Horticulture: Science and Technology

Horticulture: Science and Technology

Edited by Jamie Hanks

SYRAWOOD
PUBLISHING HOUSE
New York

Published by Syrawood Publishing House,
750 Third Avenue, 9th Floor,
New York, NY 10017, USA
www.syrawoodpublishinghouse.com

Horticulture: Science and Technology
Edited by Jamie Hanks

International Standard Book Number: 978-1-68286-373-2 (Hardback)

Cataloging-in-publication Data

Horticulture : science and technology / edited by Jamie Hanks.
p. cm.
Includes bibliographical references and index.
ISBN 978-1-68286-373-2
1. Horticulture. 2. Agricultural engineering. I. Hanks, Jamie.
SB318 .H67 2017
635--dc23

Printed in China.

TABLE OF CONTENTS

PREFACE

Horticulture is an umbrella discipline encompassing the science and technology of growing a varied array of plants such as medicinal plants, non-food crops, herbs, etc. This book is a vital source for all researching or studying the field of horticulture as it gives incredible insights into emerging trends and concepts. From theories to research to practical applications, case studies related to all contemporary topics of relevance to this field have been included in this book. Scientists and students actively engaged in this field will find this text full of crucial and unexplored topics.

The researches compiled throughout the book are authentic and of high quality, combining several disciplines and from very diverse regions from around the world. Drawing on the contributions of many researchers from diverse countries, the book's objective is to provide the readers with the latest achievements in the area of research. This book will surely be a source of knowledge to all interested and researching the field.

In the end, I would like to express my deep sense of gratitude to all the authors for meeting the set deadlines in completing and submitting their research chapters. I would also like to thank the publisher for the support offered to us throughout the course of the book. Finally, I extend my sincere thanks to my family for being a constant source of inspiration and encouragement.

Editor

Field-Evolved Resistance to Bt Maize by Western Corn Rootworm

Aaron J. Gassmann*, Jennifer L. Petzold-Maxwell, Ryan S. Keweshan, Mike W. Dunbar

Department of Entomology, Iowa State University, Ames, Iowa, United States of America

Abstract

Background: Crops engineered to produce insecticidal toxins derived from the bacterium *Bacillus thuringiensis* (Bt) are planted on millions of hectares annually, reducing the use of conventional insecticides and suppressing pests. However, the evolution of resistance could cut short these benefits. A primary pest targeted by Bt maize in the United States is the western corn rootworm *Diabrotica virgifera virgifera* (Coleoptera: Chrysomelidae).

Methodology/Principal Findings: We report that fields identified by farmers as having severe rootworm feeding injury to Bt maize contained populations of western corn rootworm that displayed significantly higher survival on Cry3Bb1 maize in laboratory bioassays than did western corn rootworm from fields not associated with such feeding injury. In all cases, fields experiencing severe rootworm feeding contained Cry3Bb1 maize. Interviews with farmers indicated that Cry3Bb1 maize had been grown in those fields for at least three consecutive years. There was a significant positive correlation between the number of years Cry3Bb1 maize had been grown in a field and the survival of rootworm populations on Cry3Bb1 maize in bioassays. However, there was no significant correlation among populations for survival on Cry34/35Ab1 maize and Cry3Bb1 maize, suggesting a lack of cross resistance between these Bt toxins.

Conclusions/Significance: This is the first report of field-evolved resistance to a Bt toxin by the western corn rootworm and by any species of Coleoptera. Insufficient planting of refuges and non-recessive inheritance of resistance may have contributed to resistance. These results suggest that improvements in resistance management and a more integrated approach to the use of Bt crops may be necessary.

Editor: Peter Meyer, University of Leeds, United Kingdom

Funding: This research was supported through the Biotechnology Risk Assessment Grant program of the United States Department of Agriculture, and through the College of Agriculture and Life Sciences at Iowa State University. The funders had no role in study design, data collection and analysis, decision to publish, or preparation of the manuscript.

Competing Interests: Aaron Gassmann has received research funding, not related to this project, from Dow, Monsanto, Pioneer, Syngenta, and Bayer CropScience. Aaron Gassmann has filed a provisional patent for the seedling based assay described in this manuscript. This does not alter the authors' adherence to all the PLoS ONE policies on sharing data and materials.

* E-mail: aaronjg@iastate.edu

Introduction

Transgenic crops engineered to produce insecticidal toxins derived from the bacterium *Bacillus thuringiensis* (Bt) were planted on more than 58 million hectares worldwide in 2010 [1]. Benefits of Bt crops include reduced use of harmful insecticides and regional suppression of some key agricultural pests [2,3,4,5,6]. Within the United States, and worldwide, more area is planted to Bt maize *Zea mays* L. than any other Bt crop [1]. The western corn rootworm *Diabrotica virgifera virgifera* LeConte (Coleoptera: Chrysomelidae) is among the most serious pests of maize within the United States, with larval feeding on maize roots causing the majority of crop losses from this pest [7]. Beginning in 2003, Bt maize was commercialized for control of western corn rootworm larvae and was rapidly adopted by farmers, constituting over 45% of maize crop in the United States during 2009 [8,9]. However, the evolution of resistance by the western corn rootworm could cut short the benefits of Bt maize.

The refuge strategy is used in the United States and elsewhere to delay pest resistance to Bt crops [10]. This strategy uses non-Bt host plants as a refuge for Bt susceptible genotypes. Mating of homozygous susceptible pests with pests that are homozygous for Bt resistance produces progeny that are heterozygous for resistance traits. The delay in resistance expected under the refuge strategy becomes greater as the dominance of resistance decreases and is greatest when resistance is completely recessive [11]. Thus, as the area planted to refuge decreases or resistance becomes more dominant, pests are predicted to evolve resistance more quickly [11,12].

The western corn rootworm has repeatedly demonstrated its ability to adapt to pest management strategies [7]. Examples include the evolution of resistance to conventional insecticides and the cultural practice of crop rotation [13,14,15]. The widespread planting of Bt maize targeting western corn rootworm raised concerns that this pest would evolve resistance to Bt. Of additional concern are data suggesting that resistance of western corn rootworm to Bt maize is not recessive [16]. Furthermore, a lack of compliance in planting of refuges has been documented among farmers that grow Bt maize in the United States [17]. Both of these factors are expected to increase the risk of western corn rootworm evolving Bt resistance.

In the present study, we compared populations of western corn rootworm that were sampled from two types of maize fields: problem fields and control fields. Farmers reported severe feeding injury by corn rootworm to Bt maize in problem fields but not in control fields planted to Bt or non-Bt maize. We report that western corn rootworm populations sampled from problem fields showed statistically significant, but not complete, resistance to the Bt maize. Resistance was found only for Cry3Bb1 maize, the type of Bt maize that had been grown historically in those fields, and was not present for maize that produced Bt toxin Cry34/35Ab1. Our results represent the first case of field-evolved resistance by the western corn rootworm to Bt maize and the first case of field-evolved resistance by a coleopteran to Bt toxin, as all previous cases of field-evolved resistance have involved Lepidoptera [18].

Methods

Corn rootworm populations were sampled during the summer of 2009 in problem fields and in control fields found within Iowa, USA (Fig. 1). Problem fields were defined as fields with severe feeding injury to Bt maize by corn rootworm and were identified by farmers that contacted the extension service of Iowa State University. Problem fields contained plants that were goose-necked (bent at the plant-soil interface) and lodged (tilted in a pronounced manner), which are characteristic of feeding by corn rootworm larvae. Additionally, farmers noted a high abundance of rootworm adults in problem fields. Upon receiving notification of a problem field, we visited the field and sampled the western corn rootworm present in the field. In all four problem fields, the vast majority of adult *Diabrotica* spp. present in the field were western corn rootworm. In one case a problem field was present on an Iowa State University research farm (Northeast Research and Demonstration Farm; site P4 in Fig. 1 and Table 1). With the exception of this Iowa State University research farm, maize roots were dug from the problem fields to evaluate rootworm feeding injury and the presence of Bt toxin was confirmed by ELISA with a kit (Envirologix, Portland, Maine). Roots were not sampled at random but were selected to confirm the presence of rootworm feeding.

Control fields were defined as fields not associated with unexpected feeding by corn rootworm on Bt maize. To allow for comparison with problem fields, only western corn rootworm were sampled from control fields. We sampled five control fields that were widely distributed throughout Iowa. Three of the control fields were located on Iowa State University research farms (C2, C4, C5 in Table 1; Fig. 1). One control field was identified based on a grower complaint of heavy rootworm injury to non-Bt corn (C3 in Table 1; Fig. 1). Another control field (C1) was identified as part of a survey of corn rootworm abundance in Iowa (M. Dunbar pers. obs.). This field was the only control field with a history of Bt maize but there was not apparent rootworm feeding as evidenced by an absence of lodging by maize plants (C1 in Table 1; Fig. 1). Maize roots were not examined in control fields, so the extent of rootworm feeding is unknown.

We interviewed farmers and farm managers to determine cropping history of fields from 2003 to 2009 (Table 1). Individuals were asked if Bt maize had been grown in the field, during which years, and what type of Bt maize (e.g., Cry3Bb1 or Cry34/35Ab1). No questions were asked about planting of refuge, size of refuge, or proximity of the refuge to the Bt field. For years in which Bt maize was not grown in a field, individuals were asked about the type of crop that was grown (e.g., maize or soybeans).

Adult western corn rootworm collected in the field were brought to Iowa State University where they were held in small cages (18 cm×18 cm×18 cm L×W×H) (Megaview Science, Taichung, Taiwan) and provided with food consisting of corn leaf tissue and an artificial diet (western corn rootworm diet, Bio-Serv, Frenchtown, New Jersey). The water source for the adult beetles was 1.5% agar solid, which was 98.5% water by mass, and provided water to the adult western corn rootworm when consumed. Cages were held in an incubator (25°C; 16/8 L/D) and individuals from

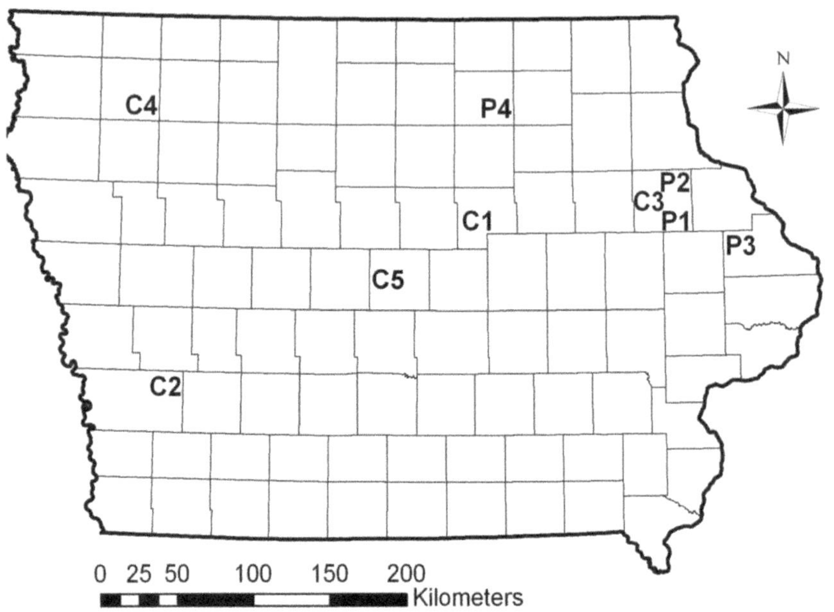

Figure 1. Distribution of sites sampled within Iowa during 2009. Sites beginning with C are control fields and were not associated with feeding injury to Bt maize, and sites beginning with P were problem fields, which were associated with farmer complaints of severe injury to Cry3Bb1 maize by corn rootworm. Codes correspond to Table 1 where a field history is provided along with the corrected survival for these populations of western corn rootworm on Cry3Bb1 maize and Cry34/35Ab1 maize.

Table 1. Sampling date in 2009, corrected survival in bioassays, and history of planting in problem fields (P1–P4) and control fields (C1–C5) from 2003 to 2009.

		Corrected Survival		**Field History**[a]						
Site	**Date Sampled**	**Cry3Bb1**	**Cry34/35Ab1**	**03**	**04**	**05**	**06**	**07**	**08**	**09**
P1	11 September	0.61+0.10	0.06±0.04	2	3	3	3	3	3	3
P2	11 September	0.61±0.06	0.03±0.02	2	2	3	5	5	5	5
P3	11 September	0.49±0.05	0.14±0.03	2	2	2	2	3	3	3
P4	14 August	0.40±0.06	0.20±0.10	2	2	2	3	3	3	3
Mean		**0.52±0.05**	**0.11±0.06**							
C1	26 September	0.32±0.11	0.25±0.06	2	2	2	3	3	3	3
C2	15 September	0.21±0.08	-------------	1	2	1	2	1	2	2
C3	11 September	0.17±0.05	0.13±0.09	2	2	2	2	2	2	2
C4	23 September	0.10±0.04	0.00±NA	1	2	1	4	2	3	2
C5	01 September	0.06±0.04	0.06±0.05	2	3	1	2	6	1	2
Mean		**0.17±0.05**	**0.11±0.06**							

[a]Field history indicates the crop that was planted in a field each year: 1 = soybean, 2 = maize lacking rootworm active Bt, 3 = Cry3Bb1 maize, 4 = Cry34/35Ab1 maize, 5 = combination of Cry3Bb1 maize and Cry34/35Ab1 maize, 6 = research plots with non-Bt maize and several Bt maize hybrids (mCry3A [46], Cry3Bb1, and Cry34/35Ab1).

each population were housed in separate cages. Adults were provided with an oviposition substrate that consisted of moist, finely sieved soil (<180 µm) placed in a 10 cm Petri dish. Eggs obtained from each population were placed separately in 45 mL plastic cups containing moistened sieved soil, and then sealed in a plastic bag and placed in a cold room at 8°C for at least 5 months to break diapause. Following exposure to cold, eggs were stored for one week at 25°C. Eggs were washed from the soil using a screen with 250 µm openings and then placed atop moistened sieved soil held in a 10 cm Petri dish. Neonate larvae began hatching approximately one week thereafter.

Neonate larvae from each population were evaluated in laboratory bioassays for their survival on two transgenic maize hybrids, each of which contained a unique Bt toxin targeting corn rootworm. One hybrid (DeKalb DKC 6169) produced Cry3Bb1. The other hybrid (Mycogen 2T789) produced Cry34/35Ab1. For both of these hybrids, we also evaluated rootworm survival on a near isogenic hybrid that lacked a gene for a rootworm active Bt toxin but otherwise was genetically similar to its respective Bt hybrid. In the case of Cry3Bb1 maize, the non-Bt hybrid was DKC 6172 (DeKalb) and for Cry34/35Ab1 maize the non-Bt hybrid was 2T777 (Mycogen).

Maize plants used in bioassays were grown in a greenhouse (25°C, 16/8 L/D) in 1 L containers made of clear plastic (Reynolds Food Packaging, Shepherdsville, Kentucky) with supplemental lighting provided with 400 W high-pressure sodium bulbs (Ruud Lighting Inc., Racine, Wisconsin). Containers were filled with 750 mL of a 1:1 ratio of Sunshine Sun Gro SB300 and Sunshine Sun Gro LC1 potting soils (Sun Gro Horticulture Canada Ltd., Vancouver, British Columbia). Seeds were planted one per container at a depth of ca. 4 cm. Beginning two weeks after planting, plants were fertilized weekly with 100 mL of Peters Excel 15-5-15 Cal-Mag Special (Everris International, Geldermalsen, The Netherlands) at a concentration of 4 mg per mL.

Maize seeds of 2T789 and 2T777 were coated with a seed treatment (CruiserMaxx 250, Syngenta, Basel, Switzerland), which contained the neonicotinoid insecticide Thiamethoxam. Prior to planting, this seed treatment was removed by washing ca. 50 seeds in a solution of 1 mL dish detergent (Ultra Palmolive Original, Colgate-Palmolive Company, New York, New York) and 250 mL deionized water. Seeds were placed in the detergent solution for 20 minutes and agitated gently using a stirring plate and magnetic stirring bar. This process was repeated three times with seeds rinsed four times with deionized water between each time they were washed. Seeds were then rinsed four times and allowed to dry for approximately 12 hours, followed by one hour of soaking in a 10% bleach solution, during which they were stirred every 15 minutes. After seeds were removed from the bleach solution, they were rinsed 10 times with deionized water and then allowed to dry for at least 24 hours. This process removed virtually all visible signs of the seed treatment. Insecticidal seed treatment was not applied to DKC 6169 and DKC 6172. However, to ensure that no residual insecticide was present, seeds were bleached following the methods used with 2T789 and 2T777.

Plants were grown in a greenhouse for three to four weeks, until they contained at least five fully formed leaves (V5 stage), and then moved to incubators for bioassays. For bioassays, plants were first trimmed to a height of 20 cm to allow for storage in incubators. Two to three leaves were left on each plant but were trimmed to 8 cm long. Recently hatched larvae (less than 24 hours old) were removed from the soil's surface within their Petri dish using a fine brush and placed at the base of a maize plant on a root that had been exposed by moving away a small amount of soil. Maize plants remained in their original 1 L containers throughout the bioassay. Between 10 and 20 neonates were placed on the base of each plant. Larvae were distributed equally between Bt and non-Bt maize plants. Cups containing plants and larvae were placed in an incubator for 17 days (25°C, 65% RH, 16/8 L/D), and plants were watered as needed.

After 17 days, the aboveground biomass of the plant was excised and the soil, containing roots and larvae, was removed from the 1 L plastic container and placed on a Berlese funnel to extract larvae from the soil. A length of 17 days was selected for bioassays because it allowed sufficient time for some of the fastest developing larvae to reach the third and final instar [19]. Root masses were held on Berlese funnels over 4 days and rootworm larvae were collected in 15 mL glass vials containing 10 mL of 85% ethanol. The average sample sizes per population were 12.7±4.8 (mean ± standard deviation) bioassay cups for Cry3Bb1 maize and for its non-Bt counterpart, and 12.8±4.8 bioassay cups for Cry34/

35Ab1 maize and for its non-Bt counterpart. We did not have sufficient western corn rootworm eggs to test one of the populations (C2 from Table 1) on Cry34/35Ab1 maize.

Data Analysis

Data on the number of field-years (i.e., planting of one field for single year) during which problem fields and control fields were planted to Cry3Bb1 maize were compared using a G test of independence with a Williams's correction [20].

For each bioassay cup, proportional survival was calculated as the quotient of the number of larvae recovered after 17 days divided by the number of neonates initially placed in a bioassay container. The mean proportional survival for each population on each type of maize was analyzed with a two-way, mixed-model analysis of variance (ANOVA) (PROC MIXED in SAS). Data for the two types of Bt maize (Cry3Bb1 maize and Cry34/35Ab1 maize) were analyzed separately. The ANOVA included the fixed factors of field type (problem field vs. control field), maize hybrid (Bt maize vs. non-Bt maize) and their interaction. Random factors in the analysis were population, which was nested within field type, and the interaction between maize hybrid and population nested within field type. Survival data were transformed by the arcsine of the square root to ensure homogeneity of variance and normality of the residuals. Pairwise contrasts were conducted using the PDIFF option in PROC MIXED.

For each population, we calculated corrected survival as the complement of corrected mortality. Corrected mortality was determined using the correction of Abbott [21], and was calculated for each population by adjusting mortality of larvae from each bioassay cup with Bt maize by the average mortality on the non-Bt near isogenic hybrid. Average corrected survival for each population was compared between control fields and problem fields for Cry3Bb1 maize and for Cry34/35Ab1 maize based on a one-way ANOVA (PROC ANOVA in SAS).

Corrected survival also was used to test the significance of three correlations among all populations sampled. We tested for the following correlations: 1) corrected survival of populations on Cry3Bb1 maize and Cry34/35Ab1 maize, 2) corrected survival for populations on Cry3Bb1 maize and the number of years populations had been exposed to Cry3Bb1 maize in the field and 3) corrected survival for populations on Cry34/35Ab1 maize and the number of years populations had been exposed to Cry34/35Ab1 maize in the field. Correlations were measured using a Pearson correlation coefficient and tested for significance against the null hypothesis of $\rho = 0$ (PROC CORR in SAS).

Results

The average level of rootworm feeding injury observed in problem fields was 1.8±0.7 nodes (mean ± standard deviation; $N = 12$) based on the Iowa State University root injury scale, which ranges from 0 nodes (no feeding injury) to 3 nodes (heavy feeding injury) [22]. All roots sampled from problem fields were from maize plants that produced Cry3Bb1 as indicated by ELISA.

Interviews with farmers indicated that problem fields were planted to Cry3Bb1 maize for at least three consecutive growing seasons while only one control field (C1) was planted to Cry3Bb1 maize for any consecutive growing seasons (Table 1). Furthermore, Cry3Bb1 maize was planted for significantly more field-years in problem fields (14 of 28 field-years) than in control fields (6 of 35 field-years) ($G = 7.68$; df = 1; $P = 0.006$). By contrast, control fields were planted to a greater diversity of crops and employed an array of management practices to control corn rootworm (Table 1). For example, control fields were planted to soybeans in 7 of 35 fields-years while problem fields were not planted to soybeans in any of the 28 field-years.

Survival in bioassays on Cry3Bb1 maize was affected by a significant interaction between field type and maize hybrid ($F = 7.06$; df = 1,7; $P = 0.03$). On the non-Bt hybrid, survival was similar, and did not differ significantly, between populations from problem fields and control fields ($P = 0.74$) (Fig. 2A). By contrast, on Cry3Bb1 maize, survival was three times higher and significantly greater for insects from problem fields than from control fields ($P = 0.011$), indicating that insects from problem fields were resistant to Cry3Bb1 maize (Fig. 2A). However, survival was significantly lower on Cry3Bb1 maize than on non-Bt maize for populations from problem fields ($P = 0.008$), indicating that problem fields contained a mixture of resistant and susceptible individuals, that resistance was incomplete, or that a combination of these factors was present. Additionally, survival was significantly lower on Cry3Bb1 maize compared with non-Bt maize for populations from control fields ($P < 0.0001$).

A different pattern emerged when populations were tested against Cry34/35Ab1 maize and its non-Bt near isogenic hybrid (Fig. 2B). The interaction between field type and maize hybrid was not significant ($F = 0.07$; df = 1,6; $P = 0.80$) and the effect of field type was not significant ($F = 0.003$; df = 1,6; $P = 0.96$). An effect of hybrid was present, with populations displaying significantly lower larval survival on Cry34/35Ab1 maize than non-Bt maize ($F = 61.48$; df = 1,6; $P = 0.0002$) (Fig. 2B). Survival was not significantly different between populations from problem fields and control fields on Cry34/35Ab1 maize ($P = 0.95$) or on non-Bt maize ($P = 0.87$). These results indicate that populations were equally susceptible to Cry34/35Ab1 maize, and that this Bt toxin significantly reduced survival.

A second set of complementary analyses were conducted with survival scores for larvae on Bt maize that were corrected for survival on the accompanying non-Bt near isogenic hybrid (Table 1). Corrected survival was significantly higher, and three times greater, for larvae from problems fields than control fields on Cry3Bb1 maize ($F = 20.61$; df = 1,7; $P = 0.003$), which indicates that populations from problem fields were resistant to Cry3Bb1 maize. However, no difference between populations from control fields and problem fields was detected on Cry34/35Ab1 maize ($F < 0.1$; df = 1,6; $P = 0.99$).

No significant correlation occurred among populations for corrected survival of larvae on Cry3Bb1 maize and Cry34/35Ab1 maize ($r = 0.068$; df = 6; $P = 0.87$), indicating an absence of cross resistance between these Bt toxins (Fig. 3A). Cry3Bb1 maize had been grown for at least one field-year in seven of the nine fields sampled (Table 1), and among populations there was a significant positive correlation between the number of years a field had been planted to Cry3Bb1 maize and survival on Cry3Bb1 maize ($r = 0.832$; df = 7; $P = 0.005$), indicating that an increased duration of exposure to Cry3Bb1 maize in the field resulted in greater resistance to this Bt toxin (Fig. 3B). Cry34/Cry35Ab1 maize had been grown in only three of the nine fields sampled and for a total of only one field-year alone and five field-years in combination with Cry3Bb1 maize (Table 1). No significant correlation was detected between frequency with which Cry34/35Ab1 maize was cultivated and survival on Cry34/35Ab1 maize ($r = -0.56$; df = 6; $P = 0.15$).

Discussion

Survival of western corn rootworm on Cry3Bb1 maize in laboratory bioassays was significantly higher for insects from problem fields where farmers reported severe root injury to

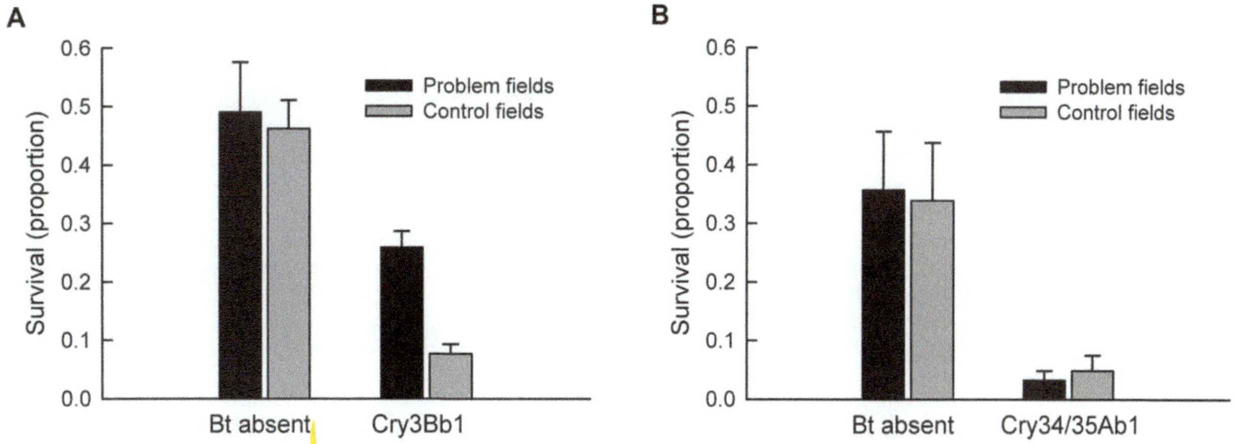

Figure 2. Survival of western corn rootworm on Bt and non-Bt maize. Data are shown for A) Cry3Bb1 maize and B) Cry34/35Ab1 maize. In both cases, survival also is shown for a non-Bt near isogenic hybrid. Bar heights are means and error bars are the standard error of the mean.

Cry3Bb1 maize than from control fields where such injury was not reported (Fig. 2A). Furthermore, there was a significant correlation between the number of years Cry3Bb1 maize had been grown in a field and survival of western corn rootworm on Cry3Bb1 maize (Fig. 3B). These data indicate that the western corn rootworm is evolving resistance to Cry3Bb1 maize in some populations in Iowa, USA. This is the first case of the western corn rootworm, or any species of beetle, evolving resistance to a Bt toxin in the field [18]. Insects collected from problem fields did not display greater survival on Cry34/35Ab1 maize (Fig. 2B). Additionally, no correlation in survival on Cry3Bb1 maize and Cry34/35Ab1 maize was observed among populations, indicating an absence of cross resistance between these Bt toxins (Fig. 3A).

One factor that may have contributed to the resistance observed here is that Bt maize producing Cry3Bb1 is not considered a high-dose event against corn rootworm [23]. High-dose events are expected to delay resistance by making the inheritance of resistance more recessive [10]. Additionally, genetic analysis of a greenhouse-selected strain found that resistance to Cry3Bb1 maize in western corn rootworm was not a recessive trait [16]. In the context of the refuge strategy, recessive inheritance of resistance caused by high-dose events will reduce survival on Bt crops for heterozygous offspring that result from mating between insects from refuge and Bt fields, thereby delaying resistance [18,24]. The ability of heterozygous resistant western corn rootworm to survive on Bt maize may have diminished the effectiveness of refuges to delay resistance [12].

A second factor that may have contributed to the evolution of resistance was insufficient refuge populations. Currently, only 50% of Bt maize planted in Midwest complies with US EPA requirements for refuge size and proximity to Bt fields [17]. Insufficient refuge populations also may have contributed to other cases of Bt resistance [25]. In general, larger populations of refuge insects will act to delay pest resistance by decreasing the proportion of homozygous resistant insects in a population, although the magnitude of this effect will depend on the

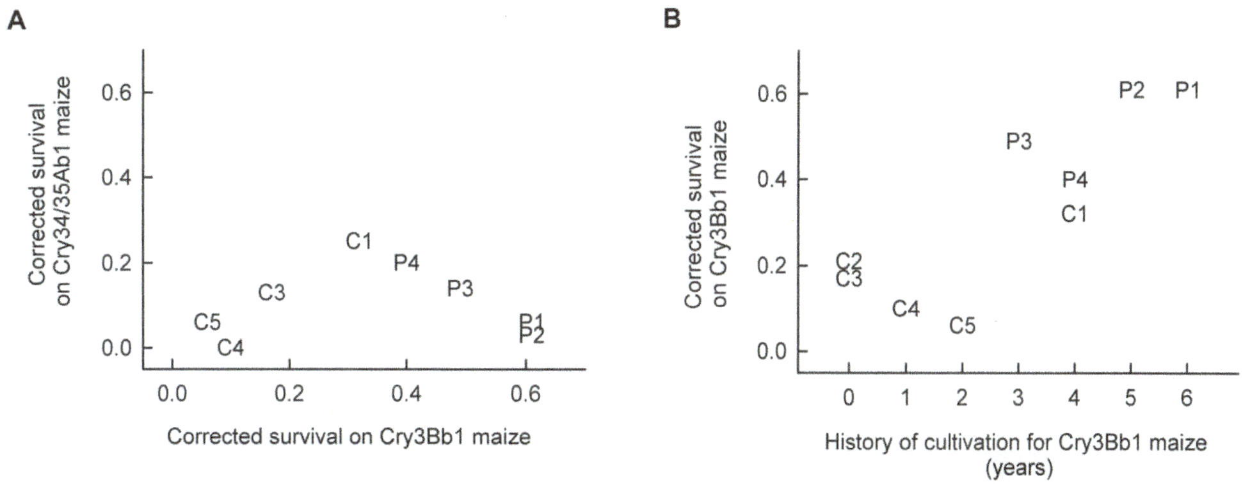

Figure 3. Correlation analysis for corrected survival of western corn rootworm. Correlations are shown for A) survival on Cry3Bb1 maize and Cry34/35Ab1 maize and B) survival on Cry3Bb1 maize and number of years Cry3Bb1 maize was planted in a field. Symbols in the graphs correspond to Table 1, which lists corrected survival for populations on Bt maize and the cultivation history of fields. For (A), no significant correlation was present between survival on Cry3Bb1 maize and Cry34/35Ab1 maize (r = 0.068; df = 6; P = 0.87). For (B), a significant positive correlation was present between corrected survival on Cry3Bb1 maize and the number of years Cry3Bb1 maize had been grown in a field (r = 0.832; df = 7; P = 0.005).

inheritance of resistance, with greater delays occurring for recessively inherited traits [24].

It is noteworthy that our findings are consistent with a greenhouse-selection experiment that found higher survival on Cry3Bb1 maize by western corn rootworm after three generations of selection [16]. In all of the problem fields we studied, Cry3Bb1 maize had been grown for at least three consecutive years (Table 1), which corresponds to three generations of selection in this univoltine pest [7]. Resistance to Cry3Bb1 maize in the greenhouse-selected strain was incomplete, with the selected strain displaying lower fitness on Bt maize than non-Bt maize [16]. In this study, populations from problem fields displayed lower survival on Cry3Bb1 maize than non-Bt maize. This may have been due to resistance in problem fields being incomplete, populations from problem fields containing a mixture of resistant individuals from those fields and susceptible migrants for neighboring fields, or a combination of these two factors.

The landscape-level effects of the resistant populations identified in this study will depend on gene flow, fitness trade-offs that accompany resistance, and selection intensity [26,27,28,29]. Current trends in planting of Bt crops suggest that intense selection for resistance in the field will continue [1]. Fitness costs could act to delay resistance, although the few data currently available suggest that costs of Bt resistance in western corn rootworm may be small [16,27]. Western corn rootworm appears to have low rates of dispersal, typically traveling less than 40 m per day, although long-distance dispersal is possible [30,31]. The tendency for short-distance dispersal may help to delay adaptation at a landscape level [29]. Taken together these data suggest that resistance to Cry3Bb1 maize in western corn rootworm should persist and intensify in localized areas, but at a landscape level, some populations may remain susceptible to Cry3Bb1 maize.

It might be the case that the observed resistance by western corn rootworm to Cry3Bb1 maize is a result of pre-adaptation rather than a response to selection. This is unlikely because of the high degree of genetic homogeneity observed among populations of this pest in the Midwest [32,33]. This genetic similarity is thought to have resulted from the recent and rapid range expansion of the western corn rootworm from the Great Plains to the East Coast [7]. Furthermore, there was a history of selection with Cry3Bb1 maize observed among problem fields (Table 1) and a significant correlation between history of selection in the field and survival on Cry3Bb1 maize in bioassays (Fig. 3b), both of which support the proposition that resistance was the result of selection rather than pre-adaptation.

Recently, Bt maize was commercialized that produces both Cry3Bb1 and Cry34/35Ab1 [34]. Pyramiding of multiple Bt toxins that target the same pest can delay the evolution of resistance to either toxin when most individuals that are resistant to one toxin are killed by the other toxin [35]. In populations where western corn rootworm populations have begun adapting to Cry3Bb1, the benefit of pyramiding two Bt toxins may be diminished [36]. However, the lack of cross resistance between these toxins (Fig. 3a) suggests that pyramiding Cry3Bb1 with Cry34/35Ab1 may still act to delay resistance in problem fields at least as long as, if not longer than, the cultivation of maize producing only a single toxin.

Although no cases of field-evolved resistance were reported during the first decade of commercialization for Bt crops, several recent cases have been reported [12,25,37,38,39,40]. Typically, there is a lag between the introduction of an insecticide and the first occurrence of resistance, which is then followed by a steady increase in the cumulative number of occurrences [41]; a trend that would clearly be undesirable for Bt crops. Stern et al. [42] outlined the foundation of integrated pest management by advocating the application of multiple methods to control pest populations, thus delaying or avoiding problems that include, but are not limited to, pest resistance [43]. To date, the widespread planting of Bt crops has resulted in pest resistance for only a small subset of all pest populations managed by this technology [18]. However, these recent cases suggest a need to develop more integrated management solutions for pests targeted by Bt crops [44,45]. A common pattern observed among problem fields in this study was the consecutive planting of the same type of Bt maize over several seasons (Table 1). Even with resistance management plans in place, sole reliance on Bt crops for management of agriculture pests will likely hasten the evolution of resistance in some cases, thereby diminishing the benefits that these crops provide.

Acknowledgments

We thank Bruce Tabashnik and Fred Gould for their comments on an earlier draft of this manuscript.

Author Contributions

Conceived and designed the experiments: AJG. Performed the experiments: JLP-M RSK MWD. Analyzed the data: AJG JLP-M. Wrote the paper: AJG JLP-M RSK.

References

1. James C (2010) Global Status of Commercialized Biotech/GM Crops: 2010. ISAAA Brief No. 42. Ithaca, New York: ISAAA.
2. Carrière Y, Ellers-Kirk C, Sisterson MS, Antilla L, Whitlow M, et al. (2003) Long-term regional suppression of pink bollworm by *Bacillus thuringiensis* cotton. Proc Natl Acad Sci U S A 100: 1519–1523.
3. Cattaneo MG, Yafuso C, Schmidt C, Huang C-y, Rahman M, et al. (2006) Farm-scale evaluation of the impacts of transgenic cotton on biodiversity, pesticide use, and yield. Proc Natl Acad Sci U S A 103: 7571–7576.
4. Huang J, Ruifa H, Rozelle S, Pray C (2005) Insect-resistant GM rice in farmers' fields: assessing productivity and health effects in China. Science 308: 688–690.
5. Wu K-M, Lu Y-H, Feng H-Q, Jiang Y-Y, Zhao J-Z (2008) Suppression of cotton bollworm in multiple crops in China in areas with Bt toxin-containing cotton. Science 321: 1676–1678.
6. Hutchison W, Burkness E, Mitchell P, Moon R, Leslie T, et al. (2010) Areawide suppression of European corn borer with Bt maize reaps savings to non-Bt maize growers. Science 330: 222–225.
7. Gray ME, Sappington TW, Miller NJ, Moeser J, Bohn MO (2009) Adaptation and invasiveness of western corn rootworm: intensifying research on a worsening pest. Annu Rev Entomol 54: 303–321.
8. James C (2009) Global Status of Commercialized Biotech/GM Crops: 2009. ISAAA Brief No. 41. Ithaca, New York: ISAAA.
9. Environmental Protection Agency (2003) Biopesticides registration action document: event MON863 Bacillus thuringiensis Cry3Bb1 Corn. Available: http://www.epa.gov/pesticides/biopesticides/ingredients/tech_docs/cry3bb1/1_%20cry3bb1_exec_summ.pdf.
10. Gould F (1998) Sustainability of transgenic insecticidal cultivars: integrating pest genetics and ecology. Annu Rev Entomol 43: 701–726.
11. Tabashnik BE, Gould F, Carrière Y (2004) Delaying evolution of insect resistance to transgenic crops by decreasing dominance and heritability. J Evol Biol 17: 904–912.
12. Tabashnik BE, Gassmann AJ, Crowder DW, Carrière Y (2008) Insect resistance to Bt crops: evidence versus theory. Nat Biotechnol 26: 199–202.
13. Meinke LJ, Siegfried BD, Wright RJ, Chandler LD (1998) Adult susceptibility of Nebraska western corn rootworm (Coleoptera: Chrysomelidae) populations to selected insecticides. J Econ Entomol 91: 594–600.
14. Parimi S, Meinke LJ, French WB, Chandler LD, Siegfried BD (2006) Stability and persistence of aldrin and methyl-parathion resistance in western corn rootworm populations (Coleoptera: Chrysomelidae). J Econ Entomol 25: 269–274.
15. Levine E, Spencer JL, Isard SA, Onstad DW, Gray ME (2002) Adaptation of the western corn rootworm to crop rotation: evolution of a new strain in response to a management practice. Am Entomol 48: 94–107.

16. Meihls LN, Higdon ML, Siegfried BD, Miller NJ, Sappington TW, et al. (2008) Increased survival of western corn rootworm on transgenic corn within three generations of on-plant greenhouse selection. Proc Natl Acad Sci U S A 105: 19177–19182.
17. Jaffe G (2009) Complancency on the Farm: Significant Noncompliance with EPA's Refuge Requirements Threatens the Future Effectiveness of Genetically Engineered Pest-protected Corn. Washington, DC: Center for Science in the Public Interest.
18. Carrière Y, Crowder DW, Tabashnik BE (2010) Evolutionary ecology of insect adaptation to Bt crops. Evol Appl 3: 561–573.
19. Nowatzki TM, Lefko SA, Binning RR, Thompson SD, Spencer TA, et al. (2008) Validation of a novel resistance monitoring technique for corn rootworm (Coleoptera: Chrysomelidae) and event DAS-59122-7 maize. J Appl Entomol 132: 177–188.
20. Sokal RR, Rohlf FJ (1995) Biometry. New York: W. H. Freeman and Company.
21. Abbott WS (1925) A method of computing the effectiveness of an insecticide. J Econ Entomol 18: 265–267.
22. Oleson JD, Park Y-L, Nowatzki TM, Tollefson JJ (2005) Node-injury scale to evaluate root injury by corn rootworms (Coleoptera: Chrysomelidae). J Econ Entomol 98: 1–8.
23. Environmental Protection Agency (2010) Biopesticides Registration Action Document. Available: http://www.epa.gov/oppbppd1/biopesticides/pips/cry3bb1-brad.pdf.
24. Roush RT (1997) Bt-transgenic crops: just another pretty insecticide or a chance for a new start in resistance management? Pestic Sci 51: 328–334.
25. van Rensburg JBJ (2007) First report of field resistance by stem borer, *Busseola fusca* (Fuller) to Bt-transgenic maize. S Afr J Plant Soil 24: 147–151.
26. Sisterson MS, Carrière Y, Dennehy TJ, Tabashnik BE (2005) Evolution of resistance to transgenic Bt crops: interactions between movement and field distribution. J Econ Entomol 98: 1751–1762.
27. Gassmann AJ, Carrière Y, Tabashnik BE (2009) Fitness costs of insect resistance to *Bacillus thuringiensis*. Annu Rev Entomol 54: 147–163.
28. Jaffe K, Issa S, Daniels E, Haile D (1997) Dynamics of the emergence of genetic resistance to biocides among asexual and sexual organisms. J Theor Biol 188: 289–299.
29. Caprio MA (1992) Gene flow accelerates local adaptation among finite populations: simulating the evolution of insecticide resistance. J Econ Entomol 85: 611–620.
30. Spencer JL, Mabry TR, Vaughn TT (2003) Use of transgenic plants to measure insect herbivore movement. J Econ Entomol 96: 1738–1749.
31. Spencer JL, Hibbard BE, Moeser J, Onstad DW (2009) Behaviour and ecology of the western corn rootworm (*Diabrotica virgifera virgifera* LeConte). Agric For Entomol 11: 9–27.
32. Coates BS, Sumerford DV, Miller NJ, Kim KS, Sappington TW, et al. (2009) Comparative performance of singel nucleotide polymorphism and microsatellite markers for population genetic analysis. J Hered 100: 556–564.
33. Kim KS, Sappington TW (2005) Genetic structuring of western corn rootworm (Coleoptera: Chrysomelidae) populations in the United States based on microsatellites loci analysis. Environ Entomol 34: 494–503.
34. Environmental Protection Agency (2009) Pesticide Fact Sheet. Available: http://www.epa.gov/oppbppd1/biopesticides/pips/smartstax-factsheet.pdf.
35. Roush RT (1998) Two-toxin strategies for management of insecticidal transgenic crops: can pyramiding succeed where pesticide mixtures have not? Philos Trans R Soc Lond B Biol Sci 353: 1777–1786.
36. Onstad DW, Meinke LJ (2010) Modeling evolution of *Diabrotica virgifera virgifera* (Coleoptera: Chrysomelidae) to transgenic corn with two insecticidal traits. J Econ Entomol 103: 849–860.
37. Storer NP, Babcock JM, Schlenz M, Meade T, Thompson GD, et al. (2010) Discovery and characterization of field resistance to Bt maize: *Spodoptera frugiperda* (Lepidoptera: Noctuidae) in Puerto Rico. J Econ Entomol 103: 1031–1038.
38. Downes S, Parker T, Mahon R (2010) Incipient resistance of *Helicoverpa punctigera* to the Cry2Ab Bt toxin in bollgard II cotton. PLoS ONE 5: e12567.
39. Bagla P (2010) Hardy cotton-munching pests are latest blow to GM crops. Science 327: 1439.
40. Dhurua S, Gujar GT (2011) Field-evolved resistance to Bt toxin Cry1Ac in the pink bollworm, *Pectinophora gossypiella* (Saunders) (Lepidoptera: Gelechiidae) from India. Pest Manag Sci 67: doi: 10.1002/ps.2127.
41. Metcalf RL (1986) The ecology of insecticides and the chemical control of insects. In: Kogan M, ed. Ecological Theory and Integrated Pest Management Practices. New York: Wiley. pp 251–297.
42. Stern VM, Smith RF, van den Bosch R, Hagen KS (1959) The integrated control concept. Hilgardia 29: 81–101.
43. Pedigo LP, Rice ME (2009) Entomology and Pest Management. Upper Saddle River: Pearson Education, Inc.
44. Tabashnik BE, Sisterson MS, Ellsworth PC, Dennehy TJ, Antilla L, et al. (2010) Suppressing resistance to Bt cotton with sterile insect releases. Nat Biotechnol 28: 1304–1307.
45. Baum JA, Bogaert T, Clinton W, Heck GR, Feldmann P, et al. (2007) Control of coleopteran insect pests through RNA interference. Nat Biotechnol 25: 1322 - 1326.
46. Walters FS, Stacy CM, Lee MK, Palekar N, Chen JS (2008) An engineered chymotrypsin/cathepsin G site domian I renders *Bacillus thuringiensis* Cry3A active against western corn rootworm larvae. Appl Environ Microbiol 74: 367–374.

2

Tillage, Mulch and N Fertilizer Affect Emissions of CO_2 under the Rain Fed Condition

Sikander Khan Tanveer, Xiaoxia Wen, Xing Li Lu, Junli Zhang, Yuncheng Liao*

College of Agronomy, Northwest A&F University Yangling, Shaanxi, P.R. China

Abstract

A two year (2010–2012) study was conducted to assess the effects of different agronomic management practices on the emissions of CO_2 from a field of non-irrigated wheat planted on China's Loess Plateau. Management practices included four tillage methods i.e. T_1: (chisel plow tillage), T_2: (zero-tillage), T_3: (rotary tillage) and T_4: (mold board plow tillage), 2 mulch levels i.e., M_0 (no corn residue mulch) and M_1 (application of corn residue mulch) and 5 levels of N fertilizer (0, 80, 160, 240, 320 kg N/ha). A factorial experiment having a strip split-split arrangement, with tillage methods in the main plots, mulch levels in the sub plots and N-fertilizer levels in the sub-sub plots with three replicates, was used for this study. The CO_2 data were recorded three times per week using a portable GXH-3010E1 gas analyzer. The highest CO_2 emissions were recorded following rotary tillage, compared to the lowest emissions from the zero tillage planting method. The lowest emissions were recorded at the 160 kg N/ha, fertilizer level. Higher CO_2 emissions were recorded during the cropping year 2010–11 relative to the year 2011–12. During cropping year 2010–11, applications of corn residue mulch significantly increased CO_2 emissions in comparison to the non-mulched treatments, and during the year 2011–12, equal emissions were recorded for both types of mulch treatments. Higher CO_2 emissions were recorded immediately after the tillage operations. Different environmental factors, i.e., rain, air temperatures, soil temperatures and soil moistures, had significant effects on the CO_2 emissions. We conclude that conservation tillage practices, i.e., zero tillage, the use of corn residue mulch and optimum N fertilizer use, can reduce CO_2 emissions, give better yields and provide environmentally friendly options.

Editor: Ben Bond-Lamberty, DOE Pacific Northwest National Laboratory, United States of America

Funding: The study was supported by the National Natural Science foundation of China (31071375 and 31171506). The funders had no role in study design, data collection and analysis, decision to publish, or preparation of the manuscript.

Competing Interests: The authors have declared that no competing interests exist.

* E-mail: yunchengliao@163.com or yunchengliao@nwsuaf.edu.cn

Introduction

Studies regarding soil CO_2 emissions have attracted significant attention because the concentration of CO_2 in the atmosphere is increasing very rapidly as a consequence of fossil fuel combustion and deforestation. The past two centuries of human activities have reportedly contributed as much as approximately half of the increase in CO_2 emissions [1], [2]. Global terrestrial ecosystems absorbed carbon at the rate of 1–4 Pg yr^{-1}, during 1980s and 1990s, which made up approximately 10–60% of the fossil fuel emissions [3], [4]. Currently, significant attention is given to CO_2 emissions from soils because this source significantly affects the global carbon cycle and the function of the terrestrial ecosystem [5]. Fluxes of greenhouse gases (CO_2, N_2O and CH_4) between the atmosphere and agricultural soils considerably influence the stock of anthropogenic greenhouse gases [6]. Agriculture is an important source of emissions for these different gases, and its contribution to climate change is approximately 20% on an annual basis [7]. It has been reported that soils have already contributed approximately 50 Pg of anthropogenic CO_2 to the atmosphere in the past, through cultivation processes [8].

Tillage is an integral part of agriculture which not only significantly affects crop production but is also considered one of the leading factors in soil degradation. This technique is a fundamental operation that has affected both the soil and the environment and is considered one of the most important sources of CO_2 emissions into the atmosphere [9] because humans have tilled the soil for crop production for thousands of years [10] and approximately 23–44% of total CO_2, is emitted into the atmosphere through soil preparation-related operations [11]. Approximately 30–50% of soil C has already been lost through the adaptations of intensive tillage practices [12], and major C losses from, soils in the form of CO_2 occur immediately after the tillage operations [13].

Agricultural management practices affect different soil processes (i.e., soil temperature, soil moisture and soil pH), and other ongoing soil decomposition processes, which ultimately result in the conversions of plant-derived C to soil organic matter and CO_2 [14]. Applications of inorganic as well as organic fertilizers [15] and different degrees of soil moisture and temperature strongly affect the fluxes of soil CO_2 [16], [17] & [18]. Similarly, the application of N fertilizer also affects soil CO_2 emissions [19]. Instead of burning crops residues, farmers, applications of inorganic fertilizers and use of green manures as well as organic manures can be of great use in maintaining soil fertility [20]. These practices can provide essential nutrients to crops and reductions in the burning of crops can reduce CO_2 emissions into the atmosphere [21].

Agricultural tillage practices can be helpful in the sequestering of atmospheric CO_2 [22], [23] and [24]. Conservation tillage has the potential to increase soil C and N [25] and other types of

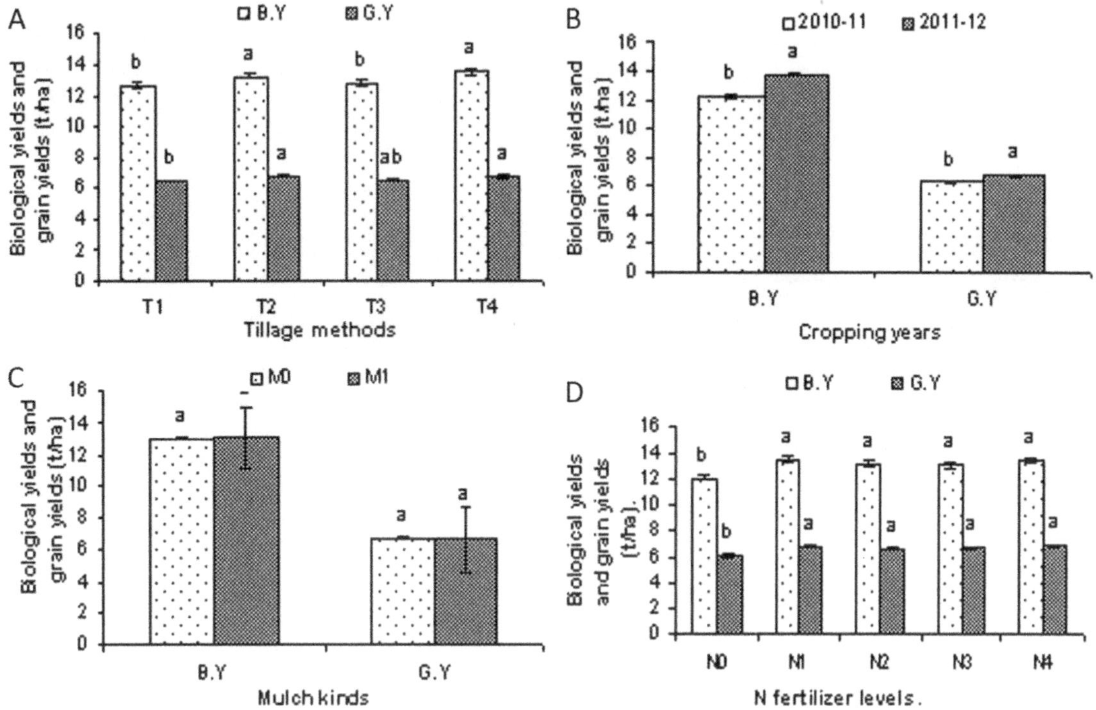

Figure 1. Wheat crop biological and grain yields as affected by different tillage methods, different mulch kinds and different N fertilizer levels during different cropping years (2010–12). (A), Wheat crop biological and grain yields as affected by different tillage methods i.e. T_1, chisel plow tillage., T_2, zero tillage., T_3, rotary tillage and T_4, mold board plow tillage., (B), Mean wheat crop biological and grain yields as affected by different tillage methods, mulch kinds and different N fertilizer levels during two cropping years (2010–12). (C), Wheat crop biological and grain yields as affected by different mulch kinds i.e. M_0, No mulch and M_1, corn residue mulch., (D), Wheat crop biological and grain yields as affected by different N fertilizer levels during two cropping years (2010–12) including, N_0, 0 kg N/ha., N_1, 80 kg N/ha, N_2, 160 kg N/ha., N_3, 240 kg N/ha and N_4, 320 kg N/ha.

conservation practices can be helpful in reducing the loss of soil organic carbon from the soils [26], [27]. Similarly, the retention of crop residue, nitrogen fertilization and no-tillage are generally supposed to enhance the soil organic carbon (SOC) stocks in the soil [28] because these farms, management practices not only increase crops biomass, but are also considered very important for the microbial decomposition of crop residues [29]. As far as N fertilization is concerned, some scientists have reported that increased N fertilization can depress CO_2 emissions [30], [31] however, others [32] have reported that N fertilization has no effect on SOC, while some other scientists [33] have reported that higher N fertilization improves the SOC of the soil.

The Loess Plateau has an area of approximately 63, 5000 km^2. It covers many provinces in China, and is home to millions of people. It is one of the most highly eroded areas of the world, and traditional agriculture, i.e., intensive tillage, is considered one of the leading man-made factors responsible for this erosion. However many crops residues are produced in this region. A small portion of these residues is used for forage or fuel consumption and the remaining residues are generally burned. Mold board plowing followed by harrowing is commonly used for the tillage operations in this region [34]. Few studies showing CO_2 emissions to the adaptation of different agronomic management practices have been previously reported from this region of China. In this area intensive tillage methods i.e. rotary tillage and mold board plow tillage methods are commonly used for land preparation. Commonly higher levels of N fertilizers are applied and crop residues are removed from the fields at the time of soil preparation.

The main aim of this two year study was to identify the effects of different tillage methods i.e. chisel plow and zero tillage in comparison with intensive tillage practices i.e. rotary tillage and mold board plow tillage methods, different N fertilizer levels and the application of corn residue mulch on CO_2 emissions. The results from this study can be of great help in improving the management of soils not only in this area of China but also in other regions of the world.

Results

Wheat crop yields

Significant variations in wheat crops biomass and grain yields were recorded during both study years (2010–12). When compared to the other tillage methods, there were better yields overall with the zero tillage planting method. No grain yields differences were recorded under different mulch treatments and similarly low yields were recorded for N_0 nitrogen fertilizer levels. Statistically equal grain yields were recorded for all of the other higher tested levels of N fertilizer (Fig. 1).

Soil CO_2 flux

All tested treatments had significant effects on the CO_2 emissions (Figs. 2, 3). The details are given below.

Tillage method effects on CO_2 emissions

Two years of combined data show that the rotary tillage and mold board plow tillage methods had their highest and statistically equal CO_2 emissions during the first week of planting the wheat

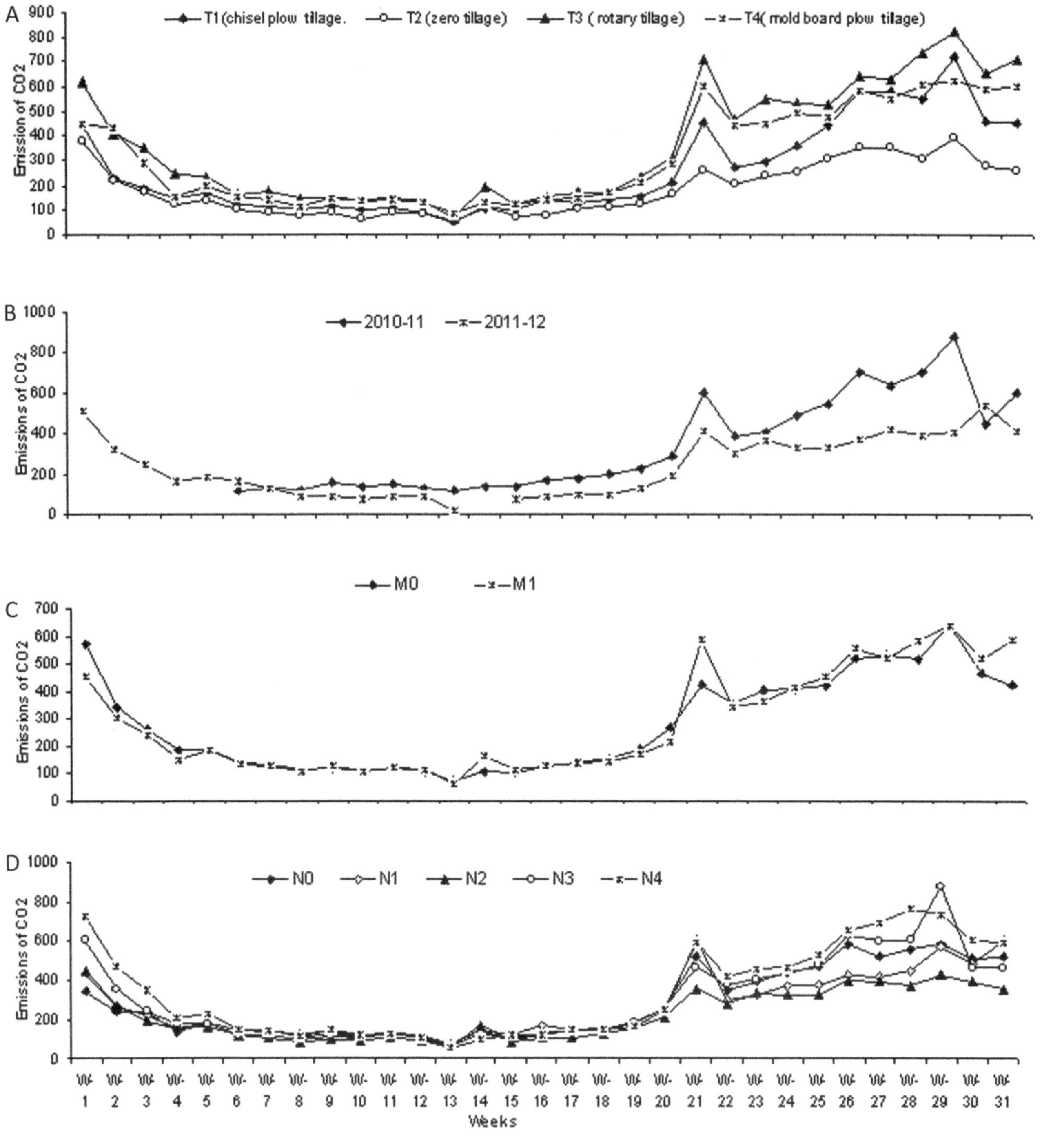

Figure 2. CO_2 emission trends as affected by different tillage methods, mulch kinds and N fertilizer levels during two cropping years (2010–12). (A). Emissions trends of CO_2 from different tillage methods i.e. T_1, chisel plow tillage, T_2, zero tillage, T_3, rotary tillage and T_4, mold board plow tillage., (B).CO_2 emission trends as affected by different kinds of tillage methods, different types of mulch and different N fertilizer levels, during two cropping years (2010–12). (C). CO_2 emission trends due to different mulch kinds i.e. M_0, no mulch and M_1, corn residue mulch., (D). CO_2 emissions trends due to different N fertilizer levels during two cropping years (2010–12) including, N_0, 0 kg N/ha., N_1, 80 kg N/ha, N_2, 160 kg N/ha, N_3, 240 kg N/ha and N_4, 320 kg N/ha.

crop, compared to the chisel plow tillage and zero tillage planting methods. When compared to all the other tillage methods, the lowest CO_2 emissions were recorded for the zero tillage planting method (Fig. 2; Fig. 3). Emission trends recorded during the whole wheat crop seasons show that although there were variations in the CO_2 emissions in response to different tillage methods, the overall highest emissions were recorded for the rotary tillage planting method, followed by mold board plow tillage. Although higher CO_2 emissions were recorded for the chisel plow tillage method, the lowest CO_2 emissions were generally recorded for zero tillage planting method (Fig. 2; Fig. 3).

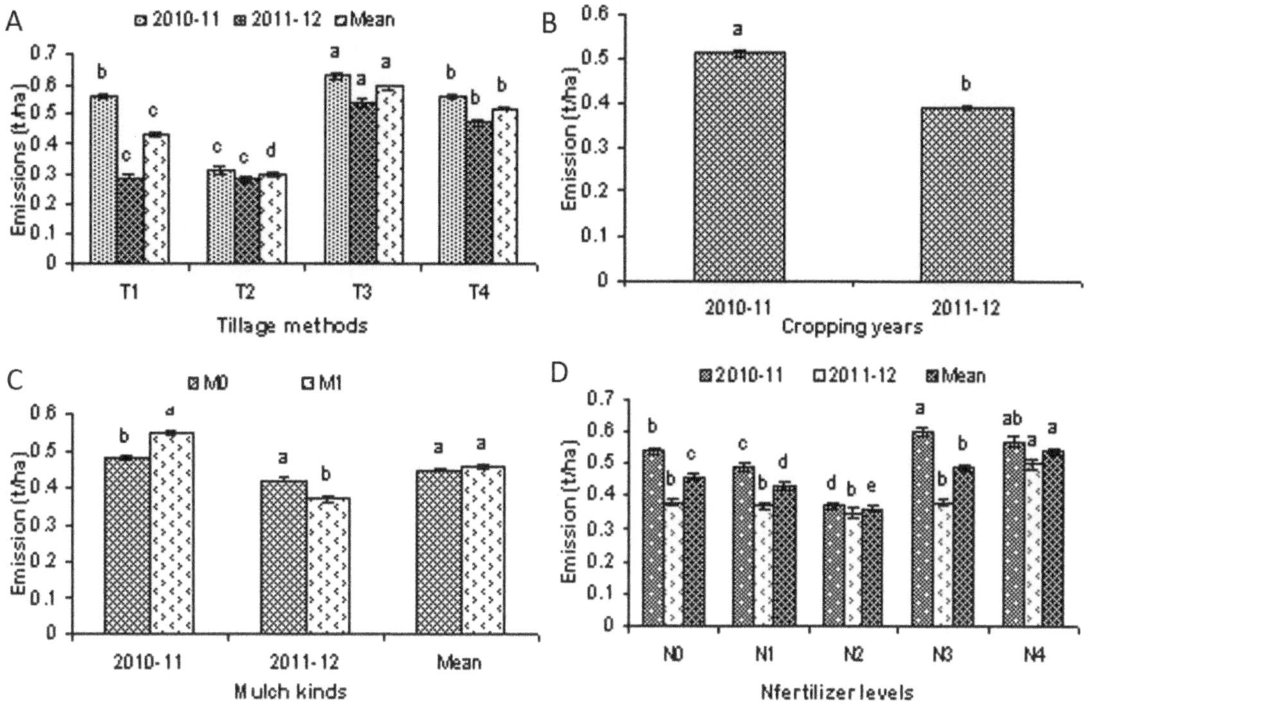

Figure 3. Total/mean emissions of CO_2 as affected by different tillage methods, mulch kinds, and N fertilizer levels during two cropping years (2010–12). (A).Emissions of CO_2 from different tillage methods i.e. T_1, chisel plow tillage., T_2, zero tillage.,T_3, rotary tillage and T_4., mold board plow tillage., (B). Total emissions of CO_2 as affected by different kinds of tillage methods, different kinds of mulch and different levels of N fertilizer, during two cropping years (2010–12)., (C), Emissions of CO_2 from the different mulch kinds i.e. M_0, no mulch and M_1, corn residue mulch., (D).Total/mean emissions of CO_2 as affected by different levels of N fertilizer, during two cropping years (2010–12) including, N_0, 0 kg N/ha., N_1, 80 kg N/ha., N_2,160 kg N/ha., N_3, 240 kg N/ha and N_4, 320 kg N/ha.

Effects of cropping years on CO_2 emissions

With the exception of one week (i.e., week 23), the highest weekly emissions of CO_2 were recorded during cropping year 2010–11 in comparison to the weekly emissions of CO_2 during cropping year 2011–12 (Figs. 2, 3).

Mulch effects on CO_2 emissions

Weekly CO_2 emissions during both wheat crops growing seasons (2010–11 and 2011–12) and the mean of the two years (2010–12) of data (Figs. 2, 3) show that there were more CO_2 emissions recorded from the corn residue-mulched treatments during cropping year 2010–11 than from the non-residue mulched treatments (Figs. 2, 3). However, fewer CO_2 emissions were recorded from the corn residue mulched treatments during cropping year 2011–12 and in comparison to the non residue-mulched treatments, over all less emissions of CO_2 were recorded from the corn residue mulched treatments (Figs. 2, 3). However the two year mean data show mixed types of emissions were recorded on a weekly basis following the applications of corn residue mulch or no mulch (Fig. 2).

N fertilizer level effects on CO_2 emissions

CO_2 emissions data recorded during both wheat cropping seasons, i.e., 2010–11 and 2011–12 and the two year (2010–12)) mean data show that there were significant differences in the weekly CO_2 emissions in response to different N fertilizer levels (Fig. 2; Fig. 3). This finding also indicates that the lowest CO_2 emissions were recorded for the N_0, N fertilizer level at the start of the wheat crop growing seasons compared to all the other higher N fertilizer levels (Fig. 2). During the winter months the CO_2 emissions decreased under all treatments but when the temperatures rose again, higher CO_2 emissions were recorded (Fig. 2). However, when compared to all the other N fertilizer level treatments, the lowest overall CO_2 emissions were recorded for 160 kg N/ha (Fig. 2). CO_2 emission fluxes increased with the increase in crop growth and temperatures, so during the last weeks of wheat crop growth equal CO_2 emissions were recorded for all of the N fertilizer treatments (Fig. 2).

Cumulative CO_2 emissions

Two years (2010–12) of CO_2 emissions data show that on a mean cumulative basis, except in the case of corn residue mulch treatments, significant differences in the emissions of CO_2 were recorded for all of the different tillage methods and different N fertilizer level treatments (Fig. 3). Statistically significant variations in the total and mean CO_2 emissions were recorded for all of the tillage methods, and the emissions trend for the different tillage methods was $T_3>T_4>T_1>T_2$ (Fig. 3). Different N fertilizer levels had significant effects on the total and mean CO_2 emissions. The emissions trend for the different nitrogen fertilizer levels was $N_4>N_3>N_0>N_1>N_2$ (Fig. 3). On a cumulative basis, more CO_2 emissions were recorded during the cropping year 2010–11 than during cropping year 2011–12 (Fig. 3), and on the whole approximately 30% more CO_2 emissions were recorded during cropping year 2010–11 than during 2011–12 (Fig. 3).

CO_2 emissions varied for all of the tillage methods and N fertilizer levels. For the chisel plow tillage treatment, using 160 kg N/ha reduced the emissions of CO_2 (data not shown) and the CO_2

emissions trend in case of the chisel plow tillage method and different N fertilizer levels was $N_4>N_3>N_0>N_1>N_2$ (data not shown) (Table 1). For the Zero tillage planting method and N fertilizer level interactions, the CO_2 emission varied but the emissions trend was $N_0>N_3>N_2>N_1>N_4$ (data not shown) (Table 1). For the rotary tillage planting method and N fertilizer level interactions, the CO_2 emissions trend was $N_4>N_3>N_1>N_0>N_2$ (data not shown) (Table 1). Similarly for the mold board plow tillage method and N fertilizer level interactions, the CO_2 emissions trend was $N_0>N_4>N_1>N_3>N_2$ (data not shown) (Table 1).

With the exception of the mold board plow tillage method, corn residue mulch applications increased the CO_2 emissions in all of the other three tillage methods. Lower CO_2 emissions were recorded for all the tillage methods during cropping year 2011–12 in comparison to cropping year 2010–11 and throughout cropping year 2010–11, corn residue mulch applications increased the CO_2 emissions to 16.5% compared to the non residue mulched treatments. Similarly during cropping year, 2011–12, 12.6% fewer CO_2 emissions were recorded for the corn residue mulched treatments in comparison to the non-residue mulched treatments (data not shown) (Table 1).

For the N_0 (0 kg N/ha), N_1 (80 kg N/ha), N_2 (160 kg N/ha) and N_4 (320 kg N/ha), treatments, the use of corn residue mulch increased the CO_2 emissions by approximately 6.2%, 5.5%, 0.6% and 35.6%, respectively, and for N_4 (240 kg N/ha), the application of corn residue mulch reduced the CO_2 emissions by approximately 35.6% (data not shown) (Table 1).

Soil temperatures versus CO_2 emissions

Temperature changes for the top 5 cm of soil from the different treatments during the two years (2010–12) of study are given in Fig. 4. Although with the passage of time, there were variations in soil temperatures for the different tillage methods, however intermediary types of temperature changes were recorded following zero tillage planting in comparison to the other three tillage methods (Fig. 4). Similarly, although statistically comparable temperatures were recorded for both types of mulch treatments, slightly increased temperatures were recorded in the corn residue mulched treatments relative to the non-residue mulched treatments (Fig. 4). However, no soil temperature differences were recorded for the different N fertilizer level treatments (Fig. 4). Generally speaking, higher soil temperatures were recorded during the cropping year 2010–11 than during cropping year 2011–12 (Fig. 4). CO_2 emission trends showed that CO_2 emissions increased with an increase in soil temperatures and vice versa (Figs. 2, 4).

Soil moisture versus CO_2 emissions

When compared to all of the other tillage methods, the highest soil water contents were recorded in response to the rotary tillage planting method (Fig. 5). Similarly, higher water contents were recorded during the cropping year 2011–12 for the different crop growth stages relative to cropping year 2010–11 (Fig. 5). Corn residue mulch increased the soil moisture contents of the crop revival stage and on the booting stage compared to the non-residue mulched treatments (Fig. 5). Lower soil moisture contents were recorded for the 80 kg N/ha (N_1) and 160 kg N/ha (N_2), treatments compared to all of the other N fertilizer level treatments (Fig. 5). Two years of mean data show that the CO_2 emissions were lower in those tillage methods or N fertilizer levels treatments that had lower water contents (Fig. 2; Fig. 4).

Table 1. ANOVA (Mean Square Values) of biological yields, grain yields, soil organic carbon, and cumulative emissions of CO_2 during two cropping years (2010–12).

Source	D.F	B.Y	G.Y	SOC	CO_2
Tillage methods	3	6593271.9***	122260.74**	32.12145866***	9605737865***
Planting years	1	13496603.4***	29504910.64***	97.90534734***	8211670776***
Mulch kinds	1	132977.5NS	20786.14NS	3.74523078***	93546720 NS
N Fertilizer levels	4	1470235.5***	4549170.27***	3.78932599***	21321766841***
Tillage methods X Mulch kinds	3	6780289.1***	1690508.29***	26.55366456***	1455896530***
Tillage methods X N Fertilizer levels	12	1443071.8 NS	302485.89NS	5.42559025***	1418530119***
Tillage methods X planting years	3	3443481.8NS	1276931.92*	12.85179871***	1455478300***
Planting years X Mulch kinds	1	66692.7NS	418751.64NS	0.67995212 NS	2547564904***
Planting years X N Fertilizer levels	4	10758689.1***	3983648.62***	3.45638215***	720140735***
Mulch kinds X N Fertilizer levels	4	975872.1NS	461889.16NS	3.23681715***	3518560905***
Tillage methods X Planting years X Mulch kinds	3	1162751.4NS	190004.99NS	12.61344035***	406313591***
Tillage methods X Planting years X N fertilizer levels	12	12108232.1NS	369168.67NS	4.50848299***	959919429***
Tillage methods X Mulch kinds X N fertilizer levels	12	1785389.2NS	420989.67NS	6.20244131***	2881830882***
Tillage methods X Mulch kinds X Planting years X N fertilizer levels	16	2029198.1 NS	782603.91***	4.05394235***	1326809061***

*Significant at 0.05 probability levels.
**Significant at 0.01 probability levels.
***Significant at 0.001 probability levels.
B.Y, Biological yields., G.Y, Grain yields., SOC, Soil organic carbon.
X, indicates interactions between different factors i.e. Tillage's X Mulches indicates interactions between different tillage methods and mulch kinds.

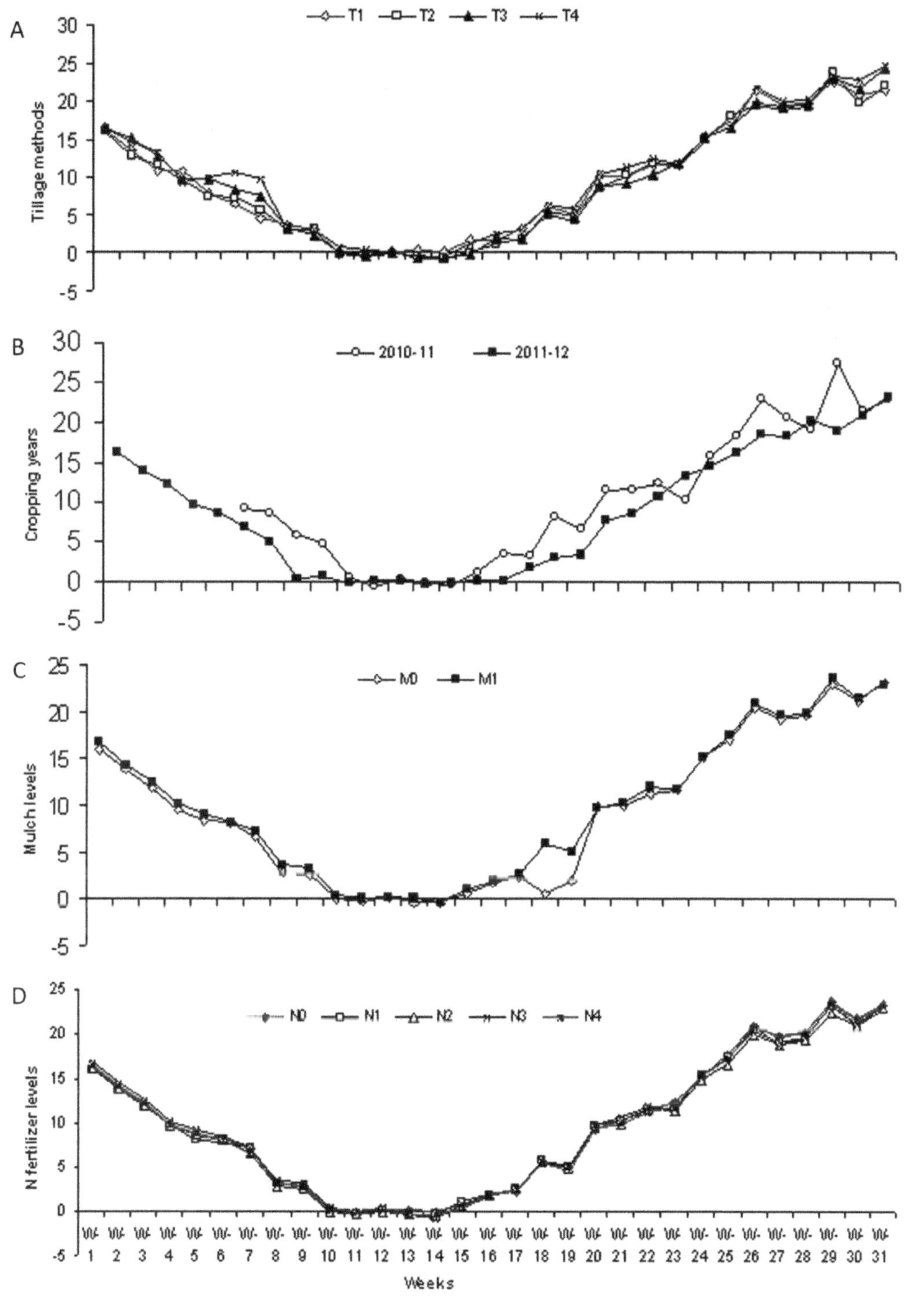

Figure 4. Changes in soil temperatures (0–5 cm depth) due to different tillage methods, corn residue mulch and N fertilizer levels during two cropping years (2010–12). (A). Changes in soil temperatures (0–5 cm depth) due to different tillage methods i.e. T_1, chisel plow tillage, T_2, zero tillage, T_3, rotary tillage and T_4, mold board plow tillage method., (B) Changes in soil temperatures (0–5 cm depth), as affected by different tillage methods, different mulch kinds and different N fertilizer levels during two cropping years (2010–12)., (C), Changes in soil temperatures (0–5 cm depth) during two cropping years (2010–12) due to different mulch kinds i.e. M_0, no mulch and M_1, corn residue mulch., (D). Changes in soil temperatures (0–5 cm depth) due to different N fertilizer levels during two cropping years (2010–12) including, N_0, 0 kg N/ha., N_1, 80 kg N/ha., N_2, 160 kg N/ha., N_3, 240 kg N/ha and N_4, 320 kg N/ha.

Soil organic carbon versus CO_2 emissions

Two years of mean data show that the tillage methods had significant effects on the SOC from 0–10 cm soil depths and that compared to all of the tillage methods, the highest SOC was recorded following chisel Plow tillage (Fig. 6). Higher SOC contents were recorded for all of treatments during cropping

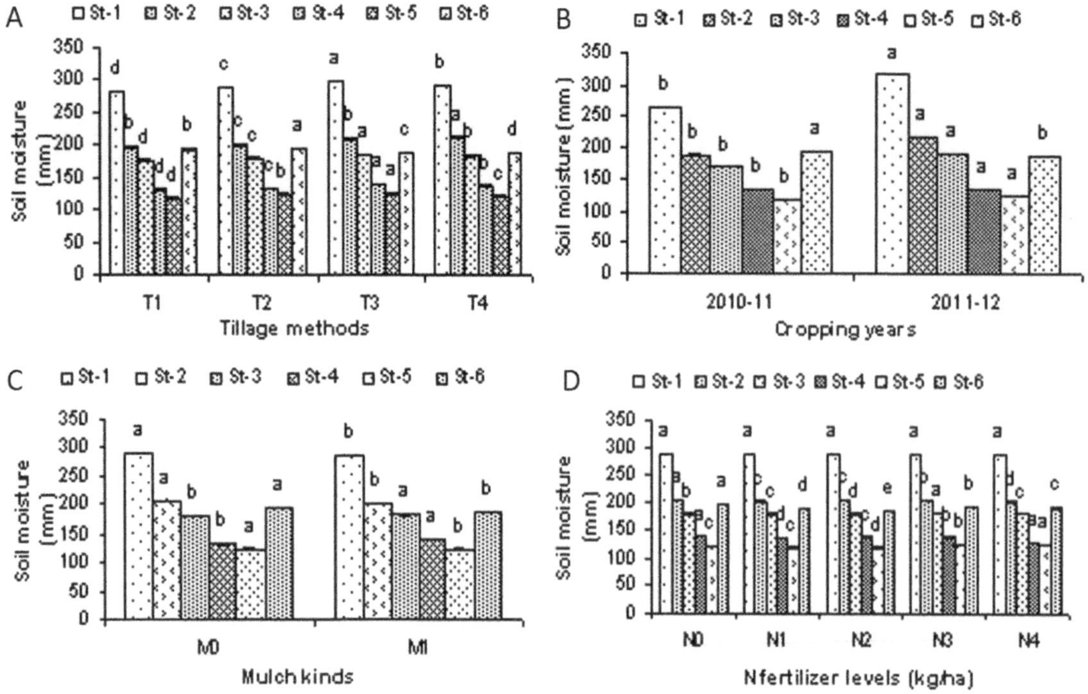

Figure 5. Total soil moisture contents (0–100 cm soil depth) as affected by different tillage methods, mulch kinds, and N fertilizer levels during two cropping years (2010–12). (A). Total soil moisture contents (0–100 cm soil depth) due to different tillage methods i.e. T_1, chisel plow tillage., T_2, zero tillage., T_3, rotary tillage and T_4, mold board plow tillage method., (B).Total soil moisture contents (0–100 cm soil depth) as affected by different tillage methods, different mulch kinds and different N fertilizer levels, during two cropping years (2010–12).(C), Total soil moisture contents (0–100 cm soil depth) as affected by different mulch kinds during the two cropping years (2010–12) i.e. M_0, no mulch and M_1, corn residue mulch., (D). Total soil moisture contents (0–100 cm soil depth) as affected by different N fertilizer levels, during two cropping years (2010–12) including, N_0, 0 kg N/ha., N_1, 80 kg N/ha, N_2, 160 kg N/ha., N_3, 240 kg N/ha and N_4, 320 kg N/ha., (*) Stage-1, (Crop sowing stage), Stage-2, (Crop revival stage), Stage-3, (Stem elongation stage), Stage-4, (Booting stage), Stage-5, (Grain formation stage) and Stage-6 (Crop harvesting stage).

year 2011–12 in comparison to cropping year 2010–11 (Fig. 6). The use of corn residue mulch increased the SOC compared to the non-mulched treatments (Fig. 6), and higher over all SOC contents were recorded in the cases of N_1 (80 kg N/ha) and N_2 (160 kg N/ha) in comparison to all of the other N fertilizer levels treatments (Fig. 6). The data show that CO_2 emissions were lower in those treatments that had higher SOC contents (Fig. 3; Fig. 6).

Effects of seasonal variations on CO_2 emissions

Seasonal temperature variations had significant effects on the CO_2 emissions. Because there were normal temperatures when the wheat crops were sown, higher CO_2 emissions were recorded, but reduced emissions were recorded with the decline in temperature during the winter seasons. With increased of crop growth and ascending of seasonal temperatures higher CO_2 emissions were recorded during both years (Figs. 2, 3, 4, 7, 8,).

Discussion

Soil CO_2 fluxes

Variations in seasonal temperatures had significant effects on soil temperatures, which ultimately affected the CO_2 emissions (Figs. 2, 3, 4, 7, 8). Higher CO_2 fluxes were recorded immediately after the tillage operations, which continued for a few days, and these emissions decreased with the passage of time (Fig. 2). Our results are in agreement with other findings [35]. Other investigators have reported changes in CO_2 emissions with seasonal variations. According to these investigators, seasonal variations in CO_2 emissions are controlled not only by the soil temperatures and soil moistures but also by the tillage practices. Changes in CO_2 emissions from seasonal variations have been reported for almost all ecosystems. These emissions mainly depend both on the type of climate and the ecosystem [36].

Similar findings regarding variations in CO_2 fluxes with changes in soil temperatures and crop growth stages have been reported for rice crops [37]. In our study, the air and soil temperatures had significant effects on CO_2 emissions. Thus with the decrease in soil temperatures during the winter months, the CO_2 emissions decreased, and the CO_2 emissions again increased with the rise in soil temperatures during the summer months (Figs. 2, 3, 4, 7, 8).

Generally, the seasonal CO_2 emissions variations found in our experiment were similar to other findings [38]. Other investigators reported that these emission variations might be related to variations in autotrophic and in heterotrophic respiration because both are involved in soil CO_2 emissions. In addition a large amount of CO_2 is released from plant roots, during the continuation of the plant-energy system. Microbial and root respiration can also significantly contribute to CO_2 emissions.

Tillage method effects on CO_2 emissions

Higher CO_2 emissions were recorded immediately after the tillage operations, which continued for a few days (Fig. 2). Our results are in agreement with findings from other researchers, who reported that CO_2 emissions following tillage increased up to 2–15 times [39], [40], [41], [42] and [43]. According to these findings [39], instead of microbial activity, the basic reason for higher CO_2 emissions immediately after the tillage was actually the release of entrapped CO_2 from the soil pores as a result of physical

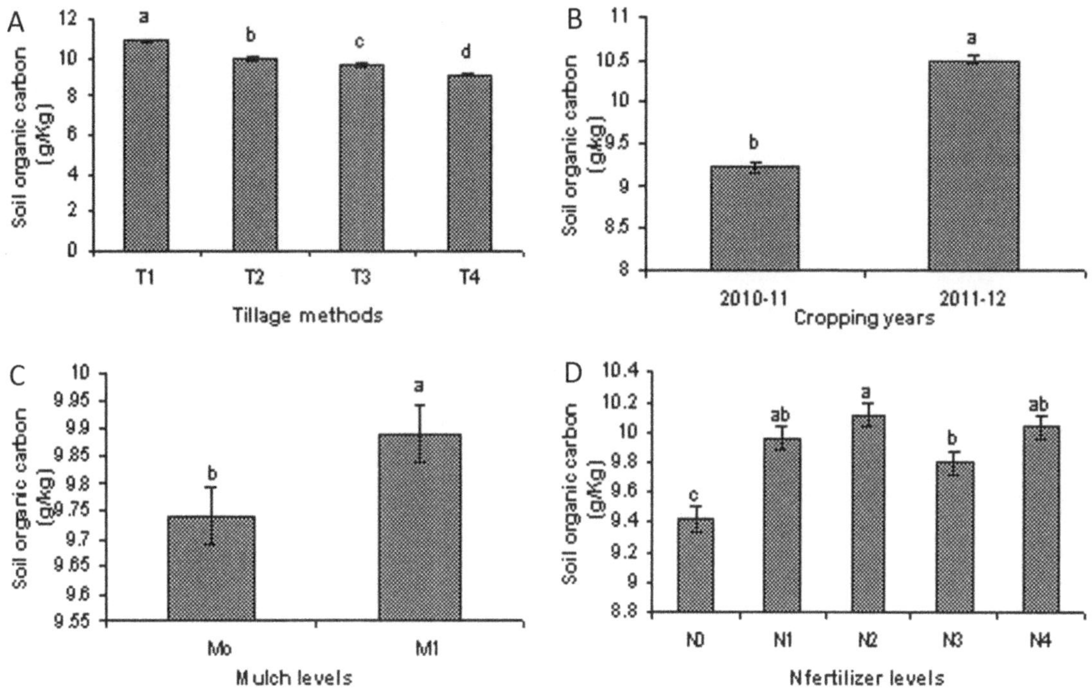

Figure 6. Soil organic carbon (SOC) contents (g/kg) as affected by different tillage methods, different mulch kinds and different N fertilizer levels during different cropping years (2010–12). (A), Soil organic carbon (SOC) contents as affected by different tillage methods i.e. T_1, chisel plow tillage., T_2, zero tillage., T_3, rotary tillage and T_4, mold board plow tillage method., (B), Mean soil organic carbon (SOC) contents as affected by different tillage methods, mulch kinds and different N fertilizer levels during two cropping years (2010–12)., (C), Soil organic carbon (SOC) contents as affected by different mulch kinds i.e. M_0, no corn residue mulch and M_1, corn residue mulch., (D), Soil organic carbon (SOC) contents as affected by different N fertilizer levels during two cropping years (2010–12) including, N_0, 0 kg N/ha., N_1, 80 kg N/ha., N_2, 160 kg N/ha., N_3, 240 kg N/ ha and N_4, 320 kg N/ha.

operations. The other reasons for these higher emissions might be that (1) tillage operations break soil aggregates and expose their organic matter to microbial attack [44], [45]; (2) tillage operations encourage the mineralization of soil organic matter by incorporating crops residues into the soil [46]; and (3) tillage operations enhance soil aeration [43]. In our study, the tillage methods had significant effects on the CO_2 emissions and, overall, the rotary tillage and mold board plow tillage methods led to higher CO_2 emissions compared to the chisel plow and zero tillage planting methods (Fig. 2 and Fig. 3). Similar findings under different tillage systems have been previously reported [13]. These researchers reported significantly more CO_2 emission fluxes from the fields tilled by a mold board plow relative to the fields prepared by the chisel plow methods. According to these researchers, the basic reason for the greater emission fluxes from the mold board plow tillage method compared to the chisel plow tillage method was the depth and extent to which the soil was disturbed by using the different tillage implements. In our experiment, higher soil temperatures in the top 5 cm depth (Fig. 4) and generally higher moisture contents (Fig. 5) in the rotary tillage and mold board plow methods might be responsible for the higher emissions, in addition to the soil preparation depths. The tillage depths also resulted in a reduction of the SOC in the top 0–10 cm soil layers for the rotary tillage and mold board plow methods, compared to the chisel plow tillage and zero tillage planting methods (Fig. 6).

Corn residue mulch effects on CO_2 emissions

Weekly CO_2 emissions data (Figs. 2, 4) show that during cropping year 2010–11, the application of corn residue mulch caused an overall increase in CO_2 emissions compared to the non-residue mulched treatments, but during the cropping year 2011–12, fewer CO_2 emissions were recorded in response to the application of corn residue mulch relative to the non-residue mulched treatments (Fig. 2 and Fig. 4). Although more CO_2 emissions fluxes were recorded in the corn residue mulched treatments in comparison to the non-residue mulched treatments, these might be due to the more microbial activities in the corn residue mulched treatments, which might have increased the SOC with the passage of time (Fig. 6). However, the two year mean data show that statistically non significant differences were recorded for CO_2 emissions following the use of the corn residue mulched or non mulched treatments. These findings might be explained by noting that both years CO_2 collecting chambers were fixed before the application of corn residue mulch. As a result there were fewer corn residues within the CO_2 collecting chambers from the corn residue mulched treatments. This smaller amount might be the reason why no CO_2 emission differences were recorded following the applications of different corn residue mulch treatments. The other reason might be that during the cropping year 2010–11, the application of corn residue mulch increased the CO_2 emissions but this mulch also increased the SOC contents of the corn residue mulched treatments, possibly resulting in the lower emission of CO_2 from the corn residue mulched treatments during the year 2011–12 (Fig. 3). A modeling study reported [47] that instead of applying higher rates of fertilizers, the use of crop residues or manure amendments would mitigate GHG emissions more efficiently. Similarly, it has been reported [48] that applications of straw increased the SOC sequestration in the soil which ultimately influenced the temporal patterns of CO_2 emissions from the soil.

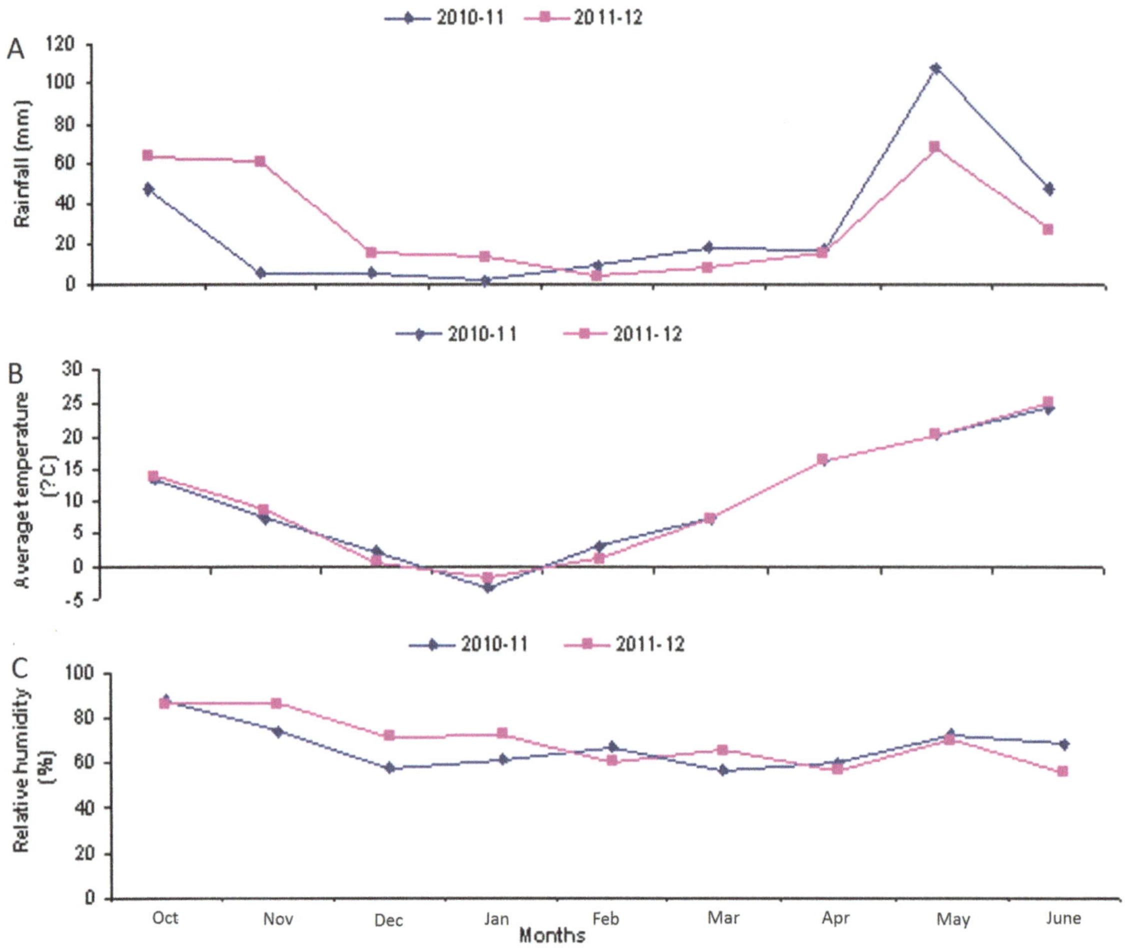

Figure 7. Monthly rainfalls, average temperatures and average relative humidity's of the study area during two wheat crop growing seasons (2010–12). (A).Rainfalls of the study area during two wheat crop growing seasons (2010–12), (B) Average temperatures of the study area during two wheat crop growing seasons (2010–12) (C) Average relative humidities of the study area during two wheat crop growing seasons (2010–12). (*). R.F-1, Rainfalls during cropping year 2010–11., R.F-2, rainfalls during cropping year 2011–12., Av.temp-1, Monthly average temperatures during the cropping year 2010–11., Av.temp-2, Monthly average temperatures during the cropping year 2011–12., R.H (%)-1, Average monthly relative humidities during the cropping year 2010–11., R.H (%)-2, Average monthly relative humidities during the cropping year 2011–12.

N fertilizer level effects on CO_2 emissions

Very few studies regarding CO_2 emissions in relation to the different tillage methods, corn residue mulch and N fertilizer levels have previously been reported in this region of China. It is expected that the application of inorganic N fertilizers along with organic materials will affect the mineralization of soil organic matter and crop productions, which will ultimately affect CO_2 emissions [8].

However variations in CO_2 emissions following fertilizer applications have been reported for different areas of China. Some scientists [15] have reported that the fertilizer applications suppresses CO_2 emissions and others [49] have reported that fertilizer application enhances CO_2 emissions. Moreover, some other scientists [19] have reported that fertilizer applications have no effects on CO_2 emissions.

Our study shows that the use of 80 kg and 160 kg N/ha, suppressed CO_2 emissions when compared with the 0 N fertilizer level, but that further increases in N fertilizer application rates enhanced CO_2 emissions. Higher emissions from the $N_{0,}$ nitrogen fertilizer treatments might be explained by noting that plants under unfertilized N treatments are considered to respond to a relative shortage of N by increasing the plant's carbon allocation to its structures and functions, which are responsible for N acquisition [50]. In our study, the use of different levels of N fertilizer relative to the no nitrogen fertilizer level significantly increased crop yields (Fig. 1).This result shows that the higher CO_2 emission fluxes in response to higher levels of N fertilizer might be explained by the increased use of C for microbial growth [51] and it might also be explained by the less efficient use of carbon by the microbial biomass, which resulted in a greater proportion of carbon loss in the form of CO_2 fluxes [52].

In our case, the use of corn residue mulch in combination with different levels of N fertilizer increased CO_2 emissions following the chisel plow, zero tillage and rotary tillage planting methods (Table 1). Our results regarding CO_2 emissions following the combination of N fertilizer and organic amendments are in agreement with the results of other scientists [53], [38] and [54]. These investigators reported a higher CO_2 emission flux from the treatments that utilized both fertilizers and organic manures. This finding shows that up to a certain extent the use of higher N fertilizer levels suppresses CO_2 emissions but in our case, higher levels of N fertilizer (i.e., 240 and 320 kg N/ha) enhanced CO_2 emissions. These results are contrary to the findings of other

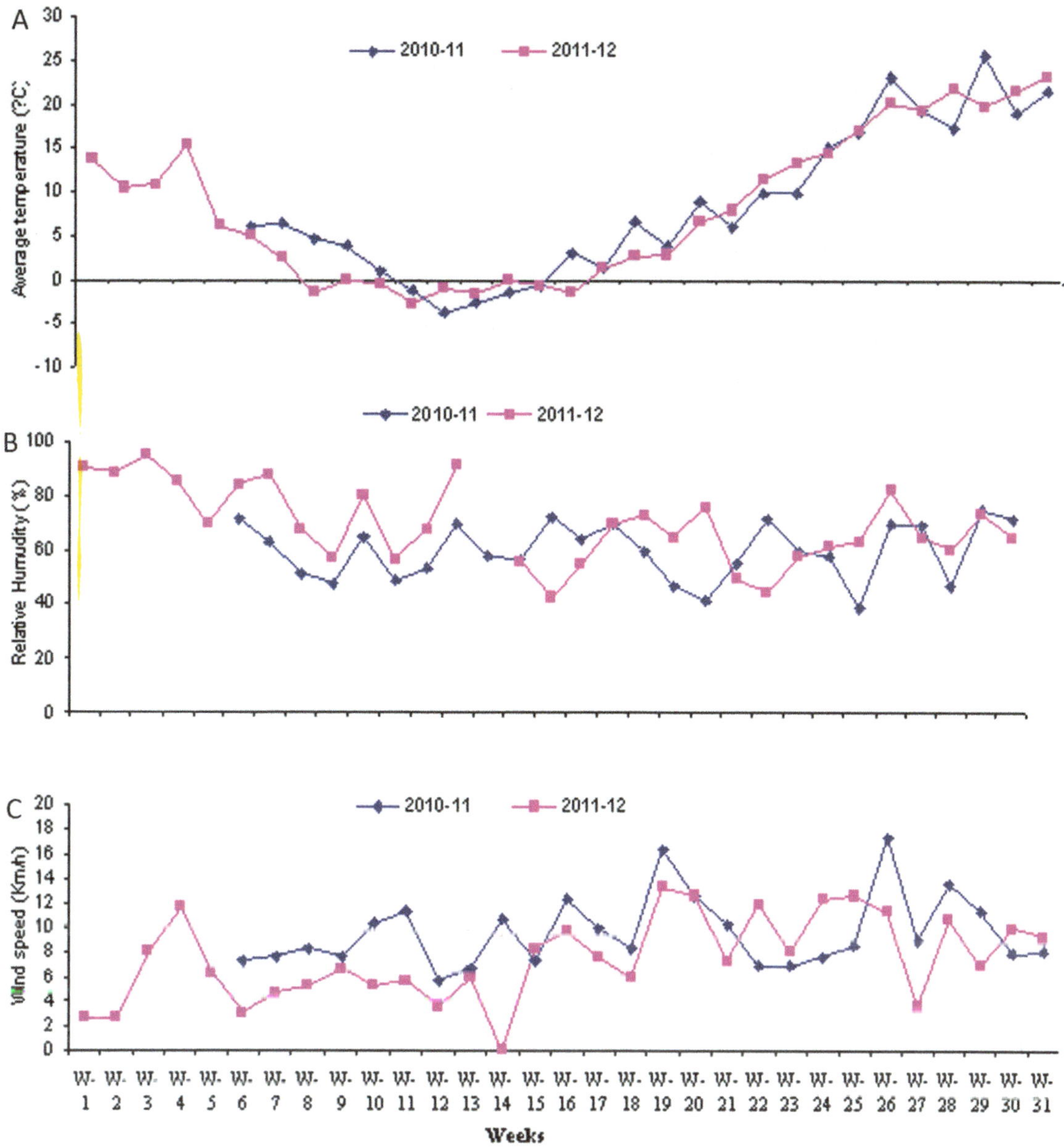

Figure 8. Average weekly temperatures, relative humidity's and wind speeds during the two wheat crop growing seasons (2010–12). (A) Average weekly temperatures of the study area during the two wheat crop growing seasons (2010–12), (B).Average weekly relative humidity's of the study area during the two wheat crop growing seasons (2010–12), (C) Average weekly wind speeds of the study area during the two wheat crop growing seasons (2010–12).

scientists [30] and [31]. These investigators reported no clear reasons for the CO_2 emission reductions.

Cropping year effects on CO_2 emissions

Two years of data on a weekly as well as on a cumulative basis show that there were more CO_2 emissions from all of the treatments during cropping year 2010–11 compared to the CO_2 emissions during cropping year 2011–12 (Figs. 2, 3). The main reasons for these lower emissions during cropping year 2011–12 were, on the whole, lower air temperatures, higher relative humidities and higher over all rainfall concentrations during the early periods of wheat crop growth relative to cropping year 2010–11 (Fig. 7). These factors all ultimately reduced the soil temperatures, which resulted in reduced CO_2 emissions during the cropping year 2011–12 compared to cropping year 2010–11 (Figs. 4, 7, 8). Another reason for the lower emissions might be an increase in the SOC from the corn residue mulched treatments, which might have ultimately helped to reduce the CO_2 emissions during cropping year 2011–12 in comparison to cropping year 2010–11 (Figs. 2, 3, 6).

Effects of soil temperatures, soil moisture and soil organic carbon on CO_2 emissions

Respiration of ecosystem mainly depends on both the heterotrophic (microbe) and autotrophic (plant) activities and both of these factors are controlled by the prevailing environmental conditions (basically temperature and water availability), availability of carbohydrates and substrates and others [55], [56] and

Table 2. Q10 values of the different treatments during the different cropping years (2010–12).

Treatments	chisel plow		zero tillage		rotary tillage		mold board plow tillage	
	2010–11	2011–12	2010–11	2011–12	2010–11	2011–12	2010–11	2011–12
	$Y = 69.220e^{0.0516x}$	$Y = 47.81\ e^{0.0898x}$	$Y = 73.371e^{0.0736x}$	$Y = 74.836e^{0.807x}$	$Y = 140.49e^{0.0427x}$	$Y = 67.628\ ^{0.0749x}$	$Y = 287.91e^{0.0707x}$	$Y = 78.941e^{0.1001x}$
N_0	$R^2 = 0.7076$	$R^2 = 0.8064$	$R^2 = 0.0823$	$R^2 = 0.7213$	$R^2 = 0.7472$	$R^2 = 0.7372$	$R^2 = 0.6076$	$R^2 = 0.7439$
	Q10 = 1.68	Q10 = 2.43	Q10 = 2.09	Q10 = 1.90	Q10 = 2.54	Q10 = 2.55	$Y = 287.91e^{0.0707x}$	Q10 = 2.72
	$Y = 79.76e^{0.0744x}$	$Y = 73.979e^{0.0816x}$	$Y = 57.341e^{0.0845x}$	$Y = 62.738e^{0.0996x}$	$Y = 142.39e^{0.0691x}$	$Y = 93.744e^{0.092x}$	$Y = 247.36e^{0.0385x}$	$Y = 129.37e^{0.0818x}$
N_1	$R^2 = 0.6672$	$R^2 = 0.7158$	$R^2 = 0.0869$	$R^2 = 0.752$	$R^2 = 0.8187$	$R^2 = 0.8074$	$R^2 = 0.5688$	$R^2 = 0.736$
	Q10 = 2.10	Q10 = 2.26	Q10 = 2.33	Q10 = 2.71	Q10 = 1.63	Q10 = 2.11	Q10 = 2.03	Q10 = 2.72
	$Y = 80.442e^{0.0827x}$	$Y = 0.787e^{0.1012x}$	$Y = 114.12e^{0.0509x}$	$Y = 57.382e^{0.0891x}$	$Y = 90.813e^{0.0616x}$	$Y = 122.76e^{0.0669x}$	$Y = 108.97e^{0.0634x}$	$Y = 79.756e^{0.0873x}$
N_2	$R^2 = 0.5944$	$R^2 = 0.8201$	$R^2 = 0.528$	$R^2 = 0.7701$	$R^2 = 0.790$	$R^2 = 0.6534$	$R^2 = 0.8134$	$R^2 = 0.7563$
	Q10 = 2.29	Q10 = 2.75	Q10 = 1.66	Q10 = 2.70	Q10 = 2.00	Q10 = 2.51	Q10 = 1.47	Q10 = 2.27
	$Y = 125.25e^{0.1085x}$	$Y = 70.222e^{0.0931x}$	$Y = 82.409e^{0.0478x}$	$Y = 55.391e^{0.1006x}$	$Y = 176.74e^{0.097x}$	$Y = 149.95e^{0.0859x}$	$Y = 103.92e^{0.0614x}$	$Y = 69.581e^{0.0843x}$
N_3	$R^2 = 0.7977$	$R^2 = 0.7054$	$R^2 = 0.7439$	$R^2 = 0.7937$	$R^2 = 0.6429$	$R^2 = 0.7876$	$R^2 = 0.852$	$R^2 = 0.6457$
	Q10 = 2.96	Q10 = 2.47	Q10 = 2.33	Q10 = 2.02	Q10 = 1.85	Q10 = 1.95	Q10 = 1.89	Q10 = 2.32
	$Y = 55.139e^{0.0653x}$	$Y = 61.6e\ ^{0.0904x}$	$Y = 43.884e^{0.0903x}$	$Y = 60.312e^{0.8051x}$	$Y = 90.419e^{0.058x}$	$Y = 221.26e^{0.0624x}$	$Y = 116.85e^{0.0613x}$	$Y = 96.28e^{0.1086x}$
N_4	$R^2 = 0.6443$	$R^2 = 0.7961$	$R^2 = 0.7996$	$R^2 = 0.9015$	$R^2 = 0.6951$	$R^2 = 0.6654$	$R^2 = 0.7312$	$R^2 = 0.7969$
	Q10 = 1.02	Q10 = 2.18	Q10 = 1.61	Q10 = 2.16	Q10 = 2.64	Q10 = 2.32	Q10 = 1.85	Q10 = 2.96

N_0, 0 kg N/ha., N_1, 80 kg N/ha., N_2, 160 kg N/ha., N_3, 240 kg N/ha and N_4, 320 kg N/ha.

Table 3. Q10 values of the different treatments during the different cropping years (2010–12).

Treatments	chisel plow		zero tillage		rotary tillage		mold board plow tillage	
	2010–11	2011–12	2010–11	2011–12	2010–11	2011–12	2010–11	2011–12
	$Y=43.137e^{0.1316x}$	$Y=87.49e^{0.0778x}$	$Y=163.72e^{0.0363x}$	$Y=61.295e^{0.09x}$	$Y=62.946e^{0.1111x}$	$Y=99.78e^{0.0961x}$	$Y=183.66e^{0.0518x}$	$Y=55.625e^{0.0947x}$
N_0+M	$R^2=0.8849$	$R^2=0.6632$	$R^2=0.4067$	$R^2=0.8372$	$R^2=0.7113$	$R^2=0.8096$	$R^2=0.753$	$R^2=0.8216$
	Q10 = 3.72	Q10 = 2.69	Q10 = 2.47	Q10 = 2.10	Q10 = 1.79	Q10 = 2.96	Q10 = 1.85	Q10 = 2.58
	$Y=254.21e^{0.0484x}$	$Y=0.0988e^{0.0516x}$	$Y=84.882e^{0.0413x}$	$Y=49.272e^{0.0965x}$	$Y=175.25e^{0.0557x}$	$Y=76.273e^{0.0858x}$	$Y=112.84e^{0.0703x}$	$Y=51.454e^{0.0957x}$
N_1+M	$R^2=0.7689$	$R^2=0.8367$	$R^2=0.7109$	$R^2=0.7881$	$R^2=0.5183$	$R^2=0.8149$	$R^2=0.695$	$R^2=0.7376$
	Q10 = 1.62	Q10 = 2.52	Q10 = 1.44	Q10 = 2.22	Q10 = 3.04	Q10 = 2.58	Q10 = 1.68	Q10 = 2.60
	$Y=72.653\ ^{0.0809x}$	$Y=45.736e^{0.0926x}$	$Y=80\ 339e^{0.0722x}$	$Y=61.262e^{0.0907x}$	$Y=163.02e^{0.0519x}$	$Y=104.06e^{0.083x}$	$Y=93.668e^{0.0666x}$	$Y=46.012e^{0.1166x}$
N_2+M	$R^2=0.8237$	$R^2=0.7309$	$R^2=0.8477$	$R^2=0.7992$	$R^2=0.4896$	$R^2=0.7173$	$R^2=0.6503$	$R^2=0.816$
	Q10 = 2.25	Q10 = 2.87	Q10 = 1.52	Q10 = 2.18	Q10 = 1.95	Q10 = 2.36	Q10 = 2.02	Q10 = 3.17
	$Y=1.015e^{0.0898x}$	$Y=46.994e^{0.1054x}$	$Y=164.85e^{0.0502x}$	$Y=30.192e^{0.1197x}$	$Y=163.9e^{0.405x}$	$Y=40.179e^{0.0983x}$	$Y=75.02e^{0.0742x}$	$Y=72.535e^{0.1075x}$
N_3+M	$R^2=0.8604$	$R^2=0.8214$	$R^2=0\ 4050$	$R^2=0.8784$	$R^2=0.6464$	$R^2=0.84$	$R^2=0.7363$	$R^2=0.8221$
	Q10 = 2.45	Q10 = 1.11	Q10 = 2.06	Q10 = 2.18	Q10 = 1.68	Q10 = 2.29	Q10 = 1.95	Q10 = 2.93
	$Y=8.69e^{0.10179x}$	$Y=69.932e^{0.0859x}$	$Y=51.822e^{0.0641x}$	$Y=36.672e^{0.1073x}$	$Y=287.97e^{0.0937x}$	$Y=121e^{0.1050x}$	$Y=127.88e^{0.0702x}$	$Y=73.455e^{0.1177x}$
N_4+M	$R^2=0.8342$	$R^2=0.7264$	$R^2=0.7247$	$R^2=0.8364$	$R^2=0.6465$	$R^2=0.854$	$R^2=0.7656$	$R^2=0.8821$
	Q10 = 2.76	Q10 = 2.36	Q10 = 1.65	Q10 = 2.74	Q10 = 1.50	Q10 = 2.67	Q10 = 2.10	Q10 = 3.24

N_0+M, 0 kg N/ha + corn residue mulch., N_1+M, 80 kg N/ha + corn residue mulch., N_2+M, 160 kg N/ha + corn residue mulch., N_3+M, 240 kg N/ha + corn residue mulch., N_4+M, 320 kg N/ha + corn residue mulch.

[57]. Many studies have shown that seasonal variations in CO_2 emissions were mainly caused by the soil temperature, soil moisture or the combination of both these factors [58], [36].

Our study also indicates that soil temperature was an important driving force for the increased CO_2 emissions, which is also supported by the Q10 values given in Table 2 and Table 3. It has also been reported that the CO_2 evolution rate significantly increases with an increase in temperature and moisture [37]. Our results are also in agreement with many other field studies, which have shown strong relationships between soil temperatures and CO_2 flux rates [59], [60], [17] and [18]. Additionally, our results agree with the findings of [61], who reported a stronger polynomial for temperature and moisture interaction ($r^2 = 0.89$) than for temperature alone ($r^2 = 0.47$). Many previous studies have reported that changes in crop management practices, i.e., the appropriate use of tillage operations, proper fertilization, crop residue applications and crop rotations can be helpful for managing soil organic matter, e.g.,. [62].

Our study also shows that the results at the end of cropping year 2010–11 revealed more SOC in the top 0–10 cm of soil than that at the end of cropping year 2011–12. This result shows that the adaptation of different management practices, i.e., the application of crop residues, increased the SOC contents, especially following chisel plow tillage and zero tillage planting compared with rotary tillage and mold board plow tillage. SOC sequestration is a long term processes and various results have been previously reported, i.e. [63] and [64], have reported that conservation tillage is a recommended management practice for agricultural ecosystems that can enhance the pool of soil organic carbon (SOC), in the soil. Following an analysis of global data, NT reportedly sequestered carbon at an average rate of 0.57 Mg C ha^{-1} compared with the mold board plow [65]. It has also been reported that increases in the SOC pools can be credited to either reductions in the CO_2 efflux from the soil or to increases in the C inputs [65]. When comparing the soil surface across different tillage systems, conservation tillage systems retain more crop residues, which ultimately result in the formation of more SOC [66] and [67]. In addition to this finding, the decomposition process of surface applied plants residues as a part of conservation tillage is slow compared to conventional tillage systems because of lower contact with the soil.

In our study a negative but highly significant correlation coefficient (r) value i.e. −0.19403** was recorded between the SOC and cumulative CO_2 emission. This finding might help in the reduction of CO_2 emissions from the chisel plow tillage and zero tillage planting methods compared to the rotary tillage and mold board plow methods (Figs. 2 and 3). In our experiment, increases in the SOC contents from different tillage methods, especially chisel plow tillage and zero tillage, might be related to the lower disturbance of the soil, retention of more crop residues on the soil surface and reductions in the efflux of CO_2 from the soil. Similar results have been reported by [65], [66] and [67].

Conclusions

Intensive tillage and higher N fertilization are not only detrimental to the soil but are also destructive for the entire environment. Adaptations of appropriate tillage methods, crop residue applications and proper fertilization are beneficial for the soil as well as for the environment. These practices are also economically beneficial for resource-poor farmers. The findings from our study clearly indicate that the tillage methods significantly affected CO_2 emissions and the zero tillage planting method emitted the lowest CO_2 compared with the other three tillage methods. No significant differences in CO_2 emissions were recorded for the applications of corn residue mulch, but the applications of corn mulch significantly improved the soil organic carbon (SOC) contents of the soil for all of the tillage systems. In addition, corn residue mulch application reduced the weed infestation by up to 40% (data not shown). Therefore, the application of corn residue mulch can be helpful for reducing the use of herbicides, which will also be helpful in establishing a healthy environment. Applications of different N fertilizer levels also significantly affected CO_2 emissions and, overall, the lowest emissions were recorded for the 160 kg N/ha treatment. Higher CO_2 emissions were recorded immediately after the tillage operations. This study also indicated that both soil temperatures and moistures strongly affected CO_2 emissions and that compared with the other tillage methods, zero tillage planting gave better grain yields. The lowest N fertilizer use gave equal yields, for the two year mean, as did the application of higher N fertilizer levels. These results clearly indicate that proper changes in farm management practices i.e., the adoption of zero tillage, crops residue application and optimum use of N fertilizers can reduce CO_2 emissions from soils. Therefore this type of long term study can be further helpful in reducing the emissions of CO_2 from soils, which will be helpful in reducing the use of inorganic fertilizers. These practices will be helpful in reducing production costs and will be beneficial for the entire environment.

Materials and Methods

Experimental site

A two-year (2010–12)), field study was conducted at the experimental area of Northwest A&F University, Shaanxi Province, northwestern China (latitude of 34°20° N, longitude of 108° 04°E and elevation of 466.7 m above sea level) on the Eum-Orthrosols (Chinese soil Taxonomy) soil, with a mean bulk density of approximately 1.29 g cm^3. The soil in the top 40 cm had an SOC of approximately 14.26 g/kg, total nitrogen 0.74 g/kg and the pH was approximately 7.85. This area is under the corn-wheat rotation system. During both years, the wheat crop was planted after harvesting the corn crop. Both fertilizers, i.e., phosphorous in the form of calcium phosphate (Ca_2 $(PO_4)_3$ with 16% P and nitrogen in the form of urea with N≥46%, were applied to the corn crop at the rate of 750 kg/ha and 375 kg/ha, respectively. The total rainfall during the wheat crop growing season (October-June) was 231.6 mm and 242.7 mm during, cropping years 2010–11 and 2011–12 respectively (Fig. 7).

Experimental design and treatments

A factorial experiment having a strip split-split arrangement, with tillage methods in the main plots, mulch levels in the sub plots and N-fertilizer levels in the sub-sub plots with three replicates, was used for this study. Different tillage methods, i.e., chisel plow tillage (T_1), zero tillage (T_2), rotary tillage (T_3) and mold board plow tillage (T_4) methods, were kept in the main plots, different mulch kinds, i.e., M_0 (no residue mulch) and M_1 (corn residue mulch), in the sub plots, while different nitrogen fertilizer rates, i.e., 0, 80, 160, 240 and 320 kg N/ha, were kept in the sub-sub plots. The area was uniform in terms of fertility. The total experimental plot area (3300 m^2) was equally divided into four main tillage treatments. The area of each tillage treatment (33 m×25 m) was further sub-divided into sub-plots for mulch treatments, and finally the sub plots were further divided into sub-sub plots, and each sub-sub individual plot had an area of 3 m×25 m. Treatments were randomized within each sub-plot.

Chisel plow tillage (T_1), was performed using a chisel plow. Following fertilizer applications, a chisel plow with a shank spacing of approximately 40 cm apart and 30–35 cm deep was used once. Later on, fertilizers were mixed by using the rotavator for up to a 5 cm depth. For zero tillage (T_2) fertilizers were mixed by using the rotavator up to a 5 cm depth due to the lack of availability of a proper zero tillage drill. For rotary tillage (T_3), the seed bed was prepared by using the rotavator up to a depth of 15–20 cm, and in the case of the mold board. plow tillage method (T_4), the soil was plowed up to a 20–30 cm depth by using the mold board plow, followed by the rotavator for the final seed bed preparation.

Urea fertilizer with N≥46%, was used as the source of the nitrogen, and phosphorous (P) fertilizer in the form of calcium phosphate ($Ca_2\ (PO_4)_3$ with 16% P was equally applied to all of the treatments at a rate of 750 kg/ha at the time of soil preparation. The treatments arrangements were kept the same during both years (2010–12) of study. Previously harvested air-dried corn crops residues were used as the source of corn residue mulch during both years. When the wheat crop was at the 3–4 leaf stage, mulch was applied at a rate of 750 g/m^2. The field was flat with a uniform topography. This area is rain fed, and a wheat-corn rotation is the main cropping system. No irrigation was applied to either crop. No changes were made to the areas with different tillage treatments, and a corn crop was planted after the wheat crop harvest by using the same tillage methods. The wheat crop was harvested using a combine harvester, and after harvesting the wheat crop, the corn crop was planted using a corn planter.

Winter wheat (C.V Shaan mai −139) was planted on October 17, 2010 and October 18, 2011 by using wheat drills. The line to line distance was maintained at approximately 16 cm apart. The seed had a 13% moisture contents and a 85% germination rate during both years. During cropping season 2010–11, a seed rate of about @ 190–200 kg/ha was used, while during cropping season 2011–12, a seed rate of approximately 205–210 kg/ha was employed. Experimental treatments were separated from each other by making boundaries between the treatments. Both years an herbicide application i.e. carfentrazone-ethyl ($C_{15}H_{14}C_{12}F_3N_3O_3$), was used to control the weeds. At physiological maturity, which occurred on June 8, 2011 and on June 10, 2012, three samples were randomly selected from each treatment and manually harvested by using the 1 m^2 quadrants to calculate the grain yields. Finally the wheat crop was harvested using a combine harvester.

Measurements

Meteorological factors. Meteorological data for the study area are given in Fig. 7 and Fig. 8, which indicate that during cropping year 2011–12, more rains were concentrated during the early period of crop growth relative to 2010–11, during which more rains were concentrated in the later crop growth stages. Higher average seasonal temperatures were recorded during the year 2010–11 than in 2011–12. However, more relative humidity was recorded during 2011–12 than in 2010–11. Weekly meteorological data showed higher, average temperatures during the year 2010–11 than during 2011–12. On the other hand, higher weekly relative humidities were recorded during the cropping year 2011–12 than during 2010–11. Similarly, higher weekly basis more wind speeds were recorded during the cropping year 2010–11 than during 2011–12.

Monitoring CO_2 emissions. Because of the high number of treatments, both years of CO_2 emissions data were recorded three times per week and one day of CO_2 emissions data was used as one replicate for statistical analysis. CO_2 emissions data were recorded using a GXH-3010E1 portable gas analyzer. This gas analyzer is made by the Beijing Huayun Yiqi Company, and the CO_2 emission was recorded by using the method described by Gao et al [68]. During both years (2010–12) CO_2 emissions data from all of treatments was recorded 6840 times, which included 3120 times during the cropping year 2010–11 and 3720 times during cropping year 2011–12.

One round PVC chamber (21 cm in diameter and 13.5 cm in height), having total area of approximately 0.0047 m^3 was permanently fixed in the center of each treatment plot. The chamber was completely fixed in the soil up to a 4.5 cm depth. The plants growing with in the chambers were removed. As a consequence of some technical problems with the Gas analyzer, during wheat growing season 2010–11, the CO_2 emissions data were recorded after the wheat had been planted for one month, and during cropping season, 2011–12, CO_2 emissions were recorded starting during the first week of wheat planting. However, due to severe snow fall, data were not recorded on the 14th week during this year. Taking data from wheat planting until harvest for 2010–11, the CO_2 emissions data were recorded for 26 weeks, and during cropping year 2011–12, these data were recorded for 31 weeks. Every week, the data were collected 3 times depending upon the environmental condition, i.e., if the field was too wet from rainfall, then the data recording was stopped. Each time the data collection started at 900 a.m., and each sequence of CO_2 flux measurements took at least 4 hours. Due to the large number of treatments the data were randomly collected from different treatments. The main purpose of this randomization was to minimize the effects of different days as well as changing soil temperatures on the emissions of CO_2. The GXH-3010E1 gas analyzer was attached to the data collector chambers with intake and an outtake tubes. These tubes were made up of soft plastic pipes and each one was approximately one meter in length. At the time of data recording, first the CO_2 data, i.e., X_1 were recorded without closing/covering the chamber, and then the chamber top was tightly closed with a cover that had a fixed small fan. The gas within the chamber was mixed for three minutes with this fan. After this CO_2 emission, data (X_2) were recorded using the gas analyzer.

Chambers tops were closed for only three minutes at the time of data recording. To avoid any chemical change in the soil, these chambers were kept open for the whole remaining time. Along with CO_2 data, soil temperatures data were also recorded from each treatment from depths of 5 cm. For this purpose, thermometers were permanently fixed in each plot each year for the whole crop growth period. The CO_2 emission rate was calculated by using equation (1) as described by Gao et al. [68].

$$F = k(X_2\text{-}X_1)H/\Delta t \tag{1}$$

Where F is the CO_2 emission in mg/(m^2.h); K is a constant with a value of 1.80 (25°C) and X_1 and X_2 are the CO_2 emissions rates from the chambers before and after covering of the chambers. H is the height of the chambers in meters and Δt is the time in hours (h). The cumulative emission of CO_2 was calculated using the following relationship, as described by Wilson H.M and Alkaisi W.W (2008) [69],

$$CO_2 - C\left(kgha^{-1}\right) = \sum_{i=first}^{n=last} X_i + X_{i+1} * N + X_{i+2} * N + \ldots + X_{i+n} * N \tag{2}$$

where (i) is the first week of the growing season when the first CO2 emission rate was taken, (n) is the last week of the growing season when the last CO2 emission rate was taken, X is the CO2 rate (Kg ha−1 day−1), and N is the number of days between two consecutive CO2 emission rates measurements. Finally, these CO2 emission rates for the whole wheat crop growing period (taken between 9.00 A.M until approximately 1.00 P.M) were converted into tons/ha.

Soil moisture measurements. Both years soil moisture contents were measured during different crop growth stages, i.e., before planting the wheat crops, at the 5 leaf stage (Zadoks stage 1.0–1.9), stem elongation stage (Zadoks stage 3.0–3.9), booting stage (Zadoks stage 4.0–4.9), grain formation stage (Zadoks stage 7.0–7.9) and harvesting stage. For this purpose, soil samples were collected from each treatment from 0–100 cm soil depths with three replicates, with increments of 0–10 cm, 10–20 cm, 20–30 cm. 30–40 cm, 40–60 cm, 60–80 cm and 80–100 cm. Soil samples were collected in aluminum boxes using a hand auger, and fresh soil samples were immediately transported to the laboratory. After recording the fresh weights, these samples were dried in an oven at 105°C for at least 48 hours, and the soil water contents were then measured using equation (3) given below.

$$\text{Soil water content (\%)} = \frac{\text{fresh weight of soil} - \text{dry weight of soil}}{\text{dry weight of soil}} \times 100 \quad (3)$$

Soil water contents (mm) were calculated as gravimetric moisture contents using the equation (soil water ×B.D× thickness of soil layer). Soil bulk densities from the different tillage treatments and from the different soil depths were measured using the core method. Total soil water contents in the 0–100 cm soil depths of the different treatments were measured on the basis of the bulk densities of these different soil layers.

Soil organic carbon (SOC) measurements. Both year soil SOC samples were collected from the top 0–10 cm soil depth after the wheat harvest. Soil samples were collected from two randomly chosen points from each plot using a hand augur. The samples from each treatment were then mixed together to make a composite sample of each treatment. These samples were then air-dried at room temperature, crushed gently and passed through a 2 mm sieve for further chemical analysis. Soil organic carbon was determined by the oxidation method with $K_2Cr_2O_7$ -H_2SO_4. For chemical analysis, 0.5 grams of soil was digested with 5 mL of 1 M $K_2Cr_2O_7$ and 5 mL of concentrated H_2SO_4 and was heated at 175°C for 5 minutes followed by titration of the digests with $FeSO_4$ [70].

Statistics

Annual data collected for the CO_2 emission rates and for other related parameters over the whole 2-year period were subjected to an analysis of variance (ANOVA) by using the factorial experiment with the strip-split-split arrangement having the tillage methods in the main plots, mulch in the sub plots and N-fertilizer levels in the sub-sub plots. The SAS analytical software package GLM (8.01) was used for the analyses. Mean values and standard errors (SE) were calculated for each treatment, and an ANOVA was used to assess the treatment effects on the measured variables. Means were declared statistically significant at a 0.05 probability level, or P≤(0.05), using the DUNCAN test (DNMRT).

Author Contributions

Conceived and designed the experiments: SKT YL. Performed the experiments: SKT XLL. Analyzed the data: SKT JLZ. Wrote the paper: SKT XXW YL.

References

1. Post WM, Peng TH, Emanuel WR, King AW, Dale VH, et al (1990) The global carbon cycle. Am Sci 78: 310–326.
2. Houghton RA, Skole DL (1990) Carbon, In: Turner BL, Clark WC, Kates RW, Richards JF, Mathews JT, Meyer WB, editors. The earth as Transformed by Human Action. Cambridge University Press. 393–408.
3. Houghton RA (2007) Balancing the global carbon budget. Ann Rev Earth Planet Sci 35: 313–47.
4. Denman KL, Brasseur G, Chidthaisong A, Ciais P, Cox PM, et al. (2007) In Climate Change 2007: The Physical Science Basis. Contribution of Working Group 1 to the Fourth Assessment Report of the International Panel on Climate Change. Solomon, S., et al, editors. Cambridge Univ. Press. 499–587.
5. Valentini R, Matteucci G, Dolman AJ, Schulze ED, Rebmann C, et al. (2000) Respiration as the main determinant of carbon balance in European forests. Nature 404: 862–864.
6. IPCC (2007) Climate Change 2007: The Physical Science Basis. Cambridge University Press. 996.
7. Cole CV, Duxbury J, Freney J, Heinemeyer O, Minami K, et al. (1997) Global estimates of potential mitigation of greenhouse gas emissions by agriculture. Nut. Cycl. Agroecosyst. 49: 221–228.
8. Paustian K, Collins HP, Paul EA (1997) Management controls on soil carbon. In: Paul EA, Paustian K, Elliot ET, Cole CV, editors. Soil Organic Matter in Temperate Agro ecosystems (Long-term Experiments in North America). CRC Press, Boca Raton, FL. 15–49.
9. Lal R (1997) Long-term tillage and maize monoculture effects on a tropical alfisol in western Nigeria.1. Crop yield and soil physical properties. Soil and Tillage Res 42: 145–160.
10. Ahmadi H, Mollazade K (2009) Effect of plowing depth and soil moisture content on reduced secondary tillage. Agricultural Engineering International: The CIGR E Journal 11: 1–9.
11. Koga N, Tsuruta H, Tsuji H, Nakanoa H (2003) Fuel consumption – derived CO_2 emission under conventional and reduced tillage cropping systems in northern Japan, Agriculture Ecosystems and Environment 99: 213–219.
12. Schlesinger WH (1986) Changes in soil carbon storage and associated properties with disturbance and recovery. In: Trabalka JR, Reichle DE, editors. The changing Carbon Cycle- A global analysis. Springer Verlag, New York. 194–220.
13. Reicosky DC, Lindstrom MJ (1993) Fall tillage method: Effect on short – term carbon dioxide flux from soil. Agron J 85(6): 1237–1243.
14. Franzluebbers AJ, Hons FM, Zubberer DA (1995) Tillage induced seasonal changes in soil physical properties affecting soil CO_2 evolution under intensive cropping. Soil Tillage Res 34: 41–60.
15. Ding W, Cai Y, Cai Z, Zheng X (2006) Diel pattern of soil respiration in N-amended soil under maize cultivation. Atmos Environ 40: 3294–3305.
16. Ren X, Wang Q, Tong C, Wu J, Wang K, et al. (2007) Estimation of soil respiration in a paddy ecosystem in the subtropical region of China. Chin Sci Bull 52 (19): 2722–2730.
17. Iqbal J, Ronggui H, Lijun D, Lan L, Shan L, et al. (2008) Differences in soil CO_2 flux between different land use types in mid-subtropical China. Soil Bio Biochem 40: 2324–2333.
18. Liu H, Zhao P, Lu P, Wang YS, Lin YB, et al. (2008) Greenhouse gas fluxes from soils of different land-use types in a hilly area of south China. Agric Ecosyst Environ 124(1–2): 125–135.
19. Lee DK, Doolittle JJ, Owens VN (2007) Soil carbon dioxide fluxes in established switch grass land managed for biomass production. Soil Biol Biochem 39: 178–186.
20. Ladd JN, Amato M, Zhou LK, Schultz JE (1994) Differential effects of rotation, plant residue and nitrogen fertilizer on microbial biomass and organic matter in an Australian alfisol. Soil Biol Biochem 26: 821–831.
21. Edmeades DC (2003) The long-term effects of manures and fertilizers on soil productivity: a review. Nutr Cycl Agroecosys 66: 165–180.
22. Kern JS, Johnson MG (1993) Conservation tillage impacts on natural soil and atmospheric carbon levels. Soil Sci SOC Am J 57: 200–210.
23. Reeves DW (1997) The role of soil organic matter in maintaining soil quality in continuous cropping system. Soil Tillage Res 43: 131–167.
24. Smith P, Powlson DS, Glendining MJ, Smith JU (1998) Preliminary estimates of potential for carbon mitigation in European soils through no-till farming. Global Chang Biol 4: 679–685.
25. Schlesinger WH (1999) Carbon sequestration in soils. Science 248: 2095.
26. Johnson MG, Levine ER, Kern JS (1995) Soil organic matter, distribution, genesis, and management to reduce greenhouse gas emissions. Water Air Soil Pollut 82: 593–615.
27. Lal R, Follett RF, Kimble J, Cole CV (1999) Managing US crop land to sequester carbon in soil. J. Soil Water Conserv 54: 374–381.

28. Paustian K, Andren O, Janzen H, Lal R, Smith P, et al. (1997) Agricultural soils as a C sink to offset CO_2 emissions. Soil Use Manage 13: 230–244.
29. Green CJ, Blackmer AM, Horton R (1995) Nitrogen effects on conservation of carbon during corn residue decomposition in soil. Soil Sci Soc Am J 59: 453–459.
30. Kowallenko CG, Ivarson KG, Cameron DR (1978) Effect of moisture content, temperature and nitrogen fertilization on carbon dioxide evolution from field soils. Soil Biol Biochem J 10: 417–423.
31. Fogg K (1988) The effect of added nitrogen on the rate of organic matter decomposition. Biol Rev 63: 433–472.
32. Halvorson AD, Wienhold BJ, Black AL (2002) Tillage, nitrogen, and cropping system effects on soil carbon sequestration. Soil Sci Soc Am J 63: 906–912.
33. Liang BC, Mackenzie AF (1992) Changes in the soil organic carbon and nitrogen after six years of corn production. Soil Sci 153: 307–313.
34. Wang XB, Cai DX, Hoogmoed WB, Oenema O, Perdok UD (2007) Developments in conservation tillage in rainfed regions of North China. Soil & Tillage Res 93: 239–250.
35. Zhang H, Wang X, Feng Z, Pang J, Lu F, et al. (2011) Soil temperature and moisture sensitivities of soil CO_2 efflux before and after tillage in a wheat field of Loess Plateau, China. J Environ Sci 23: 79–86.
36. Lou YZ, Zhou X (2006) Soil respiration and the environment. Academic, San Diego. 110–111.
37. Iqbal J, Hu R, Lin S, Hatano R, Feng M, et al. (2009) CO_2 emission in a subtropical red paddy soil (Ultisol) as affected by straw and N- fertilizer applications: A case study in Southern China. Agriculture Ecosyst and Environ 131: 292–302.
38. Zou J, Huang Y, Zheng X, Wang Y, Chen Y (2004) Static opaque chamber – based technique for determination of net exchange of CO_2 between terrestrial ecosystem and atmosphere. Chin Sci Bull 49 (4): 381–388.
39. Reiccosky DC, Dugas WA, Torbert HA (1997) Tillage- induced soil carbon dioxide loss from different cropping system. Soil Tillage Res 41: 105–118.
40. Calderon FJ, Jackson LE, Scow KM, Rolston DE (2001) Short term dynamics of nitrogen, microbial activity, and phospholipid fatty acids after tillage. Soil Sci Soc Am J 65: 118–126.
41. Reiccosky DC (2002) Long term effect of moldboard plowing on tillage-induced CO_2 loss. In: Kimble JM. Lal R, Follett RF editors. Agricultural practices and policies for carbon sequestration in soil. CRC Press, Boca, 87–96.
42. La Scala N Jr, Lopes A, Panosso AR, Camara FT, Periera GT (2005) Soil CO_2 efflux following rotary tillage of a tropical soil. Soil Tillage Res 84: 222–225.
43. Alvaro-Fuentes J, Cantero-Martinez C, Lopez MV, Arrue JL (2007) Soil carbon dioxide fluxes following tillage in semiarid Mediterranean agro-ecosystems. Soil Tillage Res 96: 331–341.
44. Beare MH, Cabrera ML, Hendrix PF, Coleman DC (1994) Aggregate–protected and unprotected organic matter pools in conventional and no-tillage soils. Soil Sci Soc Am J 58: 787–795
45. Kladivko EJ (2001) Tillage systems and soil ecology. Soil Tillage Res 61: 61–76.
46. Zhang HX, Zhou X, Lu F, Pang J, Feng Z, et al (2011) Seasonal dynamics of soil CO_2 efflux in a conventional tilled wheat field of the Loess Plateau, China. Ecol Res. 26: 735–743. DOI 10. 1007/s11284-011-0832-5.
47. Li H, Qiu J, Wang L, Tang H, Li C, et al. (2010) Modelling impacts of alternative farming management practices on greenhouse emissions from a winter wheat – maize rotation system in China. Agriculture, Ecosyst and Environ 135: 24–33.
48. Jacinthe PA, Lal R, Kimble JM (2002) Carbon budget and seasonal carbon dioxide emission from a Central Ohio Luvisol as influenced by wheat residue amendment. Soil & Tillage Res 67: 147–157.
49. Xiao Y, Xie G, Lu G, Ding X, Lu Y (2005) The value of gas exchange as a service by rice paddies in suburban Shanghai, PR China. Agric Ecosyst Environ 109: 273–283.
50. Chapin FS III (1991) Effects of multiple environmental stress upon nutrient availability In: Mooney HA, Winner WA, Pell EJ, editors. Integrated Responses of Plants to Environmental Stress. Academic New York. 67–86.
51. Fisk MC, Fahey TJ (2001) Microbial biomass and nitrogen cycling responses to fertilization and litter removal in young northern hardwood forests. Biogeochemistry 53: 201–223.
52. Anderson TH (1994) Physical analysis of microbial communities in soil: applications and limitations. In: Ritz K, Dighton J, Giller KE, editors. Beyond the Biomass: Compositional and Functional Analysis of Soil Microbial Communities. John Wiley and Sons. New York. 67–76.
53. Galantini J, Rosell R (2006) Long-term fertilization effects on soil organic matter quality and dynamics under different production systems in semiarid Pampean soils. Soil Tillage Res 34: 41–60.
54. Ding W, Meng L, Yin Y, Cai Z, Zheng X (2007) CO_2 emission in an intensively cultivated loam as affected by long-term application of organic manure and nitrogen fertilizers. Soil Biol Biochem 39: 669–679.
55. Raich JW, Schlesinger WH (1992) Modeling soil organic matter in organic amended and N-fertilized, long-term plots. Soil Sci Soc Am J 56: 476–488.
56. Davidson EA, Belk E, Boone RD (1998) Soil water content and temperature as independent or confounded factors controlling soil respiration in a temperate mixed hardwood forest. Glob Chang Biol 4: 217–227.
57. Reichstein M, Tenhunen JD, Roupsard O, Ourcival JM, Rambal S, et al. (2002) Ecosystem respiration in two Mediterranean evergreen Holm oak forests: drought effects and decomposition dynamics. Funct Ecol 16: 27–39.
58. Buchmann N (2000) Biotic and abiotic factors controlling soil respiration rates in *Picea abies* stands. Soil Biol Biochem 32: 1625–1635.
59. Saiz G, Black K, Reidy B, Lopez S, Farrell EP (2007) Assessment of soil CO_2 efflux and its components using a process-based model in a young temperate forest site. Geoderma 139: 79–89.
60. Bauer J, Herbst M, Huisman JA, Weihermuller L, Vereecken H (2008) Sensitivity of simulated soil heterotrophic respiration to temperature and moisture reduction functions. Geoderma 145: 17–27.
61. Bowden RD, Newkirk KM, Rullo GM (1998) Carbon dioxide and methane fluxes by a forest soil under laboratory-controlled moisture and temperature conditions. Soil Biol. Biochem 30: 1591–1597.
62. Haynes RJ (2000) Labile organic matter as an indicator of organic matter quality in arable and pastoral soils in New Zealand. Soil Biol Biochem 32: 211–219.
63. Lal R, Kimble JM (1997) Conservation tillage for carbon sequestration. Nutr. Cycling Agroecosyst 49: 243–253.
64. Lal R, Kimble JM, Follet RF, Cole CV (1998) The potential for US crop land to Sequester Carbon and Mitigate the Greenhouse Effect. Chelsea: Ann Arbor Science.
65. West TO, Post WM (2002) Soil organic carbon sequestration rates by tillage and crop rotation: a global data analysis. Soil Sci. Soc. Am J 66: 1930–1946.
66. Drury CF, Tan CS, Welacky TW, Oloya TO, Hamill AS, et al. (1999) Red clover and tillage influence on soil temperature, water content and corn emergence. Agron J 91: 101–108.
67. Hutchinson JI, Campbell CA, Desjardins RL (2007) Some perspectives on carbon sequestration in agriculture. Agric Meteorol 142: 288–302.
68. Gao C, Sun X, Cao J, Luan Y, Hao H, et al. (2008) A method and apparatus of measurement of carbon dioxide flux from soil surface from situ [J]. Journal of Beijing Forestry University 30 (2): 102–105. (In Chinese with English abstract).
69. Wilson HM, Al-kaisi MM (2008) Crop rotation and nitrogen fertilization effect on soil CO_2 emissions in Central Lova. Applied Soil Ecology 39: 264–270.
70. Bao SD (2008) Soil Agricultural Chemistry Analysis [M]. Beijing: China Agricultural press (In Chinese).

3

Limited Fitness Advantages of Crop-Weed Hybrid Progeny Containing Insect-Resistant Transgenes (*Bt/CpTI*) in Transgenic Rice Field

Xiao Yang[1], Feng Wang[2], Jun Su[2], Bao-Rong Lu[1]*

1 Ministry of Education Key Laboratory for Biodiversity Science and Ecological Engineering, Department of Ecology and Evolutionary Biology, Fudan University, Shanghai, China, **2** Fujian Province Key Laboratory of Genetic Engineering for Agriculture, Fujian Academy of Agricultural Sciences, Fuzhou, China

Abstract

Background: The spread of insect-resistance transgenes from genetically engineered (GE) rice to its coexisting weedy rice (*O. sativa f. spontanea*) populations *via* gene flow creates a major concern for commercial GE rice cultivation. Transgene flow to weedy rice seems unavoidable. Therefore, characterization of potential fitness effect brought by the transgenes is essential to assess environmental consequences caused by crop-weed transgene flow.

Methodology/Principal Findings: Field performance of fitness-related traits was assessed in advanced hybrid progeny of F_4 generation derived from a cross between an insect-resistant transgenic (*Bt/CpTI*) rice line and a weedy strain. The performance of transgene-positive hybrid progeny was compared with the transgene-negative progeny and weedy parent in pure and mixed planting of transgenic and nontransgenic plants under environmental conditions with natural *vs.* low insect pressure. Results showed that under natural insect pressure the insect-resistant transgenes could effectively suppress target insects and bring significantly increased fitness to transgenic plants in pure planting, compared with nontransgenic plants (including weedy parent). In contrast, no significant differences in fitness were detected under low insect pressure. However, such increase in fitness was not detected in the mixed planting of transgenic and nontransgenic plants due to significantly reduced insect pressure.

Conclusions/Significance: Insect-resistance transgenes may have limited fitness advantages to hybrid progeny resulted from crop-weed transgene flow owning to the significantly reduced ambient target insect pressure when an insect-resistant GE crop is grown. Given that the extensive cultivation of an insect-resistant GE crop will ultimately reduce the target insect pressure, the rapid spread of insect-resistance transgenes in weedy populations in commercial GE crop fields may be not likely to happen.

Editor: Guy Smagghe, Ghent University, Belgium

Funding: This research was supported by the Natural Science Foundation of China (Grant no. 30730066 and 30871503), and National Program of Development of Transgenic New Species of China 2011ZX08011-006. The funders had no role in study design, data collection and analysis, decision to publish, or preparation of the manuscript.

Competing Interests: The authors have declared that no competing interests exist.

* E-mail: brlu@fudan.edu.cn

Introduction

The commercial application of genetically engineered (GE) crops in agricultural production has aroused great biosafety concerns worldwide. The potential environmental impacts caused by the cultivation of the GE crops are the most debated issues [1–3]. Transgene flow from a GE crop into populations of wild or weedy relatives and its potential ecological risks is considered as a key environmental problem [4–6]. (Trans) gene flow from a crop to its wild relatives has been widely documented in the last decades [7–9]. However, our knowledge on the role of introgressed transgenes that confer novel traits with a strong selective advantage in changing the evolutionary process of wild or weedy populations is still limited. Therefore, assessing potential environmental risks caused by the extensive cultivation of GE crops prior to their commercialization becomes a common practice [1,5,8]. The study of potential ecological consequences created by transgene flow to wild relatives particularly the coexisting and conspecific weeds will provide solid bases for environmental risk assessment [10–13]. The fate of weedy populations that acquired transgenes through gene flow is largely different, depending on the fitness effect of the introgressed transgenes under given environmental conditions [5,14]. If the introgressed transgenes can increase fitness, the transgenes will enhance the competitiveness and invasiveness of the weedy populations, leading to the rapid spread of the transgenes in the weedy populations, and *vice versa*. Thus, estimating fitness effect of transgenes on weedy populations is essential.

Rice is an important world crop providing staple food for nearly one half of the global population [15]. Research and development of GE rice has been extensively practiced in China, and consequently a many GE rice lines with novel traits such as insect-resistance, herbicide-tolerance, and improved grain quality have been developed [5,11,16]. In 2009, Chinese authority granted biosafety certificates for two insect-resistant rice lines containing a *Bt* transgene (*Bacillus thuringiensis*) [17], meaning that GE rice may enter commercial production in near future. Transgene spread from insect-resistant GE rice to its coexisting weedy rice populations through gene flow becomes a major environmental biosafety concern. Weedy rice (also known as red rice, *O. sativa f. spontanea*) is a noxious weed that causes significant losses of rice yield and quality worldwide [18,19]. The introgression of transgenes with selective advantages may largely enhance the spread of the weedy rice, causing more serious weed problems. Weedy rice is an annual weed conspecific to cultivated rice [20], and transgene flow from cultivated rice to weedy rice cannot be avoided owing to their similar flowering phenology [21]. It is therefore valid to determine the fitness effect of insect-resistance transgenes on weedy rice populations under different insect pressure for risk assessment.

Fitness of an insect-resistance transgene is largely associated with the environment in which insect pressure can largely be variable [22–25]. The relationship between fitness effect brought by an insect-resistance transgene and the ambient insect pressure has not been well described. A number of studies of crop-weed hybrids showed variable fitness effect brought by insect-resistant transgenes under different insect occurrences [25,26]. It is apparent based on these studies that the enhanced fitness benefit of insect-resistant transgenes was always associated with the high target herbivore pressure, whereas under low herbivore pressure the benefit brought by the insect-resistant transgenes was reduced. Therefore, to analyzed fitness effect brought by an insect-resistance transgene under designed experimental conditions with different insect pressure should be the key to detect the fitness effect of such a transgene.

Studies have already revealed the effects of *Bt* transgene in crop-wild or crop-weed hybrid progeny under different environments [23,25,26–28]. Our previous studies of F_1–F_3 crop-weed hybrids and their derived lineages from insect-resistant rice revealed increased fecundity brought by *Bt/CpTI* transgenes when target insects were abundant [25,29]. However, the relationship between variation in different insect pressure and the corresponding fitness change was not well addressed by a properly designed experiment. In this study, we intended to address the following questions: (i) Are the insect-resistant transgenes (*Bt/CpTI*) still effective on resisting rice target herbivores in the transgenic F_4 populations derived from a GE crop-weed hybrid under natural insect pressure? (ii) Does the transgenic F_4 population have significantly higher fitness than the nontransgenic F_4 population and weedy rice parent under natural insect pressure? (iii) Does the mixed planting of transgenic plants significantly reduce the ambient insect pressure of the nontransgenic plants, and consequently, reduce the potential fitness benefit brought by insect-resistance transgenes in the experimental fields? The answer of above questions is essential for assessing the long-term ecological impacts caused by insect-resistant transgene flow from to its weedy rice populations.

Methods

Experimental Materials and their Sources

Our experiment included two sets of F_4 populations, with or without transgenes derived from a crop-weed hybrid, and its weedy rice parent. The F_4 populations were generated by continued self-pollination from the F_1 hybrids of an insect-resistant GE rice line crossed with a weedy rice strain (used as the pollen recipient) [25,29]. The GE rice line produced by Fujian Academy of Agricultural Sciences in Fuzhou of China had the *Bt/CpTI* (*Bacillus thuringiensis* (*cryIAc*) and cowpea trypsin inhibitor) transgenes in a double insertion, tightly linked with the selectable marker gene *hyg* (hygromycin resistance), transformed by an agrobacterium method [30]. This GE rice line was produced to control lepidopteran pests such as rice stem borers (*Scirpophaga incertulas*, *Chilo suppressalis*, and *Sesamia inferens*) and rice leaf-folders (*Cnaphalocrocis medinalis*) [24]. The weedy rice strain (coded as W) was donated by Dr. H. S. Suh of Yeungnam University from S Korea [29]. The identification and separation of the transgene homozygous population (coded as TP) from non-transgene homozygous population (TN) were made from screening the seedlings derived from seeds of F_2 plants using the hygromycin-B water solution treatment [25]. The selected transgene-positive and transgene-negative F_3 populations were self-pollinated for one more generation. Therefore, plants of both transgene-positive and transgene-negative F_4 populations were homozygous at the transgenic locus.

Experimental Design

The field experiment was conducted in 2010 in the environment with a natural insect occurrence at the designated Biosafety Assessment Center in Fuzhou, Fujian Province, China. The experimental plots were allocated in two separate blocks (natural insect *vs.* low insect pressure) isolated from each other for >100 m. The block with natural insect pressure was free of any insecticide treatment, whereas the block with low insect pressure was achieved by spraying different insecticides (Methamidophos, Folimat, Buprofezin, and Monosultap) every 7–10 days starting at the tillering stage. The main comparison was made among the weedy parent, transgenic and nontransgenic hybrid lineages under the different insect pressure. Therefore, under each of the insect-pressure environmental conditions (natural *vs.* low), the following treatments were designated: (1) pure planting of weedy rice parent, transgene-positive, and transgene-negative F_4 populations; (2) mixed planting between transgene-positive and -negative F_4 populations, or transgene-positive F_4 population and weedy rice parent, or transgenic-negative F_4 population and weedy rice parent. Consequently, there were a total of twelve treatments included in the experiment.

For each treatment, four replicates (plots) were included. In the pure planting design, each plot included 64 plants in an 8×8 grid with 20 cm spacing between plants. In the mixed planting design, each plot included 32 plants of either type of the experimental materials grown alternatingly in an 8×8 grid with 15 cm spacing between plants. The field layout of plots was arranged in a complete randomized design in the two blocks. Seeds of all the plant materials were germinated in 50×100 cm pots filled with paddy soil from fields and placed in a green house. Seedlings were transferred to a nursery bed 20 days after seed germination, and then transplanted into field plots ~40 days after seed germination. To avoid accidental seed shattering, panicles of all plants were enclosed by nylon mesh-bags 10–15 days after the flowering.

Data Collection and Analysis

For estimating insect damage and fitness-related traits, all plants in a plot were characterized at various vegetative and reproductive stages, except for those in the border rows that were excluded to avoid the edge effect. As a result, a total 36 plants from a pure-planting plot and 18 plants for each of the two types of populations from a mixed-planting plot were characterized. Blasted tillers (by rice stem borers) and folded leaves (by leaf-folder) at the beginning of flowering stage were measured to estimate the insect damage. An insect index was used to estimate the insect pressure that was calculated as the sum of percentage of both blasted tiller and

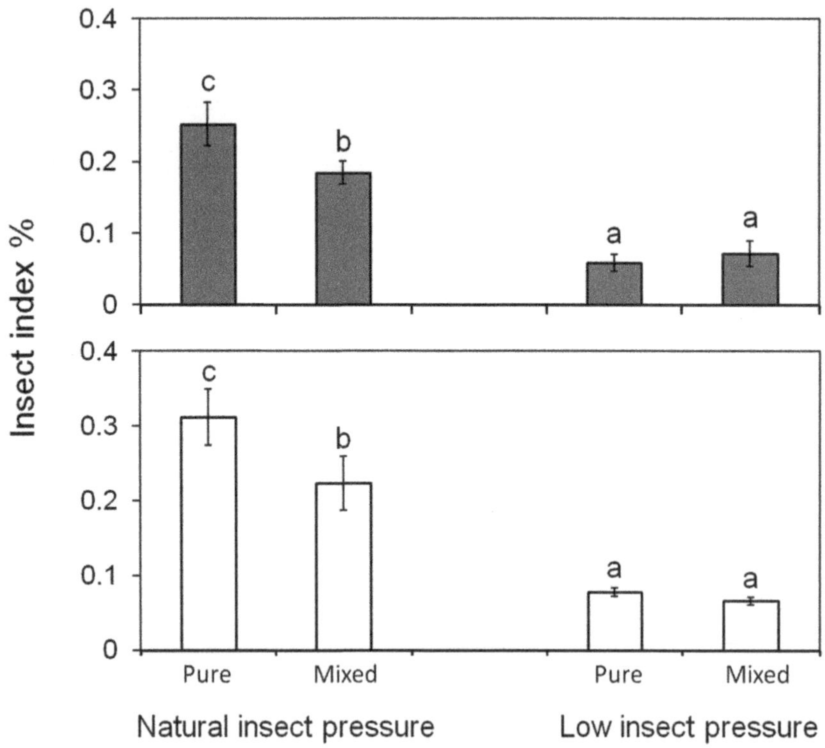

Figure 1. Insect index (%) calculated based on the ratio of blasted tillers and folded leaves on weedy parent (above) and transgene-negative hybrid progeny (below) in different cultivation modes (pure *vs.* mixed) under natural or low insect pressure. The comparison was made among pure and mixed planting under natural insect and low insect pressure. Different letters above the columns indicate significant differences according to Duncan's multiple range tests after Bonferroni correction. "Pure" indicates pure planting, "Mixed" indicates mixed planting. Bars represent standard error. Levels of significance: $P<0.05$.

folded leaves in the plots of weedy parent and nontransgenic F_4 populations. The following fitness-related traits were included for measurement: plant height, number of tillers, number of panicles, and number of filled seeds per plant, seed set, and 1000-seed weight at various growth stages.

One way ANOVA (Duncan's multiple range test) was used to examine significant differences for insect index obtained from weedy rice and transgene-negative plants in pure and mixed planting under natural and low insect pressure, and to test significant differences in insect damage and fitness-related traits among weedy parent, transgene-positive, and transgene-negative populations in pure planting, followed by the Bonferroni correction. The paired *t*-test was used to examine significant differences between transgene-positive and transgene-negative, transgene-positive and weedy parental, or transgene-negative and weedy parental populations in mixed planting. All statistical analyses were performed using the software package IBM SPSS Statistics ver. 19.0 for Windows (SPSS Inc., IBM Company Chicago, IL, USA, 2010).

Results

Insect Pressure Under the Two Environmental Conditions

The insect index calculated from the insect damage of blasted tillers and folded-leaves demonstrated a significant difference between the two blocks with natural insect pressure and low insect pressure (Fig. 1; Table 1 and 2). As estimated from the weedy parental and transgene-negative F_4 populations in pure planting, the insect index was significantly higher under the natural insect pressure (25.3% and 31.5%) than that the under low insect pressure (5.8% and 7.8%; $P<0.001$). The significant differences of insect pressure allowed us to determine the net benefit caused by the transgenes when all the plants were exposed to herbivores, and the net cost caused by the transgenes when herbivores were absent or reduced to a low level.

Noticeably, the insect index was significantly lower in the mixed planting of transgene-positive populations than in the pure planting under natural insect pressure (Fig. 1), indicating that the mixed planting with transgene-positive plants will generally reduce the ambient insect pressure. However, such reduction of insect damage was not observed in the mixed planting of transgene-positive plants under low insect pressure.

Performance of F_4 and Weedy Parental Populations in Pure Planting

In general, transgene-positive F_4 populations with insect-resistance transgenes showed significantly less insect damage than transgene-negative F_4 populations and weedy parents, under natural insect pressure (Table 1). Transgene-positive F_4 populations had a better performance for most of the fitness-related traits than transgene-negative F_4 populations and weedy parents, under natural insect pressure (Table 1). For example, transgene-positive plants produced a significantly higher number of panicles (25% increase) and ratio of seed set (17% increase) than their transgene-negative plants (Table 1). Particularly, the number of seeds produced by transgene-positive plants was significantly higher than both transgene-negative plants (44% increase) and weedy parents (31% increase) (Fig. 2). Under low insect pressure, nearly all the fitness-related traits did not show significant differences

Table 1. Average values and standard errors (±) of insect damage and fitness-related traits in weedy rice parent, transgene-positive, and transgene-negative crop-weed hybrid progeny (F_4) in pure planting under natural (above) vs. low (below) insect pressure.

Trait	Plant material		
	Weedy parent	Transgene-positive	Transgene-negative
Natural insect			
Blasted tillers %	13.4±1.1 b	7.0±1.3 a	15.6±2.0 b
Folded leaves %	11.8±2.1b	1.5±0.3 a	15.6±3.1 b
Plant height (cm)	91.9±4.0 a	111.9±4.2 b	112.3±2.6 b
No. of tillers	17.3±1.1 a	18.2±0.3 a	16.0±1.2 a
No. of panicles	10.4±1.0 a	15.0±0.5 b	12.0±0.9 a
Seed set (%)	50.0±1.5 b	47.8±2.5 b	40.8±0.9 a
1000-seed weight (g)	20.3±0.2 a	21.4±0.3 b	20.7±0.2 ab
Low insect			
Blasted tillers %	4.8±1.2 a	5.0±0.8 a	6.5±0.7 a
Folded leaves %	1.0±0.0 b	0.2±0.2 a	1.3±0.2 b
Plant height (cm)	107.3±1.7 a	111.0±2.5 a	120.5±2.7 b
No. of tillers	13.9±0.6 a	13.3±0.4 a	13.0±0.9 a
No. of panicles	13.6±0.7 b	10.6±0.7 a	11.0±0.4 a
Seed set (%)	51.3±1.3 ab	52.2±1.4 b	48.0±0.8 a
1000-seed weight (g)	19.7±0.2 a	22.1±0.2 b	21.6±0.3 b

Different letters following the average values in the same rows indicate significant differences according to Duncan's multiple range tests after Bonferroni correction (N = 4 plots).

Table 2. Average values and standard errors (±) of insect damage and fitness-related traits in weedy rice parent (W), transgene-positive (TP), and transgene-negative (TN) crop-weed hybrid progeny (F_4) in mixed planting under natural (above) vs. low (below) insect pressure.

Trait	Plant material					
	TP *vs.* TN		TP *vs.* W		TN *vs.* W	
Natural insect						
Blasted tillers %	11.4±2.0	17.3±0.9*	9.6±2.1	13.0±2.3	12.9±1.4	18.1±3.0
Folded leaves %	2.9±1.7	5.0±3.0	4.2±1.3	5.4±1.2	13.6±2.7	10.3±3.7
Plant height (cm)	101.4±2.6	103.4±4.8	96.9±2.9	93.1±3.0	107.9±6.4	99.4±6.7
No. of tillers	11.0±1.0	9.9±0.8	13.2±1.2	12.0±1.0	11.8±0.5	9.8±1.0
No. of panicles	8.9±0.4	8.6±0.8	9.7±1.0	8.3±0.4	8.9±0.7	8.2±0.9
Seed set (%)	42.5±4.2	37.1±0.6	43.5±1.7	49.3±2.6	37.0±2.0	47.8±3.8*
1000-seed weight (g)	20.4±0.6	20.5±0.2	20.6±0.1	19.6±0.3*	21.1±0.6	19.1±0.3*
Low insect						
Blasted tillers %	5.4±0.3	5.8±0.4	5.3±1.5	6.5±1.9	5.7±2.2	8.2±1.7
Folded leaves %	0.4±0.3	0.8±0.3	0.4±0.1	0.7±0.2	2.3±0.4	0.7±0.3*
Plant height (cm)	112.3±5.0	116.7±3.4	104.5±4.7	98.5±3.4	111.4±2.8	104.1±1.4
No. of tillers	7.6±0.1	8.2±0.3	8.5±0.5	7.8±0.5	8.8±0.4	8.3±0.5
No. of panicles	6.6±0.4	7.2±0.5	6.8±0.4	6.2±0.3*	7.3±0.6	7.2±0.3
Seed set (%)	53.9±0.9	49.0±3.2	49.9±5.7	61.9±2.4	51.6±2.6	55.7±2.9
1000-seed weight (g)	22.3±0.8	21.9±0.5	21.9±0.7	19.3±0.4*	21.7±0.2	18.8±0.4**

Level of significance was calculated based on the paired *t*-test between TP and TN, TP and W, or TN and W (N = 4 plots).
*P<0.05,
**P<0.01.

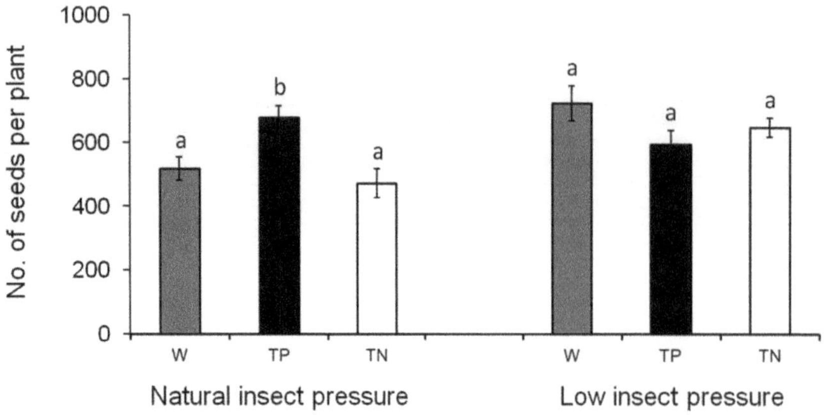

Figure 2. The average number of filled seeds per plant produced by weedy rice parent (W), transgene-positive (TP), and transgene-negative (TN) hybrid progeny under natural vs. low insect pressure in pure cultivation. The comparison was made among weedy rice parent, transgene-positive, and transgene-negative hybrid progeny. Different letters above the columns indicate significant differences according to Duncan's multiple range tests after Bonferroni correction. Bars represent standard error. Levels of significance: $P<0.05$.

between transgene-positive and transgene-negative populations (Table 1; Fig. 2) in both pure and mixed planting, suggesting no obvious fitness cost from the transgenes.

Noticeably, transgene-negative F_4 populations showed significantly higher values of plant height and 1000-seed weight than their weedy rice parental populations under both natural and low insect pressures (Table 1). This indicated that crop-weed hybridization may bring benefit through the enhanced performance of some traits of hybrids and hybrid lineages in later generations.

Performance of F_4 and Weedy Parental Populations in Mixed Planting

In contrast to the pure planting, transgene-positive and transgene-positive F_4 populations showed apparently less pronounced differences in insect damage in mixed planting than in pure planting under natural insect pressure (Table 2). The same trend was observed between transgene-positive F_4 and weedy parental populations (Table 2) in mixed planting. As a result, transgene-positive F_4 populations did not show significant differences for nearly all the fitness-related traits, compared with transgene-negative F_4 and weedy parental populations under natural insect pressure (Table 2). These results indicated that the mixed-planting plots with transgene-positive plants would result in a reduced insect damage to transgene-negative and weedy rice plants, and consequently reduce the differences in fitness-related traits between transgene-positive and transgene-negative plants.

Similarly, under low insect pressure, nearly all the fitness-related traits did not show significant differences among transgene-positive F_4, transgene-negative F_4 and weedy parental populations (Table 2; Fig. 3). However, it is worthy of notice that the 1000-seed weight of transgene-positive and transgene-negative F_4 plants was significantly higher than that of weedy parental plants, suggesting again the benefit of hybridization brought to hybrid progeny compared with their weedy parents in terms of seed weight, regardless of the presence of transgenes.

Discussion

The key point of assessing potential ecological risks caused by an insect-resistant transgene introgressed to weedy populations is to analyze the increased fitness of transgenic plants in comparison with their nontransgenic counterparts derived from the same crop-weed hybridization under controlled environmental conditions [11,22,24,25,31]. Our data from this study indicated significant differences in insect pressure (as measured by the insect index, Fig. 1) between the two rice planting environments with natural *vs.* low insect pressure, achieved by artificial spray of insecticides. The insect pressure in both pure and mixed planting plots of weedy rice parents and transgene-negative F_4 populations was significantly higher under the natural-insect condition than that under the controlled low-insect condition. These results indicate the effectiveness of insect control in our experiment. In addition, as indicated by the insect damage on weedy parents and transgene-negative plants, significantly higher insect index was also detected in pure planting plots than in the mixed-planting plots of nontransgenic plants (including parents) with transgenic plants under natural insect. In contrast, the mixed-planting plots with transgenic plants did not show such significant differences in insect index under low insect pressure. All together, these results demonstrate that the differential environment allowed us to analyze the fitness effect of insect-resistance transgenes in weedy rice plants against their nontransgenic counterparts, and that the mixed planting of transgenic plants in a field plot can significantly reduce the ambient insect pressure.

In pure planting, transgene-positive crop-weed F_4 plants showed significantly greater values of fecundity-related traits, such as the number of panicles and well-filled seeds and the ratio of seed set than their transgene-negative plants under natural insect pressure. Particularly, the number of panicles and filled seeds of transgene-positive F_4 plants showed significantly higher values than the weedy rice parents. However, such significant differences between the transgenic plants and nontransgenic plants were not detected in these traits under the low insect pressure. The increased fecundity of transgenic crop-weed F_4 plants is most likely brought by the introgression of insect-resistance transgenes that significantly reduced insect attack to the plants. The increased fecundity in the transgenic F_4 plants demonstrates that the introgressed insect-resistant transgenes (*Bt/CpTI*) are still effective on controlling rice target herbivores in the advanced generation of crop-weed hybrid populations. This result is consistent with our previous study in which the F_2–F_3 lineages derived from the crop-weed hybrids

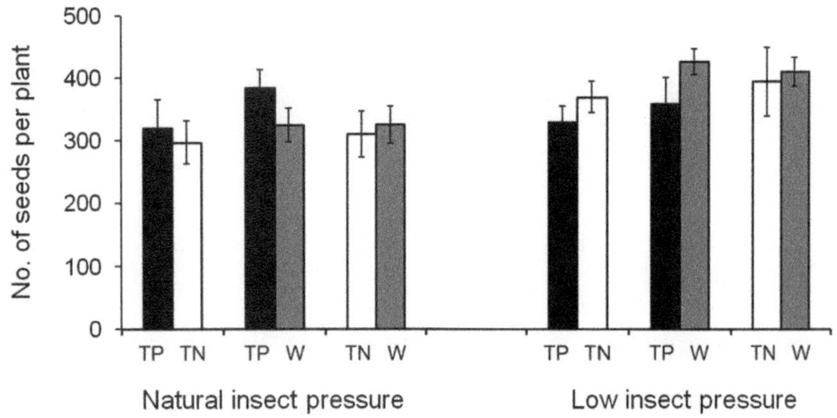

Figure 3. The average number of filled seeds per plant produced by weedy rice parent, transgene-positive (W), and transgene-negative (TP) hybrid progeny under natural vs. low insect pressure in mixed cultivation. The comparison was made between transgene-positive and transgene-negative, transgene-positive and weedy rice parent, or transgene-negative and weedy rice parent, based on the paired *t*-tests. Bars represent standard error. Levels of significance: $P<0.05$.

with the same transgenic event showed effective control of insect damages to the hybrid lineage compared with both transgene-negative hybrid lineages and the weedy parents, under a high level of insect pressure [25,29]. Many previous studies on crop-weed and crop-wild hybrid lineages also demonstrated that insect-resistance transgenes can dramatically reduce the target herbivores and increase the fecundity in hybrids and their advanced generation of hybrid populations [23,25,29]. These findings are consistent with our results in this study. It seems possible to make a conclusion from the field experimental data of above studies including ours that the introgressed insect-resistance transgenes will maintain their strong ability to control herbivores and increase fecundity of transgenic populations derived from crop-weed or crop-wild transgene introgression. The transgenic plants may have a strong ability to compete with their nontransgenic counterparts including nontransgenic hybrid and parental populations [23,25,28].

However, it is necessary to point out that in the mixed-planting plots the fecundity traits (particularly seed production) of the transgene-positive plants did not show significantly increased values, compared to the transgene-negative plants and weedy rice parents under natural insect pressure. This result suggests apparent losses of fitness benefit that should have been brought by the insect-resistance transgenes in mixed planting under natural insect pressure. The similar phenomenon was also observed in a number of studies in which the fitness benefit and cost of insect-resistant GE rice lines and their nonGE parental lines was analyzed under natural and low insect pressure, respectively [22,24]. Associated with the data of insect index from this study where significantly lower insect pressure was recorded in the mixed planting plots than in the pure planting plots under the same environmental condition of natural insect (Fig. 1), we think that the loss of fitness benefit to insect-resistant transgenic crop-weed hybrid populations in mixed planting is most likely owing to the considerably reduced insect pressure. In other words, the generally low ambient insect pressure in the experimental field caused by the mixture of insect-resistant transgenic plants will significantly reduce the potential fitness advantages that should have been brought by insect-resistance transgenes. As a result, the expected long-term persistence and rapid spread of insect-resistance transgenes in weedy rice populations caused considerably increased fitness advantages following transgene flow from a GE rice variety may not happen in the realistic situation if such fitness advantages are extremely limited. Under the actual situation where an insect-resistant GE rice variety is cultivated, the ambient insect pressure in an extensive field area should be reduced to a much greater extent than that in our experimental plots. The spread of transgenes in weedy rice populations might be considerably limited owning to the negligible fitness increase in insect-resistance GE rice fields where the target insect pressure is significantly reduced, although the issue of transgene flow from an insect-resistant GE rice variety to its coexisting weedy rice populations should not be neglected.

Our finding has its important implications for the risk assessment of transgene flow from insect-resistant GE crop to its wild relatives, and to the conspecific weedy populations in particular. Given a determined frequency of transgene flow, the magnitude of potential environmental risks should largely depend on the fitness effect of a transgene. It is generally recognized that fitness effect of an insect-resistance transgene is determined by the ambient insect pressure in the environment where wild/weedy populations occur. Therefore, the assessment of environmental risks caused by transgene flow from an insect-resistant GE crop to wild populations should first consider the pressure of target insects in the concerned environment. This principle may also applied to the risk assessment of transgene flow to wild populations, considering that insect pressure in natural habitats is significantly lower than in agriculture habitats [32]. For the crop conspecific weedy populations co-occurring with a crop (and GE crop), crop-to-weed transgene flow cannot be avoided. The expected environmental impact from transgene flow from insect-resistant GE crop could be large because hypothetically insect-resistance transgenes will bring fitness benefit to the weedy populations. However, the fitness advantages might be limited due to the fact that weedy plants will be surrounded by insect-resistant plants in a GE crop field, reducing the ambient pressure of target herbivores significantly. Consequently, the spread of the transgenes in weedy populations would be limited. In reality, the extensive commercial cultivation of an insect-resistant GE crop will largely reduce target herbivores in a GE deployed area [33]. Under such a circumstance, the environmental impacts caused by the crop-to-weed transgene flow from an insect-resistant GE crop will also be limited.

Author Contributions

Conceived and designed the experiments: BRL XY. Performed the experiments: XY JS. Analyzed the data: BRL XY FW. Contributed reagents/materials/analysis tools: FW JS. Wrote the paper: XY BRL.

References

1. Andow DA, Zwahlen C (2006) Assessing environmental risks of transgenic plants. Ecol Lett 9: 196–214.
2. Conner AJ, Glare TR, Nap JP (2003) The release of genetically modified crops into the environment - Part II. Overview of ecological risk assessment. Plant J 33: 19–46.
3. Snow AA, Palma PM (1997) Commercialization of transgenic plants: Potential ecological risks. Bioscience 47: 86–96.
4. Haygood R, Ives AR, Andow DA (2003) Consequences of recurrent gene flow from crops to wild relatives. P Roy Soc B-Biol Sci 270: 1879–1886.
5. Lu BR, Yang C (2009) Gene flow from genetically modified rice to its wild relatives: Assessing potential ecological consequences. Biotechnol Adv 27: 1083–1091.
6. Stewart CN, Halfhill MD, Warwick SI (2003) Transgene introgression from genetically modified crops to their wild relatives. Nat Rev Genet 4: 806–817.
7. Ellstrand NC, Prentice HC, and Hancock JF (1999) Gene flow and introgression from domesticated plants into their wild relatives. Annu Rev Ecol Evol S 30: 539–563.
8. Ellstrand NC (2003) Current knowledge of gene flow in plants: implications for transgene flow. Philos T R Soc B 358: 1163–1170.
9. Raybould AF, Gray AJ (1993) Genetically modified crops and hybridization with wild relatives: A UK Perspective. J Appl Ecol 30: 199–219.
10. Chen LJ, Lee DS, Song ZP, Suh HS, Lu BR (2004) Gene flow from cultivated rice (*Oryza sativa*) to its weedy and wild relatives. Ann Bot 93: 67–73.
11. Lu B-R, Snow AA (2005) Gene flow from genetically modified rice and its environmental consequences. BioScience 559: 669–678.
12. Gealy DR, Mitten DH, Rutger JN (2003) Gene flow between red rice (*Oryza sativa*) and herbicide-resistant rice (*O. sativa*): implications for weed management. Weed Technol 17: 627–45.
13. Warwick SI, Légère A, Simard MJ, James T (2008) Do escaped transgenes persist in nature? The case of an herbicide resistance transgene in a weedy *Brassica rapa* population. Mol Ecol 17: 1387–1395.
14. Hails RS, Morley K (2005) Genes invading new populations: a risk assessment perspective. Trend Ecol Evol 20: 245–252.
15. FAO (2004) The state of food and agriculture 2003–2004. Agricultural biotechnology: meeting the needs of the poor? http://www.fao.org/docrep/006/Y5160e/Y5160e00.HTM (Accessed on April 22, 2012).
16. Huang JK, Hu RF, Rozelle S, Pray C (2005) Insect-Resistant GM Rice in Farmers' Fields: Assessing Productivity and Health Effects in China. Science 308: 688–690.
17. MOA (Ministry of Agriculture, China) (2009) The second batch of genetically modified organisms that are granted biosafety certificates in 2009. (in Chinese) http://www.stee.agri.gov.cn/biosafety/spxx/P020091127591594596689.pdf. (Accessed on April 22, 2012).
18. Delouche JC, Burgos NR, Gealy DR, Zorilla-San MG, Labrada R, et al (2007) Weedy rices: origin, biology, ecology and control. FAO of the United Nations, Rome. 144. http://www.fao.org/docrep/010/a1023e/a1023e00.htm (Accessed on April 22, 2012).
19. Londo JP, Schaal BA (2007) Origins and population genetics of weedy red rice in the USA. Mol Ecol 16: 4523–4535.
20. Harlan JR, De Wet J (1971) Toward a rational classification of cultivated plants. Taxon 20: 509–517.
21. Cohen MB, Arpaia S, Lan LP, Chau LM, Snow AA (2008) Shared flowering phenology, insect pests, and pathogens among wild, weedy, and cultivated rice in the Mekong Delta, Vietnam: implications for transgenic rice. Environ Biosafety Res 7: 73–85.
22. Chen LY, Snow AA, Wang F, Lu BR (2006) Effects of insect-resistance transgenes on fecundity in rice (*Oryza sativa*, Poaceae): A test for underlying costs. Am J Bot 93: 94–101.
23. Snow AA, Pilson D, Rieseberg LH, Paulsen MJ, Pleskac N, et al (2003) A Bt transgene reduces herbivory and enhances fecundity in wild sunflowers. Ecol Appl 13: 279–286.
24. Xia H, Chen LY, Wang F, Lu BR (2010) Yield benefit and underlying cost of insect-resistance transgenic rice: Implication in breeding and deploying transgenic crops. Field Crop Res 118: 215–220.
25. Yang X, Xia H, Wang W, Wang F, Su J, et al (2011) Transgenes for insect resistance reduce herbivory and enhance fecundity in advanced generations of crop-weed hybrids of rice. Evol Appl 4: 672–684.
26. Halfhill MD, Sutherland JP, Moon HS, Poppy GM, Warwick SI, et al (2005) Growth, productivity, and competitiveness of introgressed weedy *Brassica rapa* hybrids selected for the presence of *Bt cry1Ac* and *gfp* transgenes. Mol Ecol 14: 3177–3189.
27. Mason P, Braun L, Warwick SI, Zhu B, Stewart CN (2003). Transgenic *Bt*-producing *Brassica napus*: *Plutella xylostella* selection pressure and fitness of weedy relatives. Environ Biosafety Res 2: 263–276.
28. Vacher C, Weis AE, Hermann D, Kossler T, Young C, et al (2004) Impact of ecological factors on the initial invasion of Bt transgenes into wild populations of birdseed rape (*Brassica rapa*). Theor Appl Genet 109: 806–814.
29. Cao QJ, Xia H, Yang X, Lu BR (2009) Performance of hybrids between weedy rice and insect-resistant transgenic rice under field experiments: implication for environmental biosafety assessment. J Integr Plant Biol 51: 1138–1148.
30. Rong J, Song ZP, Su J, Xia H, Lu BR, et al (2005) Low frequency of transgene flow from *Bt/CpTI* rice to its nontransgenic counterparts planted at close spacing. New Phytol 168: 559–566.
31. Jenczewski E, Ronfort J, Chèvre AM (2002). Crop-to-wild gene flow, introgression and possible fitness effects of transgenes. Environ Biosaf Reserv 2: 9–24.
32. Chen YH, Welter SC (2002) Abundance of a native moth *Homoeosoma electellum* (Lepidoptera: Pyralidae) and activity of indigenous parasitoids in native and agricultural sunflower habitats. Environ Entomol 31: 626–636.
33. Wu KM, Lu YH, Feng HQ, Jiang YY, Zhao JZ (2008) Suppression of cotton bollworm in multiple crops in China in areas with Bt toxin-containing cotton. Science 321: 1676–1678.

4

New Genes in Traditional Seed Systems: Diffusion, Detectability and Persistence of Transgenes in a Maize Metapopulation

Joost van Heerwaarden[1,2]*, Diego Ortega Del Vecchyo[2,5], Elena R. Alvarez-Buylla[2,3], Mauricio R. Bellon[4]

1 Biometris, Plant Sciences Group, Wageningen University and Research Center, Wageningen, The Netherlands, **2** Instituto de Ecología, Universidad Nacional Autónoma de México, México D.F., México, **3** Centro de Ciencias de la Complejidad, Universidad Nacional Autónoma de México, México D.F., México, **4** Bioversity International, Rome, Italy, **5** Interdepartmental Ph.D. Program in Bioinformatics, University of California Los Angeles, Los Angeles, California, United States of America

Abstract

Gene flow of transgenes into non-target populations is an important biosafety concern. The case of genetically modified (GM) maize in Mexico has been of particular interest because of the country's status as center of origin and landrace diversity. In contrast to maize in the U.S. and Europe, Mexican landraces form part of an evolving metapopulation in which new genes are subject to evolutionary processes of drift, gene flow and selection. Although these processes are affected by seed management and particularly seed flow, there has been little study into the population genetics of transgenes under traditional seed management. Here, we combine recently compiled data on seed management practices with a spatially explicit population genetic model to evaluate the importance of seed flow as a determinant of the long-term fate of transgenes in traditional seed systems. Seed flow between farmers leads to a much wider diffusion of transgenes than expected by pollen movement alone, but a predominance of seed replacement over seed mixing lowers the probability of detection due to a relative lack of homogenization in spatial frequencies. We find that in spite of the spatial complexities of the modeled system, persistence probabilities under positive selection are estimated quite well by existing theory. Our results have important implications concerning the feasibility of long term transgene monitoring and control in traditional seed systems.

Editor: Christos A. Ouzounis, The Centre for Research and Technology, Hellas, Greece

Funding: Funding was provided by the Rockefeller Foundation (grants FS013 & FS185), UC MEXUS-CONACYT CN08-187, SEMARNAT-CONACYT0538/A-1 (ucmexus.ucr.edu), CONACYT 124909 "Red Complejidad, Ciencia y Sociedad," and a UC MEXUS-CONACYT doctoral fellowship. The funders had no role in study design, data collection and analysis, decision to publish, or preparation of the manuscript.

Competing Interests: The authors have declared that no competing interests exist.

* E-mail: joost.vanheerwaarden@wur.nl

Introduction

Since the introduction of genetically modified (GM) crops in the 1990s, gene flow into non-target populations has been a cause of concern [1–2]. The case of GM maize in Mexico has been of particular interest [3–4] because of the country's status as center of origin and diversification of this important food crop [5–6]. In spite of a national moratorium on the planting of GM maize imposed in 1998, several studies have reported the presence of genetic elements of transgenic origin in populations of Mexican landraces [7–10].

Whereas biosafety issues such as introgression into wild relatives [11–13] and contamination of conventional food and seed supplies [14–15] were recognized early on, these reports have drawn attention to a hitherto unappreciated risk: gene flow into the traditional seed system that is typical of smallholder agriculture in the developing world. Unlike commercial farmers, most smallholders recover seed from the previous harvest, and frequently acquire seed from other farmers through a well-structured traditional system of rules, expectations and practices based on family and local social networks [16–19]. This traditional system effectively links all individual maize populations, or seed lots, into a single evolving metapopulation [20] in which the fate of an escaped transgene is subject to the long-term effects of population genetic processes such as drift, gene flow and selection.

All of these processes are determined to a large extent by seed management [21], the collection of farmer practices related to seed sourcing and use, including seed selection, planting, and acquisition [22] [17][23].

Seed acquisition, denoted here as seed flow, is particularly important, as it strongly affects both drift and gene flow in the metapopulation [21]. Previous work has distinguished two forms of seed flow [23]: replacement (i.e. planting a field exclusively with seed obtained from another farmer) and mixing (i.e. planting a field with a mixture of seed saved from the farmer's previous harvest and seed obtained from another farmer), which occur at different frequencies and have contrasting population genetic effects [21]. Replacement is the most common form of seed flow and mainly occurs in response to complete seed loss or when planting for the first time. Mixing is done in response to partial seed loss or as a form of experimentation [23]. Both mixing and replacement can involve seed from either inside or outside the farming community, although the former is much more common [24].

Although the importance of seed flow to the issue of transgene escape has been recognized [3] [10], research on agricultural biosafety in maize has been mostly limited to assessing the risk of cross-pollination between adjacent fields [25–26]. Empirical research is currently strongly restricted by biosafety regulations and can only address local and short-term consequences of transgene escape. Modeling studies therefore present an important alternative for exploring the issue of seed mediated gene flow and its implications by simulating future scenarios under contrasting assumptions and conditions on large spatial scales. So far there have been very few simulation studies addressing the effect of seed flow on the parameters relevant to biosafety policy (see [9–10][27] for recent examples).

Seed flow may affect biosafety in a number of ways but here we focus on three aspects that we consider particularly important – namely diffusion, detectability and persistence of a transgene. Diffusion into non-target populations is at the heart of most biosafety concerns. Proposals for release of GM crops in Mexico have been aimed at release in regions of low maize diversity, implicitly assuming that long-distance seed dispersal can be ignored. Another key aspect for GM control is detectability. It has recently been shown that a very localized distribution of transgene frequencies lowers the probability of detection [9], prompting the need to evaluate the effects of seed flow on the spatial frequency distribution. Finally, persistence in response to natural selection is perhaps the most important long-term issue to be evaluated. Transgenes are designed to be agronomically superior in their target system, which, depending on the type of gene, may extend to traditional farming systems as well. While some of the currently available transgenic traits such as herbicide resistance may not be relevant to smallholder farmers if herbicides are not used, others may be, such as transgenes designed for biotic or abiotic stress resistance, like BT genes [3]. Such an advantage would become an evolutionary benefit within traditional seed systems. In classical population genetic models, positive selection increases the probability of a gene escaping extinction caused by genetic drift, leading to indefinite persistence and eventual fixation in the metapopulation. In classical models of undivided populations, this persistence (i.e. fixation) probability is determined by the initial gene frequency and the amount of genetic drift experienced by the metapopulation [28]. The extent to which crop metapopulations deviate from classical models in this respect remains to be established.

In this study, we provide a first example of using data on farmer-mediated seed flow to model the diffusion, detectability and persistence of a transgene. As an illustrative case study, we evaluate the long-term spatial dynamics of localized transgene introduction by modeling the recurrent growing of GM maize at a single location over 100 planting cycles, with the exception of the evaluation of persistence for which a one-time introduction is assumed. We use recently-compiled data [24] on traditional maize seed management in Mexico to parameterize a spatially-explicit, stochastic model describing the population genetics of a maize metapopulation subject to pollen and seed flow. This metapopulation consists of individual fields/seed lots cultivated within villages of which only a single central village is a cite of primary planting of GM maize (Figure 1). We consider only unconscious diffusion by excluding primary GM maize as a seed source, restricting the initial influx of transgenes into the metapopulation to be pollen mediated.

We evaluate the effect of substantial variations in seed flow parameters, with and without positive selection, by simulating five contrasting scenarios, one of which corresponds to the observed patterns of seed management in our sample of farmers and four that deviate from them in one or more seed flow parameters (Table 1). These scenarios assume constant pollen flow and combine three rates of seed mixing, two of seed replacement and two proportions of between-village seed flow, thereby providing a broad but manageable range of seed management conditions.

We intend to demonstrate the value of explicit modeling of seed management when making predictions about the fate of escaped transgenes in landrace populations under proposed scenarios of introduction. We hope that our results will contribute to improved risk assessment and biomonitoring at centers of crop origin and in regions where traditional agriculture and diverse landraces dominate crop production systems.

Results

Diffusion

Recurrent planting of GM maize is expected to lead to the accumulation of transgenes in the metapopulation with a spatial distribution determined by the patterns of gene flow. In the absence of selection, the increase in transgene frequency is relatively insensitive to the specifics of seed flow (Figure 2). Under all five seed flow scenarios, gene frequency increases linearly at a rate somewhat inferior to the theoretical maximum expected for an infinite population undergoing recurrent influx of transgenic pollen (See methods). Under positive selection the effect of differences in seed flow parameters become visible, particularly for the highest selection coefficient of 0.05. Including seed flow leads to a faster increase in transgene frequency compared to the scenario without seed flow (scenario 1, Table 2). The introduction of seed mixing (scenario 2, Table 2) has a limited effect compared to the introduction of seed replacement (scenario 3–5, Table 2), probably due to the relatively low frequency of occurrence of the former. Increasing the frequency of introduction of seed from outside the village (scenario 4, Table 2) also leads to faster accumulation of transgenes, at the same total level of mixing and replacement. For the scenario with the highest frequencies of mixing, replacement and introduction of non-local seed (scenario 5, Table 2) the rate of accumulation approaches the theoretical maximum expected in an infinite population under constant influx and positive selection (See methods).

The number of affected farmers (i.e. whose fields contain more than 1% transgenic plants) responds more strongly to changes in seed flow parameters, regardless of the selection coefficient (Figure 2). In terms of both frequency and number of affected farmers, the difference between the zero seed flow scenario (1, Table 2) and the scenario with only seed mixing (scenario 2, Table 2) is small, however, again suggesting that at realistic rates seed replacement plays a more dominant role in transgene diffusion.

As expected, seed mixing and replacement strongly affect the spatial distribution of transgenes (Figure 3). Whereas under the zero seed flow scenario (scenario 1, Table 2) transgene presence beyond three km from the focus of introduction after 100 years is negligible, the introduction of seed mixing and especially replacement leads to a much more even frequency distribution throughout the metapopulation, particularly when positive selection is strong. For scenario 3 (corresponding to realistic parameter values, Table 2) under the three selection coefficients (0, 0.01, 0.05), respective transgene frequencies of $2\ 10^{-5}$, $3\ 10^{-5}$ and $5\ 10^{-4}$, and fractions of affected farmers of $6\ 10^{-4}$, $1\ 10^{-3}$ and $3\ 10^{-2}$, are found as far as 50 km from the release site (Figure 3.). In terms of absolute numbers, more than ten million transgenic plants would grow at 50 or more kilometers from the point of introduction at the highest selection level, with more than 900 fields containing in excess of 1% transgenic plants.

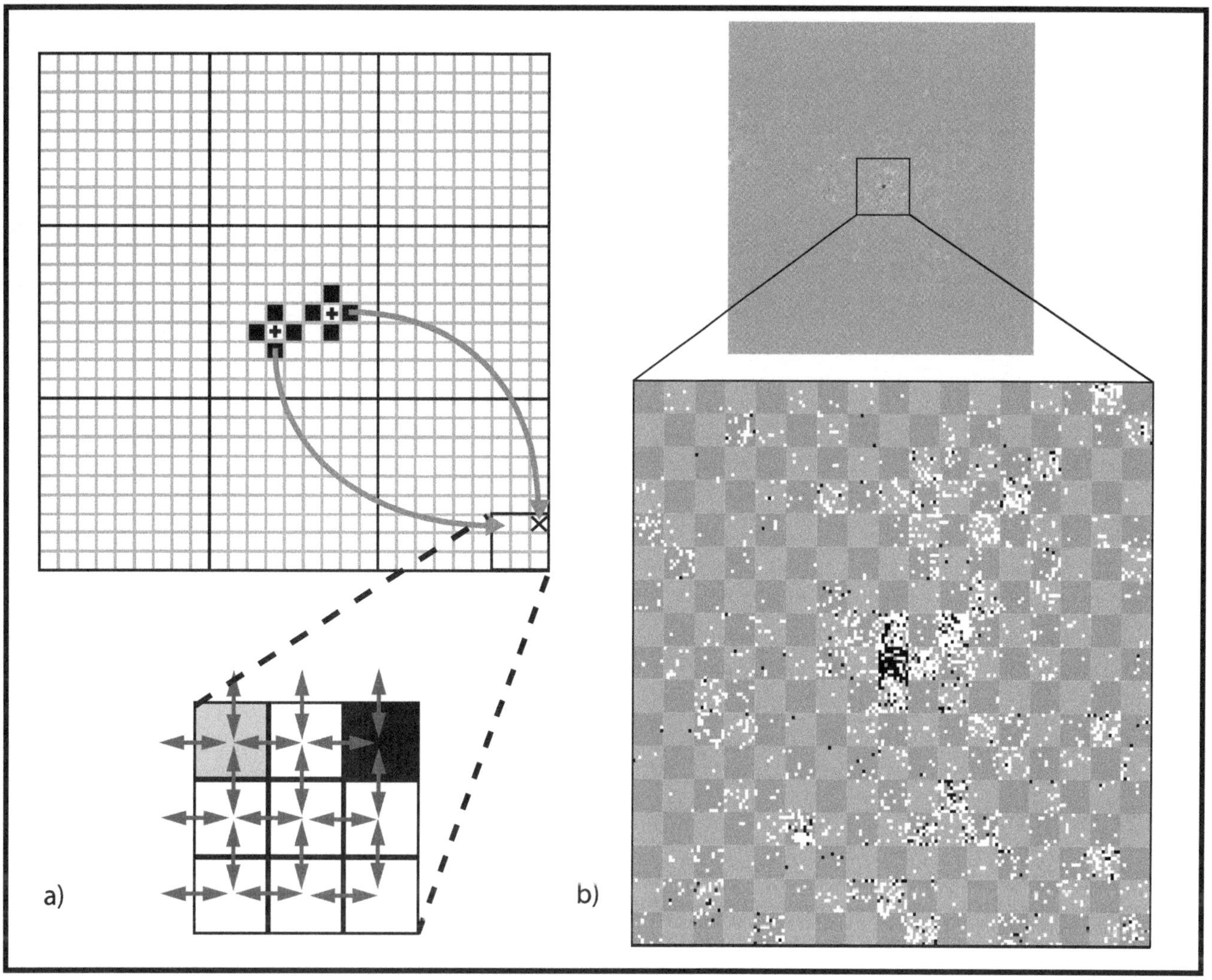

Figure 1. Diagram representing the metapopulation model (a) and example model output (b). The larger panel in (a) shows six villages with two positive farmers (black crosses) with seed flow indicated by gray arrows. Enlarged area shows a block of nine fields surrounding two fields that received contaminated seed. The field on the left mixed seed from a contaminated source, resulting in a frequency below that of the source field. The field on the right undergoes complete replacement and now has the same transgene frequency as the seed source. Panel (b) presents the spatial distribution of the transgene after 100 generations in the entire metapopulation (top) and in an enlarged area close to the focus of introduction (bottom). Transgene presence is marked in white and frequencies above 0.5 percent in black.

Table 1. Observed values for seed management data.

Parameter	**Observed values**
Number of ears used for planting (*F*)	290
Kernels per ear (*K*)	400
Probability of seed replacement (ε)	0.28
Probability of seed mixing (μ)	0.014
Proportion of seed used for mixing (*m*)	0.5
Probability of seed flow from non-local sources (*1 - V*)	0.14
Proportional contribution of pollen from a single neighbor to the gene pool of a field (m_g)	0.015
Average number of pollen neighbors (ç)	1

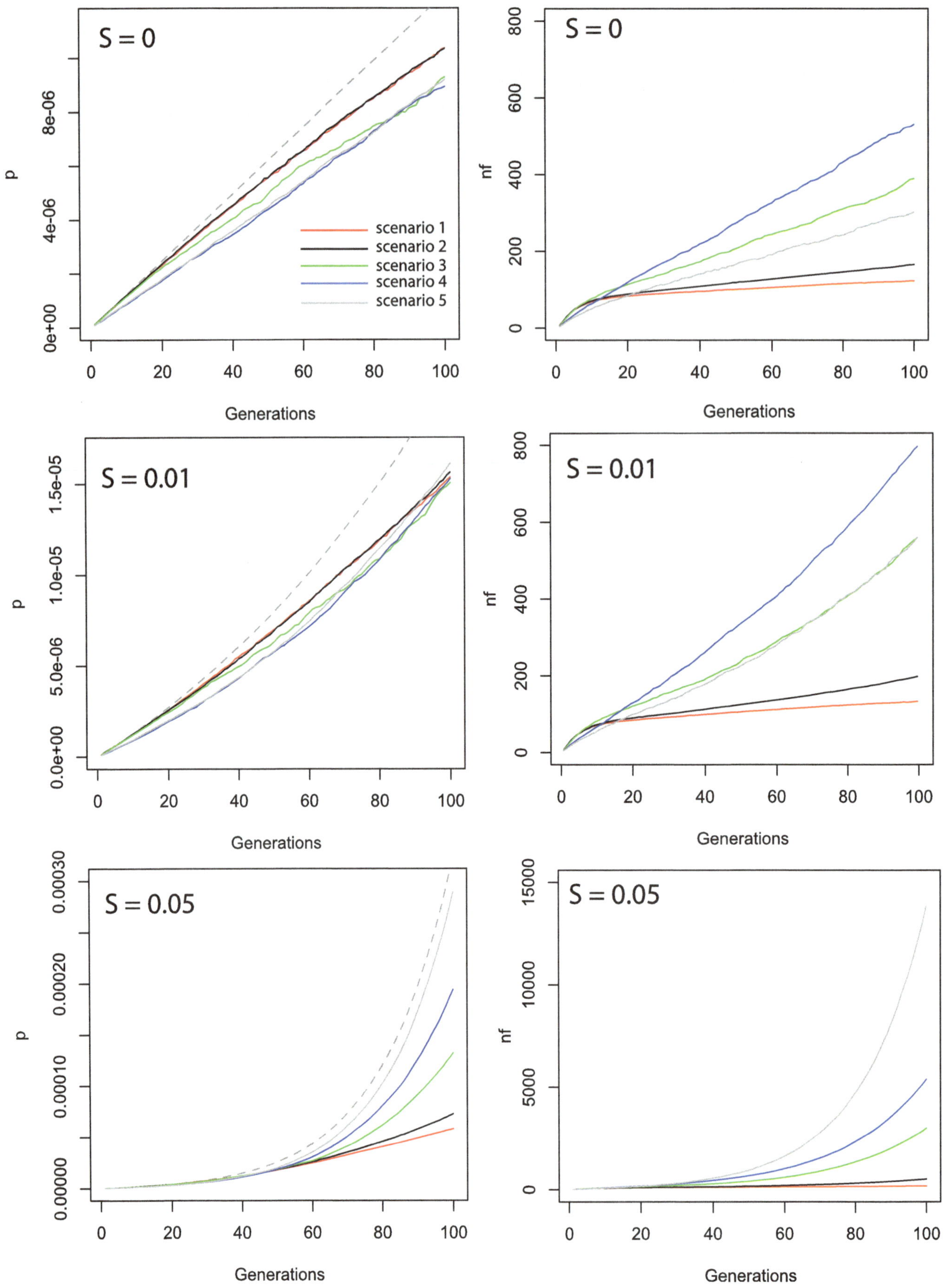
S = 0
p
0e+00
4e-06
8e-06
0
20
40
60
80
100
Generations
scenario 1
scenario 2
scenario 3
scenario 4
scenario 5
S = 0
nf
0
200
400
600
800
0
20
40
60
80
100
Generations
S = 0.01
p
0.0e+00
5.0e-06
1.0e-05
1.5e-05
0
20
40
60
80
100
Generations
S = 0.01
nf
0
200
400
600
800
0
20
40
60
80
100
Generations
S = 0.05
p
0.00000
0.00010
0.00020
0.00030
0
20
40
60
80
100
Generations
S = 0.05
nf
0
5000
10000
15000
0
20
40
60
80
100
Generations

Figure 2. Accumulation of transgene frequency (p) and number of farmers with frequencies above 0.5 percent (nf) at s = 0 (top), s = 0.01 (middle) and s = 0.05 (bottom) for the five scenarios (red = 1, black = 2, green = 3, blue = 4, gray = 5). The gray dotted line shows expectations for p.

Detectability

Using a published biomonitoring scheme, consisting of 3 fields sampled for 50 random villages [29], we take repeated samples from the simulated metapopulation at 50 planting cycles after the cessation of 100 cycles of GM maize cultivation. We then measure the probability of transgene detection as the fraction of samples that contains at least one transgene. Detection results for the five seed flow scenarios without selection confirm earlier suggestions that seed flow may affect detectability of transgenes at a given frequency by increasing spatial homogeneity of transgene frequencies [9] (Table 3). As shown by the diffusion results above, in the absence of seed flow the spatial distribution of transgenes is highly skewed, meaning that a small number of fields close the source of introduction will contain high frequencies while those further away will have extremely low frequencies (Figure 3). This skewed distribution causes the observed detection probability to fall far below that expected based on conventional estimates that assume a homogenous population [29] (see methods). As seed mixing, replacement and between villages seed flow is introduced, the probability of detection also increases. However, only for the high seed flow scenario with 28 percent mixing does the detection rate begin to approach conventional expectations. Importantly, under realistic patterns of seed mixing (scenario 3, Table 2), the detection rate is only 0.14 instead of 0.74 as expected, reflecting the limited amount of homogenization associated with seed replacement compared to seed mixing.

Persistence

Positive selection will inevitably lead to the long term persistence of a transgene unless random fluctuations in frequency, as a result of finite population size, are strong enough to cause the extinction of the gene. Starting from the low transgene frequency after a single cycle of planting of GM maize, the five scenarios and three selection levels show considerable variation in the probability of the transgene to persist indefinitely in the metapopulation (Table 4). Under a classical model of a single undivided population, such variation should reflect differences in metapopulation effective size (N_e), a determinant of the amount of genetic drift [28]. Recent theory may be used to estimate N_e for the five scenarios [21] (see methods) and substituting these estimates into the classical model indeed leads to rather precise prediction of the persistence probabilities observed for our model.

As expected, the highest estimated N_e is found in the absence of mixing and replacement (Scenario 1, Table 2), translating into probabilities of persistence close to 100 percent due to the low levels of genetic drift. Conversely, the scenarios with 28 percent seed replacement and 2 percent mixing (Scenarios 3 and 4, Table 2) have the lowest N_e and therefore relatively low probabilities of persistence. It is interesting that of the latter two scenarios the one corresponding to the observed seed management data (Scenario 3) deviates somewhat more from theoretical predictions than the scenario with higher rates of between-village seed flow (Scenario 4), in spite of having the same predicted N_e. This points to a small but unexpected effect of the simulated village structure that cannot be accounted for by existing theory.

Discussion

Predicting and managing the spread of transgenes beyond the target seed supply is of great importance for biosafety. Recent developments expanding the use of transgenes in crop plants to include the production of pharmaceuticals and other chemicals [30] have only increased the need for adequate evaluation of biosafety risks. Past experiences have shown that even within the commercial seed sector, preventing the intermixing of transgenic and conventional seed stock can be difficult [15]. The challenges of managing transgene flow in traditional seed systems in centers of crop origin are particularly daunting. Our study presents the first attempt to evaluate the potential importance of seed flow for the long-term fate of a transgene in a large crop metapopulation. By combining quantitative observations on traditional maize seed systems with a spatially explicit model, we are able to predict how deviations from observed practice may change the outcome of the introduction of a novel gene in terms of its diffusion, detectability and persistence. Our approach is a significant improvement over previous modeling studies that did not consider population genetics or seed management practices [31] [10].

As expected [3], seed flow is an important determinant of the population genetics of a transgenic element. Although in the absence of selection, seed mixing and replacement have a limited effect on the total frequency of the transgene, observed levels of replacement greatly affect the spatial distribution of the gene and the number of affected farmers. When selection is present, the frequency of mixing and replacement and the use of non-local seed sources notably influence the expected frequency of the transgene. Replacement, being the most common form of seed flow, thereby having the strongest effect.

Even though farmers predominantly rely on local seed sources, infrequent long-distance flow causes transgenes to spread much further than would be expected in the absence of seed flow. It is

Table 2. Seed flow scenarios used in the simulations.

Scenarios	Pollen flow (m_g)	Seed mixing probability (μ)	Seed replacement probability (ε)	Probability of between-village seed flow ($1- V$)	Estimated N_e
1	0.015	0	0.0	0.0	1,184,818,534
2	0.015	0.019	0.0	0.14	844,993,687
3	0.015	0.019	0.28	0.14	37,298,109
4	0.015	0.019	0.28	0.9999	37,298,109
5	0.015	0.28	0.28	0.9999	153,791,555

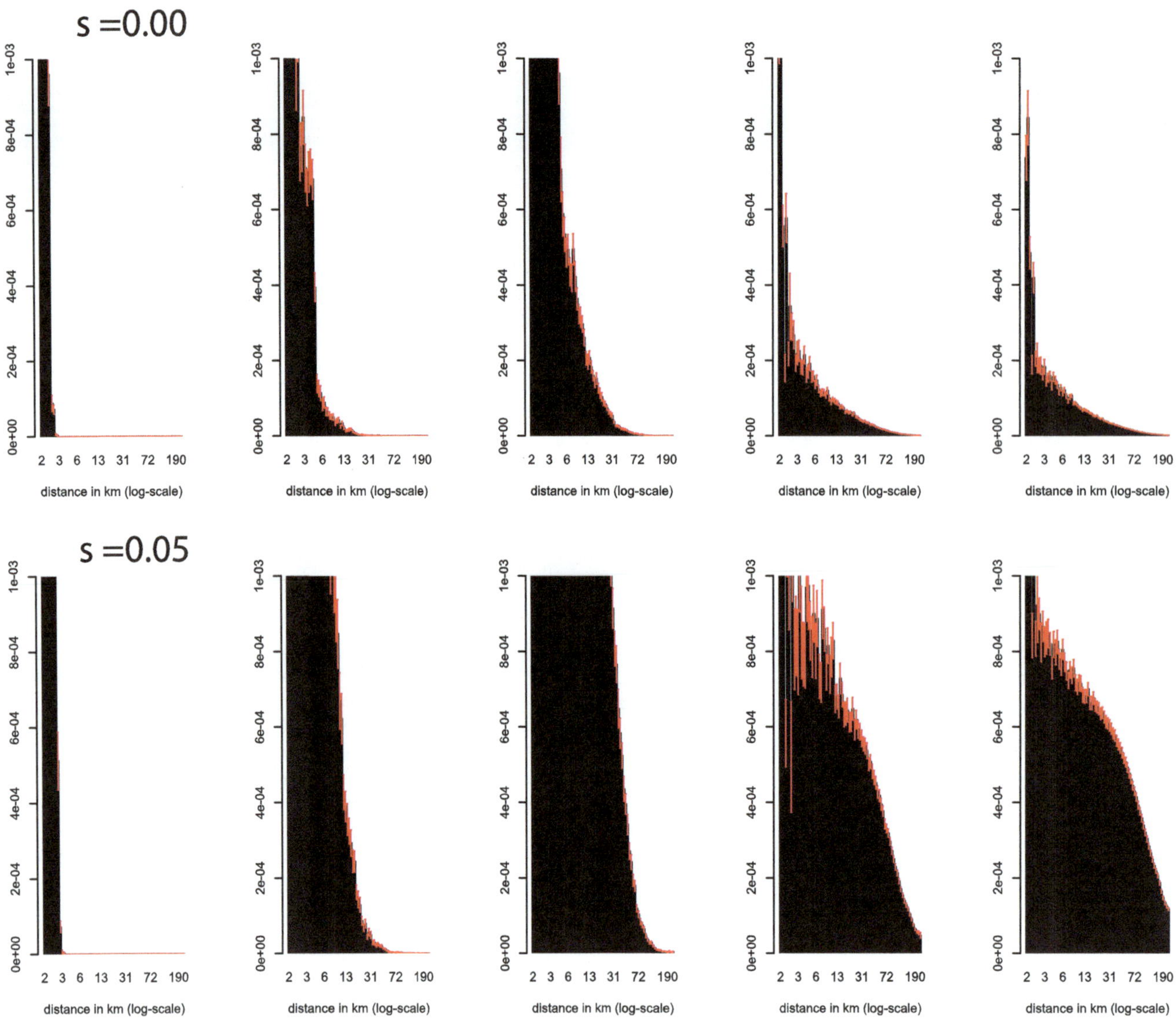

Figure 3. Frequency of the transgene vs. the distance from the focus of introduction (where the transgene is being introduced recurrently) at s = 0.00 (top) and s = 0.05 (bottom) for scenarios 1–5. Red whiskers indicate the standard error. Displayed values are truncated at p = 0.001.

Table 3. Detection probabilities of the transgene for each of the five scenarios.

Scenario	1	2	3	4	5
Expected	0.83	0.83	0.74	0.76	0.77
Observed	0.01	0.03	0.14	0.19	0.46

The observed probabilities refer to the average frequency of trials in which the positive transgene was detected among ten repetitions. Expected probabilities are simple binomial probabilities based on the assumption of spatially homogenous transgene frequencies (see methods).

Table 4. Persistence of the transgene, based on a starting frequency of 0.00001.

	selection coefficient	0.005		0.01		0.05	
Scenario	N_e	**Exp**	**Obs**	**Exp**	**Obs**	**Exp**	**Obs**
1	11,848,115	0.947	0.923	0.997	0.989	1.000	1.000
2	8,450,219	0.876	0.854	0.985	0.980	1.000	1.000
3	374,069	0.088	0.123	0.169	0.257	0.604	0.843
4	374,069	0.088	0.085	0.169	0.180	0.604	0.732
5	1,538,842	0.317	0.276	0.533	0.461	0.978	0.947

Both the observed and expected persistence probabilities are shown for each of the six scenarios using two different selection coefficients. The observed probability of extinction in each scenario is estimated as the frequency of 100 simulations where the transgene frequency drops down to zero.

noteworthy that after 100 planting cycles, even under our relatively conservative scenario of 33 fields undergoing recurrent planting of GM maize, as many as 3% of farmers at 50 km from the original release sites are affected if the selection coefficient exceeds 0.05. This should be considered when evaluating biosafety of new releases based on distance from areas where landraces grow (see recent maps by CONABIO: http://www.biodiversidad.gob.mx/genes/proyectoMaices.html).

Although seed flow greatly increases the spatial diffusion of transgenes, the observed predominance of seed replacement over seed mixing translates into a limited rate of homogenization at the metapopulation level. In practical terms this means that true detection probabilities will be much lower than expected based on conventional calculations that assume a homogenous distribution [29]. In our case, only the scenario where seed mixing was equally frequent as replacement yielded a detection probability approaching theoretical expectations.

An important result from our simulations is that in spite of the complex spatial structure represented by our metapopulation, the probability of persistence of a transgene of known frequency and selective advantage is relatively well predicted by existing theory [28]. For the parameter ranges studied here, most of the variation in persistence probability at a given level of selection is explained by effective size as predicted from the model parameters using a recent mathematical crop metapopulation model [21]. The slight discrepancy observed between scenarios 3 and 4, in spite of their identical predicted population sizes, is the only instance where our model predictions deviate from theoretical expectations. Scenario 4 assumes unlimited seed flow between villages as in the mathematical model, while between-village seed flow is restricted in scenario 4. The fact that persistence in scenario 4 better fits theoretical predictions thus suggests that spatial restricted seed flow may cause an increase in effective size.

These results show that it is possible to make quantitative statements on the fate of transgenes, albeit under admittedly unrealistic assumptions of long-term persistence of the traditional maize production system. For example, given an approximate number of about 3 million small-scale maize farmers in Mexico [32], a rough theoretical estimate of the effective population size would be on the order of 100 million. This means that for a transgene with a selective advantage of 0.01, frequencies would have to remain below 10^{-8} to have a less than 5% chance of persistence. Detecting a single transgene at such frequencies with 95% probability would require testing around 750,000 maize ears under the assumption of homogeneity. Considering that true detection probabilities are much lower due to the uneven distribution of transgenes, it is unlikely that monitoring and controlling a selectively favored transgene would be feasible.

In all model predictions, selection is a key parameter. Although our simulated selection levels are within the range of published estimates for maize genes [33–34], we know very little about the potential fitness effects of different transgenes. Since transgenes are designed to be agronomically favorable and of large effect it is not unlikely that some are subject to strong positive selection, although others may turn out to be effectively neutral within traditional agro-ecosystems. Determining selective coefficients of particular transgenes under realistic conditions will thus be pivotal for designing sound detection and biosafety schemes and regulations and should be a research priority for an evidence-based biosafety system.

In this paper we have explored a limited and rather conservative example of transgene presence and diffusion to demonstrate the interaction between farmer practice and the population genetics of a new and rare genetic element. We have not tried to represent more extreme scenarios of transgene introduction such as may be occurring now or in the near future. Our analysis of the metapopulation dynamics of newly-introduced genes in traditional seed systems is by no means exhaustive. There are obviously many more scenarios and parameter combinations that may be explored, and there may be relevant aspects of seed management and that we have not included. Although the basic patterns of seed flow, such as localized exchange and predominance of seed replacement, will be similar for most traditional farming systems, it is likely that aspects of seed management vary by geography and environment [35] [24]. Some regions may for example have a tighter integration of traditional and modern seed systems than we have assumed [36], which would imply the occurrence of seed mixing between conventional and transgenic stock and faster transgene diffusion than predicted here. Nonetheless, we hope to have shown that knowledge on seed management is of great relevance for predicting the possible outcome of proposed scenarios of transgene release and is essential for establishing a rigorous and quantitative framework for biosafety policies.

Materials and Methods

Model Description

Our computational model extends recent population genetic models of traditional seed systems [9] [21] by providing spatially explicit representation of villages as well as long distance seed dispersal with a user-determined dispersal kernel. We simulate the temporal dynamics of a single, bi-allelic locus in a square grid of n fields assigned to k villages (Figure 1) where neighboring fields are connected by pollen and seed flow that can take place both within and between villages. Fields subject to primary planting of GM maize are assigned a starting frequency, $p = 0.5$, of the transgenic allele. Recurrent planting of GM maize, such as assumed for the evaluations of diffusion and detectability, is modeled by resetting their frequency to 0.5 in each generation. The frequency change for each field in each generation is then modeled in four consecutive stages, where in each stage the starting frequency is designated by p and the new frequency by p^*:

1. Pollen flow

Pollen flow is assumed to be deterministic. Fields receive a proportion m_g of its genes from their η neighbors. The number of neighbors may be trimmed for the sake of realism. Fields at the corners and borders of the grid have two and three neighbors respectively. For a single field, transgene frequency after pollination is given by:

$$p^* = p(1 - \eta m_g) + m_g \sum_i^{\eta} p_i$$

with $\eta \leq 4$.

2. Selection

Selection is also modeled deterministically under the assumption of complete dominance. Gene frequency after selection becomes:

$$p^* = p + \frac{sp(1-p)^2}{1 - s(1-p)^2}$$

3. Drift

Each field is planted with F ears bearing K kernels, totaling $N=FK$ diploid plants. We assume random mating within fields and $F<<N$ so that the genetic sampling process is appropriately described as taking male and female gametes with replacement. Gene frequency in the paternal contribution to the next generation is given by:

$$p_m^* \sim \frac{B(N,p)}{N}$$

Whereas the frequency in the female contribution is:

$$p_f^* \sim \frac{B(N,Q)}{N},\ Q \sim \frac{B(2F,p)}{2F}$$

The double sampling describes the generation of N haploid gametes from F diploid ears. The post-drift gene frequency for each field is simply:

$$p^* = \frac{p_m^* + p_f^*}{2}$$

4. Seed flow

Seed flow occurs by mixing and replacement, respectively. For both types of seed flow there is a single seed source for each receiving field in each generation. Replacement occurs with a probability ε and involves replacement of the entire seed lot with that of another farmer. Transgenic frequency after replacement is given by $p^* = p_i$. Where p_i is the frequency in the source field. Mixing takes place with a probability μ and involves supplementing a fraction m of the native seed lot with seed from a another farmer. Frequency after mixing thus becomes:$p^* = p(1-m) + mp_i$. Seed sources are chosen randomly from non-replaced seed lots from within or from outside the village with probabilities V and (1-V) respectively. Each source outside the village is defined by a random angle δ from a uniform distribution and a random distance d (according to a chosen dispersal kernel). If a replaced or primary GM maize seed lot is chosen or if the coordinates fall within the same village or outside the grid, a new random source is selected. The exclusion of GM maize seed lots as seed sources, imposed to model unconscious diffusion, can obviously be lifted if so desired. Source code of our model is available upon request.

Information on Seed Systems and Parameterization

Seed management data. To obtain estimates for seed related model parameters we used raw data from a recently published socioeconomic survey [24] that was carried out in 2003 with the heads (i.e. the person making decisions on domestic and farm issues) of 400 randomly-selected maize-producing households at 20 sites (20 households per site) randomly selected among 46 municipalities comprised in four transects across an altitudinal gradient from 10 to 2980 masl in five states in Central Mexico.

All maize farmers interviewed relied on traditional seed systems and planted mostly landraces, although some of them had experimented with improved varieties as well. In addition to socio-economic data, this survey contained questions specific to maize seed management such as quantities, geographic origins and history of the seed planted. Estimates of pollen flow were taken from the literature [26]. Because most parameters differed substantially between maize-growing environments and in order to represent a single environment, only data for the highland environment was used, with the exception of seed source distances, which were obtained from the full dataset to assure sufficient sample size. The parameter estimates derived from these data are presented in Table 1. The reported distances of seed sources confirmed a prevalence of local seed flow with rare instances of long-distance seed flow, as reported in other studies. We fit a lognormal distribution to observed distances using the *fitdistr* routine contained in the MASS package of the R statistical software [37]. Seed flow distances between villages in our model were set to correspond to real between-village distances by scaling the average between-village distance in our simulations to match the average neighbor distance in a Delaunay triangulation of the coordinates of 911 mapped highland villages (>300 inhabitants) in a rectangular section of the study region (information: INEGI, http://www.inegi.org.mx).

Parameterization and Model Scenarios

Our simulated metapopulation contained one million fields divided equally over 10,000 villages, with a single village in the center containing 33 fields planted with GM maize. We explored five different scenarios of seed flow (Table 1), covering a wide range of seed management types ranging from the complete absence of seed flow to a high seed flow scenario of 28 percent seed replacement and mixing and a 99.99 percent probability of between-village flow. Scenario number three represented the observed seed management parameters as estimated from our survey data, i.e. infrequent seed mixing, frequent seed replacement and predominantly local seed flow. For the evaluation of transgene diffusion, continuous planting by the same 33 farmers was simulated for 100 generations and the results averaged over 100 replications. We tested selection coefficients of 0, 0.01 and 0.05. The expected maximum rate of transgene frequency increase in an undivided population of infinite size is simply the deterministic increase caused by pollen flow from positive fields and positive selection, represented by:

$$\Delta p = n_p \eta m_g n^{-1} + sp.$$

For the detectability evaluation the five scenarios were simulated in the absence of selection ($s=0$), for 150 generations with planting of GM maize by 33 farmers ceased after 100 generations. For each replication, 10,000 samples were simulated, consisting of 50 randomly-selected villages, three fields per village, seven ears per field and 165 seeds per ear. This sampling scheme was designed to be quantitatively similar to Ortiz-Garcia *et al.* [29] in their 2003–2004 biomonitoring survey of Mexican landraces. Detection was modeled as a set of Bernoulli trials, in which each sampled field has a probability of detection of $1-(1-p)^n$, where $n=(2F+FK)$ [9], and p is the transgene frequency in that field. The expected overall detection probability under the assumption of homogeneous frequencies is simply given by the above formula with F and K replaced by the total number of sampled ears and kernels respectively and by replacing p by the mean transgene frequency in the metapopulation [29].

Theoretical expectations for persistence are represented by $u=\frac{1-e^{-4N_e sp}}{1-e^{-4N_e s}}$ [28], where u is the persistence (or fixation) probability and N_e is the variance effective population size. Given p and s, higher values of N_e lead to higher persistence probabilities because of a lower chance of losing the new gene due to drift. N_e was calculated as the long-term N_e derived from the mean coalescent time T ($N_e = 2T$), given by the crop metapopulation model by van Heerwaarden et al. 2010 [21], with model parameters set as in Table 1.

For the evaluation of persistence a smaller metapopulation of only 100 villages was simulated to limit computational time. In this case 1000 replications were run of a single planting cycle of GM maize by 33 farmers, leading to a starting frequency of $p = 1.24 \times 10^{-5}$. Selection was set to 0.005, 0.01 and 0.05. Simulations were run until transgene frequency reached either 0 (extinction) or 0.001 (persistence). The threshold value of 0.001 for persistence was used to limit computational time and based on the fact that for this frequency the probability of fixation approaches unity under realistic values of N_e.

Acknowledgments

The authors wish to thank 4 anonymous reviewers for valuable comments on earlier versions of the document.

Author Contributions

Conceived and designed the experiments: JvH. Performed the experiments: JvH. Analyzed the data: JvH DODV. Contributed reagents/materials/analysis tools: MRB. Wrote the paper: JvH EAB MRB.

References

1. Ellstrand NC, Hoffman CA (1990) Hybridization as an avenue of escape for engineered genes. BioScience 40: 438–442.
2. Snow AA, Andersen B, Jørgensen RB (1999) Costs of transgenic herbicide resistance introgressed from Brassica napus into weedy B. rapa. Molecular Ecology 8: 605–615.
3. Bellon MR, Berthaud J (2004) Transgenic maize and the evolution of landrace diversity in Mexico. The importance of farmers' behavior. Plant Physiology 134: 883–888.
4. Mercer K, Wainwright J (2008) Gene flow from transgenic maize to landraces in Mexico: An analysis. Agriculture, Ecosystems & Environment 123: 109–115.
5. Matsuoka Y, Vigouroux Y, Goodman MM, Buckler ES, Sanchez G (2002) A single domestication for maize shown by multilocus microsatellite genotyping. Proceedings of the National Academy of Sciences of the United States of America 99: 6080–6084.
6. van Heerwaarden J, Doebley J, Briggs WH, Glaubitz JC, Goodman MM, et al. (2011) Genetic signals of origin, spread, and introgression in a large sample of maize landraces. Proceedings of the National Academy of Sciences of the United States of America 108: 1088–1092.
7. Quist D, Chapela IH (2001) Transgenic DNA introgressed into traditional maize landraces in Oaxaca, Mexico. Nature 414: 541–543.
8. Serratos-Hernández JA, Gomez-Olivares JL, Salinas-Arreortua N, Buendia-Rodriguez E, Islas-Gutiérrez F, et al. (2007) Transgenic proteins in maize in the soil conservation area of Federal District, México. Frontiers in Ecology and the Environment 5: 247–252.
9. Piñeyro Nelson A, Van Heerwaarden J, Perales HR, Serratos-Hernández JA, Rangel A, et al. (2009) Transgenes in Mexican maize: molecular evidence and methodological considerations for GMO detection in landrace populations. Molecular Ecology 18: 750–761.
10. Dyer GA, Serratos-Hernández JA, Perales HR, Gepts P, Piñeyro-Nelson A, et al. (2009) Dispersal of transgenes through maize seed systems in Mexico. PloS one 4: e5734.
11. Ellstrand NC, Prentice HC, Hancock JF (1999) Gene flow and introgression from domesticated plants into their wild relatives. Annual Review of Ecology, Evolution, and Systematics 30: 539–563.
12. Chapman MA, Burke JM (2006) Letting the gene out of the bottle: the population genetics of genetically modified crops. New Phytologist 170: 429–443.
13. Schoen DJ, Reichman JR, Ellstrand NC (2008) Transgene escape monitoring, population genetics, and the law. BioScience 58: 71–77.
14. Bucchini L, Goldman LR (2002) Starlink corn: a risk analysis. Environmental health perspectives 110: 5.
15. Mellon M, Rissler J (2004) Gone to seed: transgenic contaminants in the traditional seed supply. Union of Concerned Scientists.
16. Bellon MR, Brush SB (1994) Keepers of maize in Chiapas, Mexico. Economic Botany 48: 196–209.
17. Louette D, Charrier A, Berthaud J (1997) In situ conservation of maize in Mexico: genetic diversity and maize seed management in a traditional community. Economic Botany 51: 20–38.
18. Badstue LB, Bellon MR, Berthaud J, Ramirez A, Flores D, et al. (2007) The dynamics of seed flow among small-scale maize farmers in the Central Valleys of Oaxaca, Mexico. World Development 35: 1579–1593.
19. Chambers KJ, Brush SB (2010) Geographic Influences on Maize Seed Exchange in the Bajio, Mexico. The Professional Geographer 62: 305–322.
20. Louette D (2000) Traditional management of seed and genetic diversity: what is a landrace? In: Brush SB, editor. Genes in the field: on-farm conservation of crop diversity. IDRC (International Development Research Centre).
21. van Heerwaarden J, van Eeuwijk FA, Ross-Ibarra J (2010) Genetic diversity in a crop metapopulation. Heredity 104: 28–39.
22. Bellon MR (1991) The ethnoecology of maize variety management: a case study from Mexico. Human Ecology 19: 389–418.
23. Smale M, Aguirre A, Bellon MR, Mendoza J, Rosas IM (1999) Farmer management of maize diversity in the Central Valleys of Oaxaca, Mexico: CIMMYT/INIFAP 1998 baseline socioeconomic survey. Economics Working Papers.
24. Bellon MR, Hodson D, Hellin J (2011) Assessing the vulnerability of traditional maize seed systems in Mexico to climate change. Proceedings of the National Academy of Sciences 108: 13432–13437.
25. Ma BL, Subedi KD, Reid LM (2004) Extent of cross-fertilization in maize by pollen from neighboring transgenic hybrids. Crop Science 44: 1273.
26. Messeguer J, Penas G, Ballester J, Bas M, Serra J, et al. (2006) Pollen-mediated gene flow in maize in real situations of coexistence. Plant Biotechnol J 4: 633–645.
27. Dyer GA, Gonzalez C, Lopera DC (2011) Informal "seed" systems and the management of gene flow in traditional agroecosystems: the case of cassava in Cauca, Colombia. PLoS One 6: e29067.
28. Kimura M (1962) On the probability of fixation of mutant genes in a population. Genetics 47: 713.
29. Ortiz-Garcia S, Ezcurra E, Schoel B, Acevedo F, Soberon J, et al. (2005) Absence of detectable transgenes in local landraces of maize in Oaxaca, Mexico (2003–2004). Proceedings of the National Academy of Sciences of the United States of America 102: 12338.
30. Stewart CN (2008) Pharming in crop commodities. Nature biotechnology 26: 1222–1223.
31. Serratos-Hernández JA, Islas-Gutiérrez F, Buendía-Rodríguez E, Berthaud J (2004) Gene flow scenarios with transgenic maize in Mexico. Environmental Biosafety Research 3: 149–157.
32. Food and Agriculture Organization (2004) The state of agricultural commodity markets. FAO.
33. Wang RL, Stec A, Hey J, Lukens L, Doebley J (1999) The limits of selection during maize domestication. Nature 398: 236–239.
34. Hufford MB, Xu X, van Heerwaarden J, Pyhäjärvi T, Chia J-M, et al. (2012) Comparative population genomics of maize domestication and improvement. Nature Genetics DOI:10.1038/ng.2309.
35. Brush SB, Perales HR (2007) A maize landscape: Ethnicity and agro-biodiversity in Chiapas Mexico. Agriculture, Ecosystems & Environment 121: 211–221.
36. van Heerwaarden J, Hellin J, Visser RF, van Eeuwijk FA (2009) Estimating maize genetic erosion in modernized smallholder agriculture. Theoretical and Applied Genetics 119: 875–8dyer8.
37. R DCT (2009) R: A Language and Environment for Statistical Computing.

5

Quantitative Digital Imaging of Banana Growth Suppression by Plant Parasitic Nematodes

Hugh Roderick[1], Elvis Mbiru[2], Danny Coyne[2], Leena Tripathi[2], Howard J. Atkinson[1]*

1 Centre for Plant Sciences, University of Leeds, Leeds, United Kingdom, 2 International Institute of Tropical Agriculture, Kampala, Uganda

Abstract

A digital camera fitted with a hemispherical lens was used to generate canopy leaf area index (LAI) values for a banana (*Musa* spp.) field trial with the aim of establishing a method for monitoring stresses on tall crop plants. The trial in Uganda consisted of two cultivars susceptible to nematodes, a plantain, Gonja manjaya and an East African Highland banana, Mbwazirume, plus a nematode resistant dessert banana, Yangambi km5. A comparative approach included adding a mixed population of *Radopholus similis*, *Helicotylenchus multicinctus* and *Meloidogyne* spp. to the soil around half the plants of each cultivar prior to field planting. Measurements of LAI were made fortnightly from 106 days post-planting over two successive cropping cycles. The highest mean LAI during the first cycle for Gonja manjaya was suppressed to 74.8±3.5% by the addition of nematodes, while for Mbwazirume the values were reduced to 71.1±1.9%. During the second cycle these values were 69.2±2.2% and 72.2±2.7%, respectively. Reductions in LAI values were validated as due to the biotic stress by assessing nematode numbers in roots and the necrosis they caused at each of two harvests and the relationship is described. Yield losses, including a component due to toppled plants, were 35.3% and 55.3% for Gonja manjaya and 31.4% and 55.8% for Mbwazirume, at first and second harvests respectively. Yangambi km5 showed no decrease in LAI and yield in the presence of nematodes at both harvests. LAI estimated by hemispherical photography provided a rapid basis for detecting biotic growth checks by nematodes on bananas, and demonstrated the potential of the approach for studies of growth checks to other tall crop plants caused by biotic or abiotic stresses.

Editor: Sunghun Park, Kansas State University, United States of America

Funding: This work was supported by funding from the BBSRC (grant number BB/F004001/1; http://www.bbsrc.ac.uk/home/home.aspx) and IITA (http://www.iita.org/). The funders had no role in study design, data collection and analysis, decision to publish, or preparation of the manuscript.

Competing Interests: The authors have declared that no competing interests exist.

* E-mail: h.j.atkinson@leeds.ac.uk

Introduction

Establishing damage levels from biotic stresses and the resultant economic loss to crops requires optimisation between assuring the reliability of the information gained and the effort and resources invested in its acquisition. The plant parasitic nematodes that cause severe losses to global banana and plantain (*Musa* spp.) production provide a specific example of the difficulties of damage assessments. The yield losses they cause are a combination of reduced plant growth that extends the production cycle, reduced harvested banana bunch weight and toppling of plants with damaged root systems in storms [1,2]. They are often controlled in commercial plantations by periodic application of pesticides [3] but chemical control is rarely affordable or available to small producers in African countries where losses may be up to 71±16% [4].

Damage and economic thresholds based on nematode densities recovered from roots have been determined for *Radopholus similis*, which is widely damaging in commercial plantations, but the values vary for different geographical regions [1,5]. A second approach measures the extent of root necrosis caused by this nematode. The resultant disease index is readily obtained and can be related to economic thresholds [5–7]. However, this approach has limitations when additional economically damaging nematode species are present, as frequently occurs in Uganda [8]. In that country *Helicotylenchus multicinctus* and *Meloidogyne* spp. are also present in tropical lowland ecologies with *Pratylenchus goodeyi* present at cooler higher altitudes [8]. Work in Costa Rica on the dessert banana cultivar Grand Naine ranked *R. similis*, *P. coffeae*, *M. incognita* and *H. multicinctus* in descending order of severity of root damage by equal densities of each nematode species but recorded *M. incognita* as causing the largest reduction in bunch weight [9]. Similar work with the plantain cultivar Apantu-pa in Ghana identified that *P. coffeae* caused the most severe necrosis and yield reductions but both *M. javanica* and *H. multicinctus* caused losses of 30% and 26% respectively with relatively less necrosis [10]. Consequently root necrosis may not be a reliable basis for assessing nematode losses to yield when *Meloidogyne* spp. are one of the economic species present [11] or generally when *R. similis* is not the only damaging species present.

This work uses leaf area index (LAI) to measure banana plant growth checks by nematodes more directly than inferring them from nematode root density or damage. LAI is widely used to measure the growth of crop plants and can be defined as the amount of leaf surface area (single side) per-unit area of the ground. It relates to the capacity of the banana canopy to intercept solar radiation and fix carbon and varies with location, planting density and additional factors such as season [12]. LAI is important because leaves normally far outweigh the contributions from other plant parts to the yield of most crops [13]. A range of methods have been used to measure LAI. Those applied to crops include direct measurement of the lengths and widths of leaves [14] and approaches that use ground based instruments that measure the fraction of transmitted radiation that passes through a plant canopy [15,16]. The approach taken in this work is based on hemispherical photography using a camera with a 180° fisheye lens directed from the ground to the sky zenith to capture an image. This is then analysed to determine the LAI for those plants in the field of view. Hemispherical photography has been shown to be an effective tool for estimating LAI in various settings, including for coniferous forests [17] and single urban trees [18]. The method has also been validated against other indirect methods for monitoring LAI [18–20]. The approach has benefited from the development of digital photography, avoiding the need to develop film camera images. Proprietary software for a personal computer rapidly processes digital images to LAI estimates.

The current study explores the use of digital photography and LAI to detect growth checks to banana plants using a replicated field trial design in Uganda imposed by challenge with a combination of *R. similis*, *H. multicinctus* and *Meloidogyne* spp. Two nematode susceptible banana cultivars were used together with Yangambi km5 that has known resistance against *R. similis* [21]. The differences detected that are attributable to nematode challenge were related to bunch weights and root necrosis over two successive cropping cycles, to provide a comprehensive assessment of the technique. The results establish the potential of the approach for studies of growth checks to banana and other tall crop plants including those caused by biotic or abiotic stresses.

Methods

Study Site

The trial was conducted at Sendusu (328340E, 08320N), 1150 m above sea level, 28 km north of Kampala, Uganda. The site receives mean annual rainfall of 1300 mm in a bimodal pattern and annual minimum and maximum mean temperatures of 16°C and 28°C. The trial site land had been fallow for one year following cultivation of yam and cassava. No endangered or protected species were involved in the described field study and no specific permits were required for the studies, which were carried out on land designated for field trials at the research station of the International Institute for Tropical Agriculture (IITA).

Experimental Design

The trial included two factors: banana cultivar and nematode treatment. Three cultivars were used, the East African highland banana (EAHB) Mbwazirume (AAA-EA genotype), the plantain Gonja manjaya (AAB genotype) and the dessert banana Yangambi km5 (AAA genotype) which were selected as a representatives of different banana genotypes and nematode host status. All planting material was sourced from farms in close proximity to the Sendusu site. Suckers were pared and hot water treated for 30 sec at 98°C for disinfection of plant parasitic nematodes and weevil larvae [22], then potted into steam-sterilised sandy loam soil in 10 litre polythene bags. Nematode treatment comprised either no supplementation of the natural population or addition of approximately 5,500 nematodes to soil around plants prior to planting. Nematodes were added using 20 g of fresh banana root segments containing 77% *R. similis*, 17% *H. multicinctus* and 6% *Meloidogyne* spp. The inoculum roots were collected from an infested banana site at Sendusu and placed into a shallow excavation around each plant and mixed into the soil. Estimated nematode densities were calculated from five 5 g root segment sub-samples immediately prior to inoculation. Root pieces were bulked, chopped (~0.5 cm) and sub-samples removed, macerated in a blender for 10 s and the nematodes extracted overnight using a modified Baermann technique [23]. Each plant not given added nematodes received 20 g of autoclaved root segments. All plants were planted into the field twenty days later on 18th May 2009. A standard split-plot design [24] was used with plot replicates of 36 banana plants spaced 3 m apart in a 6 by 6 grid. Each cultivar was replicated in eight plots, four of which were infected with nematodes giving a total of 144 plants for each cultivar/treatment combination. The cultivation followed local practises with 10 kg of dry cow manure added per planting hole and fresh elephant grass (*Pennisetum purpureum*) mulched four times annually around each plant. Water was applied daily for 1 week to ensure transplanted plants established. The trial was surrounded by a row of guard plants transplanted with the trialled plants.

Crop Data Collection

Plant growth was assessed from 106 days post planting (dpp) every 14 days between 10 am and noon by hemispherical photography. Images were captured with an 8 mm full circle fisheye lens (EX-DG, Sigma, Kanagawa, Japan) on a full-frame digital camera (EOS 5D, Canon, Tokyo, Japan) attached to a self levelling mount (SLM7, Delta-T, Cambridge, UK) and a monopod. The camera lens front was placed 95 cm above the ground in the centre of four plants and directed at the sky zenith. A single exposure was taken while the observer knelt below the field of view. The majority of the canopy recorded in each image was dominated by the four plants surrounding the camera. Nine images per plot were taken, so that each plant was dominant in only one image. Mature bunch weight and date of harvest was recorded for each plant which spanned 350–520 dpp and 530–900 dpp for the first and second harvests respectively.

The crop was continually monitored by field observation and during analysis of hemispherical images for the foliar diseases Black Sigatoka, Fusarium wilt and bacterial *Xanthomonas* wilt and for the presence of the banana weevil *Cosmopolites sordidus*.

Analysis of Hemispherical Photographic Images

The images collected by digital camera were imported into proprietary software (HemiView Version 2, Delta-T, Cambridge, UK) calibrated to the lens used from pre-loaded settings. The software classifies images to distinguish visible and obscured sky directions and so define canopy openings relative to banana foliage. The sky visibility and obstruction is then defined as a function of sky direction and canopy indices are calculated from this information to provide results including LAI [25]. The algorithms estimate LAI as half of the total leaf area per unit ground area by means of an iterative inversion of Beer's Law [12]. LAI generated for each image was exported into Excel (Microsoft, Redmond, WA) before further statistical analysis. The mean of the measurements across four plots, 36 images in total, was calculated for each cultivar/treatment combination on each day of measurement.

Nematode Damage Assessment

At harvest, the percentage root necrosis for each plant was derived from five randomly selected functional primary roots. They were cut into 10 cm lengths, sliced length-wise and the proportion of cortex necrosis estimated for each root piece [11]. The overall percentage root necrosis comprised the average of the five pieces. The five functional root segments used to score for root necrosis were cut into 0.5 cm sections, thoroughly mixed, and a 5 g sub-sample used for nematode extraction. Each sub-sample was macerated in a blender for 10 s and the nematodes extracted overnight as above. Nematode population densities were estimated from three 2 ml aliquots from a 25 ml suspension. Nematodes were identified to species level and population densities estimated per 100 g root fresh weight. At the first harvest nematode counts were made for all plants, at the second harvest nematode counts were made on three samples per plot. Nematode densities were log transformed before analysis.

Statistical Analysis

All data were analysed using a standard statistical package (SPSS v18; IBM Corporation Armonk, New York, USA; http://www-01.ibm.com/software/analytics/spss/) installed on a personal computer with choice of method informed by a standard text [26].

Results

The Leaf Area Index of Growing Plants

Figure 1 provides mean canopy LAI values based on 36 images for each data point. Images were collected over two growth cycles for plants with or without nematode addition before their transfer to the field. The images in the figure indicate the range of LAI values that were obtained. In the first cropping cycle, the highest grand mean value for LAI occurred at 340 dpp for Gonja manjaya and Mbwazirume. The mean (± SEM) for the images on this day for plants receiving nematodes before transplanting was 74.8±3.52% for Gonja manjaya and 71.1±1.88% for Mbwazirume of LAI values of 0.652±0.0154 and 0.587±0.0148, respectively, of plants not challenged by added nematodes. The highest grand mean LAI for the same two cultivars before the second harvest occurred at 540 dpp. When nematodes had been added to plants before transplanting, the mean values were 69.2±2.15% for Gonja manjaya and 72.2±2.74% for Mbwazirume of mean LAI values of 0.791±0.0406 and 0.743±0.0240, respectively, when no nematodes were added. A small opposite effect of adding nematodes was observed for the nematode resistant Yangambi km5 plants. Mean values at the same dpp as used for the other two cultivars were 118.3%±5.52% and 113.4±4.16% for the first and second crop cycles relative to LAI values when nematodes were not added of 0.787±0.0307 and 1.048±0.0471 respectively. The difference in LAI between adding and not adding nematodes was statistically significant for all three cultivars ($P<0.001$, for both occasions of Gonja and Mbwazirume and $P<0.05$ for both Yangambi km5 comparisons; $n = 36$ for each mean, Oneway Anova with *apriori* contrasts).

Nematode Densities

For the two susceptible cultivars, the addition of nematodes increased the number of *R. similis* and *H. multicinctus* and decreased the number of *Meloidogyne* spp. recovered at both harvests relative to those plants not receiving that treatment. A low level of invasion occurred in plants not receiving a nematode treatment prior to transfer to the field from the nematode population in the field soil at the site (Figure 2). The density of *R. similis* relative to *H. multicinctus* increased from the first to second harvests on both susceptible cultivars. Nematode addition also led to a higher density of nematodes recovered from the resistant cultivar, Yangambi km5, at both harvests compared to the plants not receiving additional nematodes, but less than was seen on the susceptible cultivars. The values are statistically significant ($P<0.001$) with the exception of the non-abundant *H. multicinctus* on Yangambi km5 at the first harvest and *Meloidogyne* spp. on Mbwazirume and *R. similis* on Yangambi km5 at the second harvest.

Leaf Area Index and Nematode Densities

The relationship between LAI and nematode densities was described by a cubic regression curve fitted to LAI values and nematode densities at harvest (Figure 3). There were similar decline curves for LAI values with increasing nematode density at both harvests for cultivars Gonja and Mbwazirume ($P<0.001$ in both cases; not shown). The data were pooled for the two cultivars and summarised with a cubic regression using mean LAI values at logarithmic mean nematode densities for a range of density classes. The first significant fall in LAI to 92.7±2.89% of 0.565±0.0111 in the absence of recovered nematodes occurred at a mean density of 4,166±348 nematodes/100 g roots for Mbwazirume ($P<0.05$; Figure 3). A significant decline in LAI for Gonja was established at the next higher density class with a mean LAI of 80.0±1.58% of 0.504±0.0172 at a mean density of 28,248±970 nematodes/100 g roots ($P<0.001$; Figure 3). The resistant cultivar Yangambi km5 showed no significant decrease from a LAI of 0.840±0.0307 as the nematode density increased and so this cultivar is not included in Figure 3.

The Influence of Nematodes on Banana Bunch Weight

Bunch weights were lower for Gonja manjaya and Mbwazirume for both harvests (univariate ANOVA) on plants receiving added nematodes compared to plants that did not (Figure 4). Bunch weights were lower when nematodes had been added by 23.5±1.82% and 24.0±2.34% for Gonja manjaya and 14.8±2.13% and 38.0±1.42% for Mbwazirume, for the first and second harvests respectively. Yangambi km5 showed a slight increase in yield when nematodes had been added of 7.0±2.45% and 1.3±2.08% at the first and second harvests respectively.

Necrosis of Roots

There was only limited necrosis observed on roots of plants not receiving nematodes before field planting and for Yangambi km5 roots with or without nematodes added (Figure 5). In contrast considerable necrosis was observed on both Gonja manjaya and Mbwazirume plants that had received nematodes. It was more severe on Gonja manjaya than Mbwazirume over both harvests ($P<0.001$; univariate analysis). Coefficients of the linear equation between necrosis and the nematode density were obtained by a stepwise entry of eligible variables. *R. similis* contributed most to the correlation coefficient (0.596), which increased to 0.628 when *H. multicinctus* was also entered with both additions being statistically significant ($P<0.001$). The partial correlation coefficient for each nematode for its linear effect on necrosis when the contribution of the other nematodes was removed was also statistically significant at 0.512 and 0.246 respectively ($P<0.001$ in both cases). The small contribution made by *Meloidogyne* spp. was not significant and so was excluded from the final model.

The Incidence of Toppling

During the first cropping cycle 41 plants of Mbwazirume and Gonja manjaya plots to which nematodes had been added toppled. No plants toppled in plots without added nematodes for these two cultivars and no Yangambi km5 plants toppled irrespective of treatment (Table 1). In the second cropping cycle plants from all three cultivars toppled, mostly those to which nematodes had been added and less so for Yangambi km5, which had the same level for both treatments. Toppling incidence was similar when nematodes had been added for the two cropping cycles for Mbwazirume but increased significantly ($P<0.01$) for Gonja manjaya in the second cycle.

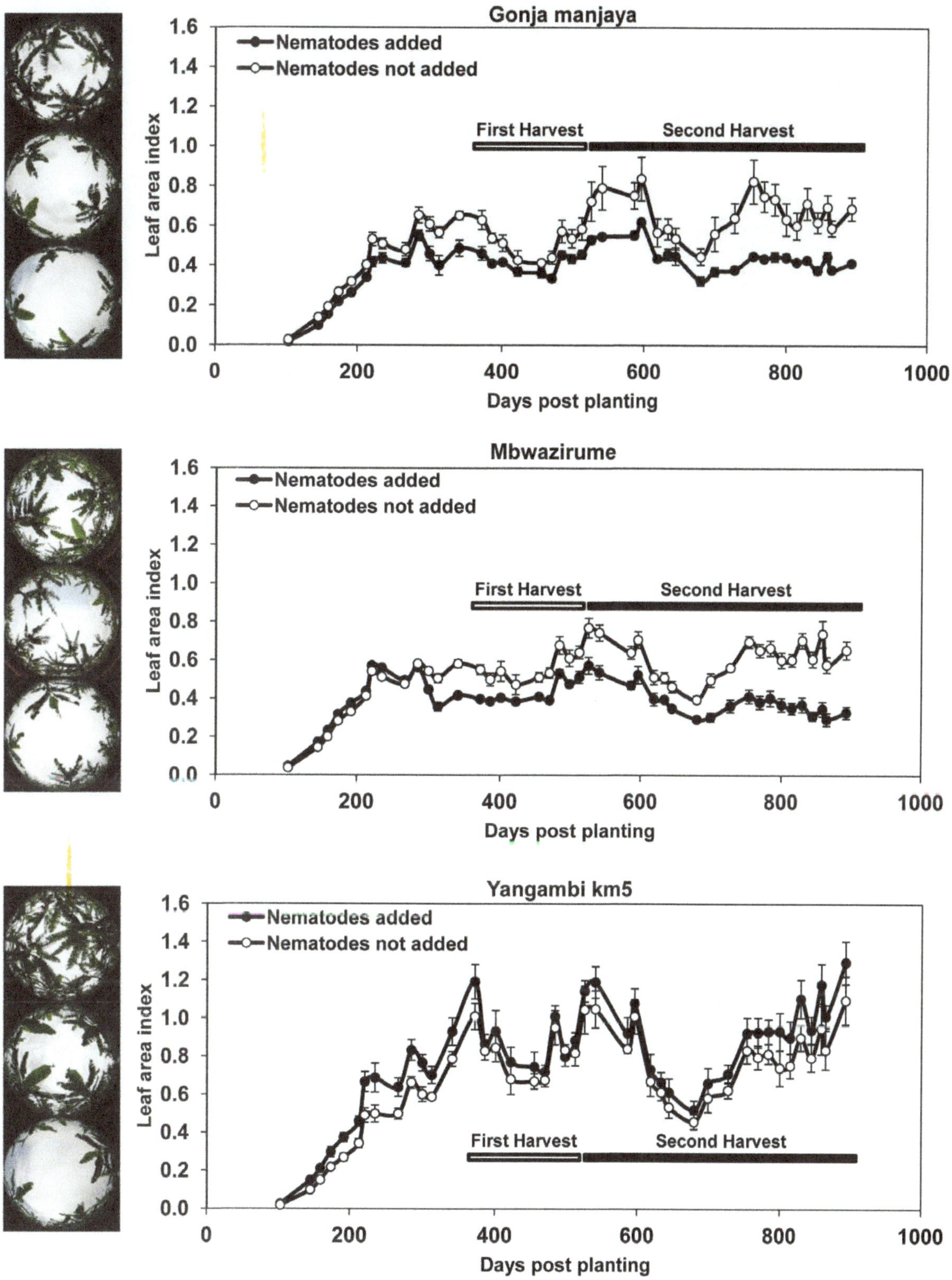

Figure 1. Leaf area index (LAI) measured by hemispherical digital photography. Mean canopy LAI measurements of four replicate plots of 36 banana plants per treatment of the susceptible cultivars Gonja manjaya and Mbwazirume and the resistant cultivar Yangambi km5 across two growth cycles with or without nematode addition of a mixed population of 5,500 nematodes per plant before transfer to the field. Error bars indicate ± SEM (n = 36) and the first and second harvest periods are indicated by open and closed bars, respectively. Adjacent images indicate example LAI values at the minimum, median and maximum values for each banana cultivar.

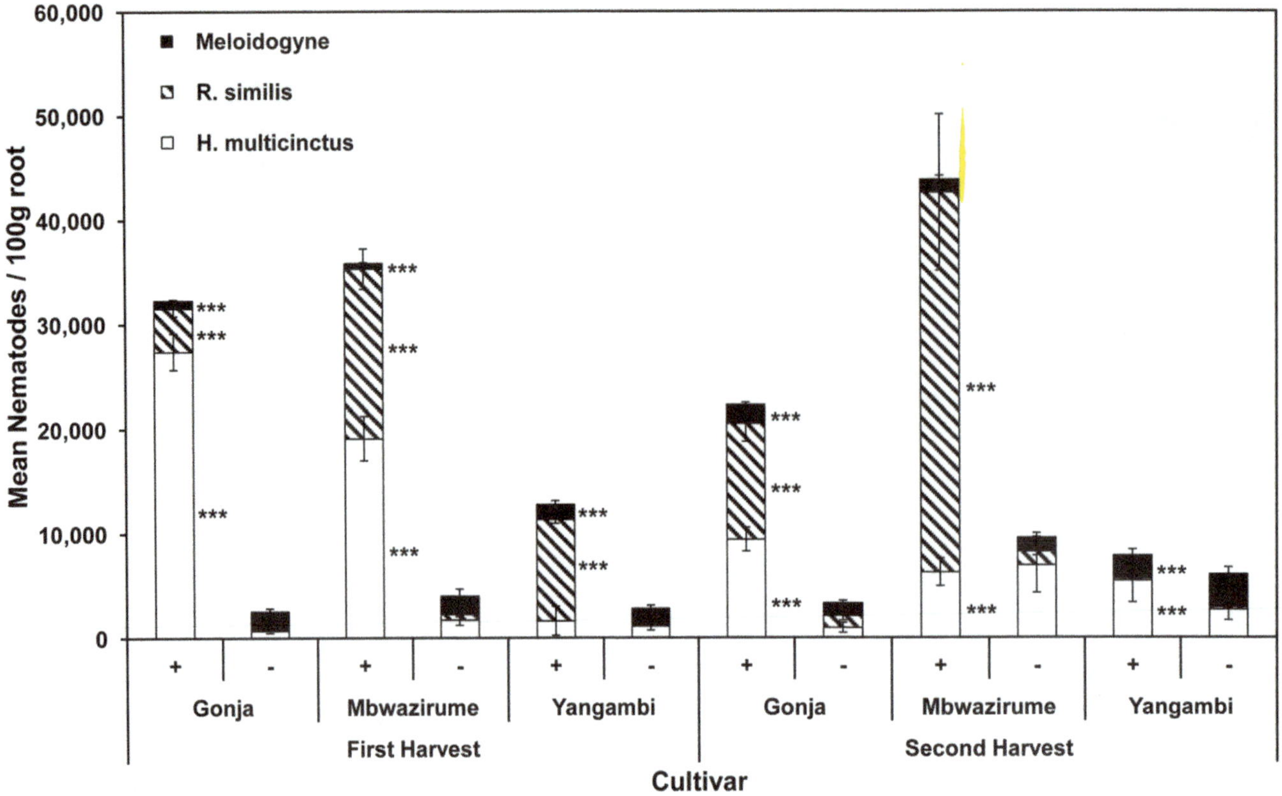

Figure 2. Nematodes recovered at harvest. Mean numbers per 100 g root of each of the three added nematode species recovered at first and second harvest on each of the three banana cultivars. Error bars indicate ± SEM (Harvest 1 n = 144; Harvest 2 n = 12). Univariate analysis was used to detect differences between nematodes added (+) and not added (–) treatments (***–P<0.001).

Plant Height, Girth and Time to Harvest

A significant reduction in plant height at harvest was caused by adding nematodes before planting for both harvests with Gonja manjaya and Mbwazirume but the effect was only 5% and 12–15% for the first and second harvests respectively. The additional nematode presence also resulted in a 10–16% reduction in girth of Gonja manjaya at both harvests and Mbwazirume for the second harvest only. Similarly, there was a small, significant reduction in days to the first harvest for Mbwazirume and an increase for Yangambi km5 but no significant difference between treatments for Gonja manjaya (Table 2).

The Incidence of Other Biotic Stresses

Plants were examined in the field for foliar diseases on each day that hemispherical images were captured for LAI analysis and again during the analysis of the images. Initial symptoms of Bacterial wilt were noticed for only 15 of the 766 (2%) standing plants during the second crop cycle. These plants were immediately destroyed on detection of the disease. Black Sigatoka and Fusarium wilt were absent or present at such a low level that damage to leaves was not observed. Weevil damage was detected for just 5 plants but not until close to the second harvest. The low incidence of these biotic stresses is likely due to the deliberate sighting of the trial away from established banana plantations and the use of land that had been fallow for a year following cultivation of cassava and yam to ensure LAI differences detected could be attributed to nematodes.

Discussion

Measurement of LAI by hemispherical digital photography provided a rapid method of assessing banana growth. The approach provides a basis for the continual measurement of a wide range of stresses on the growth of tall crops such as banana under field conditions. Its potential was established in this work using the example of nematode challenge, which is a known major biotic stress for banana crops. The blocked and replicated field trial design ensured variation in growth between plants in the same treatment did not prevent the effects under study reaching statistical significance. The design enabled the decrease in LAI values recorded with high nematode density to be related to the biotic stress the animals imposed. It incorporated a comparative approach which revealed a lower LAI for those plants with a substantially higher nematode density at harvest after their addition prior to planting relative to those plants challenged only by the lower natural population in the soil. In addition, the nematode resistant cultivar Yangambi km5 was included and it showed no decrease in LAI with high nematode density in contrast to the two susceptible cultivars Mbwazirume and Gonja manjaya. There was also only a low incidence of the other major biotic stresses in Uganda, bacterial *Xanthomonas* wilt, Fusarium wilt, Black Sigatoka and weevils, such that their contribution to mean and variability of the LAI values was limited.

The LAI values increased as the plants grew in an asymptotic manner to a plateau phase over the first cropping cycle with the increasing leaf area being completed by 340 dpp for both Mbwazirume and Gonja manjaya. Yangambi km5 grew vigorously and its LAI values reached a higher value than the other two

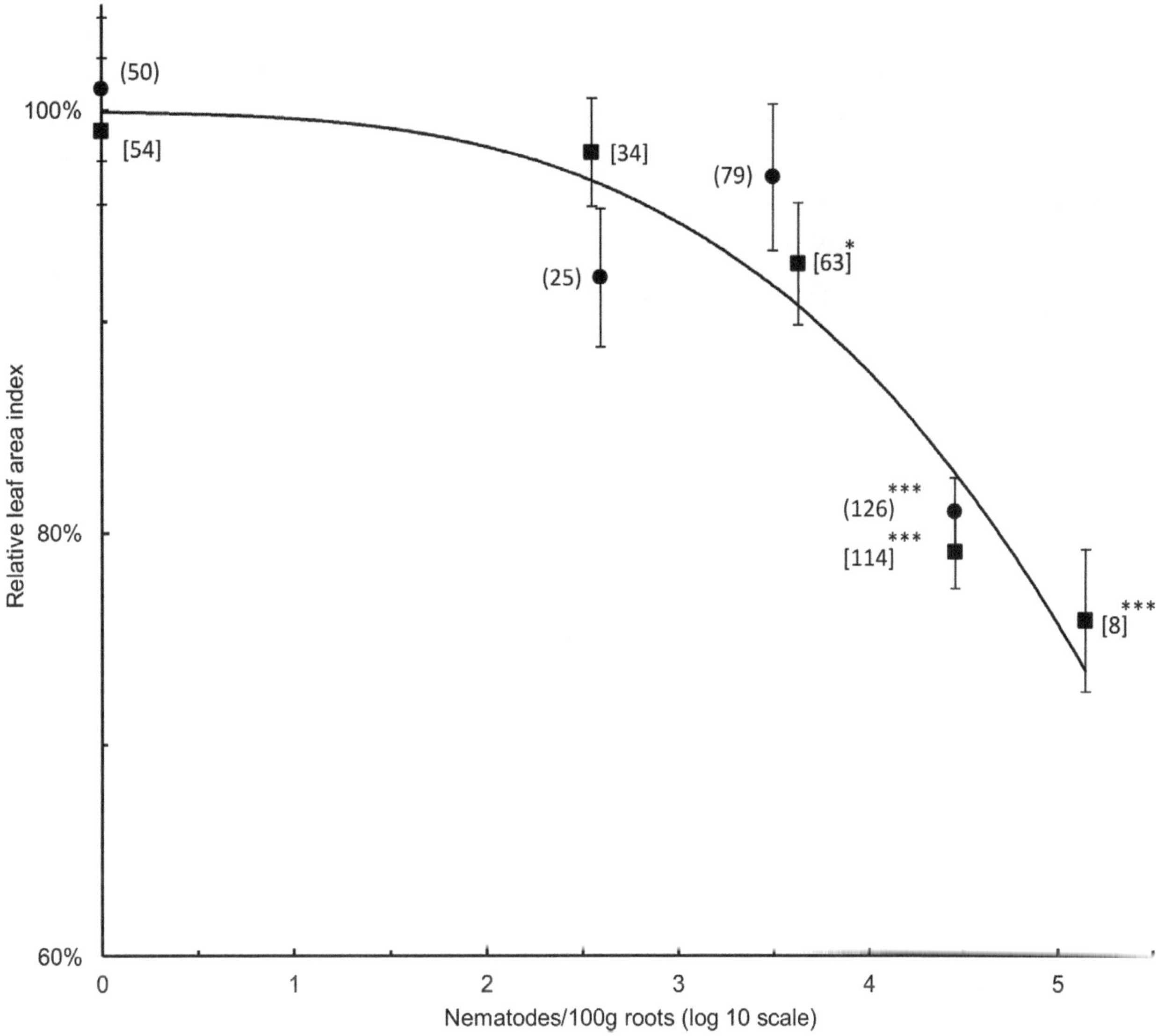

Figure 3. Relationship between LAI and nematode density. A cubic regression curve is fitted for the relationship for the cultivars Gonja (●) and Mbwazirume (■) between LAI and mean nematode density for values within one of up to five nematode density classes (0; 1–<1,000; 1,000–<10,000; 10,000–<100,000 and 100,000 or more/100 g roots) for two harvests for both plants to which nematodes were and were not added prior to planting. All nematode density classes for which there is significant percentage reduction in LAI relative to when no nematodes were recovered from roots is shown for both cultivars (*–P<0.05; ***–P<0.001, Oneway Anova with *apriori* contrasts). The values in parenthesis are the number of values associated with each mean; (n), Gonja and [n] Mbwazirume.

cultivars. The plant growth pattern measured by LAI was less evident for the second crop of the cultivars as second generation plants had grown considerably before the first harvest removed the contribution of mother plants to the values. The LAI also fluctuated with the impact of climatic factors such as storms. The addition of nematodes to soil around roots before planting out to the field reduced the subsequent LAI for both Mbwazirume and Gonja manjaya from about 300 dpp relative to those not receiving this treatment. The difference in the LAI for these cultivars between the two treatments increased across the two cropping cycles, indicating that a progressive suppression of growth was occurring. Yangambi km5 showed a slightly increased LAI in the presence of added nematodes over both cropping cycles, indicating that in this study this cultivar responds to nematode infection with stimulated growth. Such an effect has previously been reported under certain circumstances such as low nematode densities [27].

The suppression of peak LAI by 25–31% for Mbwazirume and Gonja manjaya, when nematodes were added before planting, was greater than the change of 5–11% found in plant height at harvest. Pseudostem girth has been used to predict yields [14] but no significant reduction was recorded at first harvest of Mbwazirume in this work, although its yield was reduced significantly by the added nematode challenge. The results suggest LAI is superior to both these other measures of plant size for detecting the impact of nematodes on banana plant growth.

The LAI values are lower than recorded previously in other parts of the world but similar to previous values for EAHB obtained by measuring individual leaf dimensions at the same locality as this trial [14]. This relatively low value is partly because the trial planting density was typical of plantations in Uganda which are less than in many commercial plantations elsewhere. Presumably they are optimal for EAHB based on grower

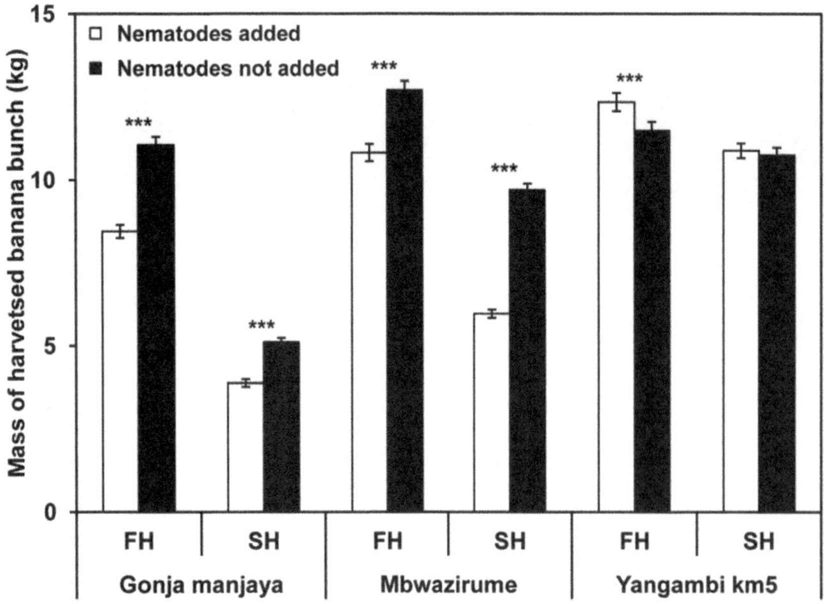

Figure 4. Harvested banana bunch weights. Mean values for each treatment for the three banana cultivars at first and second harvest. Error bars indicate ± SEM (Harvest 1 n≥120; Harvest 2 n≥95). Univariate analysis was used to detect significant differences (***–P<0.001) between bunch weights of nematodes added and not added treatments at first harvest (FH) and second harvest (SH).

experience given a cultivation history in East Africa of at least 1,000 years [14].

The relationship between growth suppression, as measured by LAI, and nematode numbers fitted a polynomial curve (Figure 3) as used previously to relate nematode density to root necrosis of banana [7,9] and conformed in shape to that derivable on a theoretical basis [28]. It was validated by comparing LAI values at different levels of nematode challenge and by measurement of root necrosis at harvest. The addition of nematodes resulted in considerable necrosis of the root systems of Gonja manjaya and Mbwazirume but not Yangambi km5 at both harvests. The nematodes were responsible for initiating much of the necrosis as it was much less evident when they were not added to the soil around plants. Correlations from linear regression indicate *R. similis* contributes more to necrosis than *H. multicinctus* with *Meloidogyne* not contributing significantly to this effect. The necrosis associated with *R. similis* causes structural damage to roots leading to an incidence of toppling [29]. In the current study, necrosis of Mbwazirume roots increased proportionally with *R. similis* densities over the two cropping cycles. However, considerable necrosis was associated with more *H. multicinctus* than *R. similis* in roots of Gonja manjaya by first harvest. This damage did not

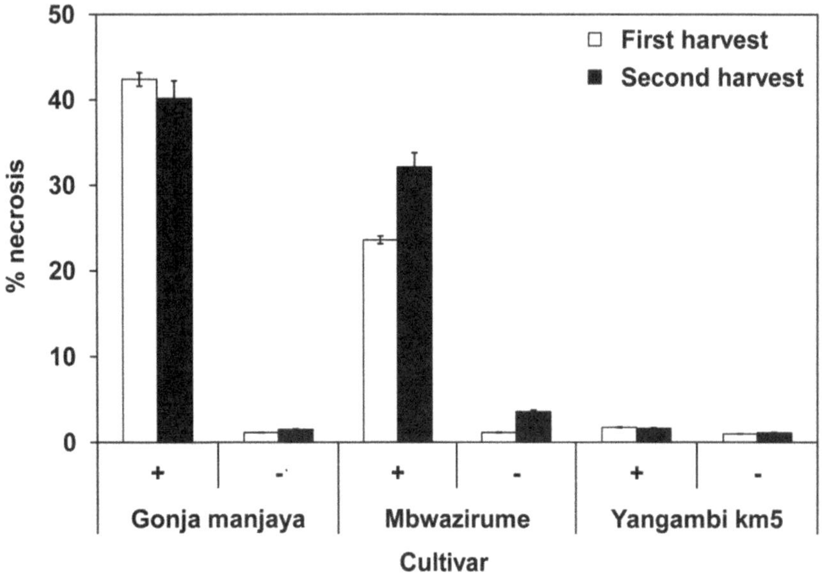

Figure 5. Percentage root necrosis at harvest. Mean values for each treatment for the three banana cultivars at first and second harvest. Values presented are back transformed from log necrosis values. Error bars indicate ±SEM (Harvest 1 n≥120; Harvest 2 n≥95).

Table 1. The impact of the addition of nematodes prior to transplantation into the field on the number of standing (S) and toppled (T) banana plants during two crop cycles.

	First crop cycle						Second crop cycle					
	Nematodes added		Nematodes not added		χ^2 P Value		Nematodes added		Nematodes not added		χ^2 P Value	
Cultivar	S	T	S	T	added vs not	*cf.* Mbw+Gonja	S	T	S	T	added vs not	*cf.* Mbw+Gonja
Mbwazirume	120	24	144	0	<0.001	NS	115	29	140	4	<0.001	<0.01
Gonja manjaya	127	17	144	0	<0.001		95	49	136	8	<0.001	
Yangambi km5	144	0	144	0	NS		140	4	140	4	NS	
Total	391	41	432	0			350	82	416	16		

become more severe by the second harvest although an increase in the density of *R. similis* was recorded. This may indicate that the *H. multicinctus* population caused much of the necrosis of Gonja manjaya roots and indicate variability in responses of different cultivars to complex populations of parasitic nematodes.

The nematode density recovered from Gonja manjaya at harvest of the second crop cycle was also less than the first. This effect may relate to the influence on nematode densities of the root-carrying capacity of the damaged root system as a crop grows, as studied for *Pratylenchus zeae* on rice [30]. Possibly the high density of nematodes present towards the end of the first cropping cycle of this plantain suppressed early root growth in the second cycle so reducing the subsequent root-carrying capacity for nematodes. In contrast, Mbwazirume had greater nematode densities associated with its roots at harvest of the second cycle than the first. It may have offered a larger root system than Gonja manjaya, which consequently reduced toppling incidence and allowed a subsequent build up of the nematode population to levels that resulted in a higher loss of yield by the second harvest.

It has been suggested that in Uganda a disease index in excess of 10, the equivalent of 273 *R. similis*/100 g root, can result in economic damage at an altitude of 1200 m above sea level [5]. This is a considerably lower nematode density than obtained in the current work at harvest even when no nematodes were added and limited necrosis occurred. Disease indices based on root necrosis alone may be an unreliable predictor of economic damage, particularly in the complex situations such as occurs in Uganda. A combination of percentage dead roots, number of large lesions and the nematode population density was used to measure *Musa* genotype field response to *R. similis* and *H. multicinctus* [31]. That approach is labour intensive and requires more specialist expertise than required to measure LAI as in the current study. In our work we selected three distinct banana genotypes to optimise the outcome but the comparison of planting types needs to be expanded. For instance, there is a considerable range of EAHB cultivars with five distinct genotypes [11] that may have differing responses to nematode presence in roots.

We have shown that LAI provides a continual measure of the impact of a plant stress on banana growth at a locality for particular genotypes but like other measurements of crop growth and yield in the field it is an integrative measure. As such studies need to be designed so that it can be demonstrated that the stress considered is the major factor in the outcome recorded. Therefore the cause of the effect on growth, such as the density of nematodes, should be confirmed to define the one or more plant stresses principally involved. This is, however, a requirement of all methods of assessing nematode induced losses to banana in field conditions as none provide measures that are independent of the consequences of other factors and stresses. For instance the level of necrosis is sometimes not reflected in yield loss in the absence of banana toppling for both *R. similis* in the Cameroon [32] and *P. goodeyi* in Rwanda [33]. In otherwise favourable conditions, the sum of all stresses on these plants may not check growth. The proportion of root mass undamaged by nematodes may be sufficient for normal physiological functions but inadequate to prevent the plant from toppling during storms and high winds. More work may establish that LAI at a specific crop growth stage offers a reliable approach for defining damage thresholds for banana crops that can be related to even complex stresses, such as a combination of nematode species.

Although the yield loss contributions due to toppling and bunch weight loss of standing plants differed for Gonja manjaya and Mbwazirume the overall values were similar. The combined yield loss including toppled plants on plots receiving nematodes over the

Table 2. The impact of the addition of nematodes prior to transplantation into the field on days to harvest, plant height at harvest and pseudostem girth at harvest for two crop cycles.

Cultivar	Treatment	Days to harvest		Plant height (cm)		Pseudostem girth (cm)	
		First harvest	Second harvest	First harvest	Second harvest	First harvest	Second harvest
Gonja manjaya	Nematodes added	442.3 ±3.34	714.9 ±8.55	255.4 ±3.72**	218 ±4.69***	43.8 ±0.542***	37.2 ±0.850***
	Nematodes not added	440.3 ±3.05	712.8 ±6.72	269.4 ±2.28	257.6 ±4.77	48.4 ±0.389	44.5 ±0.824
Mbwazirume	Nematodes added	363.6 ±2.48 **	630.7 ±7.38	231.3 ±1.54***	231.2 ±2.93***	47.9 ±0.450	46.8 ±0.718***
	Nematodes not added	376.4 ±3.03	617.1 ±6.46	243.7 ±1.44	264.1 ±2.92	53.9 ±3.63	52.5 ±0.750
Yangambi km5	Nematodes added	411.3 ±3.12 ***	613.1 ±5.62*	198 ±1.81	250.7 ±3.38	38.1 ±0.453	43.5 ±0.801
	Nematodes not added	431.9 ±4.01	636 ±7.18	197.7 ±1.90	245.1 ±3.98	38.7 ±0.401	42.3 ±0.753

***$P<0.001$;
**$P<0.01$;
*$P<0.05$; t-test comparison between the two treatments for each cultivar at harvest. Each mean (± SEM) is based on 83–138 and 27–137 plants for the first and second harvests, respectively.

second cropping cycle was significantly larger for Gonja manjaya and Mbwazirume than the losses over the first cycle. This demonstrates progressive and severe nematode induced losses in perennial banana plantations of susceptible cultivars that can be inflicted within the first few crop cycles. Loss estimates in excess of 50% have been reported from several studies [4] and emphasises the need for improved nematode control of this important food security crop in sub-Saharan Africa.

The results establish that LAI has potential for continual measurement of stress on the growth of tall crops, such as banana under field conditions. We will next apply this approach in a contained field trial to assess yield responses to nematode challenge of recently developed transgenic plantains with resistance to *R. similis* and *H. multicinctus* [34]. This will establish whether LAI is a reliable diagnostic tool for assessing nematode resistance in such plants.

Acknowledgments

The authors thank the International Institute for Tropical Agriculture, Uganda for providing the land for the trial and laboratory facilities.

Author Contributions

Conceived and designed the experiments: HR DC HA. Performed the experiments: EM. Analyzed the data: HR EM HA. Contributed reagents/materials/analysis tools: HR DC LT HA. Wrote the paper: HR DC HA.

References

1. Bridge J, Price NS, Kofi P (1995) Plant parasitic nematodes of plantain and other crops in Cameroon, West Africa. Fundam Appl Nematol 18: 251–260.
2. Sarah JL, Pinochet J, Stanton J (1996) The burrowing nematode of bananas *Radopholus similis* Cobb 1913. *Musa* Pest Fact Sheet No. 1. Montpellier, France: INIBAP.
3. Gowen SC, Quénéhervé P, Fogain R (2005) Nematode parasites of bananas and plantains. In: Luc M, Sikora RA, Bridge J, editors. Plant parasitic nematodes in subtropical and tropical agriculture. Second edition. Wallingford, UK: CAB International. 611–643.
4. Atkinson HJ (2003) Strategies for resistance to nematodes in *Musa* spp. In: Atkinson HJ, Dale J, Harding R, Kiggundu A, Kunert K, Muchwezi JM, Sagi L, Viljoen A, editors. Genetic transformation strategies to address the major constraints to banana and plantain production in Africa. Montpellier, France: INIBAP. 74–107.
5. Pattison AB, Stanton JM, Cobon JA, Doogan VJ (2002) Population dynamics and economic threshold of the nematodes *Radopholus similis* and *Pratylenchus goodeyi* on banana in Australia. Int J Pest Manage 48: 107–111.
6. Bridge J, Gowen SC (1993) Visual assessment of plant parasitic nematodes and weevil damage on banana and plantain. In: Gold CS, Gemmill B, editors. Biological and Integrated Control of Highland Banana and Plantain Pests and Diseases: Proceedings of a Research Coordination Meeting, 12–14 November 1991, Cotonou, Benin. Ibadan, Nigeria: IITA. 147–154.
7. Moens T, Araya M, De Waele D (2001) Correlation between nematode numbers and damage to banana (*Musa* AAA) roots under commercial conditions. Nematropica 31: 55–65.
8. Kashaija IN, Speijer PR, Gold CS, Gowen SC (1994) Occurrence, distribution and abundance of plant parasitic nematodes of bananas in Uganda. Afr Crop Sci J 2: 99–104.
9. Moens T, Araya M, Swennen R, De Waele D (2006) Reproduction and pathogenicity of *Helicotylenchus multicinctus*, *Meloidogyne incognita* and *Pratylenchus coffeae*, and their interaction with *Radopholus similis* on *Musa*. Nematology 8: 45–58.
10. Brentu CF, Speijer PR, Green K, Hemeng B, De Waele D, et al. (2004) Micro-plot evaluation of the yield reduction potential of *Pratylenchus coffeae*, *Helicotylenchus multicinctus* and *Meloidogyne javanica* on plantain cv. Apantu-pa (*Musa* spp., AAB-group) in Ghana. Nematology 6: 455–462.
11. Speijer PR, De Waele D (1997) Screening of *Musa* germplasm for resistance and tolerance to nematodes. Montpellier, France: INIBAP. 47 p.
12. Turner DW (1997) Ecophysiology of bananas: the generation and functioning of the leaf canopy. Acta Hortic 490: 211–222.
13. Guan J, Nutter FW (2002) Relationships between defoliation, leaf area index, canopy reflectance, and forage yield in the alfalfa-leaf spot pathosystem. Comput Electron Agr 37: 97–112.
14. Nyombi K, Van Asten PJA, Leffelaar PA, Corbeels M, Kaizzi CK, et al. (2009) Allometric growth relationships of East Africa highland bananas (*Musa* AAA-EAHB) cv. Kisansa and Mbwazirume. Ann Appl Biol 155: 403–418.
15. Bréda NJJ (2003) Ground-based measurements of leaf area index: a review of methods, instruments and current controversies. J Exp Bot 54: 2403–2417.
16. Kreye C, Bouman BAM, Castañeda AR, Lampayan RM, Faronilo JE, et al. (2009) Possible causes of yield failure in tropical aerobic rice. Field Crops Res 111: 197–206.
17. Chen JM, Rich PM, Gower ST, Norman JM, Plummer S (1997) Leaf area index of boreal forests: Theory, techniques, and measurements. J Geophys Res 102: 29429–29443.
18. Peper PJ, McPherson EG (1998) Comparison of five methods for estimating leaf area index of open-grown deciduous trees. J Arboriculture 24: 98–111.
19. White MA, Asner GP, Nemani RR, Privette JL, Running SW (2000) Measuring fractional cover and leaf area index in arid ecosystems: digital camera, radiation transmittance, and laser altimetry methods. Remote Sens Environ 74: 45–57.
20. Leblanc SG, Fernandes R, Chen JM (2002) Recent advancements in optical field leaf area index, foliage heterogeneity, and foliar angular distribution measurements. In: Proceedings of IGARSS 2002, Toronto, Canada, 24–28 June.
21. Fogain R, Gowen SC (1998) Yangambi km5 (*Musa* AAA, Ibota subgroup): a possible source of resistance to *Radopholus similis* and *Pratylenchus goodeyi*. Fundam Appl Nematol 21: 75–80.
22. Coyne DL, Wasukira A, Dusabe J, Rotifa I, Dubois T (2010) Boiling water treatment: a simple, rapid and effective technique for nematode and banana weevil management in banana and plantain (*Musa* spp.) planting material. Crop Prot 29: 1478–1482.

23. Coyne DL, Nicol J, Claudius-Cole A (2007) Practical plant nematology: a field and laboratory guide. Ibadan, Nigeria: IITA. 82 p.
24. Little TM, Hills FJ (1978) Agricultural experimentation: Design and analysis. New York, USA: Wiley. 350p.
25. Rich PM, Wood J, Vieglais DA, Burek K, Webb N (1999) Guide to HemiView: software for analysis of hemispherical photography. Cambridge, England: Delta-T Devices. 79p.
26. Snedecor GW, Cochran WG (1989) Statistical methods, eighth edition. Ames, USA: Iowa State University Press. 503p.
27. Barker KR, Olthof THA (1976) Relationships between nematode population densities and crop responses. Annu Rev Phytopathol 14: 327–353.
28. Seinhorst JW (1965) The relationship between nematode density and damage to plants. Nematologica 11: 137–154.
29. Tixier P, Salmon F, Chabrier C, Quénéhervé P (2008) Modelling pest dynamics of new crop cultivars: the FB920 banana with the *Helicotylenchus multicinctus-Radopholus similis* nematode complex in Martinique. Crop Prot 27: 1427–1431.
30. Prot JC, Savary S (1993) Interpreting upland rice yield and *Pratylenchus zeae* relationships: correspondence analyses. J Nematol 25: 277–285.
31. Hartman JB, Vuylsteke D, Speijer PR, Ssango F, Coyne DL, et al. (2010) Measurement of the field response of *Musa* genotypes to *Radopholus similis* and *Helicotylenchus multicinctus* and the implications for nematode resistance breeding. Euphytica 172: 139–148.
32. Fogain R (2000) Effect of *Radopholus similis* on plant growth and yield of plantains (*Musa*, AAB). Nematology 2: 129–133.
33. Gaidashova SV, van Asten PJA, Delvaux B, De Waele D (2010) The influence of the topographic position within highlands of Western Rwanda on the interactions between banana (*Musa* spp. AAA-EA), parasitic nematodes and soil factors. Sci Hortic-Amsterdam 125: 316–322.
34. Roderick H, Tripathi L, Babirye A, Wang D, Tripathi J, et al. (2012) Generation of transgenic plantain (*Musa* spp.) with resistance to plant pathogenic nematodes. Mol Plant Pathol 13: 842–851.

Multiple Routes of Pesticide Exposure for Honey Bees Living Near Agricultural Fields

Christian H. Krupke[1]*, Greg J. Hunt[1], Brian D. Eitzer[2], Gladys Andino[1], Krispn Given[1]

1 Department of Entomology, Purdue University, West Lafayette, Indiana, United States of America, **2** Department of Analytical Chemistry, The Connecticut Agricultural Experiment Station, New Haven, Connecticut, United States of America

Abstract

Populations of honey bees and other pollinators have declined worldwide in recent years. A variety of stressors have been implicated as potential causes, including agricultural pesticides. Neonicotinoid insecticides, which are widely used and highly toxic to honey bees, have been found in previous analyses of honey bee pollen and comb material. However, the routes of exposure have remained largely undefined. We used LC/MS-MS to analyze samples of honey bees, pollen stored in the hive and several potential exposure routes associated with plantings of neonicotinoid treated maize. Our results demonstrate that bees are exposed to these compounds and several other agricultural pesticides in several ways throughout the foraging period. During spring, extremely high levels of clothianidin and thiamethoxam were found in planter exhaust material produced during the planting of treated maize seed. We also found neonicotinoids in the soil of each field we sampled, including unplanted fields. Plants visited by foraging bees (dandelions) growing near these fields were found to contain neonicotinoids as well. This indicates deposition of neonicotinoids on the flowers, uptake by the root system, or both. Dead bees collected near hive entrances during the spring sampling period were found to contain clothianidin as well, although whether exposure was oral (consuming pollen) or by contact (soil/planter dust) is unclear. We also detected the insecticide clothianidin in pollen collected by bees and stored in the hive. When maize plants in our field reached anthesis, maize pollen from treated seed was found to contain clothianidin and other pesticides; and honey bees in our study readily collected maize pollen. These findings clarify some of the mechanisms by which honey bees may be exposed to agricultural pesticides throughout the growing season. These results have implications for a wide range of large-scale annual cropping systems that utilize neonicotinoid seed treatments.

Editor: Guy Smagghe, Ghent University, Belgium

Funding: This research was funded by an award from the North American Pollinator Protection Campaign to CK, GH and BE and by an award from the Managed Pollinator Coordinated Agricultural Project (United States Department of Agriculture National Institute of Food and Agriculture 2009-8511805718) to GH. The funders had no role in study design, data collection or analysis, decision to publish or manuscript preparation.

Competing Interests: CK has received funding in the past from agrochemical companies that manufacture and distribute a wide range of pesticides, including many of those mentioned in this manuscript. These entities include the Bayer Corporation, BASF and Syngenta. None of these entities were involved with this work or are aware of it in any way, financial or otherwise. This does not alter the authors' adherence to all the PLoS ONE policies on sharing data and materials.

* E-mail: ckrupke@purdue.edu

Introduction

Pollinator health is receiving increased attention as both managed pollinators (i.e. honey bees) and native pollinator populations decline worldwide [1–3]. Several causal mechanisms (including viral pathogens, parasitic mites and pesticides) have been proposed and investigated as contributing causes [4]. Pesticide exposure has received significant attention and recently-published analyses of pollen from managed bees located near agricultural environments demonstrated that many agricultural chemicals (including insecticides, miticides, fungicides and herbicides) are detectable in honey bee wax and pollen samples [5,6]. Of the many compounds detected, the neo-nicotinoid group has arguably received the most attention. These compounds act as nicotinic acetylcholine receptor agonists in insects, causing persistent excitation of these receptors and eventually death [7]. As a group, neonicotinoids possess several key attributes that have facilitated heavy adoption in both agricultural and urban environments, including low vertebrate toxicity [8] and the ability to be translocated by plants. Neonicotinoids are also persistent, offering the potential for a large window of activity. The half-lives of these compounds in aerobic soil conditions can vary widely, but are best measured in months (148–1,155 days for clothianidin) [8]. Among the largest single uses of these compounds is application to maize seed; production of maize for food, feed and ethanol production represents the largest single use of arable land in North America. Maize planting reached unprecedented levels in the US in 2010 (35.7 million hectares [9], and is expected to increase. Virtually all if the maize seed planted in North America (the lone exception being organic production = 0.2% of total acreage [9]) is coated with neonicotinoid insecticides. There are 2 major compounds used: clothianidin and thiamethoxam; the latter is metabolized to clothianidin in the insect. The current application rates for these compounds range from 0.25 to 1.25 mg/kernel. These compounds are highly toxic to honey bees: a single kernel contains several orders of magnitude of active ingredient more than the published LD50 values for honey bees (defined as the amount of material that will kill 50% of exposed individuals), which ranges from 22–44 ng/bee for clothianidin (contact toxicity) [8,10]. Maize seeds are typically planted at a rate ca. 12,500 kernels/hectare, making it essential that any potential routes for honey bee exposure to these compounds be quantified as thoroughly as possible. This study was initiated in response to reports of bee kills at Indiana apiaries in spring of 2010. These reports coincided with the peak period of maize planting in the area [11]. Results of analyses of these bees and pollen from the hives revealed that both clothianidin and thiamethoxam were present on dead bees and in pollen collected from a single hive. These compounds were also present in dead bees from other hives but not in bees from hives that did not show mortality. Also found was atrazine, a herbicide that is commonly used in maize production and is relatively non-toxic to honey bees [12].

These preliminary results prompted additional experiments to determine how honey bees may be gaining exposure to clothianidin and other pesticides commonly applied to either maize seed or to plants later in the season. We collected samples from a variety of potential exposure routes near agricultural fields and analyzed them to determine whether pesticides were present. We sampled soils, pollen (both collected by honey bees and directly from plants), dandelion flowers, and dead and healthy bees. We also examined waste products produced during the planting of treated seed. Sowing of maize seed in North America is accomplished using tractor-drawn planters that employ a forced air/vacuum system and a perforated disc to pick up individual seeds and drop them into the planting furrow at the selected spacing. Because maize kernels are treated with neonicotinoids and other compounds (usually fungicides) they do not flow readily and may stick to one another, causing uneven plant spacing. To remedy this, talc (a common mineral composed of hydrated magnesium silicate) is typically added to seed boxes to reduce friction and stickiness and ensure smooth flow of seed during planting. Much of the talc is exhausted during planting, either down with the seed or behind the planter and into the air using an exhaust fan. We sampled waste talc following planting to determine whether this material was contaminated with pesticides abraded from treated seeds. This waste is a mixture of the talc that has been in contact with treated maize kernels and minute pieces of the seeds themselves.

Results

Soil collected from areas near our test site revealed that neonicotinoid insecticide residues were present in all samples tested (Table 1), with clothianidin occurring in each field sampled. Herbicide residues were also found in these samples. Sampling of the waste talc from planting activities revealed that extremely high concentrations of clothianidin were found in talc exposed to treated seed (Table 2). Fungicides applied to the seed were also found. Analysis of talc used to plant untreated seed contained low quantities of the same pesticides, this is likely due to contamination and reflects the difficulties associated with thorough cleaning of equipment between plantings. Direct sampling of anthers revealed that many of the same compounds were present in maize grown from treated seed, albeit in far lower concentrations (Table 3). Collection of pollen from traps in the field demonstrated that thiamethoxam was present in 3 of 20 samples, while pollen containing clothianidin was present in 10 of 20 samples (Table 4). Fungicides were also frequently detected: azoxystrobin and propiconazole were found in all pollen samples, while trifloxystrobin was found in 12 of the 20 samples analyzed. Maize pollen was frequently collected by foraging honey bees while it was available: maize pollen comprised over 50% of the pollen collected by bees, by volume, in 10 of 20 samples.

The samples collected in 2011 revealed some similar trends (Table 5); clothianidin was found on all the dead and dying bees we sampled, while the apparently healthy bees we sampled from the same locations did not contain detectable levels of clothianidin. Atrazine and metolachlor were also found, providing further evidence that these bees were foraging near agricultural fields; as these herbicides are commonly applied prior to or during maize planting. When we sampled the contents of wax combs removed from two hives at the same apiary, we found both clothianidin and thiamethoxam in pollen removed from both hives. Nectar did not contain either compound. The miticide coumaphos was found at low levels in each nectar and pollen sample as well. Both soil and dandelion flowers obtained from the fields closest to the affected apiary (soybeans in 2010, unplanted when sampled in 2011) contained clothianidin (Table 6), therefore clothianidin in/on the dandelions could have resulted from translocation from the soil to the flower, from surface contamination of the flowers from dust, or a combination of these two mechanisms. Dandelion flowers growing far from agricultural areas served as controls; no neonicotinoids were detected.

Discussion

The results we present here more clearly define some potential intersections between foraging honey bees and some of the seed treatments used during planting of maize. These results demonstrate that honey bees living and foraging near agricultural fields are exposed to neonicotinoids and other pesticides through multiple mechanisms throughout the spring and summer. The potential for greatest exposure (and the period when mortality was noted), occurs during planting time when there is potential for exposure to extremely high concentrations of neonicotinoids in waste talc that is exhausted to the environment during and after planting. Furthermore, we show that bees living in these environments will forage for maize pollen and transport pollen

Table 1. Pesticide concentrations found in soil samples taken from production fields surrounding study area, all concentrations shown are expressed as parts per billion.[1,2]

FIELD HISTORY	THIAMETHOXAM LOD = 1.0	CLOTHIANIDIN LOD = 2.0	IMIDACLOPRID LOD = 1.0	METOLACHLOR LOD = 2.0	ATRAZINE LOD = 0.5
MAIZE-MAIZE*	ND	**6.3**	**2.9**	**5.9**	**52.0**
SOY-SOY	ND	**9.6**	**7.3**	**11.1**	**7.8**
MAIZE-SOY	ND	**4.9**	**ND**	**6.1**	**8.5**
SOY-MAIZE	ND	**2.1**	**ND**	**ND**	**22**

[1]ND = Not detected.
[2]* = Experimental field where hives were placed in 2010.

Table 2. Pesticide concentrations found in talc samples removed from planter following planting with treated and untreated maize seed, all concentrations shown are expressed as parts per million.[1,2]

SEED TYPE	THIAMETHOXAM LOD = 0.02	CLOTHIANIDIN LOD = 0.04	METALAXYL LOD = 0.04	TRIFLOXYSTROBIN LOD = 0.04
TALC ONLY	ND	ND	ND	ND
COMMERCIALLY TREATED MAIZE SEED 1	735	3400	116	66
COMMERCIALLY TREATED MAIZE SEED 2**	68	10000	92	3.8
COMMERCIALLY TREATED MAIZE SEED 3	13240	4900	263	503
COMMERCIALLY TREATED MAIZE SEED 4	70	15030	131	4.4
COMMERCIALLY TREATED MAIZE SEED 5	588	11413	116	189
UNTREATED MAIZE SEED***	12	12	4	4

[1]ND = Not detected.
[2]**, *** = Talc samples taken following planting of treated and untreated sections of experimental fields, respectively, in 2010.

containing neonicotinoids to the hive. Pollen contaminated with levels of neonicotinoids similar to those shown in our results has been known to impair pollinator health [13,14,15]. Although we anticipated that planter dust may cause higher pesticide concentrations in samples taken immediately after planting our plots, we found the opposite trend: Pollen collected just prior to our planting period contained the highest levels of neonicotinoids detected (and not detected) in samples from both the treated and untreated fields. This may reflect the high variability in the types of pollen being brought back to the hive. Most of the maize in our study area was shedding pollen before and during our atypically late planting period (mid-July). After sampling anthers directly and identifying maize pollen in our samples, we know that pollen originating from treated seed does contain clothianidin, although not at the levels found in some of the bee-collected pollen samples, indicating the likelihood of additional pathways or sources. The levels of clothianidin in bee-collected pollen that we found are approximately 10-fold higher than reported from experiments conducted in canola grown from clothianidin-treated seed [16].

Detection of agricultural pesticides (and neonicotinoids in particular) in hives (including honey, pollen and wax) has been documented in the past: Bee-collected pollen was found to contain the neonicotinoid imidacloprid [17] in one study, although no adverse effects upon adults or brood were found [18]. However, a more recent study found that rearing brood in comb contaminated with pesticides (including the neonicotinoids found in our study, thiamethoxam and clothianidin) led to delayed worker development [6]. A field study examining the effects upon honey bees of clothianidin-treated canola found low levels of clothianidin in both pollen and nectar (0.93ppb and 2.59 ppb, respectively), but also found no significant effects upon honey bee populations [16]. In studies in maize, guttation droplets produced by plants grown from neonicotinoid treated seed were shown to have from 10–100 mg/L of the pesticides and were found to cause paralysis and eventual death when fed to honey bees [19], while other studies have found traces of the seed treatment imidacloprid on vegetation near maize plantings and have hypothesized that sowing treated seed can cause dispersal of dust containing insecticide [20]. Further evidence of detrimental effects of planting treated maize seed was noted by researchers in Italy, who found that honey bee mortality increased on the day seeds were planted and that numbers of foragers declined in the days following planting [21]. A subsequent study demonstrated that bees that were induced to fly near a maize planter in Europe showed up to 100 ng of clothianidin/bee upon analysis. Interestingly, however, these bees did not die unless they were kept in conditions of high humidity [22].

Detection of clothianidin in pollen, both in stored pollen in cells and in pollen traps is a critical finding because clothianidin is even more toxic when administered to bees orally, with an LD50 of 2.8–3.7 ng/bee [23,24]. Given an average weight of 80–100 mg/bee, some of our pollen sample concentrations exceed the oral LD50. This, combined with the result that our samples of dead and dying honey bees consistently demonstrated the presence of clothianidin, suggests that the levels of both clothianidin and thiamethoxam found in our sampling of stored pollen in May of 2011 may have contributed to the deaths of the bees we analyzed. However, our analytical methods do not allow us to determine what fraction of the pesticide is on the surface of bees (contact toxicity, due to drift of soil or planter exhaust) vs. inside the body (oral toxicity, due to ingestion of contaminated pollen or guttation droplets). A combination of these exposure modalities is not unlikely.

Our results also demonstrate that clothianidin is present in the surface soil of agricultural fields long after treated seed has been

Table 3. Pesticide concentrations found in pollen removed from maize anthers at anthesis. Samples were taken from the experimental field where hives were placed. All concentrations shown are expressed as parts per billion.[1]

	THIAMETHOXAM LOD = 0.2	CLOTHIANIDIN LOD = 0.5	METALAXYL LOD = 0.5	TRIFLOXYSTROBIN LOD = 0.5
TREATED MAIZE SEED 2	1.7	3.9	3.1	1.7
UNTREATED MAIZE SEED	ND	0.3	4.0	5.5

[1]ND = Not detected.

Table 4. Pesticide concentrations found in pollen samples removed from returning foragers of hives placed adjacent to maize fields at planting, all concentrations shown are expressed as parts per billion.[1]

HIVE NUMBER	DATE	% MAIZE POLLEN	THIAMETHOXAM LOD = 0.5	CLOTHIANIDIN LOD = 1.0	TRIFLOXYSTROBIN LOD = 0.4	AZOXYSTROBIN LOD = 0.5	PROPICONAZOLE LOD = 2.0
HIVE #5: Untreated	**2 d pre-planting**	**82.7**	**ND**	**ND**	**1.6**	**22**	**9.3**
	1 d pre-planting	**78.5**	**ND**	**20.1**	**3.9**	**23**	**12.5**
	At planting	**54.7**	**ND**	**ND**	**17.9**	**9.6**	**15.0**
	1 d post planting	**16.4**	**ND**	**ND**	**18.0**	**9.2**	**9.6**
	2 d post planting	**56.5**	**ND**	**ND**	**1.6**	**82**	**41.0**
HIVE #7: Untreated	**2 d pre-planting**	**58.8**	**ND**	**1.1**	**ND**	**1.4**	**7.3**
	1 d pre-planting	**73.4**	**ND**	**13.1**	**ND**	**17.9**	**10.9**
	At planting	**47.9**	**ND**	**ND**	**0.4**	**18.9**	**5.6**
	1 d post planting	**8.8**	**ND**	**6.7**	**0.9**	**12.9**	**14.3**
	2 d post planting	**52.9**	**ND**	**3.4**	**0.4**	**26.9**	**15.8**
HIVE #6: Treated	**2 d pre-planting**	**69.1**	**ND**	**ND**	**0.5**	**12.4**	**5.6**
	1 d pre-planting	**76.3**	**ND**	**ND**	**ND**	**5.4**	**7.6**
	At planting	**59.5**	**ND**	**ND**	**1.8**	**7.1**	**4.8**
	1 d post planting	**18.2**	**ND**	**ND**	**ND**	**66**	**23.8**
	2 d post planting	**18.7**	**ND**	**ND**	**1.2**	**7.5**	**6.5**
HIVE #8: Treated	**2 d pre-planting**	**43.0**	**7.4**	**88**	**9.8**	**30.5**	**9.2**
	1 d pre-planting	**24.0**	**2.3**	**25**	**3.3**	**11.3**	**9.5**
	At planting	**35.4**	**1.2**	**10**	**2.6**	**6.7**	**6.6**
	1 d post planting	**2.6**	**ND**	**12**	**2.1**	**7.5**	**9.1**
	2 d post planting	**13.5**	**ND**	**3.9**	**2.6**	**4.3**	**3.2**

[1]ND = Not detected.

planted in that field. All soil samples we collected contained clothianidin, even in cases where no treated seed had been planted for 2 growing seasons. During the spring planting period, dust that arises from this soil may land on flowers frequented by bees, or possibly on the insects themselves. Of potentially greater concern are the very high levels of neonicotinoids (and fungicides) found in the talc that has been exposed to treated seed, since part of this highly mobile material is exhausted to the outside environment during planting and after planting. The large areas being planted with neonicotinoid treated seeds, combined with the high persistence of these materials and the mobility of disturbed soil and talc dust, carry potential for effects over an area that may exceed the boundaries of the production fields themselves. A key mechanism for honey bee exposure may occur during the period

Table 5. Pesticide concentrations found in/near apiary colonies during planting period in 2011, all concentrations shown are expressed as parts per billion.[1]

SAMPLE TYPE	Sample wt. (g)	THIAMETHOXAM LOD = 0.5	CLOTHIANIDIN LOD = 1.0	METOLACHLOR LOD = 2.0	ATRAZINE LOD = 0.5	AZOXYSTROBIN LOD = 0.5	COUMAPHOS LOD = 1.0
Dead/dying bees	2.96	**ND**	**6.9**	**3.0**	**6.5**	**ND**	**ND**
Dead/dying bees	2.47	**ND**	**10.8**	**1.7**	**3.9**	**ND**	**ND**
Dead/dying bees	1.32	**ND**	**3.8**	**5.5**	**9.5**	**ND**	**ND**
Dead/dying bees	2.57	**ND**	**4.9**	**0.8**	**4.6**	**ND**	**ND**
Dead/dying bees	1.62	**ND**	**13.3**	**1.1**	**3.9**	**ND**	**ND**
Healthy bees	0.59	**ND**	**ND**	**ND**	**5.9**	**ND**	**ND**
Nectar hive 1 (healthy)	5.78	**ND**	**ND**	**ND**	**0.5**	**0.6**	**1.1**
Nectar hive 2 (sick)	5.72	**ND**	**ND**	**ND**	**ND**	**0.3**	**4.7**
Pollen hive 1 (healthy, 2 samples)	5.05	**6.2±4.9**	**2.9±1.3**	**28.5±3.5**	**16±1.4**	**28.5±3.5**	**1.3±0.4**
Pollen hive 2 (sick, 2 samples)	5.08	**20.4±2.7**	**10.7±2.3**	**81.5±0.7**	**36.5±3.5**	**0.8±0.3**	**2.7±0.3**

When two aliquots of the same sample were analyzed the results are expressed as ± the standard deviation of the two analyses.
[1]ND = Not detected.

Table 6. Pesticide concentrations found in unplanted fields near apiary during planting period in 2011, all concentrations shown are expressed as parts per billion.[1]

SAMPLE TYPE	Sample wt. (g)	THIAMETHOXAM LOD = 1.0	CLOTHIANIDIN LOD = 1.0	METOLACHLOR LOD = 0.5	ATRAZINE LOD = 0.2	AZOXYSTROBIN LOD = 0.2	COUMAPHOS LOD = 1.0
Soil, unplanted field 1, Soybeans 2010 (2 samples)	5.15, 5.01	ND	6.0±0.3	1014±14	771±170	0.2±0.1	ND
Soil, unplanted field 2, Soybeans 2010 (2 samples)	5.28, 5.43	ND	8.9±0.1	8.3±0.7	160±15	26±17	ND
Dandelions near maize field	2.96	ND	1.4	49	677	ND	ND
Dandelions near maize field	3.81	1.6	5.9	64	1133	ND	ND
Dandelions near maize field	4.51	1.3	3.1	28	522	ND	ND
Dandelions near maize field	4.05	2.9	1.1	60	269	ND	ND
Dandelions near maize field	3.10	1.1	1.6	5.7	125	ND	ND
Dandelions near maize field	3.44	ND	9.4	295	1004	ND	ND
Dandelion, CAES (non-agricultural area)	3.93	ND	ND	ND	0.3	ND	ND

When two aliquots of the same sample were analyzed the results are expressed as ± the standard deviation of the two analyses.
[1]ND = Not detected.

when maize is typically planted across much of the Midwest (mid-April through early May). At this time, the energetic requirements of honey bee colonies are increasing rapidly and pollen and nectar resources are being gathered for colony growth. Talc and soil dusts from planting are mobile and have the potential to contaminate any flowering plants that are commonly found in or near agricultural fields and are visited by honey bees, including dandelion (*Taraxacum officinale*), which has been shown to be a preferred pollen and nectar source for honey bees during this period, when floral resources are relatively limited [25].

Later in the season, when planting is largely complete, we found that honey bees will collect maize pollen that contains translocated neonicotinoids and other pesticides from seed. Translocation of neonicotinoids into pollen has previously been reported for maize grown from imidacloprid-treated seed [26], although the degree to which honey bees in our study gathered maize pollen was surprising. The finding that bee-collected pollen contained neonicotinoids is of particular concern because of the risks to newly-emerged nurse bees, which must feed upon pollen reserves in the hive immediately following emergence. Pollen is the primary source of protein for honey bees, and is fed to larvae by nurse bees in the form of royal jelly. A bee will consume 65 mg of pollen during the 10 day period it spends as a nurse bee [27], therefore a concentration of 20 ng/g (ppb) in pollen would correspond to a dose of 1.3 ng (65 mg×20 ng/g) or almost 50% of the oral LD50 of ca. 2.8 ng/bee [23]. Some of our pollen concentrations were even higher, although it is important to note that LD50 is measured as a one-time dose, while exposure through contaminated pollen would be spread out over the 10 d period and that there is likely substantial metabolic decay of the compounds during this time. Lethal levels of insecticides in pollen are an obvious concern, but sublethal levels are also worthy of study as even slight behavioral effects may impact how affected bees carry out important tasks such as brood rearing, orientation and communication.

Also potentially important are the three fungicides found in bee-collected pollen samples (trifloxystrobin and azoxystrobin and propiconazole). Azoxystrobin and trifloxystrobin are frequently used in maize seed treatments as protectants and all three of these compounds are also widely applied to maize in North America, even in the absence of disease symptoms [28]. These compounds are typically applied using aerial application during anthesis. Propiconazole has been shown to synergize toxicity of some neonicotinoids (thiacloprid and acetamiprid) to honey bees in the laboratory, although the same results have not been shown in field studies [10,29]. Although these fungicides are not acutely toxic to honey bees [5], the fact that they are routinely applied to areas that bees will frequent (i.e. maize plants at anthesis) coupled with the difficulties and uncertainties in assessing the toxicity of pesticide mixtures [30], indicate that they should be considered in future work.

In evaluating our results, it is important to bear in mind that toxicity is only one variable in addressing pesticide risks to pollinators – the intersection between toxicity and exposure is key in determining how much risk is posed by a toxicant to a given organism. These components are assessed by regulators in developing a "risk cup" which combines these parameters to assess the cumulative risks of a given toxicant to an organism [31]. In the case of honey bees, the toxicity of the neonicotinoid seed treatments used for large acreage field crops has been well-established [14,31], although when assessing the overall threat to posed to honey bee populations, calculations are complicated even further by the observation that sublethal doses of insecticides can weaken bees and increase susceptibility to key parasites or pathogens [32].

Because we found these compounds in pollen, oral LD50 is a relevant parameter in discussing toxicity to honey bees. In terms of acute toxicity (based on the oral LD50 of 2.8 ng/bee [23]), the amount of clothianidin on a single maize seed at the rate of 0.5 mg/kernel contains enough active ingredient to kill over 80,000 honey bees. However, the overall level of risk has been more difficult to quantify, as there has not been a clear mechanism whereby honey bees could be exposed to high levels of these compounds – once the treated seed is planted, opportunities for honey bee exposure to concentrations of neonicotinoids over a wide area should drop dramatically (although see [19]). Our results suggest that of the factors we quantified in this study, used talc exhausted during and after planting (the latter would occur during routine cleaning of planting equipment) stands out as potential routes for exposure that should be prioritized for further quantification and remediation. A recently published review of the risks posed by planting treated seeds in the E.U. estimates that measures taken there may reduce the dust generated during planting by 99% [33]. In North America, different planting equipment is used and there are currently no guidelines for disposal of waste talc, nor are there devices for filtering exhaust material from the vacuum planting systems. Producers may be largely unaware that this material is highly toxic to pollinators. However, given the unprecedented levels of maize production across the United States, coupled with the increasing adoption of neonicotinoid seed treatments in other annual crops covering a wide area, including soybeans (31.3 million ha), wheat (24.7 million ha), and cotton (4.4 million ha, all figures 2010 planting) [9], it is clear that this material presents a risk that is worthy of further investigation and possibly corrective action.

Our findings have implications both for honey bees located near these crops year-round, but also for migratory colonies (bees used to pollinate winter-flowering specialty crops in western North America, such as almonds and other fruit and nut crops). Many of these colonies reside in areas where treated seed is used extensively (i.e. the upper Midwestern United States) during the period from early spring through late fall. During this period, these bees forage on a variety of crops that may be planted using neonicotinoid treated seed, including maize, soybeans and canola. Although our study was confined to honey bees, these results are relevant for any pollinators that forage in or near agricultural fields, both in the crop itself or on other flowering plants (i.e. weeds) that are present in or near the field.

Materials and Methods

Ethics statement

No specific permits were required for the described field studies. All studies were performed at a Purdue Agricultural Center, which is comprised of agricultural land that is owned and maintained by Purdue University specifically for research trials. All products used during maize planting (seed, herbicides, fertilizer etc.) were registered for legal use and were applied in accordance with label guidelines. These field studies did not involve endangered or protected species.

We based our study design upon planting practices, which helped us to identify several risk factors: during maize planting, tractor movement and wind generate large quantities of dust (i.e. disturbed soil) that may land directly on bees and/or the flowers that they visit, making this dust one potential route of exposure for foraging bees.

Sample preparation and chemical analysis

All samples were analyzed using a modified version of the QuEChERs protocol [34]. The pesticides were extracted from the matrix (i.e. soil, talc or pollen) by agitation with water, acetonitrile and the salts magnesium sulfate and sodium acetate. After centrifuging, a portion of the supernatant acetonitrile layer containing the pesticides was further cleaned by the addition of solid phase dispersants (primary secondary amine, magnesium sulfate, C-18 silica). An aliquot of solvent was then concentrated and analyzed by liquid chromatography/mass spectrometry utilizing an Agilent 1200 LC interfaced to a Thermo –LTQ mass spectrometer. The MS was operated in a positive electrospray mode with a separate MS/MS scan function utilized for each pesticide monitored. This analytical technique provides both qualitative (retention time and mass spectra) and quantitative information. The combined techniques allow the various pesticides to be unambiguously identified at the parts per billion concentration level.

Pollen identification

A portion of each pollen sample from 2010 experiments was used to count and identify the types of pollen collected by the bees. The identification of maize pollen was accomplished using comparisons to reference slides prepared from maize pollen in our study area. The other pollen types were identified using comparisons to online images (www.polleninfo.org). The four pollen types identified were maize (*Zea mays*), common dandelion (*Taraxacum officinale*), plantain (*Plantago* spp.) and goldenrod (*Solidago* spp.). To estimate the proportion of maize pollen present, a subsample of pollen suspended in water was pipetted onto a disposable hemacytometer (DRM-700 Cell-VU CBC) and the pollen grains were counted. Fresh maize pollen grains are prolate spheroidal in shape [35]. The maize pollen volume was estimated measuring the principal diameters of five individual grains from each subsample, resulting in an average volume for each maize pollen grain of 3.69×10^{-4} mm^3.

Field experiment design, planting and sample collection methods

Our experimental field was surrounded by fields that had been planted with maize and soybeans. We sampled soil collected from the experimental field and each of the adjacent fields. Soil samples of 500 cc were taken at four equidistant locations within each field on July 7, 2010. Because we were most interested in the content of dust that may arise from disturbed soil, we confined our sampling to the top 2–3 cm of soil. The soil samples from each field were pooled and mixed prior to analysis. A 500 cc sample of soil was taken from each pooled sample for chemical analysis. On July 8 2010, we placed 8 hives surrounding an unplanted field in an area of long-term maize and soybean planting. The field dimensions were approximately 142 m wide by 148 m long. This area was divided in half, and the hives were then arranged such that each field half had a hive in the center of each border. Hive entrances faced the field interior. Pollen traps (Model #M00682, Dadant and Sons, Inc. M. Hamilton, IL) were placed in the entrances of each hive and pollen samples collected daily both before and after the field was planted. Samples were labeled and stored at $-10°C$ prior to analysis. Hives were selected for approximately equal populations. Each hive contained about 20,000 to 30,000 bees. This level of stocking is much lower than other apiaries that are kept closer to Purdue University, and not expected to result in excessive competition for resources. However, honey bees will forage up to several kilometers if necessary and individual hives will vary in their foraging needs and resource preferences. We chose this number of hives to give an adequate sampling of the agricultural fields in which they were placed, without the expectation that they would forage primarily in the experimental field.

On July 12 2010, we planted half of the field with commercial maize seed treated with 1.25 mg/kernel of clothianidin, adding talc to each seed box at the recommended rate (approx. 240 cc talc/ 75 kg of maize seed) [36]. Because untreated maize seed was not available commercially, we used maize harvested the previous year to plant the other half of the field. Talc was added to this seed at the same rate as above. Fields were planted using a 6-row John Deere 7200 MaxEmerge planter. Collection of waste talc for analysis was performed following planting by manually removing approximately 50 g of talc from the manifold of the planter vacuum system. The planter and vacuum system was exhausted thoroughly and cleaned with compressed air prior to each planting and following each collection at a location far from the experimental fields. Samples of fresh pollen from maize anthers were taken by removing the entire tassel from approximately 100 plants in the field while pollen was being shed. Samples were vigorously shaken into paper bags. The resulting mixture of pollen and anthers was spread over paper towels. Anthers and other large debris were removed so that only pollen grains remained for analysis.

2011 sampling

During the spring of 2011, we again received reports of dead and dying bees at a local apiary, located in a small wooded area near maize and soybean production fields in northwestern Indiana. As in 2010, these reports coincided with local planting and tillage activities. We collected bees from the entrances of several hives on May 10th and 12th, 2011. We also collected apparently healthy returning foragers from hives at the same apiary. We removed frames containing nectar and pollen from two colonies at this location: one frame was taken from a hive with dead bees near its entrance, the second frame was removed from a nearby hive without any dead bees visible. Pollen and nectar from these frames was removed from cells for analysis, and two separate analyses of pollen samples were conducted. Finally, we collected samples of surface soil (using the methods outlined above) and dandelion flowers (multiple areas sampled, approximately 7–10 flowers were collected/sample) from maize fields within 2 km of this apiary that were being planted at the time of our bee collections.

We also collected additional waste talc samples in 2011, using commercially available neonicotinoid treated maize seed from several different manufacturers. Because our goal was to develop a representative sample from a variety of maize hybrids used in our research area, all hybrids were selected based upon agronomic suitability for local planting. Both clothianidin and thiamethoxam treated seed was used, at application rates ranging from 0.25 mg/kernel to 1.25 mg/kernel. Talc was added to each seed box at the recommended rate (approx. 240 cc talc/75 kg of maize seed) (36). Fields were planted using a 6-row John Deere 7200 MaxEmerge planter. Collection of waste talc for analysis was performed following planting by manually removing approximately 50 g of talc from the manifold of the planter vacuum system using a scoopula. The planter and vacuum system was exhausted thoroughly and cleaned with compressed air prior to each planting and following each collection.

Acknowledgments

We appreciate the comments of anonymous reviewers whose thoughtful and thorough reviews helped improve the manuscript. We thank Alexzandra Murphy, who assisted with collation of background literature. Caitlin Race, Madeline Spigler and Nicole Parker assisted with colony maintenance and sample collection. We are also grateful for the assistance of Peter Illingworth and Larry Bledsoe who aided in planting of fields.

Author Contributions

Conceived and designed the experiments: CK GH BE. Performed the experiments: CK BE GA. Analyzed the data: CK BE. Contributed reagents/materials/analysis tools: BE CK. Wrote the paper: CK GH BE. Maintained and transported honeybee colonies used for these experiments: KG.

References

1. vanEngelsdorp D, Underwood R, Caron D, Hayes J (2007) An estimate of managed colony losses in the winter of 2006–2007: A report commissioned by the Apiary Inspectors of America. Am Bee J 147: 599–603.
2. vanEngelsdorp D, Hayes J, Underwood RM, Pettis J (2008) A survey of honey bee colony losses in the U.S., fall 2007 to spring 2008. PLoS ONE 3: 1–6 e4071.
3. Biesmeijer JC, Roberts SPM, Reemer M, Ohlemuller R, Edwards M, et al. (2006) Parallel declines in pollinators and insect-pollinated plants in Britain and the Netherlands. Science 313: 351–354.
4. Alaux C, Brunet JT, Dussaubat C, Mondet F, Tchamitchan S, et al. (2009) Interactions between Nosema microspores and a neonicotinoid weaken honey bees. Env Microbiol 12: 774–782.
5. Mullin CA, Frazier M, Frazier JL, Ashcraft S, Simonds R, et al. (2010) High levels of miticides and agrochemicals in North American apiaries: Implications for honey bee health. PLoS ONE 5: e9754.
6. Wu J, Anelli C, Sheppard W (2011) Sub-lethal effects of pesticide residues in brood comb on worker honey bee (*Apis mellifera*) development and longevity. PLoS ONE 6: e14720.
7. Jeschke P, Nauen R (2008) Neonicotinoids – from zero to hero in insecticide chemistry, Pest Mgt Sci 64: 1084–1098.
8. US Environmental Protection Agency, Office of Pesticide Programs. Fact-sheet Clothianidin (2003) EPA Publication 7501C; www.epa.gov/opprd001/factsheets/clothianidin.pdf.
9. National Agriculture Statistics Service (NASS), online database: Crop production 2010 summary (2010) Available: http://usda.mannlib.cornell.edu/MannUsda/viewDocumentInfo.do?documentID=10471. Accessed 2011 July 10.
10. Iwasa T, Motoyama N, Ambrose JT, Roe RM (2004) Mechanism for the differential toxicity of neonicotinoid insecticides in the honey bee, *Apis mellifera*. Crop Prot 23: 371–378.
11. National Agriculture Statistics Service (NASS) online database (2010) Indiana crop and weather report, report for week ending April 25, 2010. Available: www.nass.usda.gov. Accessed 2011 April 15.
12. US Environmental Protection Agency, Office of Pesticide Programs. Decision documents for atrazine (2003) Available: http://www.epa.gov/oppsrrd1/REDs/atrazine_combined_docs.pdf. Accessed 2011 April 15.
13. Decourtye A, Lacassie E, Pham-Delegue M (2003) Learning performances of honey bees (*Apis mellifera* L.) are differentially affected by imidacloprid according to the season. Pest Mgt Sci 59: 269–278.
14. Desneux N, Decourtye A, Delpuech JM (2007) The sublethal effects of pesticides on beneficial arthopods. Annu Rev Entomol 52: 81–106.
15. Mommaerts V, Reynders S, Boulet J, Besard L, Sterk G (2010) Risk assessment for side-effects of neonicotinoids against bumblebees with and without impairing foraging behavior. Ecotoxicol 19: 207–215.
16. Cutler GC, Scott-Dupree C (2007) Exposure to clothianidin seed-treated canola has no long-term impact on honey bees. J Econ Entomol 100: 765–772.
17. Chauzat MP, Faucon JP, Martel AC, Lachaize J, Cougoule N, et al. (2006) A survey of pesticide residues in pollen loads collected by honey bees in France. J Econ Entomol 99: 253–262.
18. Chauzat MP, Carpentier P, Martel AC, Bougeard S, Cougoule N, et al. (2009) Influence of pesticide residues on honey bee (Hymentoptera: Apidae) colony health. Environ Entomol 38: 514–523.
19. Girolami V, Mazzon L, Squartini A, Mori N, Marzaro M, et al. (2009) Translocation of neonicotinoid insecticides from coated seeds to guttation drops: a novel way of intoxication for bees. J Econ Entomol 102: 1808–1815.
20. Greatti M, Barbattini R, Stravisi A, Sabatini AG, Rossi S (2006) Presence of the a.i. imidacloprid on vegetation near corn fields sown with Gaucho dressed seeds. Bull Insectol 59: 99–103.
21. Tremolada P, Mazzoleni M, Saliu F, Colombo M, Vighi M (2010) Field trial for evaluating the effects on honey bees of corn sown using Cruiser and Celest xl treated seeds. Bull Environ Contam Toxicol 85: 229–234.
22. Girolami V, Marzaro M, Vivan L, Mazzon L, Greatti M, et al. (2011) Fatal powdering of bees in flight with particulates of neonicotinoids seed coating and humidity implication. doi: 10.1111/j.1439-0418.2011.01648.x.
23. Laurino D, Porporato M, Patetta A, Manino A (2011) Toxicity of neonicotinoid insecticides to honey bees: laboratory tests. Bull Insectol 64: 107–113.
24. US Environmental Protection Agency, Office of Chemical Safety and Pollution Prevention (2011) PC Code: 044309. Memorandum: Clothianidin registration of Prosper T400 seed treatment on mustard seed and Poncho/Votivo seed treatment on cotton. (2010). Available: www.epa.gov/pesticides/chem_search/cleared_reviews/csr_PC-044309_2-Nov-10_b.pdf Accessed 2011 Aug 1.
25. Ginsberg HS (1974) Ecology of bees in an old field. Ecology 64: 165–175.
26. Bonmatin JM, Marchand PA, Charvet R, Moineau I, Bengsch ER, et al. (2005) Quantification of imidacloprid update in maize crops. J Agri Food Chem 53: 5336–5341.
27. Rortais A, Arnold G, Hahm MP, Touffet-Briens F (2005) Modes of honey bee exposure to systemic insecticides: estimated amounts of contaminated pollen and nectar consumed by different categories of bees. Apidologie 36: 71–83.
28. Paul PA, Madden LV, Bradley CA, Robertson AE, Munkvold GP, et al. (2011) Meta-analysis of yield response of hybrid field corn to foliar fungicides in the US corn belt. ;doi: 10.1094/phyto-03-11-0091.
29. Schmuck R, Stadler T, Schmidt HW (2003) Field relevance of a synergistic effect observed in the laboratory between an EBI fungicide and a chloronicotinyl insecticide in the honey bee (*Apis mellifera* L, Hymenoptera). Pest Mgt Sci 59: 279–286.
30. Lydy M, Belden J, Wheelock C, Hammock B, Denton D (2004) Challenges in regulating pesticide mixtures. Ecology and Society 9(6): 1.

31. Decourtye A, Devillers J In: Steeve Herve Thany, editor. Insect Nicotinic Acetylcholine Receptors. Springer Press, New York. pp 85–96.
32. Vidau C, Diogon M, Aufauvre J, Fontbonne R, Vigues B, et al. (2011) Exposure to sublethal doses of fipronil and thiacloprid highly increases mortality of honey bees previously infected by *Nosema ceranae*. PLoS One 6: e21550.
33. Thompson HM (2010) Risk assessment for honey bees and pesticides: recent developments and 'new issues'. Pest Mgt Sci 66: 1157–1162.
34. Anastassides M, Lehotay SJ, Stajnbaher D, Schenk FJ (2003) Fast easy multiresidue method employing acetonitrile partitioning and 'dispersive solid phase extraction' for determination of pesticide residues in produce. J AOAC Int 86: 412–431.
35. Kiesselbach TA (1949) The structure and reproduction of corn. Research Bulletin 161. LincolnNE: Univ. of Nebraska, College of Agriculture.
36. Deere & Company, Publication A83046. Ready to plant guide. Available: http://www.deere.com/en_US/ag/media/pdf/ready_to_plant/rtp_guide.pdf. Accessed 2011 Jun 20.

Wheat Seedling Emergence from Deep Planting Depths and Its Relationship with Coleoptile Length

Amita Mohan, William F. Schillinger, Kulvinder S. Gill*

Department of Crop and Soil Sciences, Washington State University, Pullman, Washington, United States of America

Abstract

Successful stand establishment is prerequisite for optimum crop yields. In some low-precipitation zones, wheat (*Triticum aestivum* L.) is planted as deep as 200 mm below the soil surface to reach adequate soil moisture for germination. To better understand the relationship of coleoptile length and other seed characteristics with emergence from deep planting (EDP), we evaluated 662 wheat cultivars grown around the world since the beginning of the 20th century. Coleoptile length of collection entries ranged from 34 to 114 mm. A specialized field EDP test showed dramatic emergence differences among cultivars ranging from 0–66% by 21 days after planting (DAP). Less than 1% of entries had any seedlings emerged by 7 DAP and 43% on day 8. A wide range of EDP within each coleoptile length class suggests the involvement of genes other than those controlling coleoptile length. Emergence was correlated with coleoptile length, but some lines with short coleoptiles ranked among the top emergers. Coleoptiles longer than 90 mm showed no advantage for EDP and may even have a negative effect. Overall, coleoptile length accounted for only 28% of the variability in emergence among entries; much lower than the 60% or greater reported in previous studies. Seed weight had little correlation with EDP. Results show that EDP is largely controlled by yet poorly understood mechanisms other than coleoptile length.

Editor: Randall P. Niedz, United States Department of Agriculture, United States of America

Funding: This research was funded by Washington State University, the Washington Grain Commission, and the National Science Foundation (NSF-BREAD) grant # 0965533 (http://www.nsf.gov/funding/pgm_summ.jsp?pims_id = 503285). The funders had no role in study design, data collection and analysis, decision to publish, or preparation of the manuscript.

Competing Interests: The authors have declared that no competing interests exist.

* E-mail: ksgill@wsu.edu

Introduction

Deep planting of wheat into stored soil moisture is practiced in Mediterranean-like climate regions, including the Pacific Northwest (PNW) of the United States, parts of western and southwestern Australia, central Chile, and several countries surrounding the Mediterranean Sea. Winter wheat in the low-precipitation (<300 mm annual) region of the PNW is planted 100 to 200 mm below the surface of summer-fallowed soils with deep-furrow drills to reach adequate moisture for seedling emergence [1]. Introduction of high-yielding semi-dwarf wheat cultivars in the early 1960s spawned the "green revolution". Height reduction in semi-dwarf cultivars was due to mutations in *Rht-B1* (*Rht1*) or *Rht-D1* (*Rht2*) genes that reduce either the production or the perception of gibberellin (GA), an important plant growth hormone [2]. Semi-dwarf wheat cultivars are lodging resistant, thus allowing application of higher fertilizer inputs, and have improved harvest index compared to standard-height cultivars due to increased partitioning of assimilates to reproductive organs resulting in more fertile florets per spikelet [3–5]. However, semi-dwarf cultivars are not popular among farmers in some low-precipitation Mediterranean regions because of their relatively poor and/or slow emergence from deep planting depths [6–8]. Decreased response to endogenous GA in *Rht-B1b* and/or *Rht-D1b gene* (GA-insensitive) mutants results in reduced cell size and elongation with corresponding reduction in coleoptile length, plant height, and leaf area [9–13]. Compared to GA-insensitive mutants, GA-responsive mutants (*Rht8*) are reported to reduce plant height without reducing coleoptile length [9,14] and thus have less negative effect on emergence.

Wheat seedling emergence does not have linear relationship with coleoptile length as many other factors have been implicated in the process. Emergence and coleoptile length is reported to be influenced both by genetic background and environmental factors including soil texture, seed-zone water content, temperature, light penetration, and crop residue [15–20]. Increased seed size has been reported to positively affect early seedling vigor in many crop species [21–23], although about an equal number of reports argue against this correlation [18,24]. Similarly, reports are mixed on the effect of seed weight on coleoptile length of wheat, with no effect reported for semi-dwarfs but a positive effect on some tall durum lines [25].

The pleiotropic effects associated with GA insensitive mutants on early growth have been reported and coleoptile length was positively correlated with plant height at maturity [6,8,26]. The major phenotypic effect quantitative trait loci (QTL) for coleoptile length were mapped on the short arm of 4B and 4D chromosomes in the close proximity of dwarfing genes, thus accounting for the pleotropic effect of dwarfing genes on coleoptile length [27]. Moreover, small-effect QTLs with additive effects were identified on wheat chromosomes 1A, 2B, 2D, 3B, 3A, 5A, 6A [27,28]. Further, genetic background of the cultivar and environmental conditions influence the relation between coleoptile length with plant height. This suggests the possibility to select for a longer coleoptile in reduced plant height genotypes [6], but more detailed systematic study is needed to understand the mechanism behind the relationship of coleoptile length, emergence, and plant height.

Our study was undertaken to gain a better understanding of wheat seedling emergence from deep planting depths and its relation to coleoptile length along with plant height, seed weight, and size. Another objective of the study was to identify the fastest and best emerging wheat cultivar to serve as a donor to improve the emergence trait in new and existing PNW cultivars. We used a population of 662 wheat cultivars from the major wheat growing areas of the world to capture variation present among cultivars. The collection included both winter and spring type habit and represented all major market classes of wheat.

Materials and Methods

Ethics Statement

Field experiments were conducted at the Washington State University (WSU) Dryland Research Station located near Lind, WA. No special permissions were required to conduct the experiment. The field studies did not involve any endangered or protected species. The wheat germplasm used in the study was obtained from the Germplasm Resources Information Network (GRIN), CIMMYT, and from WSU historical collection. These are open access seed sources available for research and education purposes.

Germplasm

We collected 662 wheat cultivars grown around the world; 360 with winter and 302 with spring growth habit (Table S1). The collection included 96 cultivars from the PNW, 125 from a historical US wheat collection including cultivars released since 1871, and 80 from the International Maize and Wheat Improvement Center (CIMMYT), Mexico representing the major wheat growing regions of the world. The remaining 361 entries were US cultivars collected from major wheat growing regions (outside the PNW) that have been widely grown since 1950. The most popular cultivars during each 5-year interval since 1950 from all major breeding programs in the US were included. Our intention was to capture as much variation among cultivated wheat as possible without making the population size unmanageably large. The world collection also represented different market classes depending upon hardness, color, and kernel shape. The collection consisted of 112 soft white winter (SWW), 90 soft white spring (SWS), 101 soft red winter (SRW), 35 soft red spring (SRS), 97 hard red winter (HRW), 127 hard red spring (HRS), 9 hard white winter (HWW), 57 hard white spring (HWS), and 34 club wheat cultivars. Due to concerns about seed purity and vigor, a single seed of each entry was grown in the greenhouse to obtain single-plant seeds. The harvested single-plant seed was then multiplied in the greenhouse and used for all the experiments.

Coleoptile Length Measurement

For coleoptile length measurement, 15 uniform-sized seeds of each cultivar with no physical damage were placed in the middle of a moist germination paper (Heavy Germination paper #SD 7615L), about one centimeter apart with germ end down. The germination paper was then folded vertically in half with the seed placed in the crease, the folded half was again folded horizontally four times and placed in a plastic tray with holes at the base to drain excess water. The plastic trays were then placed inside a completely darkened box and kept in a growth chamber at a constant temperature of 22°C. After 10 days, the average coleoptile length of 10 randomly-selected seedlings was recorded to the nearest millimeter measuring from the base of the seed to the coleoptile tip. Germination percentage of all the lines was recorded.

Field Test for Emergence from Deep Planting

Emergence from deep planting was conducted at the WSU Dryland Research Station near Lind, WA during the 2009–2010 crop season. The experiment was planted on land that had been fallowed for 13 months (i.e., since the last wheat harvest). Soil type was Shano silt loam (coarse-silty, mixed, superactive, mesic Xeric Haplocambids) with a soil textural size distribution of 10% clay, 51% silt, and 39% fine sand. Slope was <2%. Soil volumetric water content in the seed zone was determined gravimetrically at time of planting on September 3. These measurements were obtained in 2-cm increments avoiding wheel tracks, to a depth of 24 cm using an incremental soil sampler. Mean water content for each increment was determined from four soil cores.

For each entry, 50 uniform-sized seeds with no physical damage were planted using a four-opener deep-furrow drill in 3-m-long rows with a 0.38-m spacing between rows. The experimental design was randomized complete block with four replicates per entry. To account for any differences among openers, each entry was planted once with each of the four openers.

Seeds were planted 150 mm below the soil surface into a seed zone having a water potential of −0.50 MPa (11.5% water by volume). The depth of the soil covering the seeds was 125 mm. Wheat seedling emergence was determined by counting individual seedlings at 24-hour intervals. Seedling emergence counts were obtained on 6, 7, 8, 9, 10, 15, and 21 DAP. No rainfall occurred at the site from time of planting through 21 DAP.

Plant Height, Seed Weight and Seed Size

Plant height of each entry was recorded at maturity by measuring the average distance from the soil surface to the top of the spike (excluding awns) of the three tallest plants. Several entries (mainly spring types) were winterkilled and, therefore, plant height could obviously not be obtained. Thousand kernel weight (TKW) for each entry was determined using electronic counter (Oldmill Company, Model 850.2) and then weighing on a digital scale. Seed size and grain hardness was determined using 200 seeds of each entry with the Single-Kernel Characterization System (Perten Instruments, Springfield, IL, Model 4100).

Statistical Analysis

Data analysis was conducted using the PROC GLM procedure of Statistical Analysis System (SAS; SAS Inst. 2000). Means and standard deviations were determined using PROC MEANS in SAS. Analysis of variance was performed using PROC MIXED (SAS Institute, 2000). Pearson's linear correlation coefficients were calculated by PROC CORR. Significance level was tested at $P<0.05$. Graphical representation of data was conducted using JMP genomics (JMP, Version 10. SAS Institute Inc., Cary, NC, 1989–2007).

Results

Coleoptile Length Variation

Mean coleoptile length among entries ranged from 34 mm to 114 mm (Figure 1A–B). The vast majority of entries had a mean coleoptile length between 41 to 90 mm. Only 10 of the 662 lines had coleoptiles shorter than 41 mm and 40 lines had coleoptiles longer than 90 mm (Figure 1A). The median coleoptile length was 62.4 mm. Among the nine coleoptile length intervals made in 10 mm increments, the interval 51 to 60 mm accounted for the largest number of entries (168). Coleoptile length of spring wheat lines ranged from 37 to 114 mm with a median of 62 compared to the winter lines with a range of 34 to 103 mm and a median of 63. Among various grain market classes, mean coleoptile length was the highest (71) for club wheat and the lowest for white classes (HWW, HWS, SWW), having a mean coleoptile length of 60 mm. The mean coleoptile length was 68 mm for SWS, 66 mm for both SRW and HRW, 65 mm for SRS and 64 mm for the HRS classes.

Variation in Emergence from Deep Planting Depth

The fastest emergence was recorded in four cultivars (Spinkcota, Regent, Pacific Bluestem, ARS95-452) that had an average of at least two emerged seedlings on 7 DAP. By 8 DAP, 43% of the entries had an average of at least one emerged seedling, and this was 96% on 9 DAP. No differences were observed for EDP between spring and winter growth habit. By 21 DAP, 658 lines (99.4%) had some emergence while four entries (0.6%) did not emerge at all. Of the entries that emerged, the extent of emergence of total seeds planted ranged from 4% to 66% by 21 DAP. Emergence in some lines was fast and uniform (e.g., Spinkcota, Indian, White Federation, and Barbee) and was completed within 2–3 days after initial emergence, whereas emergence of other lines (e.g., Golden Cross, Triumph 64 and 90451ARS) was gradual and extended over many days. Eleven lines showed >60% emergence by 21 DAP. The four entries with no emergence at all had coleoptile lengths ranging from 53 to 85 mm (Table 1).

The percent emergence at 10 DAP among coleoptile length classes ranged from 14% to 38%. There was a gradual increase in emergence percentage with increasing coleoptile length up to 90 mm, beyond which there were no statistical differences (Figure 2A). In general, emergence decreased for entries with coleoptile length >110 mm, although this interpretation is based only on two entries. The best early emergers on 7 DAP had coleoptiles ranging from 50–110 mm with no significant improvement in emergence above 60 mm (data not shown). The emergence trend was very similar at 15 and 21 DAP with no significant improvement in emergence of entries with coleoptiles longer than 90 mm (Figure 2B–C). Mean emergence percentage among entries on 10 and 15 DAP ranged from 2 to 62% and on 21 DAP from 4% to 66%. Overall, five of the nine coleoptile length classes differed significantly in their mean percent emergence (Figure 2A–C).

Table 1. Coleoptile length (CL), thousand kernel weight (TKW) and emergence percentage of selected cultivars in the world wheat collection having maximum and minimum emergence from deep planting on 15 and 21 days after planting.

Entry Name	CL (mm)	TKW (gm)	% Emergence	
			DAY15	DAY21
Klein Dragon	51.0	26	52	66
Fortuna	90.8	33.8	56	64
Canadian Red	99.1	37.2	58	64
Luft	82.3	37.2	58	64
Sonora	106.3	29.8	60	62
Okla. 61STW8637	60.5	33.3	50	62
J961051	96.7	33.5	46	62
Bounty 309	46.8	39.1	4	8
WA 64250	49.0	32.0	2	4
Chinook	74.0	39.7	0	0
Blizzard	85.4	38.7	0	0
Newton	53.4	39.0	0	0

Within each of coleoptile length class, the range of emergence percentage was wide and variable (Figure 2D–F). The entry with the greatest emergence (66%) had a coleoptile length of only 52 mm. The minimum emergence in this (51–60 mm) coleoptile length class was 12%. The emergence percentage of entries with coleoptiles >100 mm (average coleoptile length of 112 mm) was 46% and ranged from 36% to 62%. The within-coleoptile class range was wider at 10 DAP compared to that at 21 DAP, suggesting that final emergence is better correlated with coleoptile

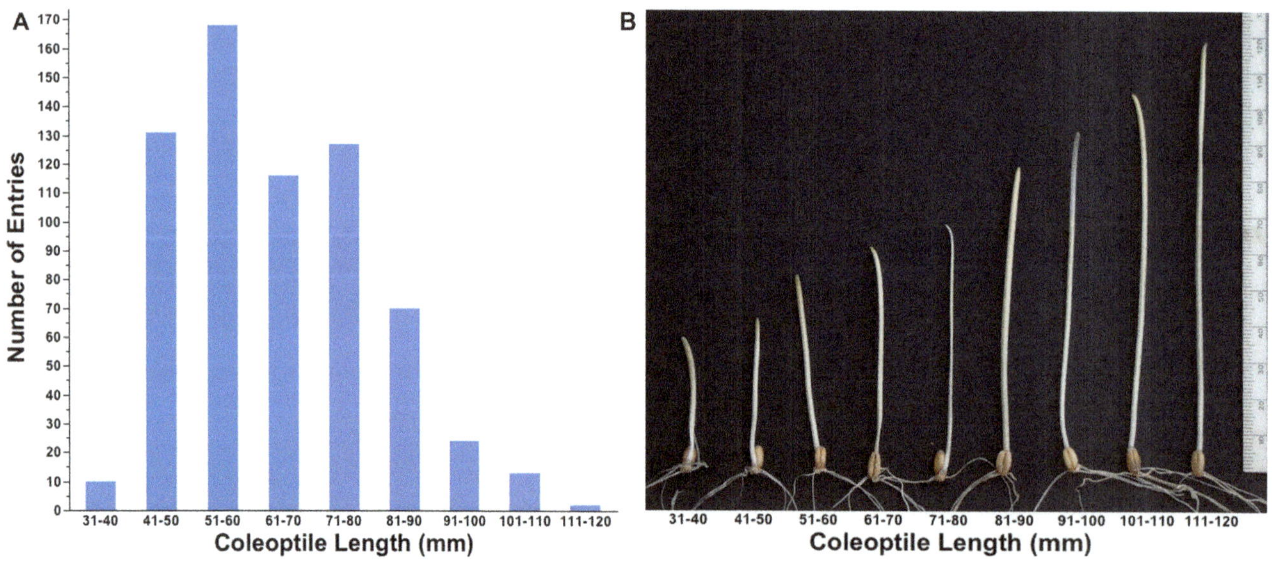

Figure 1. Coleoptile length distribution among 662 entries of the world wheat collection in 10-mm intervals (A). Representative picture of coleoptiles in each 10-mm increment class interval (B).

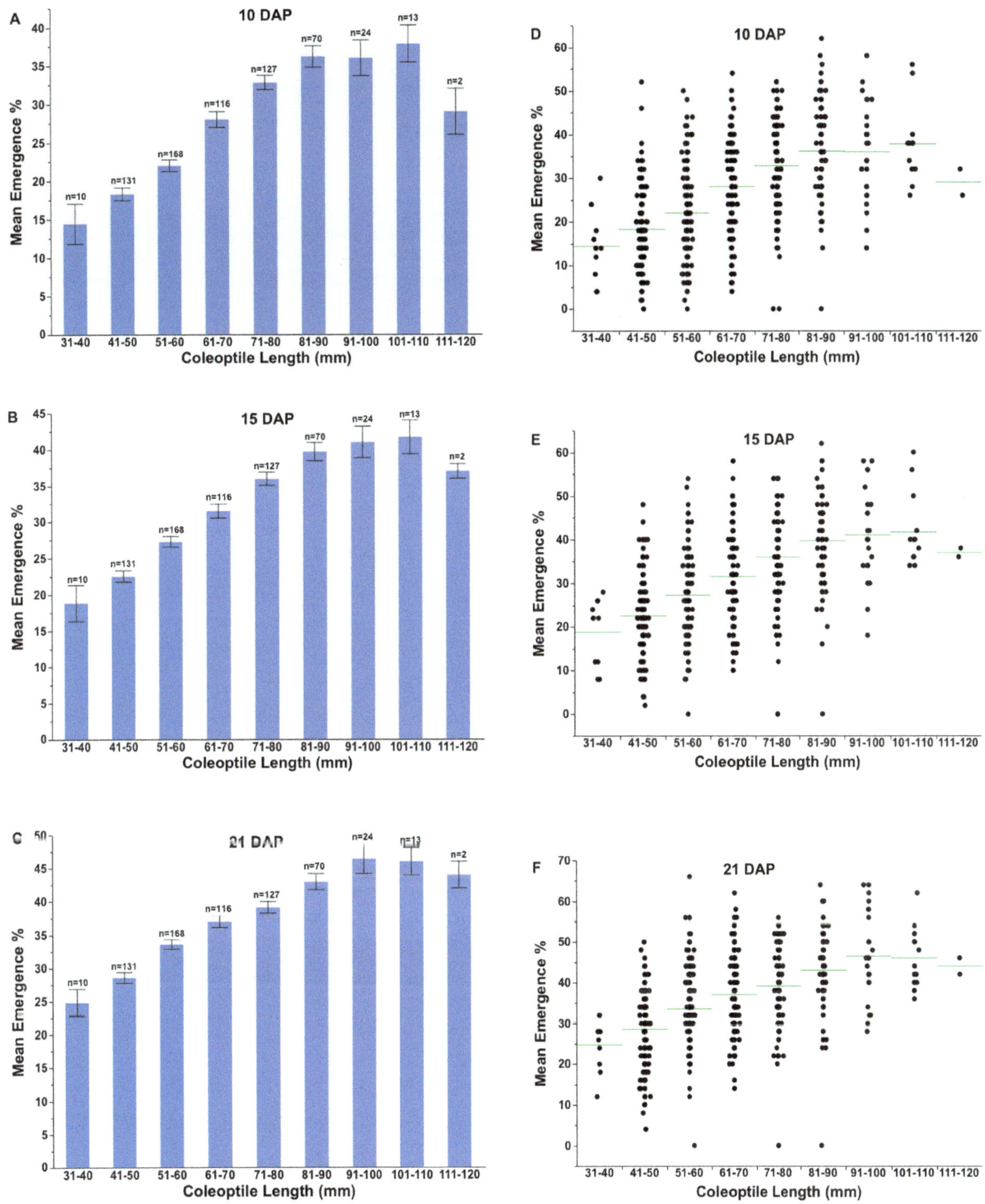

Figure 2. Mean emergence percentage among various coleoptile length classes (A–C) and, range of variation in coleoptile length classes on 10, 15, and 21 days after planting (D–F). The error bar represents the standard error of the mean. Number of lines in each coleoptile length class (n) is given above each bar. The horizontal line represents the mean in each coleoptile class (D–F).

length than to emergence speed (Figure 2E–F). Except for the four non-emerging entries that spanned four coleoptile length classes with the highest being 80 to 90 mm, minimum emergence in each coleoptile length class gradually increased with the increasing coleoptile length (Figure 2D–F). The greatest relative emergence improvements occurred for the three shortest coleoptile classes (31–40, 41–50, and 51–60 mm) and then leveled off for the next four classes (Figure 2D–F). Maximum emergence for the two longest coleoptile length classes gradually decreased from 64% in 91–100 mm class to 46% in the 111–120 mm class (Figure 2D–F).

Correlation Between Emergence and Coleoptile Length

There was an overall positive correlation between coleoptile length and emergence from deep planting (Figure 3A–B). On 8 DAP, correlation between emergence and coleoptile length was weak ($r^2 = 0.12$, $p<0.0001$), but somewhat improved on 9 DAP ($r^2 = 0.23$, $p<0.0001$). Coefficients of determination (r^2) for the relation of coleoptile length with emergence on 10, 15, and 21 DAP were 0.28, 0.29, and 0.23 respectively ($p<0.0001$) (Figure 3A, data for 10 and 21 DAP not shown).

Coefficients of determination of emergence with coleoptile length were similar among market classes and, except for the HWW class where there was no significant correlation ($p = 0.1$) on any DAP. Club wheat showed the highest coefficients of determination ($p<0.001$) on 10, 15 and 21 DAP ($r^2 = 0.43$, 0.43, 0.34, respectively).

Kernel Weight Versus Seedling Emergence

Thousand kernel weight (TKW) of the world wheat collection cultivars ranged from 15.0 to 47.4 g and averaged 32.0 g. Mean TKW for winter and spring cultivars was 33.2 and 30.8 g, respectively. There was no overall correlation between coleoptile length and TKW for spring and winter type and all the market classes except for HWS (Figure 4 A–B). The TKW of coleoptile class 31–40 and 110–120 mm entries differed significantly compared to other coleoptile classes (Figure 5A). The widest variability in TKW occurred in coleoptile classes between 41–80 mm (Figure 5B). Although, the coefficient of determination for TKW and seed size was 0.27 ($p<0.001$), seed size had negligible correlation with coleoptile length ($r^2 = 0.01$, $p<0.01$). The weak correlation held true for both winter and spring types and all market classes except for HWS, which showed the most positive correlation with coleoptile length ($r^2 = 0.22$, $p<0.001$). Thousand kernel weight did not affect emergence on any DAP nor was it associated with plant height at maturity (data not shown).

Plant Height

Plant height at maturity of 357 winter and two spring entries ranged from 46 to 116 cm with an average height of 82 cm. In general, plant height was positively correlated with coleoptile length ($r^2 = 0.25$, $p<.001$, Figure 6). There were no differences in plant height among entries with coleoptiles longer than 80 mm (Figure 7A). The tallest entry had a height of 116 cm with a coleoptile length of 74 mm. The shortest entry had a plant height of 46 cm with a coleoptile length of 54 mm. There was wide variability in plant height within each coleoptile length class (Figure 7B).

Discussion

Variation among wheat cultivars for coleoptile length and its correlation with seedling emergence has been reported from several authors [13,29,30] but the number of cultivars evaluated in those studies was relatively small. Here, we present a comprehensive analysis of variation among 662 wheat cultivars from major wheat growing regions around the world and provide a detailed analysis of the correlation between coleoptile length and seedling emergence.

Our wheat collection was carefully selected to capture the variation present in cultivated wheat while minimizing redundancy. We observed unprecedented variation in the population for the three traits studied. With a range from 34 mm to 114 mm, our collection had more than a three-fold difference in coleoptile lengths. Similarly, variation for seedling emergence from deep planting was high with some lines showing some emergence as early as 7 DAP and four lines with no emergence even on 21 DAP. Variation for seedling emergence was related to speed, extent, and uniformity of final plant stand.

It has previously been reported that the commonly used GA-insensitive semi-dwarf gene mutations (RhtB1b and RhtD1b) reduce coleoptile length by 30–40% compared to standard-height cultivars [15]. The GA insensitivity reduces cell wall extensibility, thus reduces plant height and also negatively affects growth of the coleoptile and first leaf [10,31]. Coleoptile length variation among the CIMMYT lines in our collection was very narrow with the majority having a coleoptile length between 40–60 mm. This narrow variation is probably due to the semi-dwarf nature of all the CIMMYT lines and/or due to a common genetic background. The CIMMYT lines had an average 32% (ranging from 12–54%) of seedlings emerged on 21 DAP. In addition to genetics, other factors including soil temperature, texture and moisture may also influence coleoptile length [20,25,32,33].

Emergence from deep planting is particularly important in the low-precipitation dryland regions of the world, such as the inland PNW, where wheat is planted as deep as 200 mm with deep-furrow drills and seedlings must emerge through 150 mm or more

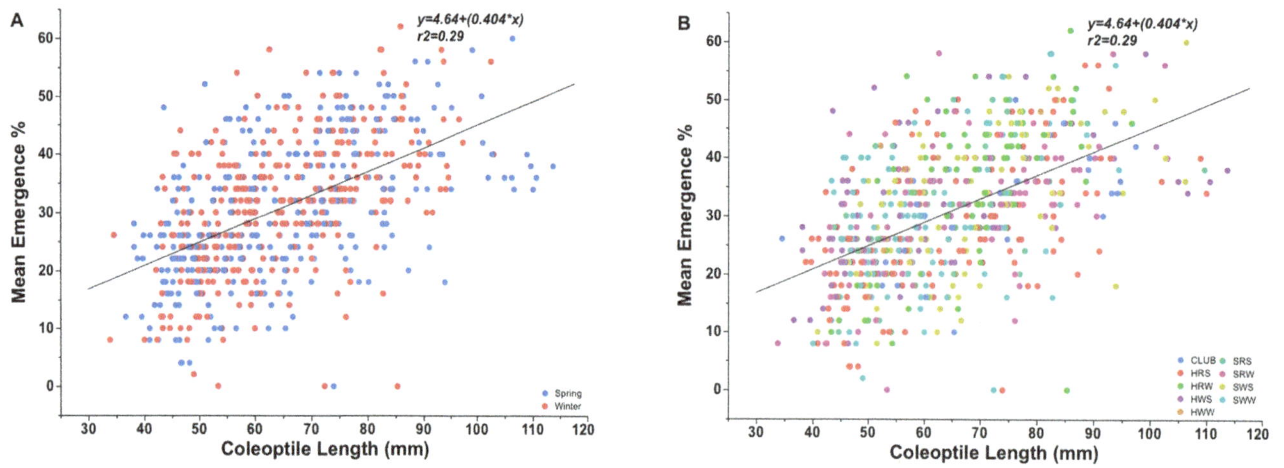

Figure 3. Mean emergence percentage of the (A) spring and winter type and (B) different market class of wheat world collection entries plotted against coleoptile length on 15 days after planting.

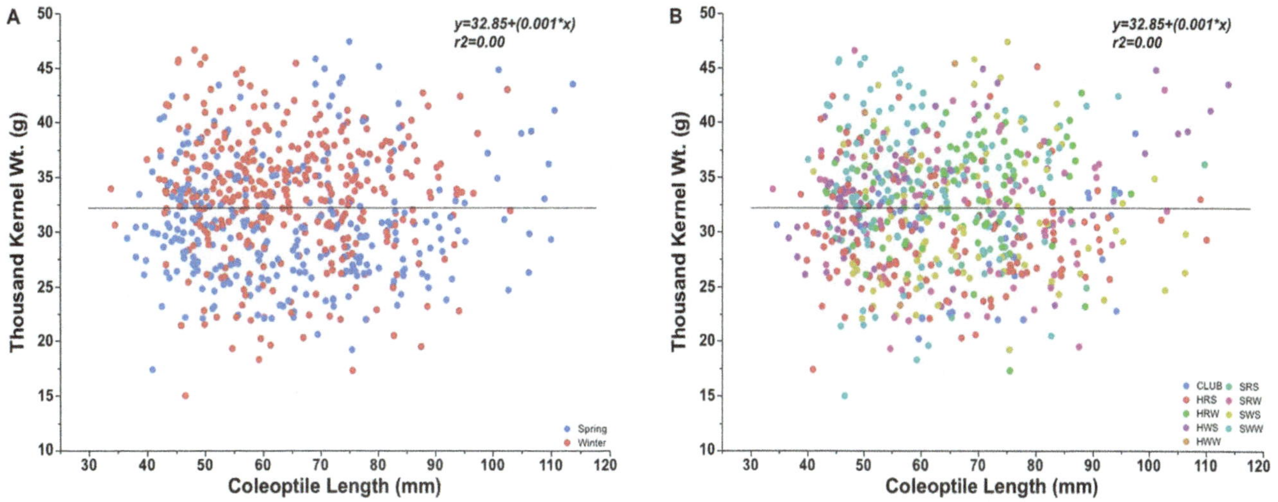

Figure 4. Thousand kernel weight of (A) spring and winter type wheat and (B) different market classes of world wheat collection entries plotted against coleoptile length.

soil cover. The coleoptile protects the emerging leaf (i.e., until the first leaf protrudes through the tip of the coleoptile while still below the soil surface) and also provides the mechanical support required for seedling emergence. Coleoptile length has previously been reported to be important for seedling emergence. In the present study, coleoptile length was significantly correlated with seedling emergence but accounted for only 28% of the variability for emergence on 10 and 15 DAP (Figure 2A and B). The emergence percentage of entries in coleoptile class 31–40 mm (25%) on 21 DAP was significantly lower than any other class (Figure 3C) with a gradual increase in mean emergence percentage up to a coleoptile length of 90 mm. Comparing various coleoptile length classes, emergence increased more dramatically with increased coleoptile length at 10 DAP compared to that at 21 DAP (Figure 3A and C). Our linear regression coefficients of determination for the relationship of coleoptile length to emergence were considerably lower compared to previous reports [6,8,13,29]. Careful evaluation of present data suggests that coleoptile length is not as critical for seedling emergence as previously thought. Our five best emerging cultivars had coleoptile length range from 51–106 mm. Further, a wide range of emergence was observed within each of the coleoptile length classes. The lower value of emergence decreased with decreasing coleoptile length, suggesting an underlying effect of coleoptile length on emergence, but the maximum emergence values did not increase with coleotiptiles longer than 90 mm, suggesting that coleoptile lengths longer than 90 mm may not be beneficial for emergence.

Several entries in our study with relatively short coleoptiles emerged well from deep planting while other entries with longer coleoptiles had poor emergence. This difference is not due to uneven germination as during the coleoptile length measurement experiment all lines germinated well. Except for the shortest class, each coleoptile length class showed a wide range of seedling

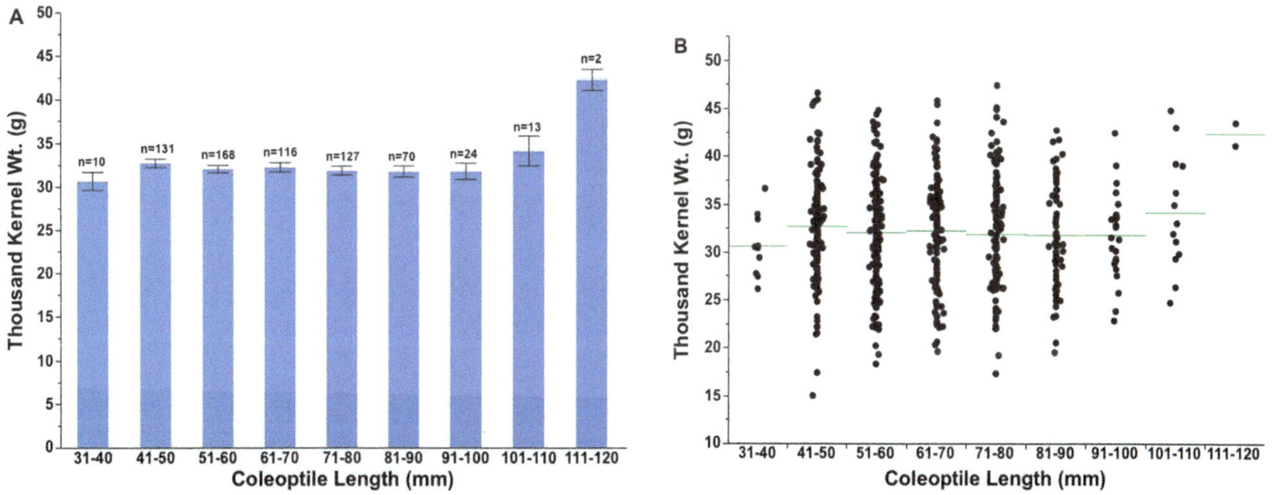

Figure 5. Thousand kernel weight distribution (A) and range variation (B) among different coleoptile length classes. Bars represent the standard error of the mean. The sample size (n) of each coleoptile class length is provided above the data bars. The horizontal line represents the mean in each coleoptile class (B).

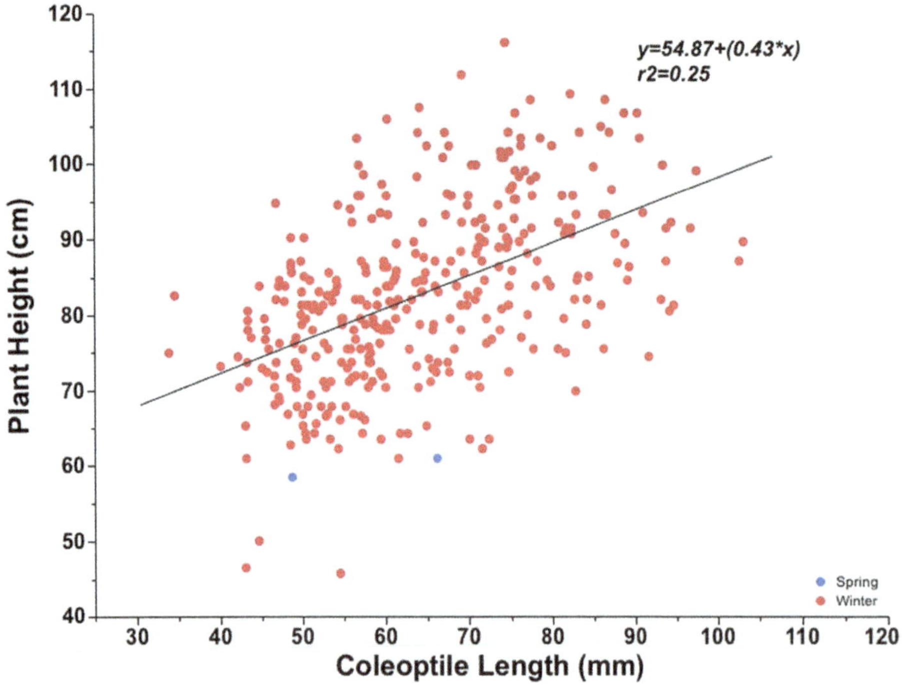

Figure 6. Mean plant height at maturity of spring and winter type cultivars in the world wheat collection entries plotted against coleoptile length.

emergence, suggesting that coleoptile length is not as strongly correlated to seedling emergence as previously reported, thus making emergence a complex trait. Similar observations were made in barley (*Hordeum vulgare L.*) where lines with widely different coleoptile lengths showed similar seedling emergence [34]. Another reason for range of coleoptile length variation within the coleoptile class might be genotype parentage which affects coleoptile length [35]. For example, club wheat had higher mean coleoptile length among the different market classes studied.

Lower dependence of emergence on coleoptile length in comparison to previous reports may partly be due to our much larger and highly diverse material providing increased precision to study this correlation. Another explanation for the weaker correlation could be that previous studies used wheat lines with limited geographic diversity, whereas our wheat lines collected from around the world may respond differently under different environmental conditions.

In general, seed weight had no effect on coleoptile length or seedling emergence. This is in agreement with most other studies,

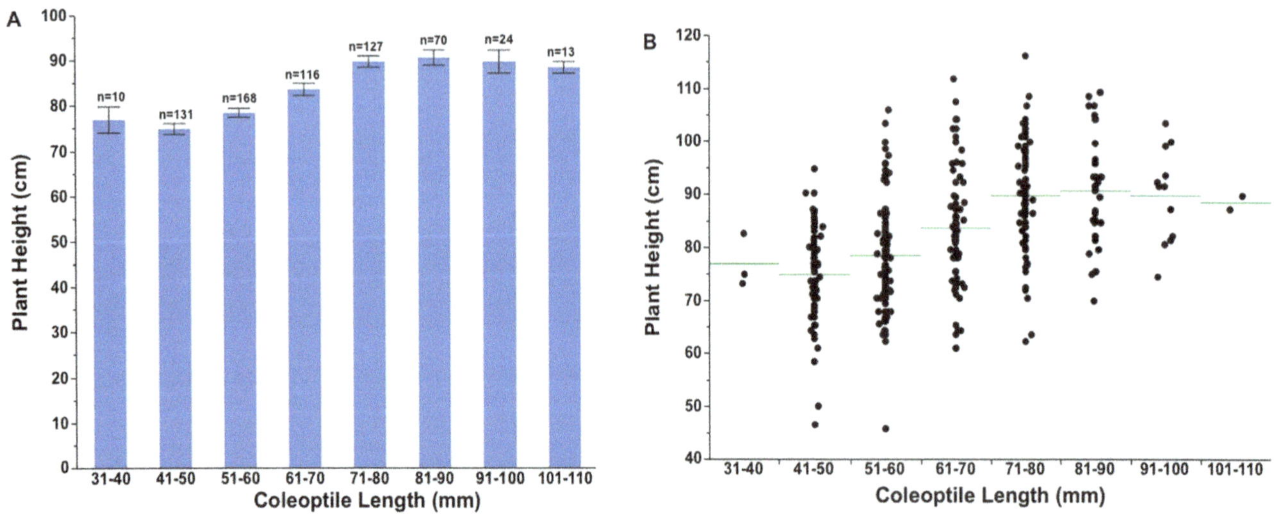

Figure 7. Mean (A) and range variation (B) in plant height at maturity in different coleoptile length classes of the world wheat collection. The bars are the standard error of the mean. The sample size (n) of each coleoptile class is provided above the data bars. The horizontal line represents the mean in each coleoptile class (B).

but there are some reports of a strong correlation between seed weight/size and coleoptile length in wheat [22,35,36]. In barley, seed weight had no effect on coleoptile length though it accounted for 31–53% of variability in coleoptile width [37]. In our study, except for the lightest and the heaviest seed weight classes, there were essentially no emergence differences. Coleoptile length and emergence of the lightest seed weight class was significantly lower than the other classes. Similarly, the highest coleoptile length class had significantly higher seed weight compared to the other coleoptile classes. In previous studies reported in the literature, essentially no effect of seed weight was reported on emergence or yield [18,20,24]. In our study the diverse genetic nature of the entries in each coleoptile length class might account for the differences in the coleoptile length and emergence within a particular seed weight range.

Coleoptile length showed a positive overall correlation with final plant height in our study (Figure 6). These results are in accordance with the earlier reports [38]. Coleoptile length and plant height are polygenic traits and many QTLs have been reported that influence traits [26,28,39–41]. We also observed a full range of plant heights in particular coleoptile class e.g. cultivar with short plant height (29 cm) with a long coleoptile (92 mm). As we do not know the nature of dwarfing genes in the world collection, these could possibly be GA sensitive genotypes that are reported to have no adverse effect on coleoptile length but reduce plant height, or possibly have favorable alleles for coleoptile length in semi-dwarf background [41]. In general, plant height was positively correlated with emergence at 10, 15, and 21 DAP. There was a general trend of increased plant height up to 90 mm coleoptile length, but no differences in the mean plant height for entries with coleoptiles longer than 90 mm.

Conclusions

A collection of 662 carefully selected wheat cultivars from around the world representing a full range of variation for the studied traits allowed us to gain improved understanding of the relation between coleoptile length and emergence from deep planting depths. Refuting the previous claims of coleoptile length explaining 60% or more of the variation for seedling emergence, our extensive study explains only 28% of the variability for seedling emergence related to coleoptile length. Contradictory to earlier reports suggesting a linear relationship between coleoptile length and seedling emergence, our study clearly showed that coleoptiles longer than 90 mm has no positive affect on seedling emergence and may actually have a detrimental effect on emergence from deep planting depths. Further, a full range of emergence percentage within each coleoptile length class clearly suggest the involvement of factors other than coleoptile length in determining the extent as well as speed of seedling emergence from deep planting depths. Results showed that except for the smallest and the largest seed weight classes that evidenced a proportional affect on coleoptile length, a similar affect was not observed for the remaining classes. We observed that the effect of coleoptile length on seedling emergence was different for different wheat market classes. Further, we observed that seed weight has no significant affect on seedling emergence from deep planting. Hence, we conclude that wheat seedling emergence from deep planting depths is a complex trait and further detailed studies are required to understand the underlying mechanisms.

Acknowledgments

The authors gratefully acknowledge the excellent technical support provided by Timothy Smith and Steven Schofstoll at the WSU Dryland Research Station at Lind and by Patrick Reisenauer in collecting coleoptile data. We also thank WSU graduate students Amandeep Dhaliwal and Ramanjot Kaur for their help with the field emergence experiment at Lind and Maninder Chahal with statistical analysis.

Author Contributions

Conceived and designed the experiments: KSG WFS AM. Performed the experiments: AM KSG WFS. Analyzed the data: AM KSG. Contributed reagents/materials/analysis tools: KSG WFS. Wrote the paper: AM KSG WFS.

References

1. Schillinger WF, Papendick RI (2008) Then and now: 125 years of dryland wheat farming in the Inland Pacific Northwest. Agronomy Journal 100: S–166–S–182. doi:10.2134/agronj2007.0027c.
2. Peng J, Richards DE, Hartley NM, Murphy GP, Devos KM, et al. (1999) "Green revolution" genes encode mutant gibberellin response modulators. Nature 400: 256–261. doi:10.1038/22307.
3. Brooking I, Kirby E (1981) Interrelationships between stem and ear development in winter wheat: the effects of a Norin 10 dwarfing gene, Gai/Rht2. Journal of Agricultural Science 97: 373–381.
4. Evans L (1993) Crop evolution, adaptation, and yield. Cambridge; New York: Cambridge University Press.
5. Miralles DJ, Slafer GA (1995) Yield, biomass and yield components in dwarf, semi-dwarf and tall isogenic lines of spring wheat under recommended and late sowing dates. Plant Breeding 114: 392–396. doi:10.1111/j.1439-0523.1995.tb00818.x.
6. Sunderman DW (1964) Seedling emergence of winter wheats and its association with depth of sowing, coleoptile length, and plant height. Agronomy Journal 56: 23–25.
7. Feather JT, Qualset CO, Vogt HE (1968) Planting depth critical for short-statured wheat varieties. California Agriculture: 12–14.
8. Fick G, Qualset C (1976) Seedling emergence, coleoptile length, and plant height relationships in crosses of dwarf and standard-height wheats. Euphytica 25: 679–684.
9. Ellis M, Rebetzke G, Chandler P, Bonnett D (2004) The effect of different height reducing genes on the early growth of wheat. Functional Plant 31: 583–589.
10. Keyes G, Sorrells M, Setter T (1990) Gibberellic acid regulates cell wall extensibility in wheat (*Triticum aestivum* L.). Plant Physiology 92: 242–245.
11. Hoogendoorn J, Rickson JM, Gale MD (1990) Differences in leaf and stem anatomy related to plant height of tall and dwarf wheat (*Triticum aestivum* L.). Journal of Plant Physiology 136: 72–77. doi:10.1016/S0176-1617(11)81618-4.
12. Matsui T, Inanaga S, Shimotashiro T (2002) Morphological characters related to varietal differences in tolerance to deep sowing in wheat. Plant Production Science 5: 169–174.
13. Rebetzke GJ, Richards RA, Fettell NA, Long M, Condon AG (2007) Genotypic increases in coleoptile length improves stand establishment, vigour and grain yield of deep-sown wheat. Field Crop Research 100: 10–23. doi:10.1016/j.fcr.2006.05.001.
14. Rebetzke GJ, Richards RA (1999) Genetic improvement of early vigour in wheat. Australian Journal of Agricultural Research 50: 291. doi:10.1071/A98125.
15. Allan R, Vogel O, Burleigh J (1962) Length and estimated number of coleoptile parenchyma cells of six wheat selections grown at two temperatues. Crop Science 2: 522–524.
16. Nebreda IM, Parodi PC (1977) Effect of seed type on coleoptile length and weight in triticale, X Triticosecale Wittmack. Cereal Research Communications 5: 387–398.
17. Jessop RS, Stewart LW (1983) Effects of crop residues, soil type and temperature on emergence and early growth of wheat. Plant and Soil 74: 101–109.
18. Chastian T, Ward K, Wysocki D (1995) Stand establishment response of soft white winter wheat to seedbed residue and seed size. Crop science 35: 213–218.
19. Rebetzke G, Richards R (1999) Breeding long coleoptile, reduced height wheats. Euphytica 106: 159–168.

20. Botwright TL, Rebetzke GJ, Condon AG, Richards RA (2001) Influence of variety, seed position and seed source on screening for coleoptile length in bread wheat (*Triticum aestivum* L.). Euphytica 119: 349–356.
21. Richards R, Lukacs Z (2002) Seedling vigour in wheat: sources of variation for genetic and agronomic improvement. Australian Journal of Agricultural Research 53: 41–50.
22. Nik MM, Babaeian M, Tavassoli A (2011) Effect of seed size and genotype on germination characteristic and seed nutrient content of wheat. Scientific Research and Essays 6: 2019–2025.
23. Lafond G, Baker R (1986) Effects of genotype and seed size on speed of emergence and seedling vigor in nine spring wheat cultivars. Crop Science 26: 341–346.
24. Mian AR, Nafziger ED (1992) Seed size effects on emergence, head number, and grain yield of winter wheat. Journal of Production Agriculture 5: 265–268.
25. Trethowan R, Singh R, Huerta-Espino J, Crossa J, Van Ginkel M (2001) Coleoptile length variation of near-isogenic Rht lines of modern CIMMYT bread and durum wheats. Field Crops Research 70: 167–176.
26. Liatukas Z, Ruzgas V (2011) Coleoptile length and plant height of modern tall and semi-dwarf European winter wheat varieties. Acta Societatis Botanicorum Poloniae 80: 197–203. doi:10.5586/asbp.2011.018.
27. Rebetzke G, Ellis M (2007) Molecular mapping of genes for coleoptile growth in bread wheat (*Triticum aestivum* L.). Theoretical and Applied Genetics 114: 1173–1183. doi:10.1007/s00122-007-0509-1.
28. Spielmeyer W, Hyles J, Joaquim P, Azanza F, Bonnett D, et al. (2007) A QTL on chromosome 6A in bread wheat (*Triticum aestivum*) is associated with longer coleoptiles, greater seedling vigour and final plant height. Theoretical and Applied Genetics 115: 59–66. doi:10.1007/s00122-007-0540-2.
29. Schillinger WF, Donaldson E, Allan RE, Jones SS (1998) Winter Wheat Seedling Emergence from Deep Sowing Depths. Agronomy Journal 90: 582–586. doi:10.2134/agronj1998.00021962009000050002x.
30. Rebetzke GJ, Bruce SE, Kirkegaard JA (2005) Longer coleoptiles improve emergence through crop residues to increase seedling number and biomass in wheat (*Triticum aestivum* L.). Plant and Soil 272: 87–100. doi:10.1007/s11104-004-4040-8.
31. Tonkinson CL, Lyndon RF, Arnold GM, Lenton JR (1995) Effect of the *Rht3* dwarfing gene on dynamics of cell extension in wheat leaves, and its modification by gibberellic acid and paclobutrazol. Journal of Experimental Botany 46: 1085–1092. doi:10.1093/jxb/46.9.1085.
32. Pereira MJ, Pfahler PL, Barnett RD, Blount AR, Wofford DS, et al. (2002) Coleoptile length of dwarf wheat isolines: Gibberellic acid, temperature, and cultivar interactions. Crop Science 42: 1483–1487.
33. Radford B, Key A (1993) Temperature affects germination, mesocotyl length and coleoptile length of oats genotypes. Australian Journal of Agricultural Research 44: 677–688.
34. Kaufmann M (1968) Coleoptile length and emergence in varieties of barley, oats, and wheat. Canadian Journal of Plant Science 361: 357–361.
35. Addae P, Pearson C (1992) Variability in seedling elongation of wheat, and some factors associated with it. Australian Journal of Experimental Agriculture 32: 377–382. doi:10.1071/EA9920377.
36. Cornish P, Hindmarsh S (1988) Seed size influences the coleoptile length of wheat. Australian Journal of Experimental Agriculture 28: 521–524.
37. Ceccarelli S, Pegiati M (1980) Effect of seed weight on coleoptile dimensions in barley. Canadian Journal of Plant Science 60: 221–225.
38. Landjeva S, Karceva T, Korzun V, Ganeva G (2011) Seedling growth under osmotic stress and agronomic traits in Bulgarian semi-dwarf wheat: comparison of genotypes with *Rht8* and/or *Rht-B1* genes. Crop Pasture Science 62: 1017–1025.
39. Yu J-B, Bai G-H (2010) Mapping Quantitative Trait Loci for long coleoptile in chinese wheat landrace Wangshuibai. Crop Science 50: 43–50. doi:10.2135/cropsci2009.02.0065.
40. Rebetzke GJ, Appels R, Morrison AD, Richards RA, McDonald G, et al. (2001) Quantitative trait loci on chromosome 4B for coleoptile length and early vigour in wheat (*Triticum aestivum* L.). Crop and Pasture Science 52: 1221–1234. doi:10.1071/AR01042.
41. Wang J, Chapman SC, Bonnett DG, Rebetzke GJ (2009) Simultaneous selection of major and minor genes: use of QTL to increase selection efficiency of coleoptile length of wheat (*Triticum aestivum* L.). Theoretical and Applied Genetics 119: 65–74. doi:10.1007/s00122-009-1017-2.

8

Alfalfa Carbon and Nitrogen Sequestration Patterns and Effects of Temperature and Precipitation in Three Agro-Pastoral Ecotones of Northern China

Shujuan Chang[1,2]⁹, Nan Liu[1]⁹, Xiaoya Wang[1], Yingjun Zhang[1]*, Yue Xie[1]

1 Institute of Grassland Science, China Agricultural University, Beijing, People's Republic of China, **2** Inner Mongolia Rangeland Survey & Design Institute, Hohhot, Inner Mongolia, People's Republic of China

Abstract

Alfalfa (*Medicago sativa* L.) is a primary and widely cultivated forage crop in China. As a perennial leguminous grass, continuous planted alfalfa may influence carbon and nitrogen sequestration in soils. We evaluated the effect of alfalfa, planted for different lengths of time, and temperature and precipitation on soil organic carbon (SOC) and total nitrogen (TN) contents, and estimated soil SOC and TN inventories from 0–60 cm in three agro-pastoral ecotones of northern China. Alfalfa SOC and TN storage patterns were significantly different with increasing soil depths between the three regions of northern China. Continuous alfalfa grassland planted had a positive effect on accumulation of both SOC and TN in the Northwest region, whereas SOC storage peaked 6–7 years after planting in the Northeast and North region. Moreover, relatively higher TN storage appeared 7 years after planting in the Northeast and North regions. This study controlled as many factors as possible, but we caution that such temporal inferences could be artifacts of site selection. The regression analysis indicated that SOC and TN accumulation was mainly dependent on temperature (≥10°C of effective total accumulated temperature) in the North region. Precipitation in the growing season was the main limiting factor for SOC storage in the Northwest region and TN accumulation in the North regions. Therefore, the different climate factors affecting SOC and TN sequestration in alfalfa occurred at a regional scale.

Editor: Ben Bond-Lamberty, DOE Pacific Northwest National Laboratory, United States of America

Funding: This research was supported by the China Forage and Grass Research System (CARS-35) and the Beijing Key Laboratory of Grassland Science. The funders had no role in study design, data collection and analysis, decision to publish, or preparation of the manuscript.

Competing Interests: The authors have declared that no competing interests exist.

* E-mail: zhangyj@cau.edu.cn

⁹ These authors contributed equally to this work.

Introduction

Soil is a vital component of the global carbon cycle [1]. Soil organic carbon (SOC) is crucially significant in sustaining soil quality, crop production and environmental quality [2]. A sustainable soil C pool will mitigate increasing CO_2 levels in the atmosphere, and consequently have a substantial influence on the C content of the atmosphere [3–5]. How to enhance the potential of the soil C pool to sequester soil carbon, and thus alleviate the rate of increase of CO_2 concentrations in the atmosphere, has become an important research area globally.

Carbon sequestration is a biotic process. Carbon fixed by photosynthesis is the primary source of SOC, with the organic litter being decomposed and converted into soil organic matter [6]. In grassland ecosystems, more than 90% of the organic carbon is present in roots and soils [7]. Many studies have shown that changes in land-use are inevitably accompanied by variations in soil carbon storage [8–10]. A change of land-use from native vegetation to agriculture resulted in sharp declines in soil organic matter [11], and the conversion of annually cultivated land to forage grasses has the potential to increase C and N sequestration [12,13]. However soil carbon inventories and turnover rates are also influenced by climate, vegetation, parent material, topography, time, and the fundamental state factors of soil formation [14,15], and so soils developed under different vegetation types and climates may have different modes of soil organic matter stabilization [16]. Climate change and their interaction with vegetation have been acknowledged as one of the main driving forces of SOC variations [17]. Many authors, studying very different environments, have found that the SOC content is directly proportional to the mean annual precipitation and inversely proportional to the mean annual temperature [17–23]. The selection and operation of appropriate land management measures for mitigating atmospheric carbon concentrations will ultimately depend on knowledge of the carbon sequestration potential under various types of land-use in different ecosystems and biomes. At the same time, we need to consider potential compromises that may arise from changes in land-use with respect to food and fiber production that support global socio-economic activities [10]. Knowledge of soil carbon sequestration by particular plants in different spatial arrangements and temporal changes in that sequestration will enable better use of the land and achievement of optimum ecological and economic benefits.

Alfalfa (*Medicago sativa* L.) is a perennial leguminous grass that is widely planted through the world. In China alfalfa has been grown for more than 2000 years, primarily as a cultivated forage crop, and is widely distributed in northern regions. In 2009, it was estimated that the alfalfa-planted area in China reached 2.80 million hectares [24]. As a principal crop in legume-cereal rotations, alfalfa plays an important ecological and economic role in the agro-pastoral ecotones of China. Alfalfa is also a primary species used to restore vegetation in the National Western Development Program [25]. In China, many studies on alfalfa grassland converted by cropland showed that this land use change in favor for soil carbon sequestration, but most of those study sites were form the semiarid Loess Plateau, Northwest of China[13,25–26]. Investigating soil carbon dynamics under continuously planted alfalfa will also contribute to the taking of suitable measures to alleviate the rate of increase of CO_2 levels in the atmosphere.

The objective of the present study was to investigate at large regional scale spatial and temporal changes in the soil carbon pool under long-term plantings of alfalfa converted from annually cultivated land. To explain the differences in dynamics of the soil C and TN storage of alfalfa grassland from three main cultivation regions, temperature and precipitation data (China Meteorological Date sharing Service System: http://cdc.cma.gov.cn/) were used in stepwise multiple regression analysis to establish which factors may be responsible for the observed differences. Based on the above content, three questions are investigated. First, what are the differences among the three regions in carbon and nitrogen sequestration in soil? Second, which are the primary climatic factors in this process? And third, whether there is a threshold in the soil C and TN storage under alfalfa grassland?

Materials and Methods

Study site

The study area is shown in Fig.1. It covers the main cultivation regions of alfalfa grassland from east to west in the agro-pastoral ecotone of the north of China. Three sites were sampled along this gradient, with each region having different climate and soil types, but a similar cultivation history. One site (the Northeast region) was located near city of Qiqihar in the northeast of China, on a sandy black loam. The second site, located near city of Hohhot, in the Inner Mongolia Autonomous Region of China (the North region), was established on a light chestnut soil. The third site was located near city of Guyuan, which is on the Loess Plateau, in Ningxia province of northwest China (the Northwest region), on a light black loessial soil and a cultivated loessial soil.

The climate of the Northeast region is typified by a windy dry spring and cold winter with an average annual temperature of 4.5°C and mean annual precipitation of 437 mm according to the statistical data from 1985 to 2010. The climate of the North region has an average annual temperature of 7.5°C, and mean annual precipitation of 392 mm. The Northwest region has a moderate semiarid climate, that with an average annual temperature of 7.09°C, and mean annual precipitation of 421 mm (Fig.2).

Samples were selected from 17 fields in 2010, comprising six alfalfa grasslands from the Northeast region, six from the North region and five from the Northwest region with the samples from each region being from fields where alfalfa had been growing for different numbers of years (Table 1). In northeast China, the six alfalfa grasslands in Qiqihar were planted in different years in an area belonging to the Institute of Animal Science in Heilongjiang province. The six alfalfa grasslands from Hohhot of the North region, all planted in different years, belong to the Shaerqin forage resource field scientific observation station of the Grassland Research, Institute of the Chinese Academy of Agricultural Sciences. The five alfalfa grasslands from Guyuan in the Northwest region, all planted in different years, are mowed grassland that was planted by local farmers. All of the grasslands are cut two or three times per year, and followed maize crops.

Soil Sampling and Preparation

Soil samples were taken at the end of the grassland growth period on September 5, 2010 in the Northeast region, September 9th in the North and the 14th in the Northwest, respectively. Five sub-samples were randomly collected using a coring tube (diameter 5 cm) and mixed as one composite sample. The soil sampling was replicated five times for each field. Soil samples were collected from depths of 0–5 cm, 5–10 cm, 10–20 cm, 20–30 cm, 30–45 cm, and 45–60 cm. After air-drying, the soils were placed in self-sealing bags and returned to the laboratory for soil organic C (SOC) and total N (TN) analyses. Another three soil samples were collected from the same depth increments in the same fields using a cutting ring (volume of 100 cm^3) for measurement of the soil bulk density.

Measurements and Soil Analyses

Soil samples from the 17 fields were analyzed for soil bulk density, SOC, and TN. The soil samples for SOC and TN analyses were sieved through a 1 mm sieve. SOC concentrations were determined by the Walkley-Black method [27]. TN concentrations were determined using the Kjeldahl wet digestion procedure [28], using a 2300 Kjeltec Analyzer Unit (FOSS, Sweden). SOC and TN concentrations were multiplied by the soil bulk density to give the SOC and TN mass per unit area per 1 cm depth for the six sampled soil layers.

Statistical Methods

Regression analyses were used to investigate the relationship between SOC or TN storage and growing years in the three regions. Multiple stepwise regression was used to investigate the correlation of SOC and TN sequestration with ≥10°C of the effective total accumulated temperature, T_e, and total precipitation (mm) in the growing reason (April to October), P_g. The temperature and precipitation data were from China Meteorological Date sharing Service System: http://cdc.cma.gov.cn/. The analysis was carried out using SPSS (SPSS, 2000).

Results

Soil Bulk Density, SOC and TN Storage Changes

There were no changes in bulk density among soil layers in the Northwest region. Moreover, there was a trend of increasing bulk density with time since planting, except for the 17-year-old field. In the Northeast region, an increase in bulk density with the increase of soil depths appeared in the top 30 cm soil layer, and then declined. The soil bulk density was in the order of North>Northeast>Northwest (Fig. 3).

In the North region, SOC and TN storage was far higher in surface soil than in sub-layer soil, and decreased with increasing of soil depth. Compared with the sharp decrease of SOC and TN with the increase of soil depth in the North, SOC and TN storage had a relative slight decrease in the Northwest. In the top 30 cm, SOC and TN storage was increased with increasing of soil depth, but decrease from 30 to 60 cm soil layer in the Northeast. The SOC and TN storage were in the order of Northeast>North>Northwest (Fig. 4).The regression model (Fig. 5) showed the relationship between the growing years and SOC and TN storage on 0–60 cm layers in alfalfa grassland. In the North region, SOC storage was increased with increasing of growing years, and reached the maximal storage after growing six

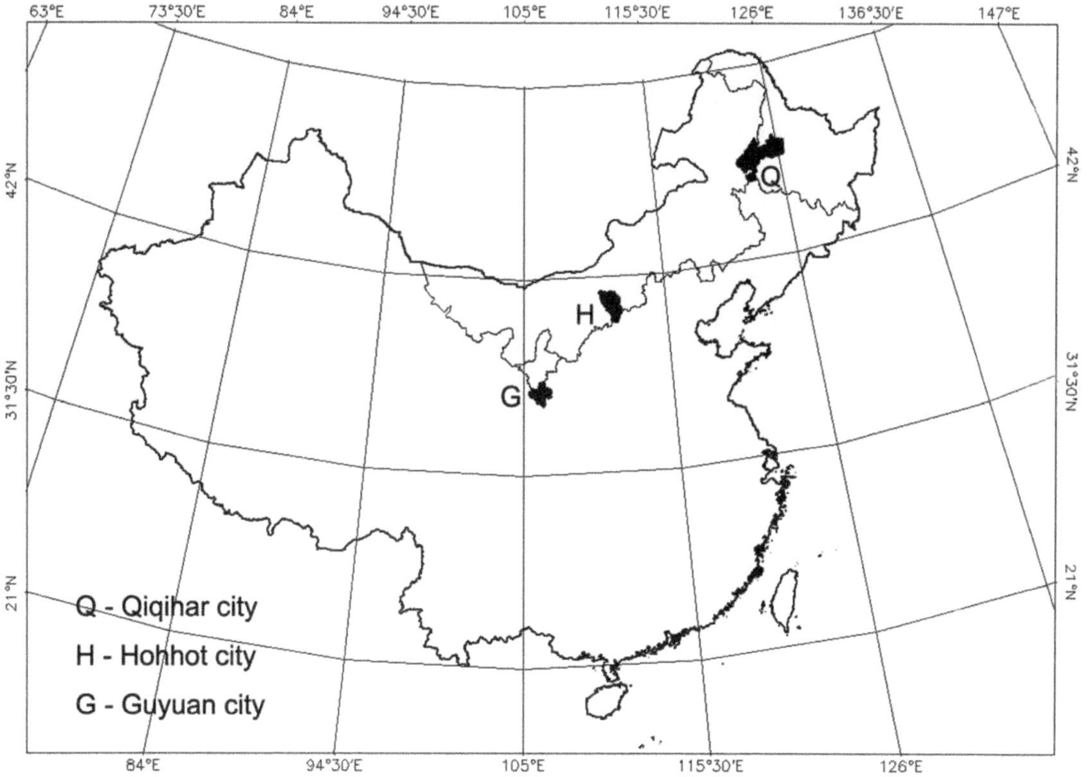

Figure 1. Location of the study areas in China.

years (41.11 Mg/ha), then begin to decline (n = 30, $P<0.05$; $R^2=0.495$). Similarly with TN storage, the maximal storage was appeared in 7th years of 5.99 Mg/ha in the North region (n = 30, $P<0.05$; $R^2=0.397$). In the Northeast region, the peak of SOC and TN storage was appeared in 7th years of 123.32 and 14.12 Mg/ha (n = 30, $P<0.05$; $R^2=0.299$ in SOC and 0.596 in TN). Soil carbon and nitrogen storage in the Northwest region had a significantly positive linear correlation relationship with the growing years (n = 25, $P<0.05$; $R^2=0.813$ in SOC and 0.775 in TN).

The Regression Model of SOC and TN Storage in Relation to Temperature and Precipitation

In the Northeast region, the stepwise regression analysis (Fig. 6) showed that SOC and TN storage cannot be quantitatively described by a combination of T_e or P_g. In the Northwest region, SOC storage increased with increasing T_e (n = 25, $R^2=0.819$, $P<0.001$), indicating that approximately 81.9% of the variability in SOC can be explained by variation in T_e. But TN storage can be quantitatively described by a linear combination of both T_e and P_g (n = 25, $R^2=0.856$, $P<0.001$). Similarly, in the North region, SOC storage was correlated with T_e in the 0–60 cm layer (n = 30, $R^2=0.484$, $P<0.05$), but TN storage was weakly correlated with P_g in the 0–60 cm layer (n = 30, $R^2=0.450$, $P=0.05$). However, the correlation coefficients are all significant. Meanwhile, the linear regression analysis indicates that soil C in the 0–60 cm layer is more sensitive to changes in temperature than TN storage in the North region soils. But the TN (0–60 cm depths) sequestration is more sensitive to changes in precipitation than effective accumulated temperature in the Northwest region.

Discussion

Dynamics of SOC and TN

SOC and TN storage patterns among the three regions were very different in alfalfa grassland, with the Northeast having the highest storage and the Northwest the lowest. Different background conditions together with differences in soil type and climate may cause the diversity of settings on SOC and TN storage in three regions. Soil carbon and nitrogen sequestration are long-term biotic processes that are affected by climate, plant species abundance and productivity, litter decomposition, soil structure, and agricultural practices [29–34]. Changes in SOC often lead to differences in soil total N, because organic forms generally account for more than 95% of the soil N [35].

With respect to differences in the length of time that alfalfa has been planted, the amount of carbon increased with an increase in the planting time, representing a considerable accumulation of SOC in these soils following conversion of annual crops to a continuous alfalfa system. But the increase trends with the increase in planting time had a threshold in the Northeast and North regions. The Northwest region had less carbon sequestration than the other regions. Meanwhile, in the Northwest region, soil bulk density was lower than the other region, with no significant difference between the six depths. The relative lower soil bulk density led to the lower capability on fertilizer and water conservation in the soil arable layer. This means that the conversion of annual cropland to the continuous alfalfa grassland in this poor soil may have greater potential on the accumulation of soil organic carbon and nitrogen. In the Northeast region, soil bulk density on 20–30 cm layers was significant higher than other layers, that caused by the plow pan profile on this layer. Long-term cultivation of a single crop in company with surface tillage will lead

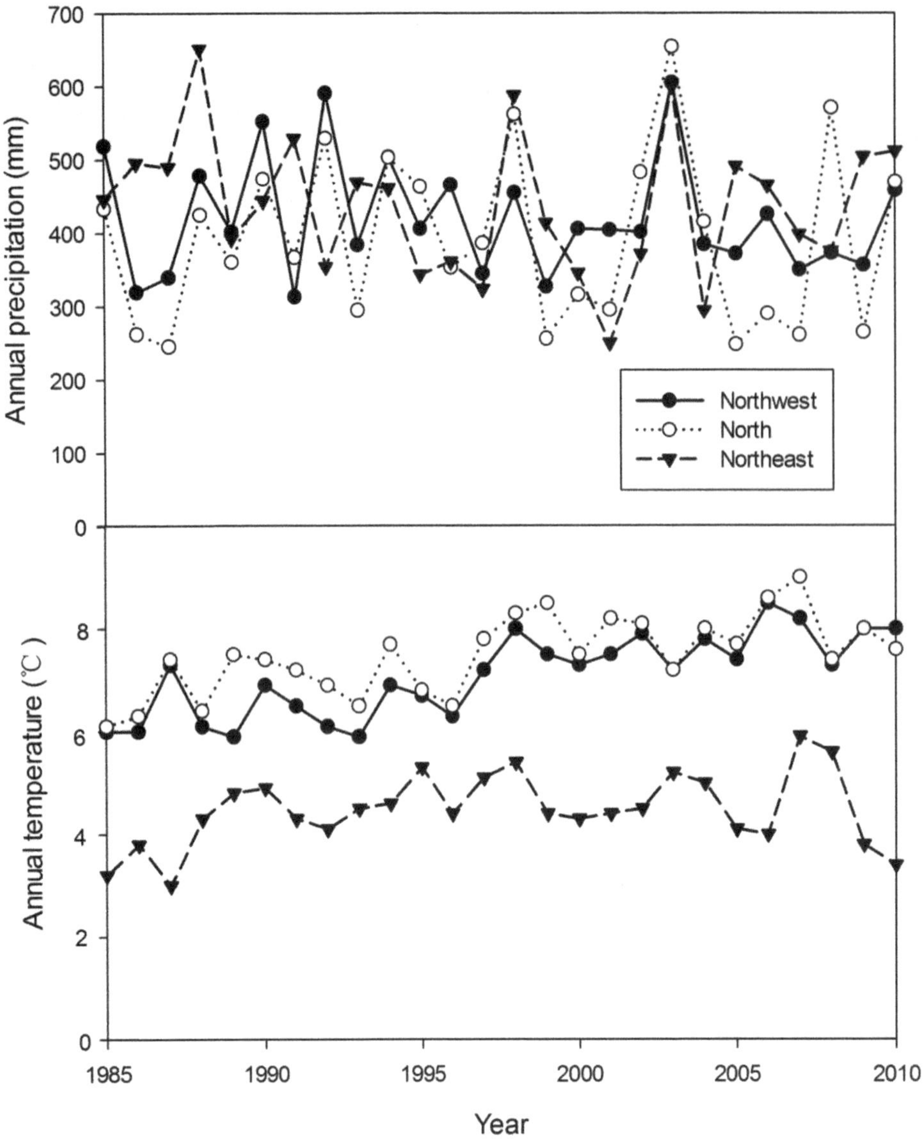

Figure 2. Annual precipitation (mm) and annual temperature (°C) in the three regions from 1985 to 2010. The data from China Meteorological Date sharing Service System: http://cdc.cma.gov.cn/.

Table 1. Details of soil samples from alfalfa grasslands growing for different numbers of years in the three regions.

(Northeast Region) Qiqihar		(North Region) Hohhot		(Northwest Region) Guyuan	
Ab.	**No. of growth years**	**Ab.**	**No. of growth years**	**Ab.**	**No. of growth years**
QM10	10-yr alfalfa grassland	HM9	9-yr alfalfa grassland	GM17	17-yr alfalfa grassland
QM8	8-yr alfalfa grassland	HM7	7-yr alfalfa grassland	GM10	10-yr alfalfa grassland
QM6	6-yr alfalfa grassland	HM5	5-yr alfalfa grassland	GM7	7-yr alfalfa grassland
QM4	4-yr alfalfa grassland	HM4	4-yr alfalfa grassland	GM4	4-yr alfalfa grassland
QM3	3-yr alfalfa grassland	HM2	2-yr alfalfa grassland	GM1	1-yr alfalfa grassland
QM2	2-yr alfalfa grassland	HM1	1-yr alfalfa grassland		
Latitude	47°15′43.70″		40°36′11.05″		36°11′41.32″
longitude	123°40′32.30″		111°45′24.45″		106°24′35.64″

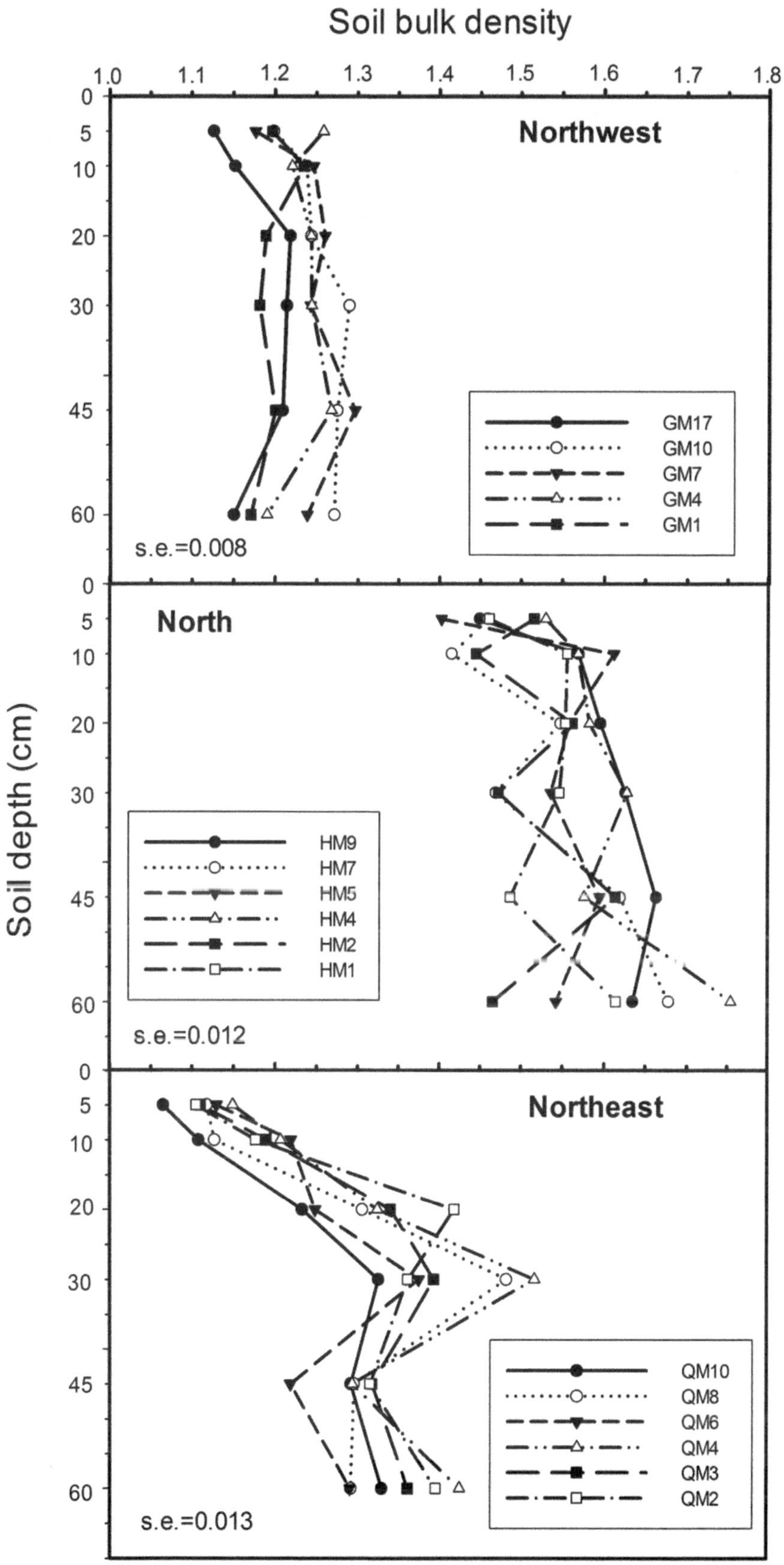

Figure 3. Mean soil bulk density in different soil layers from the three regions. The value of standard error (s.e.) for each region was stated in the graph (n = 90 for Northwest, n = 108 for North and Northeast region).

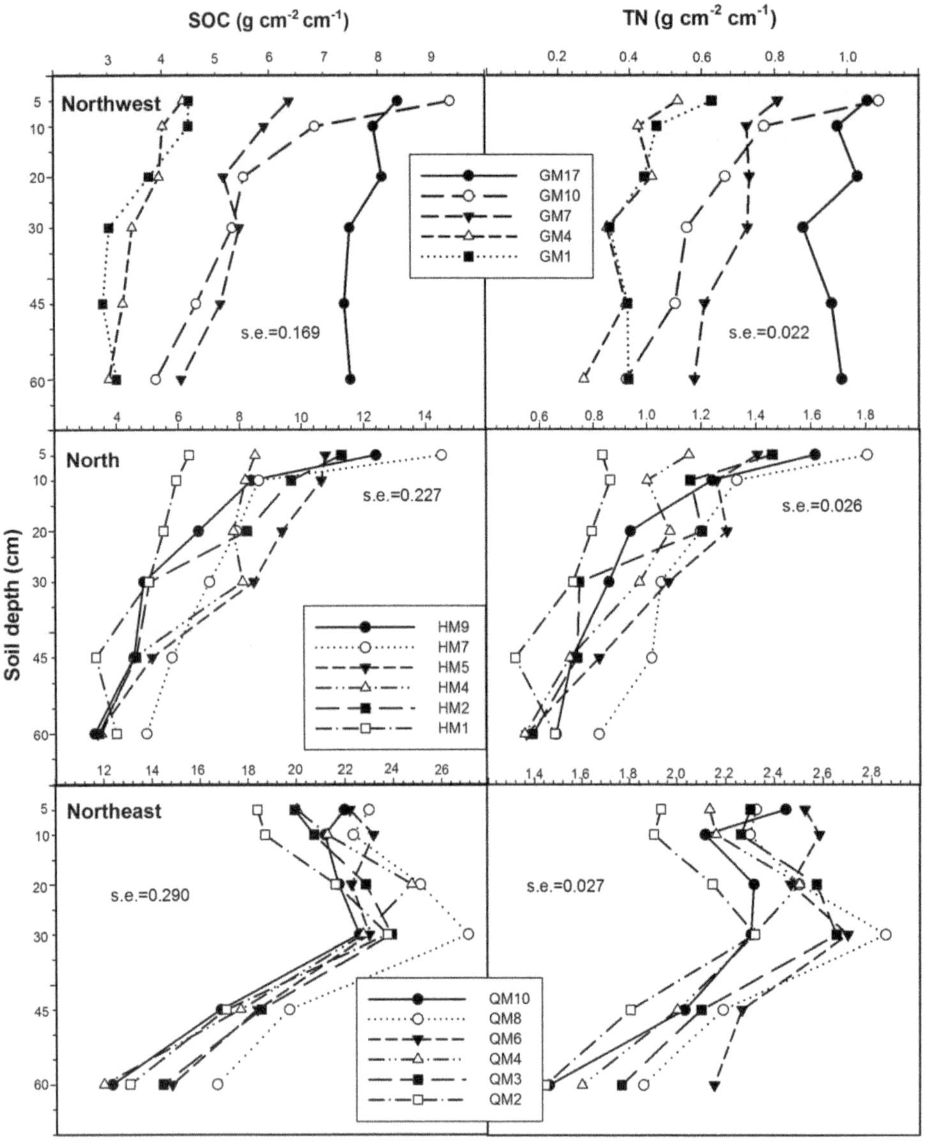

Figure 4. Mean SOC and TN storage in different soil layers from the three regions. The value of standard error (s.e.) for each region was stated in the graph (n = 150 for Northwest, n = 180 for North and Northeast region).

to thickening of the plow pan profile, soil compaction and bulk density increased [36]. Fluctuations of the SOC content in the Northeast region are gentler than in the other regions, indicated by smaller differences with increasing soil depth. In the North region, the SOC and TN contents in the 30–60 cm layer are significantly lower than in the 0–30 cm layer. This is possibly because the soil in the North region is a chestnut soil that has a calcic horizon at 30–60 cm [37]. The tight structure of calcic horizons influences root growth.

In recent years, there has been considerable research into changes in soil properties when annual cropland is converted to perennial forage. Such conversions have been shown to increase soil carbon and nitrogen contents [12–13,25]. However, Jiang et al. [26] considered that removal of large quantities of forage from alfalfa grassland systems through mowing, which is the continuous alfalfa system used in Northwest China, can exhaust soil water leading to severe soil dryness, and deplete SOC and soil nutrients in the 0–100 cm soil layer.

How to identify changes in organic matter is crucial in defining sustainable agricultural practices [38]. Changes in SOC storage in the three regions under the continuous alfalfa system are not consistent. In the North and Northeast region, SOC and TN storage changes did not show a consistent trend with the increase in the number of growing years. However, SOC storage in the Northwest region showed significant increases with increasing years of growing alfalfa. This is consistent with other research findings [13,39–40], which found considerable accumulation of SOC in soils where the land use had changed from annual crops to perennial grasses.

We used a chronosequence approach (space-for-time) to study changes of SOC and TN storage in alfalfa with different growing time. The increasing trends of increasing planting time had a threshold in the Northeast and North regions, so temporal results (e.g., strange decline in SOC after 6 years in Figure 5) are potentially confounded with site selection problems, such as spatial heterogeneity. Although we controlled for many factors, we

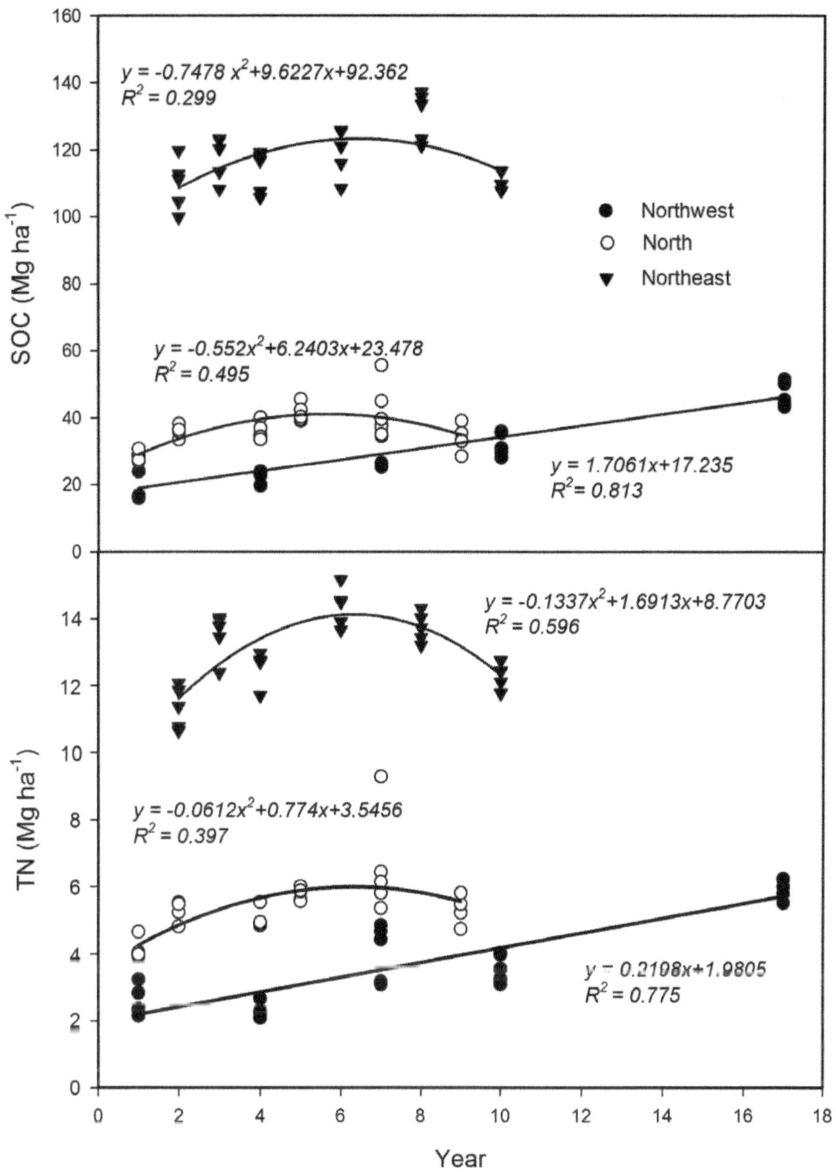

Figure 5. The relationship of soil organic carbon (SOC) and total nitrogen (TN) storage with years of growing alfalfa in North, Northeast and Northwest region of China.

caution that such temporal inferences could be artifacts of site selection. Walker et al. [41] suggested that if the date of the initial disturbance and subsequent history of the site are known, chronosequences provide the opportunity to study ecological processes over time periods that are longer than direct observation would permit. With careful site selection the chronosequence approach is thus an effective research method for evaluating.

The Relationship between Soil C and Temperature and Precipitation

Changes in temperature and precipitation may affect the soil and vegetation, with the soil affecting vegetation through its influence on water availability, elemental cycling and the soil temperature regime at an ecosystem level [42]. How to balance the effects between the temperature and the precipitation on SOC content determines what climatic factors in a specific area will be limited or decision of the soil carbon sequestration mode. In Canada, Fantappiè et al. [23] indicated that the most remarkable climate effect was that high air temperatures corresponded to low SOC values, probably because of SOC mineralization. High precipitation instead, when considered alone, had a positive effect on SOC content, but their interaction with temperature seemed to have more effect on SOC variations. In the Northeast region the accumulation of SOC and TN in the 0–60 cm layer does not agree with a regression on the regional temperature or precipitation. Precipitation is the main limiting factor in the Northwest region and the North region on TN sequestration. However, in the North region, the accumulation of SOC in the 0–60 cm layer is mainly dependent on the regional temperature. Temperature and precipitation together determine the changes in TN storage in the Northwest region. This suggests that other factors, like cultivation history, the type of cultivation, and soil properties, have more contribution to the changes of soil C and N in Northeast region. Nitrogen sequestration in soil is more sensitive

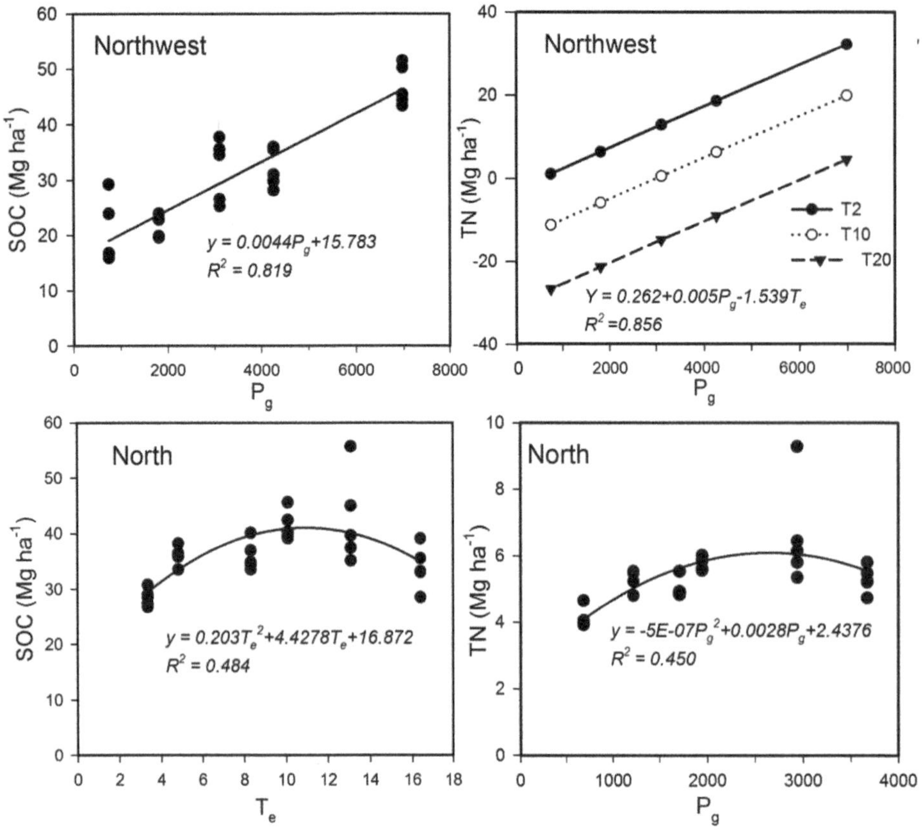

Figure 6. Relationships between SOC or TN (0-60cm) and ≥10°C of the effective total accumulated temperature (T_e) or total precipitation (mm) in the growing reason (P_g) in the North and Northwest region.

to changes in precipitation. Jenny [15] indicated a strong interdependence between climate and soil quality. Water is the key factor limiting plant growth in arid and semi-arid regions [43], and it also determines vegetation composition and distribution patterns [44,45].

Conclusion

Alfalfa SOC and TN storage patterns with increasing soil depths were significantly different between the three regions of northern China. Continuous alfalfa grassland planted had effective accumulation of both SOC and TN in the Northwest region. The peak SOC and TN storage were appeared in 6 and 7 years after alfalfa established in the North region, and both 7 years in the Northeast region. Temperature had effect on SOC storage in the North regions, whereas precipitation had more contribution to the SOC storage in the Northwest region and TN storage in the North region. Temperature and precipitation together determine the changes in TN reserves in the Northwest region. In short, SOC and TN storage has close relationship with the changes of temperature and precipitation, especially in the Northwest region.

Acknowledgments

We are grateful to the Institute of Animal Science in Heilongjiang province (Li Hong, E-mail: ; hljlihong@163.com), and the Shaerqin forage resource field scientific observation station of the Grassland Research Institute, Chinese Academy of Agricultural Sciences (Liu Lei; E-mail: ; liu4311755@163.com) for permission to use the study sites and research facilities. We are grateful to Huang YX, the owner of sites form city of Guyuan (Tel: 086-013995449140). All of study sites were got the permission. No specific permits were required for the described field studies. The location is not privately-owned or protected in any way. The field studies did not involve endangered or protected species. We wish to thank Zhou HS, Zhang XQ, Kan HM, Lin JH, Xiang M and Xu RX for their help with field sampling and data collection, and Jiang C for technical advice on the statistical analysis.

Author Contributions

Conceived and designed the experiments: YZ SC NL. Performed the experiments: SC NL. Analyzed the data: SC NL. Contributed reagents/materials/analysis tools: SC XW NL YX. Wrote the paper: SC NL.

References

1. Shi XZ, Wang HJ, Yu DS, David CW, Cheng XF, et al. (2009) Potential for soil carbon sequestration of eroded areas in subtropical China. Soil & Tillage Research 105: 322–327.
2. Doran JW, Parkin TB (1994) Defining and assessing soil quality. In: Defining Soil Quality for a Sustainable Environment, Soil Sci. Soc. Am, Madison, WI, USA. 3–21.
3. Lal R (2004) Soil carbon sequestration to mitigate climate change. Geoderma 123:1–22.
4. Bruce JP, Frome M, Haite E, Janzen H, Lal R, et al. (1999) Carbon sequestration in soils. Journal of Soil and Water Conservation 54: 382–389.
5. Upadhyay TP, Sankhayan PL, Solberg B (2005) A review of carbon sequestration dynamics in the Himalayan region as a function of land-use change and forest/soil degradation with special reference to Nepal. Agriculture Ecosystems and Environment 105: 449–465.
6. Aber J, Melillo J, Mcclaugherty C (1990) Predicting long-term patterns of mass loss, nitrogen dynamics, and soil organic matter formation from initial fine litter

chemistry in temperate forest ecosystems. Canadian Journal of Botany 68: 2201–2208.

7. Burke IC, Laurenroth WK, Milchunas DG (1997) Biogeochemistry of managed grasslands in central North America. In: Paul EA, Elliott ET, Paustian K, Cole CV, editors. Soil organic matter in temperature agroecosystems: long-term experiments in North America. CRC Press, Boca Rato, FL, 85–102.
8. Guo LB, Gifford RM (2002) Soil carbon stocks and land use change: a meta analysis. Global Change Biology 8: 345–360.
9. Grünzweig JM, Sparrow S, Yakir D, Chapin FS III (2004) Impact of agricultural land-use change on carbon storage in Boreal Alaska. Global Change Biology 10: 452–472.
10. Zhou ZY, Sun OJ, Huang JH, Li LH, Liu P, et al. (2007) Soil carbon and nitrogen stores and storage potential as affected by land-use in an agro-pastoral ecotone of northern China. Biogeochemistry 82: 127–138.
11. Wilson AT (1978) Pioneer agriculture explosion and CO2 levels in the atmosphere. Nature 273: 40–41.
12. Lal R, Follett RF, Kimble J, Cole CV (1999) Managing U.S. cropland to sequester carbon in soil. Journal of Soil and Water Conservation 54: 374–381.
13. Su YZ (2007) Soil carbon and nitrogen sequestration following the conversion of cropland to alfalfa forage land in northwest China. Soil & Tillage Research 92: 181–189.
14. Jenny H (1941) Factors of Soil Formation. McGraw–Hill, New York. 281.
15. Jenny H (1980) The Soil Resource: Origin and Behavior. Springer, New York.
16. Elliott ET (1986) Aggregate structure and carbon, nitrogen, and phosphorus in native and cultivated soils. Soil Science Society of American Journal 50: 627–633.
17. Jenny H (1980) The Soil Resource: Origin and Behavior, Ecological Studies 37. Springer-Verlag, New York. 377.
18. Kirshbaum MUF (1995) The temperature dependence of soil organic matter decomposition, and the effect of global warming on soil organic storage. Soil Biology and Biochemistry 27: 753–760.
19. Garten CT, Post WM, Hanson PJ, Cooper LW (1999) Forest soil carbon inventories and dynamics along an elevation gradient in the southern appalachian mountains. : Biogeochemistry, 45. Kluwer Academic Publishers, 115–145.
20. Jobbágy EG, Jackson RB (2000) The vertical distribution of soil organic carbon and its relation to climate and vegetation. Ecological Applications 10: 423–436.
21. Percival HJ, Parfitt RL, Scott NA (2000) Factors controlling soil carbon levels in New Zealand grasslands: is clay content important. Soil Science Society of America Journal 64: 1623–1630.
22. Lemenih M, Itanna F (2004) Soil carbon stocks and turnovers in various vegetation type and arable lands along an elevation gradient in southern Ethiopia. Geoderma 123: 177–188.
23. Fantappiè M, L'Abate G, Costantini EAC (2011) The influence of climate change on the soil organic carbon content in Italy from 1961to 2008. Geomorphology 135: 343–352.
24. Kang JM, Yang QC, GuoWS, Zhang TJ, Sun Y (2010) Yield evaluation of ten introduced alfalfa cultivars in beijing area of China. Chinese Journal of Grassland 32: 5–10.
25. Su YZ, Liu WJ, Yang R, Chang XX (2009) Changes in Soil Aggregate, Carbon, and Nitrogen Storages Following the Conversion of Cropland to Alfalfa Forage Land in the Marginal Oasis of Northwest China. Environmental Management 43: 1061–1070.
26. Jiang HM, Jin PJ, Yu J, Feng ML, Jin ZX (2006) Soil carbon pool and effects of soil fertility in seeded alfalfa fields on the semi-arid Loess Plateau in China. Soil Biology & Biochemistry 38: 2350–2358.
27. Nelson DW, Sommers LE (1982) Total carbon, organic carbon and organic matter. In: Page AL et al, editors. Methods of Soil Analysis, Part 2, 2nd edn. Agron. Monogr. ASA and SSSA Press, Madison, WI, USA, 102–129.
28. Gallaher RN, Weldon CO, Boswell FC (1976) A semi-automated procedure for total nitrogen in plant and soil samples. Soil Science Society of American Journal 40: 887–889.
29. Fu XL, Shao MA, Wei XR, Horton R (2010) Soil organic carbon and total nitrogen as affected by vegetation types in Northern Loess Plateau of China. Geoderma 155: 31–35.
30. Jimenez J, Lal R (2006) Mechanisms of C sequestration in soils of Latin America. Critical Reviews in Plant Sciences 25: 337–365.
31. Mcclaugherty CA, Pastor J, Aber JD, Melillo JM (1985) Forest litter decomposition in relation to soil–nitrogen dynamics and litter quality. Ecology 66: 266–275.
32. Posada JM, Schuur EAG (2011) Relationships among precipitation regime, nutrient availability, and carbon turnover in tropical rain forests. Oecologia 165: 783–795.
33. Post WM, Pastor J, Zinke PJ, Stangenberger AG (1985) Global patterns of soil–nitrogen storage. Nature 317: 613–616.
34. Schmitt A, Glaser B (2011) Organic matter dynamics in a temperate forest soil following enhanced drying. Soil Biology & Biochemistry 43: 478–489.
35. Stevenson FJ (1994) Humus Chemistry: Genesis, Composition, Reactions, 2nd ed. John Wiley, New York.
36. Wang CC, Chen CQ, Huang S, Deng AX (2010) Change characteristics of climatic and soil resources in Northeast China. Journal of Nanjing Agricultural University 33: 19–24.
37. Yan PJ, Wu T, Wang T (2011) Effects of calcic horizon on tree root distribution and growth in Wulanchabu city. Inner Mongolia Science Technology and Economy 235: 63–64.
38. Li XG, Li FM, Rengel Z, Bhupinderpal S, Wang ZF (2006) Cultivation effects on temporal changes of organic carbon and aggregate stability in desert soils of Hexi Corridor region in China. Soil & Tillage Research 91: 22–29.
39. Paustian K, Collins HP, Paul EA (1997) Management controls on soil carbon. In: Paul EA, Paustian KH, Elliott ET, Cole CV, editors. Soil Organic Matter in Temperate Agroecosystems: Long-Term Experiments in North America. CRC Press, Boca Raton, FL. 15–50.
40. Conant RT, Paustian K, Elliott ET (2001) Grassland management and conversion into grassland: effects on soil carbon. Journal of Applied Ecology 11: 343–355.
41. Walker LR, Wardle DA, Bardgett RD, Clarkson BD (2010) The use of chronosequences in studies of ecological succession and soil development. Journal of Ecology 98: 725–736.
42. Cheddadi R, Guiot J, Jolly D (2001) The Mediterranean vegetation: what if the atmospheric CO_2 increased? Landscape Ecology 16: 667–675.
43. Breshears DD, Myers OB, Johnson SR, Meyer CW, Martens SN (1997) Differential use of spatially heterogeneous soil moisture by two semiarid woody species: Pinusedulis and Juniperus monosperma. Ecology 85: 289–299.
44. Rutherford MC (1980) Annual plant production-precipitation relations in arid and semi-arid regions. South African Journal of Science 76: 53–56.
45. Skarpe C (1992) Dynamics of savanna ecosystems. Journal of Vegetation Science. 3: 293–400.

9

Ecoinformatics Reveals Effects of Crop Rotational Histories on Cotton Yield

Matthew H. Meisner[1,2]*, Jay A. Rosenheim[3]

1 Department of Evolution and Ecology, University of California Davis, Davis, California, United States of America, 2 Department of Statistics, University of California Davis, Davis, California, United States of America, 3 Department of Entomology and Nematology, University of California Davis, Davis, California, United States of America

Abstract

Crop rotation has been practiced for centuries in an effort to improve agricultural yield. However, the directions, magnitudes, and mechanisms of the yield effects of various crop rotations remain poorly understood in many systems. In order to better understand how crop rotation influences cotton yield, we used hierarchical Bayesian models to analyze a large ecoinformatics database consisting of records of commercial cotton crops grown in California's San Joaquin Valley. We identified several crops that, when grown in a field the year before a cotton crop, were associated with increased or decreased cotton yield. Furthermore, there was a negative association between the effect of the prior year's crop on June densities of the pest *Lygus hesperus* and the effect of the prior year's crop on cotton yield. This suggested that some crops may enhance *L. hesperus* densities in the surrounding agricultural landscape, because residual *L. hesperus* populations from the previous year cannot continuously inhabit a focal field and attack a subsequent cotton crop. In addition, we found that cotton yield declined approximately 2.4% for each additional year in which cotton was grown consecutively in a field prior to the focal cotton crop. Because *L. hesperus* is quite mobile, the effects of crop rotation on *L. hesperus* would likely not be revealed by small plot experimentation. These results provide an example of how ecoinformatics datasets, which capture the true spatial scale of commercial agriculture, can be used to enhance agricultural productivity.

Editor: Raul Narciso Carvalho Guedes, Federal University of Viçosa, Brazil

Funding: Funding Sources: 1. California State Support Committee of Cotton Incorporated. URL: http://www.ccgga.org/cotton_research/cssc.htm. 2. University of California Statewide IPM Program. URL: http://www.ipm.ucdavis.edu. 3. USDA-NRICGP (Grant 2006-01761). URL: http://www.csrees.usda.gov/funding/rfas/nri_rfa.html. 4. California Department of Pesticide Regulation. URL: http://www.cdpr.ca.gov. 5. National Science Foundation GRFP (Grant DGE-1148897). URL: http://www.nsfgrfp.org. The funders had no role in study design, data collection and analysis, decision to publish, or preparation of the manuscript.

Competing Interests: The authors have declared that no competing interests exist.

* E-mail: mhmeisner@ucdavis.edu

Introduction

Maximizing agricultural crop yield is an important goal for several reasons. First, a growing worldwide population will generate increased demand for agricultural resources [1]. Since expanding the land area devoted to agriculture is often unfeasible, or would involve the destruction of sensitive landscapes such as forests and wetlands, the only way to meet this demand will be to increase the crop yield generated from existing farmland. Second, there are substantial economic incentives for profit-seeking farmers to maximize the yield of their crops, especially given the low profit margins typical of commercial agriculture [2].

Farmers make a wide range of decisions regarding the management of their crops, involving pest management, planting/harvest dates, fertilization, irrigation, and, as we focus on in this study, crop rotation. These decisions are, along with external factors that fall outside farmers' control, such as weather, likely to affect crop performance and yield substantially. A rigorous quantitative understanding of the factors, including farmer management decisions, that affect crop yield is an essential prerequisite for developing management strategies that maximize yield.

A critical factor known to affect crop yield in a given field is the crop rotational history of that field [3]. There are several possible mechanisms by which the crops previously grown in a field can affect crop yield. First, different crops have different effects on the nutrient composition of the soil, so the identities of crops previously grown in a field can affect nutrient availability and crop yield [3]. For example, nitrogen-limited crops can benefit from rotation with nitrogen-fixing legumes [4], and phosphorus nutrition in California cotton is shaped by whether or not the previous crop received phosphorus fertilizer [5]. Second, certain crops may increase the local abundance of particular insect pests and pathogens [6–8]. Since different crops are often susceptible and resistant to different pathogens and pests, the identities of the crops recently grown in a field can affect yield. For example, if one crop increases local abundances of an insect pest that also attacks a second crop, planting the second crop immediately following the first may lead to decreased yield resulting from attack from the built up local pest population. In contrast, such a yield depression could potentially be averted if the second crop were planted following a crop that does not lead to local accumulation of the pest. In monocultures of wheat, substantial yield declines have been noted and attributed to the buildup of the soil-borne fungal pathogen *Gaeumannomyces graminis* [9]. Third, many studies have shown that a field's crop rotational history can strongly affect weed densities [10]. Numerous other mechanistic explanations for the yield effects of crop rotation have also been suggested [3].

Crop rotation has been practiced for thousands of years; evidence for its inception dates back to ancient Roman and Greek societies [11,12]. Experimental studies on the effects of crop rotation first appeared in the early 20th century, revealing that growing crops in rotation led to increased crop yields of up to 100% compared to continuous planting of a single crop [13,14]. Interest in the yield effects of crop rotation waned during the middle of the 20th century, due to the increasing availability of cheap fertilizers, insecticides, and herbicides [3,14]. However, crop rotation continues to be a relevant and important practice; low-input farming remains desirable due to the costs of fertilizers and pesticides, and fertilizer and pesticide applications can often not fully compensate for the benefits afforded by crop rotation [3]. In addition, the significant environmental and public health concerns surrounding fertilizer and pesticide use [1,15] highlight the desirability of methods of increasing crop yield through alternative methods such as crop rotation.

The effects of rotational histories on yield are well understood for some crops, such as corn, where rotation is recognized to be crticial in avoiding the buildup of corn rootworms [16]. However, for many crops, the direction, magnitude, and mechanism of the effect of crop rotational histories on crop yield remain poorly understood [3]. Cotton is one such crop. Experimental field studies of the effect of crop rotation on cotton yield have demonstrated increased cotton yield, compared to continuous cultivation of cotton, when cotton is grown in rotation with sorghum [17,18], corn [19], and wheat [20,21]. Despite these useful results, only a small subset of possible rotations has been studied, experiments have been restricted to plots significantly smaller than typical commercial cotton fields, and mechanisms for these effects remain poorly understood. To help address these limitations, we seek to expand upon this work by exploring the effects of crop rotational histories on yield in commercial cotton fields in California, using an "ecoinformatics" approach [22] capitalizing on existing observational data gathered by growers and professional agricultural pest consultants.

In recent years, there has been a surge in research and interest involving the rapidly emerging field of "big data." The big data movement has been fueled by several developments, including a dramatic increase in the magnitude of data generation, an improved ability to cheaply store, manipulate, and explore massive datasets, and the development of new analytic methods [23]. Most importantly, the movement has been driven by a growing realization that existing data, and data generated as a byproduct of our everyday lives, can be leveraged to explore key questions about nature and human behavior, even if the data were not collected for this purpose [24]. Ecoinformatics is a nascent field focused on harnessing the power of big data to address questions in environmental biology. Ecoinformatics approaches typically involve the analysis of large datasets, the synthesis of diverse data sources, and the analysis of pre-existing, observational datasets [22]. In some commercial agricultural settings, farmers, along with hired consultants, collect a great deal of regular data about their fields that are used to guide real-time crop management decisions, such as the timing of pesticide applications. By capitalizing on data that are already generated as a byproduct of commercial agriculture, ecoinformatics provides a low-cost means of obtaining a large dataset that can be used to explore key questions in agricultural biology, some of which might be too difficult or too costly to explore experimentally. Furthermore, the large size of datasets created for ecoinformatics can afford greater statistical power than could possibly be generated through experimental work.

Experimentally studying the yield effects of crop rotational histories is challenging for several reasons. There are a plethora of possible rotational histories, which means that a large number of treatments would be required to explore the space of possible rotational histories thoroughly. Furthermore, experimentally studying effects of crop rotations requires experiments spanning several growing seasons, which may be logistically challenging. Finally, in order to maintain realism and applicability to commercial fields, which are typically quite large, sizeable experimental plots would be required, especially in light of research suggesting that landscape composition as far as 20 km from a focal field can affect the densities of agricultural pests in that field [25]. While yield effects of non-mobile factors such as soil characteristics may be readily detected through small plot experimentation, the effects of highly mobile arthropods may only be detected at much larger spatial scales.

An ecoinformatics approach offers attractive solutions to these challenges. Since we analyze a large preexisting dataset that includes over a thousand records, a diversity of the possible crop rotational histories already exists in the dataset. In addition, our dataset spans 11 years of data, so the data span the temporal scale necessary to ask questions regarding effects of multi-year rotational histories. And, since the data come from the exact setting where we wish to apply our results, the data are realistic and capture the appropriate spatial scale of commercial agriculture.

First, we sought to identify which crop rotational histories are associated with increased and decreased cotton yield, and to quantify these yield effects. We then explored possible explanations for the yield effects identified in the previous step by examining the associations between crop rotational histories and pest abundance.

Materials and Methods

Dataset

The dataset was constructed by collecting existing crop records from commercial cotton fields in California's San Joaquin Valley. The data were shared by both growers and pest control advisors (PCAs), professional consultants hired to monitor field conditions and provide crop management recommendations. The dataset contains records of 1498 unique field-year instances from 566 unique fields, ranging from 1997 to 2008. Growers and PCAs collect and maintain detailed records of the conditions in their fields; numerous variables were recorded for each field-year record, and the following were used in our analyses:

1. Cotton yield. Measured once for each field-year instance, cotton lint yield was measured in bales/acre (converted to kg/ha for our analyses) and recorded for 1240 of the 1498 total records.
2. Crop rotational histories. The identity of the crop grown in the same field in previous growing seasons was recorded. For some fields, records extended back for 10 years. However, the vast majority of fields did not have records extending this far into the past. There were 15 unique crops that appeared in rotational histories: alfalfa, barley, carrots, corn, cotton, garbanzo beans, garlic, lettuce, melons, onions, potatoes, safflower, sugarbeets, tomatoes, and wheat.
3. Surrounding crops. For 1026 of the 1498 crops, we had data on the identity of the crop grown in each of the 8 fields immediately adjoining the focal field (to the North, Northeast, East, Southeast, South, Southwest, West, and Northwest).
4. Cotton variety. The database consisted of records of two different cotton species: *Gossypium barbadense* L. ("Pima cotton") and *Gossypium hirsutum* L. ("upland cotton").
5. *Lygus hesperus* densities. The plant bug *L. hesperus* is one of the most damaging pests of cotton, and a frequent target of

insecticide applications [26,27]. PCAs measured *L. hesperus* densities approximately weekly, primarily during June and July. The PCAs' sampling procedure consisted of 50 swings of a sweep net across the top of the plant canopy. Since not all PCAs sampled on the same days or at exactly the same intervals for all fields, we transformed successive samples into mean *L. hesperus* density estimates by calculating the area under the linear curve of *L. hesperus* density versus time and dividing by the number of days in the sampling interval.

Modeling approach

We employed a hierarchical Bayesian modeling approach, fitting linear mixed models to explore our questions about the effects of crop rotational histories on cotton yield. Mixed models combine the use of random effects and fixed effects, making them ideally suited for analysis of data that are structured, or clustered, in some known way, such that separate observations from within clusters are expected to be similar to one another [28]. When we model a source of clustering using a random effect, we assume that each cluster-specific parameter was drawn from a common distribution, and we estimate the parameters of this distribution from the data. We use this common distribution as the prior when calculating the posterior distribution of each cluster-specific parameter. The parameters (often called hyperparameters) of the distribution of cluster-specific parameters have posteriors that are estimated from the data, typically after assuming uninformative priors for the hyperparameters [28]. Using a common, empirical prior for all cluster-specific parameters allows pooling of information across clusters, so that data from all clusters can help inform estimates of every other per-cluster parameter. Assuming all clusters are the same introduces high bias and tends to underfit the data, whereas estimating fixed effects for each cluster introduces high variance and tends to overfit the data; however, using a random effect provides an optimal compromise between introducing bias and introducing variance [28]. In this dataset, there are several plausible sources of clustering.

1. First, we expect the data to be clustered by field, since there likely exist field-specific factors that affect yield, such as soil characteristics, local climate, and grower agronomic and pest management practices. We controlled for variable yield potential between fields by including field identity as a random effect in our models. Random effects allow pooling of information across clusters, so they are particularly useful when there are few observations from some clusters - a situation in which it is difficult to accurately estimate each per-cluster parameter with only the data from that one cluster [28]. Since there are three or fewer records for 78% of the fields in our database, we feel that including field as a random effect was preferable to trying to estimate field-specific fixed effects with very few observations per field.

   Additionally, including field as a random effect provides a straightforward way to make predictions for fields not represented in our database. Since modeling field as a random effect involves estimating a distribution of per-field parameters, we can simply sample a field-specific parameter from this distribution if we wish to make predictions about a previously unobserved field. Uncertainty in this field-specific parameter can be propagated by simulating many samples from this distribution, while simultaneously accounting for uncertainty in the parameters of this distribution. However, if we were to model field as a fixed effect, we would not estimate a distribution of field-specific parameters. We would only estimate parameters for the specific fields in our database, leaving us with no obvious way to make inferences about new fields.

2. Second, we expect that our data are clustered by year, since there is substantial between-year variability in climate, particularly in the winter and early spring. Climatic variables can affect crop performance, planting date, and insect pest populations, all of which can in turn affect cotton yield. To control for and quantify variation in yield due to year-specific factors, we included year as a random effect in our models. Our reasons for including year as a random effect are the same as those for field: there are few observations from some years, and we may wish to make predictions for future years not covered by the existing database.

All models were fit using a No-U-Turn Sampler variant of Hamiltonian Markov Chain Monte Carlo [29] implemented in Stan version 1.3.0, accessed through the rstan packing in R [30,31]. We ran three chains from random initializations, each with 10,000 samples, and discarded the first 5,000 samples from each as burn-in. Inferences were based upon the remaining 15,000 samples. We checked convergence by making sure that $\hat{R}$, an estimate of the potential scale reduction of the posterior if sampling were to be infinitely continued, was near 1 [32].

Models

Model 1. To explore the yield effects of the crop grown in the same field the previous year, we fit a linear mixed model with yield as the response variable. The predictor variable of primary interest was the identity of the crop grown in that field the previous year, which was included as a fixed effect.

Given that we are working with an observational dataset, a critical step in order to make meaningful inferences about the variable of primary interest - the crop grown the year before - was to control, to the extent possible, for potentially confounding variables that could generate spurious correlations and taint the validity of our inferences about crop rotation. To control for variable yield potential between fields and years, field and year were included in the model as random effects. The field terms control for the possibility that some fields may have higher yield potential due to their location, soil characteristics, or growing practices; the year terms control for the substantial year-to-year variation in cotton yield, which likely results from yearly weather differences. A term indicating cotton species (Pima or upland) was included in the model to account for yield differences between cotton species. Cotton species was modeled as a fixed effect, since there are only two possible categories - not enough to meaningfully estimate a random effects distribution [28]. We also included 15 real-valued fixed effect predictor variables that indicate the number of fields, out of the 8 surrounding fields, planted with each of the 15 crops we analyzed. The goal was to control for effects of the surrounding landscape, and thereby avoid spurious correlations between rotational history (which may be correlated with the crops surrounding the focal crop) and yield.

Our Bayesian modeling approach required the specification of priors for all parameters whose posteriors were estimated using MCMC. Noninformative priors (normal distributions with mean 0 and standard deviation of 100) were used for all fixed effects. The random effects for both field and year were assumed to follow a normal distribution with mean 0 (allowing means of these distributions to be estimated from the data would lead to nonidentifiability with the fixed effects for prior crop identity) and variance hyperparameters estimated from the data. Since the support of variance parameters is constrained to positive real numbers, noninformative inverse gamma distributions with shape and scale parameters set to 0.001 were used as the prior for the variance parameter of the top-level stochastic node, and as the priors for the variance hyperparameters of the field and year random effects distributions.

Model 2. To help us understand whether any effects of the crop grown in the field the previous year on cotton yield could be due to effects on *L. hesperus*, we fit the same model as Model 1, but with average June *L. hesperus* abundance as the response variable.

Model 3. Next, to formally assess whether there was an association between the effects of crop rotation on yield and the effects of crop rotation on *L. hesperus* abundance, we performed a linear regression of the estimated effects on yield (measured as the posterior means from Model 1) against the estimated effects on *L. hesperus* abundance (measured as posterior means from Model 2). Noninformative $\mathcal{N}(0,100^2)$ priors were used for the mean and intercept, and a noninformative inverse gamma distribution with shape and scale parameters set to 0.001 was used as the prior for the variance.

Model 4. A great deal of experimental evidence has demonstrated that crop rotation leads to increased yield compared to successive plantings of a single crop [3]; therefore, we explored whether or not a yield loss was incurred by cotton crops grown in fields where cotton was grown in previous years. For the 782 fields that had complete crop rotational records for the previous 4 years, we calculated the number of consecutive cotton plantings (from 1 to 4) in the 4 years preceding the focal cotton crop. We then fit a model, with yield as the response variable, using the number of consecutive prior cotton plantings as a predictor (again with the same noninformative prior of $\mathcal{N}(0,100^2)$). Field, year, and cotton type were included as they were in Models 1 and 2. Since the number of prior consecutive cotton plantings could be correlated with the number of cotton fields in the surrounding landscape during the focal year, we avoided a possible spurious correlation between consecutive cotton plantings and yield by also including a fixed effect for the number of cotton fields in the 8 fields adjacent to the focal field. We chose not to explore rotational histories of specific crops (and instead just grouped all crops into "cotton" or "not cotton") for longer than one previous year, since the number of possible rotational histories becomes very large and the number of records for each possible history becomes too small to allow for robust statistical analysis.

Model 5. To see if the number of consecutive years of cotton cultivation preceding the focal year was associated with June *L. hesperus* densities, we fit the same model as Model 4, but with June *L. hesperus* as the response variable.

Results

Model 1

Using our samples from the joint posterior of Model 1, we calculated, for each crop other than cotton, the posterior distribution of the difference in mean cotton yield in fields where that crop was grown the year before compared to mean yield in fields where cotton was grown the year before. The posterior means of these comparisons, as well as 95% highest posterior density intervals (HPDIs), are displayed in Figure 1A. Highest posterior density intervals are a Bayesian analogue of frequentist confidence intervals; they denote the narrowest region of parameter space containing 95% of the posterior probability [28]. Three crops had 95% HPDIs that did not overlap 0. Garlic (lower limit = 42.0 kg/ha, upper limit = 213.7 kg/ha), tomatoes (57.9 kg/ha 178.1 kg/ha), and melons (92.9 kg/ha, 793.7 kg/ha) had entirely positive 95% HPDIs, suggesting that previous cultivation of these crops was associated with increased cotton yield. While no crops had entirely negative 95% HPDIs, the posterior probability of safflower and sugarbeets having negative effects on yield was 96% and 95%, respectively, suggesting that cultivation of these crops the previous year was associated with decreased cotton yield. Yield was 153.0 kg/ha higher, with a 95% HPDI of (115.0 kg/ha, 192.8 kg/ha), for upland cotton than for Pima cotton.

Model 2

Using the joint posterior of Model 2, we calculated the posterior distribution of the difference in mean June *L. hesperus* densities between fields where cotton was the year grown before and where other specific crops were grown the year before. The posterior means of these comparisons, as well as 95% HPDIs, are displayed in Figure 1B. Corn (0.10 insects/sweep, 1.41 insects/sweep), onions (0.09 insects/sweep, 0.76 insects/sweep), and garlic (0.06 insects/sweep, 0.50 insects/sweep) all had 95% HPDIs that were entirely positive, suggesting that previous cultivation of these crops was associated with increased *L. hesperus* abundance. June *L. hesperus* density was 0.35 insects/sweep lower, with a 95% HPDI for this decrease of (0.26 insects/sweep, 0.44 insects/sweep), for upland cotton than for Pima cotton.

Model 3

While there were exceptions, we noticed that there was a trend for crops associated with increased pest abundances to also be associated with decreased yield. To more rigorously quantify this trend, for the 14 crops other than cotton, we regressed the posterior mean of the yield difference from cotton against the posterior mean of the *L. hesperus* difference from cotton. There was a negative slope with posterior mean −0.49 and 95% HPDI of (−1.16, 0.15) that marginally overlapped 0; the posterior probability of there being a negative slope was 93.4%. This provided evidence that crops associated with increased June *L. hesperus* densities were also associated with negative effects on yield (Figure 2).

Model 4

Model 4 suggested that every additional consecutive year of prior cotton cultivation in a field led to reduced cotton yield. Figure 3A displays the posterior distribution of the change in yield for each additional year that cotton was consecutively grown in the field prior to the focal year; the posterior mean for this change in yield was −40.9 kg/ha, with 95% HPDI (−57.5,−23.4 kg/ha). This translates to a mean of the percentage change in yield of −2.4% per year and 95% HPDI of (−1.4%,−3.4%) per year. We refit Model 4 without the term for consecutive cotton plantings; boxplots of the residuals are plotted against consecutive cotton plantings in Figure 3B, where a decreasing trend can be observed. Yield was 169.2 kg/ha higher, with a 95% HPDI of (123.4 kg/ha, 211.5 kg/ha), for upland cotton than for Pima cotton.

Model 5

Model 5 revealed a positive association between the number of preceding consecutive cotton plantings and June *L. hesperus* densities; the posterior mean of the slope regressing June *L. hesperus* on consecutive cotton plantings was 0.037 insects/sweep with a 95% HPDI that slightly overlapped 0 of (−0.007,0.079). The posterior probability of there being a positive relationship between consecutive cotton plantings and *L. hesperus* densities was 95.3%. June *L. hesperus* density was 0.32 insects/sweep lower, with a 95% HPDI of (0.20 insects/sweep, 0.43 insects/sweep), for upland cotton than for Pima cotton.

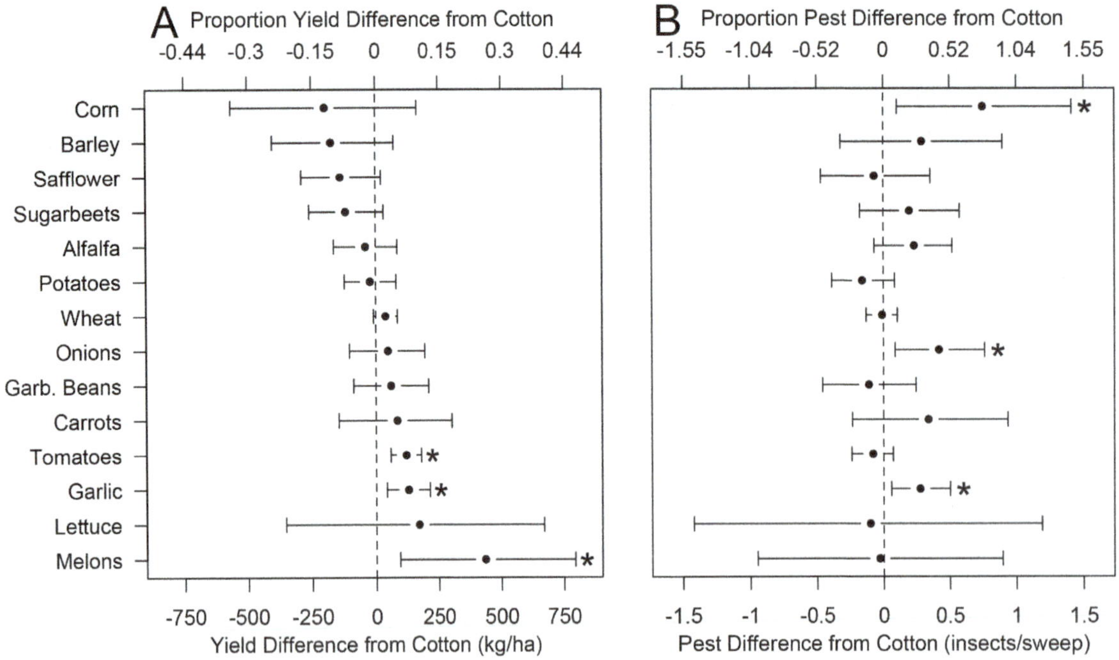

Figure 1. Means and 95% HPDIs of the differences in mean yield (A) and mean June *L. hesperus* density (B) between fields where a certain crop was grown the previous year and where cotton was grown the previous year. 95% HPDIs that do not overlap 0 are marked with a (∗).

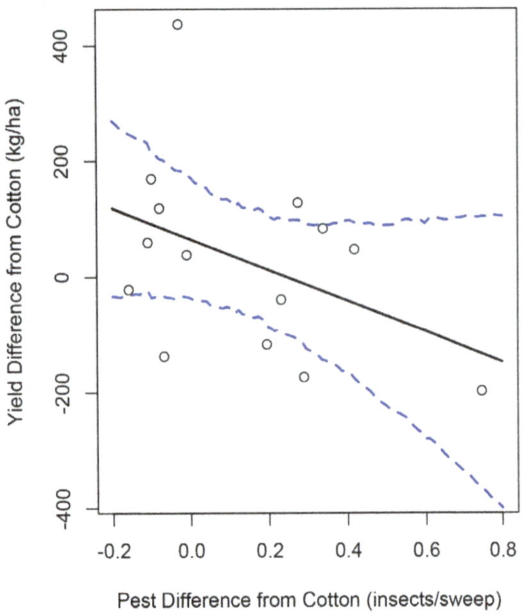

Figure 2. For each of the 14 crops other than cotton, we calculated the posterior mean of the mean difference in yield when that crop was grown in the field the year before compared to when cotton was grown in the field the year before (y-axis; these estimates are also displayed in Figure 1A). We also calculated the posterior mean of the difference in mean June *L. hesperus* densities between fields where a specific crop was grown the year before and where cotton was grown the year before (x-axis; these estimates are also displayed in Figure 1B). These estimates are plotted above (open circles). Then, we fit a linear model by regressing the mean yield differences on the mean *L. hesperus* differences. The posterior mean of the model fit (solid black) and 95% HPDI (dashed blue) are overlaid.

Discussion

Capitalizing on a large existing set of crop records from commercial cotton fields in California, we employed an ecoinformatics approach to explore the effects of crop rotational histories on cotton yield. Our hierarchical Bayesian analyses revealed evidence that several crops, when grown in the same field the year before the focal cotton planting, were associated with either decreased or increased cotton yield (Figure 1A), and either increased or decreased early season densities of the pest *L. hesperus* (Figure 1B). Furthermore, crops associated with decreased yield were generally also associated with increased *L. hesperus* densities, while those associated with increased yield were also associated with decreased *L. hesperus* densities (Figure 2).

These results suggest a possible mechanism for the observed yield effects of these rotational histories. Since *L. hesperus* preferentially attacks certain crops [33], a field cultivated with a crop that is heavily attacked by *L. hesperus* may, if *L. hesperus* disperse from the focal field, increase the abundance of *L. hesperus* in nearby fields. These populations may subsequently attack the crop planted in the focal field the following year, explaining the increase in early-season *L. hesperus* densities that we detected following certain crops. In turn, these increased *L. hesperus* populations may exert strong herbivorous pressure on focal cotton crops, possibly explaining the corresponding decrease in yield.

We believe that the effect of rotational history on early-season *L. hesperus* likely operates at a landscape scale that is larger than the within-field scale. If cotton was grown in a field the previous year, then farmers in the San Joaquin Valley are required to maintain a 90-day plant-free period prior to 10 March of the following year [27]. This prevents *L. hesperus*, which overwinter as adults on live host plants, from overwintering in a focal field where cotton was grown the year before. If a crop other than cotton was grown the previous year, then it could be possible for *L. hesperus* to overwinter in the focal field on residual plant or weed populations; however,

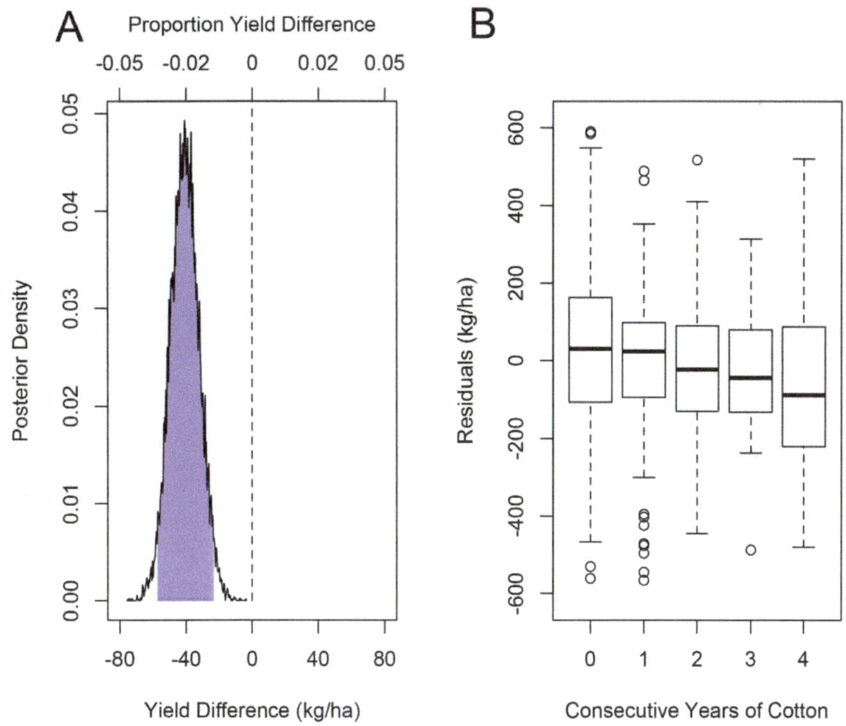

Figure 3. The posterior distribution of the change in yield for every additional year that cotton was grown consecutively in a field prior to the focal year (A). We refit the model without consecutive cotton plantings in the model and display boxplots of the residuals vs. consecutive cotton plantings, where a decreasing trend can be observed (B).

since fields are completely plowed prior to planting cotton in the spring, *L. hesperus* adults would still need to temporarily leave the focal field. Therefore, we believe that the preferred host crops for *L. hesperus* increase *L. hesperus* populations at a landscape scale. Then, when cotton, another target of *L. hesperus*, is planted in the same field the following year, the cotton field is attacked by this regional population. If regional populations are already large due to lingering effects from crops grown the previous year, *L. hesperus* populations may move into cotton early in the growing season; this could be particularly damaging given research suggesting that cotton yield is particularly sensitive to *L. hesperus* densities early in the growing season [34]. Using our data, we were not able to determine at exactly what scale the effects of rotation on *L. hesperus* likely operate. We do not believe a within-field scale is plausible, but determining a more precise spatial scale for these effects could be an interesting topic for future research.

Our findings match expectations of crop yield effects based on previous research on *L. hesperus* host crop preferences, lending support to our hypothesis that yield effects of crop rotational histories are, at least partially, mediated by effects on *L. hesperus*. Alfalfa and sugarbeets, both crops for which we found negative effects on yield and positive effects on *L. hesperus* when grown in a field the previous growing season, are all considered preferred hosts for *L. hesperus* [27], and have been shown to also increase *L. hesperus* populations in nearby cotton fields during an individual growing season [33,35]. Presumably, this effect is due to these crops supporting large *L. hesperus* populations. Large *L. hesperus* populations are known to build up in alfalfa [36], and their dispersal following alfalfa harvesting can threaten nearby cotton crops [27,37]. *L. hesperus* is also known to emigrate to nearby cotton fields when safflower begins to dry in mid-summer [38]. While the potential for nearby alfalfa [27,33] and safflower [33,35] fields to increase *L. hesperus* populations in cotton fields in a given year has been recognized, our results are the first indication that these landscape effects may extend temporally, affecting *L. hesperus* populations, and yield, in the next growing season. Tomatoes, associated with increased yield and decreased pest abundance in our data, have likewise been shown to decrease *L. hesperus* abundances in nearby cotton fields within a given year [33].

While previous experimental work has examined the effects of crop rotations on cotton yield [14,17,18,20,21], our work expands on these studies in several ways. First, we explore a much wider diversity of possible crop rotational histories, providing quantitative estimates of the cotton yield effects of cultivating 14 different crops the previous year. Second, since we analyze records from commercial cotton fields, our data have the potential to capture yield effects (such as those due to highly mobile arthropods) that could only be detected at this realistic spatial scale. Third, since we have collected data on pest abundances, not only yield, we have also been able to use our data to generate and build evidence for a hypothesized mechanistic explanation of the yield effects we identify.

We also found that farmers incurred a decline in cotton yield of about 2.4% for every additional year cotton was grown consecutively in a field preceding the focal season (Figure 3). This is consistent with previous research suggesting that continuous cultivation of cotton in the same location can reduce yield compared to interspersing cotton with other crops [14,17,18, 20,21]. We also found some evidence that the number of years cotton was grown consecutively in a field was associated with higher June *L. hesperus* densities: the posterior probability of there being a positive association was about 95%. Identifying the actual mechanism underlying this yield effect is beyond the scope of this study, but would be an interesting avenue for future research. It is possible that the yield decline is not caused by changes in *L. hesperus* densities, and instead results from the buildup of soil pathogens, especially in light of previous research showing that continuous cotton cultivation increases the densities of fungal pathogens in the soil [18].

When interpreting our results, it is important to remain cognizant of the challenges of drawing causal inferences from observational data. The key assumption required to make causal inferences from regression coefficients is that all variables that affect both the treatment assignment (crop rotation, in our analyses) and the response variable (yield and *L. hersperus* density, in our analyses) are included in the model; this ensures that the probability of receiving each treatment becomes, conditional on the predictor variables included in the model, conditionally independent of the response variable [28]. In experimental studies, the treatment assignment is typically controlled by the experimenter, so one can be confident that the only difference between treatment and control groups is in fact the treatment. However, in observational studies, it is impossible to prove definitively that there was no other factor that affected both the treatment assignment and the response variable (thus spuriously suggesting a treatment effect).

As such, we want to be very clear that our hypothesis that the effects of rotation on yield are mediated by effects on *L. hesperus* densities is exactly that - a hypothesis. While our data do support a negative *association* between effects on *L. hesperus* and effects on yield, we cannot prove with observational data that the varying effects on yield are *caused* by the varying effects on *L. hesperus*. This could be a fruitful topic for future experimental work.

Although causality is impossible to prove using observational data, ecoinformatics paves the way for implementing data-driven agricultural strategies and allows us to mine large datasets to explore important questions that are difficult to address experimentally. While by no means a replacement for experimentation, ecoinformatics can be a cost-effective and realistic complementary approach. In particular, our result identifying the effects of crop rotation on *L. hesperus* density would have been extremely difficult to reach experimentally. Since *L. hesperus* readily disperse across spatial scales of more than 1000 meters [37], an experimental study would have required massive plots comparable to the size of commercial fields in order to adequately capture their spatial dynamics.

Our results have numerous practical applications for commercial cotton growers. Growers with knowledge of the crop rotations associated with depressed cotton yield could make more informed decisions, selecting the sequence of crop cultivations that lead to maximized yield. When feasible, cotton plantings could be avoided following crops that decrease cotton yield, and instead limited to fields where crops that increase cotton yield were previously planted. In some cases, market conditions may lead a grower to plant cotton following a yield-depressing crop, even given the knowledge of likely yield loss. In those situations, our results may still be helpful, as an early warning sign of a potential pest problem in a particular field could allow the grower and PCA to focus pest detection efforts on that field and provide time to eliminate the problem before severe yield loss was incurred.

Our results suggest that the yield effects of crop rotational histories in cotton are relatively modest in magnitude: the posterior means for effects of any specific crop were mostly under 15%. However, given the tight profit margins of commercial agriculture, a 15% change in yield could translate into a far greater percentage change in profit, and could therefore be of substantial economic significance to a grower. As we seek to feed a growing worldwide population while doing minimal harm to the environment, crop management practices that increase yield while reducing the need for costly and damaging pesticides and fertilizers are of great value. Crop rotation is one such method, and we are optimistic that ecoinformatics approaches may be helpful in elucidating the details of how to optimally implement crop rotation.

Acknowledgments

We would like to sincerely thank the growers and PCAs who generously donated their data and time to help make this research possible, and R.F. Denison for valuable feedback on this manuscript.

Author Contributions

Conceived and designed the experiments: MM JR. Performed the experiments: MM JR. Analyzed the data: MM JR. Contributed reagents/materials/analysis tools: JR. Wrote the paper: MM JR.

References

1. Godfray HCJ, Beddington JR, Crute IR, Haddad L, Lawrence D, et al. (2010) Food security: the challenge of feeding 9 billion people. Science 327: 812–818.
2. Crookston RK (1984) The rotation effect: What causes it to boost yields? Crops Soils 36: 12–14.
3. Karlen DL, Varvel GE, Bullock DG, Cruse RM (1994) Crop rotations for the 21st century. Adv Agron 53: 1–45.
4. Russelle MP, Hesterman OB, Shaeffer CC, Heichel GH (1987) Estimating nitrogen and rotation effects in legume-corn rotations. In: Power JF, editor, The role of legumes in conservation tillage systems, Soil Conserv Soc Am. pp. 41–42.
5. Forbes AA, Rosenheim JA (2011) Plant responses to insect herbivore damage are modulated by phosphorus nutrition. Entomol Exp Appl 139: 242–249.
6. Benson GO (1985) Why the reduced yields when corn follows corn and possible management responses. Proc 40th Corn Sorghum Res Conf: 161–174.
7. Cook RJ (1984) Root health: Importance and relationship to farming practices. In: Bezdicek DF, editor, Organic Farming: Current Technology and its Role in a Sustainable Agriculture, Am Soc Agron. pp. 111–127.
8. Edwards JH, Thurlow DL, Eason JT (1988) Inuence of tillage and crop rotation on yields of corn, soybean, and wheat. Agron J 80: 76–80.
9. Rothamsted Research Center (2006) Guide to the classical and other long-term experiments, datasets and sample archive. Technical report, Rothamsted Research Center.
10. Liebman M, Dyck E (1993) Crop rotation and intercropping strategies for weed management. Ecol Appl 3: 92–122.
11. White KD (1970) Fallowing, crop rotation, and crop yields in Roman times. Agric Hist 44: 281–290.
12. White KD (1970) Roman farming. Ithaca: Cornell University Press.
13. Johnson TC (1927) Crop rotation in relation to soil productivity. J Am Soc Agron 19: 518–527.
14. Mitchell CC, Westerman RL, Brown JR, Peck TR (1991) Overview of long-term agronomic re-search. Agron J 83: 24–29.
15. Rosner D, Markowitz G (2013) Persistent pollutants: a brief history of the discovery of the widespread toxicity of chlorinated hydrocarbons. Environ Res 120: 126–133.
16. Peairs FB, Pilcher SD (2013) Western Corn Rootworm. Colorado State University Extension. URL http://www.ext.colostate.edu/pubs/insect/05570.html. Accessed 6 November 2013.
17. Bordovsky JP, Mustian JT, Cranmer AM, Emerson CL (2011) Cotton-grain sorghum rotation under extreme deficit irrigation conditions. Appl Eng Agric 27: 359–371.
18. Wheeler TA, Bordovsky JP, Keeling JW, Mullinix BG, Woodward JE (2012) Effects of crop ro-tation, cultivar, and irrigation and nitrogen rate on verticiullium wilt in cotton. Plant Dis 96: 985–989.
19. Mitchell CC, Delaney DP, Balkcom KS (2008) A historical summary of Alabama's Old Rotation (circa 1896): the world's oldest, continuous cotton experiment. Agron J 100: 1493–1498.
20. Constable GA, Rochester IJ, Daniells IG (1992) Cotton yield and nitrogen requirement is modi_ed by crop rotation and tillage method. Soil Tillage Res 23: 41–59.
21. Bordovsky JP, Lyle WM, Keeling JW (1994) Crop rotation and tillage effects on soil water and cotton yield. Agron J 86: 1–6.
22. Rosenheim JA, Parsa S, Forbes AA, Krimmel WA, Law YH, et al. (2011) Ecoinformatics for integrated pest management: Expanding the applied insect ecologist's tool-kit. J Econ Entomol 102: 331–342.
23. Streibich K (2013) Using big data to drive commercial advantage: excellent processes, excellent results. Database Netw J 43: 12–13.

24. Gobble MM (2013) Big data: The next big thing in innovation. Res-Technol Manage 56: 64–66.
25. O'Rourke ME, Rienzo-Stack K, Power AG (2011) A multi-scale, landscape approach to predicting insect populations in agroecosystems. Ecol Appl 21: 1782–1791.
26. Rosenheim JA, Steinmann K, Langellotto GA, Zink AG (2006) Estimating the impact of Lygus hesperus on cotton: the insect, plant, and human observer as sources of variability. Environ Entomol 35: 1141–1153.
27. Godfrey LD, Goodell PB, Natwick ET, Haviland DR, Barlow VM (2013) UC IPM pest management guidelines: cotton. University of California Division of Agriculture and Natural Resources. URL http://www.ipm.ucdavis.edu/PMG/r114301611.html. Accessed 17 September 2013.
28. Gelman A, Hill J (2009) Data analysis using regression and multilevel/hierarchical models. Cam- bridge: Cambridge University Press.
29. Hoffman MD, Gelman A (2013) The No-U-Turn sampler: Adaptively setting path lengths in Hamiltonian Monte Carlo. J Mach Learn Res: In press.
30. Stan Development Team (2013) Stan: A C++ library for probability and sampling, Version 1.3. URL http://mc-stan.org/. Accessed 17 September 2013.
31. Stan Development Team (2013) Stan modeling language user's guide and reference manual, Version 1.3. URL http://mc-stan.org/. Accessed 17 September 2013.
32. Gelman A, Rubin DB (1992) Inference from iterative simulation using multiple sequences. Stat Sci 7: 457–511.
33. Carriere Y, Goodell PB, Ellers-Kirk C, Larocque G, Dutilleul P, et al. (2012) Effects of local and landscape factors on population dynamics of a cotton pest. PLOS ONE 7: e39862.
34. Rosenheim JA, Meisner MH (2013) Ecoinformatics can reveal yield gaps associated with crop-pest interactions: a proof-of-concept. PLOS ONE 8: e80518.
35. Carriere Y, Ellsworth P, Dutilleul P, Ellers-Kirk C, Barkley C, et al. (2006) A GIS-based approach for areawide pest management: the scales of Lygus hesperus movements to cotton from alfalfa, weeds, and cotton. Entomol Exp Appl 118: 203–210.
36. Sevacherian B, Stern VM (1975) Movement of Lygus bugs between alfalfa and cotton. Environ Entomol 4: 163–165.
37. Sivakoff FS, Rosenheim JA, Hagler JR (2012) Relative dispersal ability of a key agricultural pest and its predators in an annual agroecosystem. Biol Control 63: 296–303.
38. Mueller AJ, Stern VM (1974) Timing of pesticide treatments on safflower to prevent Lygus from dispersing to cotton. J Econ Entomol 67: 77–80.

Legumes Can Increase Cadmium Contamination in Neighboring Crops

Ling Liu[1], Qian Zhang[1], Liangliang Hu[1], Jianjun Tang[1], Ligen Xu[1], Xiantian Yang[3], Jean W. H. Yong[2], Xin Chen[1]*

1 College of Life Science, Zhejiang University, Hangzhou, China, 2 Singapore University of Technology and Design, Singapore, Singapore, 3 Taizhou Academy of Agriculture Sciences, Taizhou, China

Abstract

Legumes are widely used in many cropping systems because they share their nitrogen fixation products and phosphorus mobilization activities with their neighbors. In the current study, however, we showed that co-cultivation with legumes increased cadmium (Cd) contamination in the adjacent crops. Both field and mesocosm experiments indicated that legumes increased Cd levels in edible parts and shoots of four neighboring crops and five maize varieties tested, regardless of the Cd levels in the soil. This enhanced Cd accumulation in crops was attributed to root interactions that alter the rhizosphere environment. Co-cultivation with legumes reduced soil pH, which somewhat increased the exchangeable forms of Cd. Our results have demonstrated the inevitable increases in Cd levels of crops as a direct result of co-cultivation with legumes even under situations when these levels are below the permissible threshold. With this new revelation, we need to consider carefully the current cropping systems involving legumes and perhaps to re-design the current and future cropping systems in view of avoiding food contamination by Cd.

Editor: Sunghun Park, Kansas State University, United States of America

Funding: The work was supported by the Natural Science Foundation of China (no. 30970477), the Natural Science Foundation of Zhejiang Province (no. Y307418) and the Key Laboratory of Ministry of Agriculture of the People's Republic of China (no. 278 KYJD09021). The funders had no role in study design, data collection and analysis, decision to publish, or preparation of the manuscript.

Competing Interests: The authors have declared that no competing interests exist.

* E-mail: chen-tang@zju.edu.cn

These authors contributed equally to this work.

Introduction

Legumes are good partner plants in natural plant communities because legumes can accumulate soil N and improve soil conditions for their neighbors [1]. Legume plants are widely used as nurse plants for degraded land restoration [2–4]. In agriculture, because legumes fix nitrogen and solubilize phosphorus, therefore they are often used to improve the growth of partner crops and are widely used in many intercropping systems and crop rotation cycles, especially in poor soils [5–7]. Legume-based cropping systems are considered the better way to increase productivity of the land and to maintain soil fertility [5]. In China, more than ten million hectares of legume-based intercropping or crop rotation are planted each year [7]. Legume-based cropping systems are also becoming common in other parts of the world such as India, Southeast Asia, South America, Central America, and Africa [8]. Legume-based cropping systems are also widely practiced in North America [9] and the European countries [10].

However, recent studies showed that legumes can also enhance Cd uptake of their neighboring plants. When co-planted with legumes maize can accumulate more Cd in its leaf, stem and root [11]. Legume can increase Cd concentration and total Cd accumulation of the co-cultivated tobacco [12]. Cadmium concentration in the grains was also highest in the wheat crop which was grown immediately after a crop rotation cycle with lupins as the predecessor [13]. As soil pollution by heavy metals is an inevitable problem associated with the economic and agricultural development in many countries, there are concerns that legumes enhance the risk of their neighboring crops growing in soils contaminated with varying levels of cadmium in any form of legume-based intercropping systems.

Therefore, we conducted field and greenhouse experiments to examine whether legumes increase Cd accumulation in the edible parts of their neighboring crops. We first compared Cd concentration in edible parts of four crops grown in monoculture and in mixture with three legumes through field experiments. Following which, we conducted a mesocosm experiment to further examine whether legume can increase Cd concentration in the edible parts by using three Cd levels in soil and five different maize varieties. In the microcosm experiment, we examined the changes in pH and concentration of exchangeable Cd fractions in rhizosphere when the crops are co-planted with legumes.

Results

Field study

In field experiments, yields of maize and tomato increased ($P<0.05$) but yields of cabbage and pakchoi did not changed ($P>0.05$) under mixed planting with legumes when compared to monocultures (Fig. 1).

When using Japanese clover as a partner crop, planting pattern caused significant changes ($P<0.01$) in Cd concentration of edible

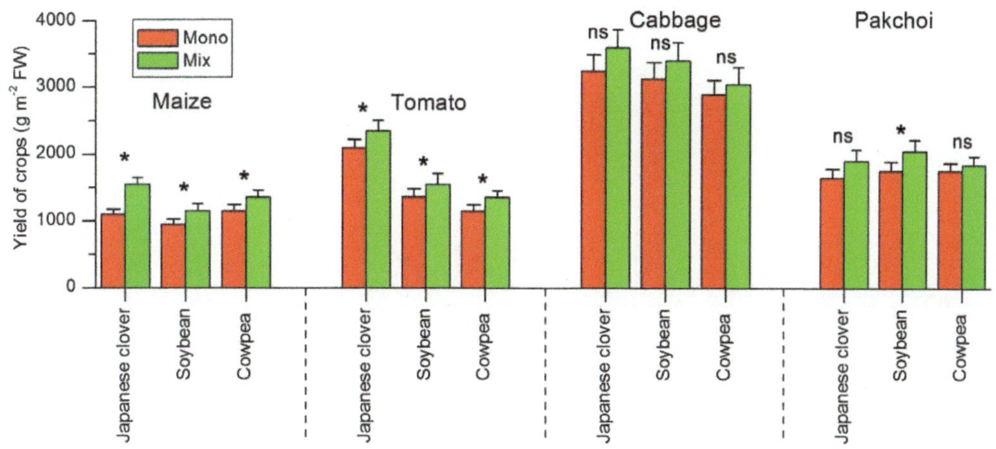

Figure 1. Yield of the edible parts of four crops grown in monoculture and mixture with Japanese clover, soybean and cowpea. Values are means (± SE) of four replicate plots. For each crop, values with asterisk (*) are significantly different (LSD at P<0.05).

parts for the four crops (maize, tomato, cabbage and pakchoi). Specifically, the levels of Cd were significantly higher when crop plants were planted with Japanese clover than when the crops were grown alone (Fig. 2A). For tomato, Cd concentration of edible part was below the maximum permissible concentration (MPC) standard of the National Standard Agency in China [14] when grown alone, but above the MPC when co-planted with legume (Fig. 2A). Using soybean as a partner crop, the mixed planting approach increased Cd concentration of the edible parts of maize, cabbage and pakchoi ($P<0.05$) with the exception for tomato ($P>0.05$), when compared to the plants in monoculture (Fig. 2A). When the main crops were grown with cowpea, the Cd concentration of edible parts of tomato and pakchoi increased ($P<0.05$), whereas Cd levels of edible parts of maize and cabbage remained at similar levels ($P>0.05$) (Fig. 2A).

Japanese clover, soybean and cowpea also significantly increased Cd accumulation in the shoots of maize ($P<0.05$, Fig. 2B). For tomato, soybean did not significantly increase ($P>0.05$), but Japanese clover and cowpea significantly increased ($P<0.05$) Cd accumulation in shoot (Fig. 2B). Effects of the three legumes on Cd accumulation in shoots of cabbage and pakchoi were similar to the effects in edible parts. (Fig. 2B).

There were no significant differences for the total and exchangeable Cd in soils between monoculture and intercropping after each experiment (Table S1).

Mesocosm study

In mesocosm experiment, soil Cd concentrations significantly impacted Cd concentration of edible part of the five maize varieties (for variety Zhengdan-958, $F_{2,20}=68.130$, $P=0.000$; for variety Suyunuo-2, $F_{2,20}=197.763$, $P=0.000$; for variety Huyunuo-2, $F_{2,20}=135.844$, $P=0.000$; for variety Jinzhumichaotian, $F_{2,20}=96.257$, $P=0.000$; for variety Shentian-1, $F_{2,20}=5.538$, $P=0.012$). Cd concentrations in edible parts increased with the soil Cd concentrations (Fig. 3A). While the planting patterns did not change Cd concentration in edible parts of variety Zhengdan-958 ($F_{1,20}=0.810$, $P=0.379$), planting patterns significantly altered the Cd concentration in edible parts of the other four varieties (for variety Suyunuo-2, $F_{1,20}=13.675$, $P=0.001$; for variety Huyunuo-2, $F_{1,20}=14.649$, $P=0.001$; for variety Jinzhumichaotian, $F_{1,20}=18.271$, $P=0.000$; for variety Shentian-1, $F_{1,20}=5.179$, $P=0.027$). Cd concentrations in edible parts of the four varieties were significantly higher under mixed planting with legume than that when growing under monoculture for all Cd treatments ($P<0.05$) (Fig. 3A).

Except for the variety Jinzhumichaotian (V4) at control and 0.8 mg kg^{-1} Cd treatment, Cd accumulation in shoots of the other maize varieties was significantly enhanced when co-planted with Japanese clover across the three soil Cd levels ($P<0.05$, (Fig. 3B).

Microcosm study

Soil pH of both crops (maize and tomato) was significantly influenced by planting patterns (for maize, $F_{3,12}=11.231$, $P=0.001$; for tomato $F_{3,12}=36.166$, $P=0.000$). Compared to monoculture, Soil pH significantly decreased when either maize or tomato were planted with Japanese clover and cowpea (Fig. 4A, $P<0.05$), but soil pH did not change when maize or tomato plants were co-planted with soybean (Fig. 4A, $P>0.05$).

Planting patterns also significantly impacted the levels of Cd exchangeable fractions in soil for both maize and tomato planting system (for maize, $F_{3,12}=10.507$, $P=0.001$; for tomato $F_{3,12}=23.700$, $P=0.001$). For maize, plants co-planted with Japanese clover and soybean ($P<0.05$) significantly increased the Cd exchangeable fraction in soil (Fig. 4B, $P<0.05$) but not with cowpea (Fig. 4B, $P>0.05$). All the three legumes increased the Cd levels of exchangeable fraction in rhizosphere soil when co-planted with tomato (Fig. 4B, $P<0.05$).

Discussion

In our field experiment when Cd concentration (0.283 mg kg^{-1}) in soil was under the permissible limiting concentration [15], the yields of all crops using the legume-crop co-planting approach generally increased when compared to monoculture. However, Cd levels of edible parts in all four crops (maize, tomato, cabbage and pakchoi) were higher when the crops were co-cultivated with legumes than when planted alone. The edible parts of tomato grown in monoculture had Cd levels which were below the maximum permissible concentration (MPC) standard of the National Standard in China [14], but these levels exceeded the MPC [14] when co-cultivated with legumes. Our mesocosm experiment further ascertained the fact that legume increased Cd accumulation in the edible parts of its neighbors. For the mesocosm experiment, co-planting with legumes increased Cd concentrations in edible parts of all five maize varieties regardless of the Cd levels (non-polluted, 0.047 mg kg^{-1} Cd; slightly

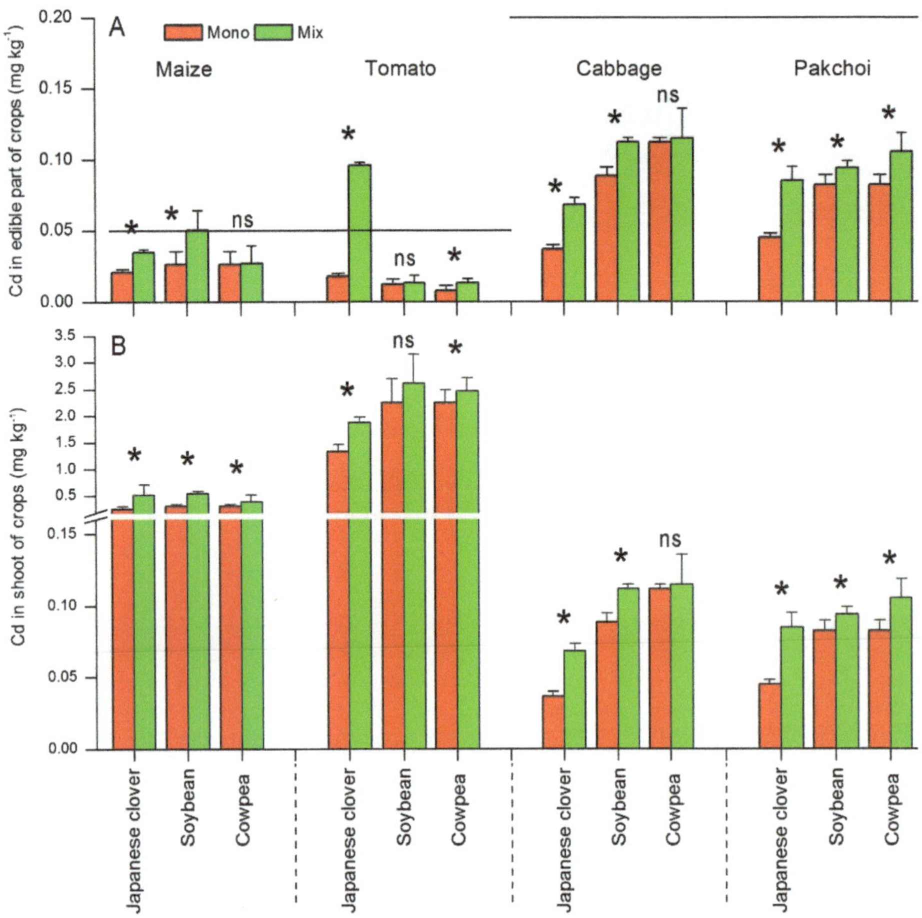

Figure 2. Cd concentrations in the edible parts (A) and shoots (B) of four crops grown in monoculture and mixture with Japanese clover, soybean and cowpea. The lower and upper horizontal lines indicate the Cd MPC for fruit-type and leafy vegetables, respectively. Values are means (± SE) of four replicate plots. For each crop, values with asterisk (*) are significantly different (LSD at $P<0.05$).

polluted, 0.3 mg kg^{-1} Cd; and moderately polluted, 0.8 mg kg^{-1} Cd) in the soil. These results suggested that legumes may enhance the risk of crop foods contaminated by Cd even though the Cd concentration in the soil is under the permissible limiting concentration.

There are two plausible mechanisms that could explain why legumes can enhance Cd accumulation in the edible parts of the neighbor crops. Firstly, legumes may increase the total Cd uptake of neighboring crops through root interaction. Secondly, legumes may alter Cd allocation to different parts of the crops [12] (Liu et al., 2011). In our field experiments, Cd accumulation in shoots of crops was significantly higher when co-planted with legumes compared to crops grown alone. Our mesocosm study on the five genotypes of maize also showed that legumes significantly increased Cd accumulation in shoots. These results suggested that the enhancement of Cd uptake by legumes played a major role in enhancing Cd accumulation in the edible parts of the crops.

Results from our microcosm study further suggested that root interactions between the co-existing crops could cause the significant increases in Cd accumulation. Plants can only absorb Cd in the exchangeable form, which usually constitutes a small portion of the total Cd content in soils [16]. However in the rhizosphere, there are two factors that could increase the exchangeable form of Cd. One is the presence of low molecular weight organic acids (e.g., acetic, oxalic, fumaric, citric, and tartaric acids) which form soluble complexes with Cd, thus releasing Cd from the soil matrix [17]. The other is attributed to the low soil pH values which desorbs the Cd from absorbed forms in soils [18,19]. Legumes are well known to reduce pH values because these plants exude large quantities of H^+ during the process of nitrogen fixation. Consequently, localized soil acidification occurs when crops are co-planted with legumes [5]. When maize or tomato was inter-planted with legumes (Japanese clover and cowpea) in our microcosm experiment, soil pH significantly decreased (Fig. 4) with a concomitantly increased in soil exchangeable Cd levels (Fig. 5). In our microcosm study, soybean did not reduce soil pH, but still increased the exchangeable Cd in rhizosphere soil of the soybean-crop system. At present, it remains unclear whether low molecular weight organic acids play a role in determining Cd exchangeability in the soil, and more research is needed to clarify this phenomenon.

Our study indicated that the enhanced Cd accumulation in neighbor crops (both the edible part and shoot) by legumes varied in specific legume-crop system. Cd accumulation in all partner crops used in the study was increased by Japanese clover. This could be attributed to the fact that Japanese clover can acidify rhizosphere soil and leading to higher levels of exchangeable Cd [12] (Fig.4 and Fig. 5). Although soybean did not reduce soil pH, it was able to enhance Cd accumulation in three crops but not for tomato. Contrary to soybean, cowpea reduced the pH in soils of all

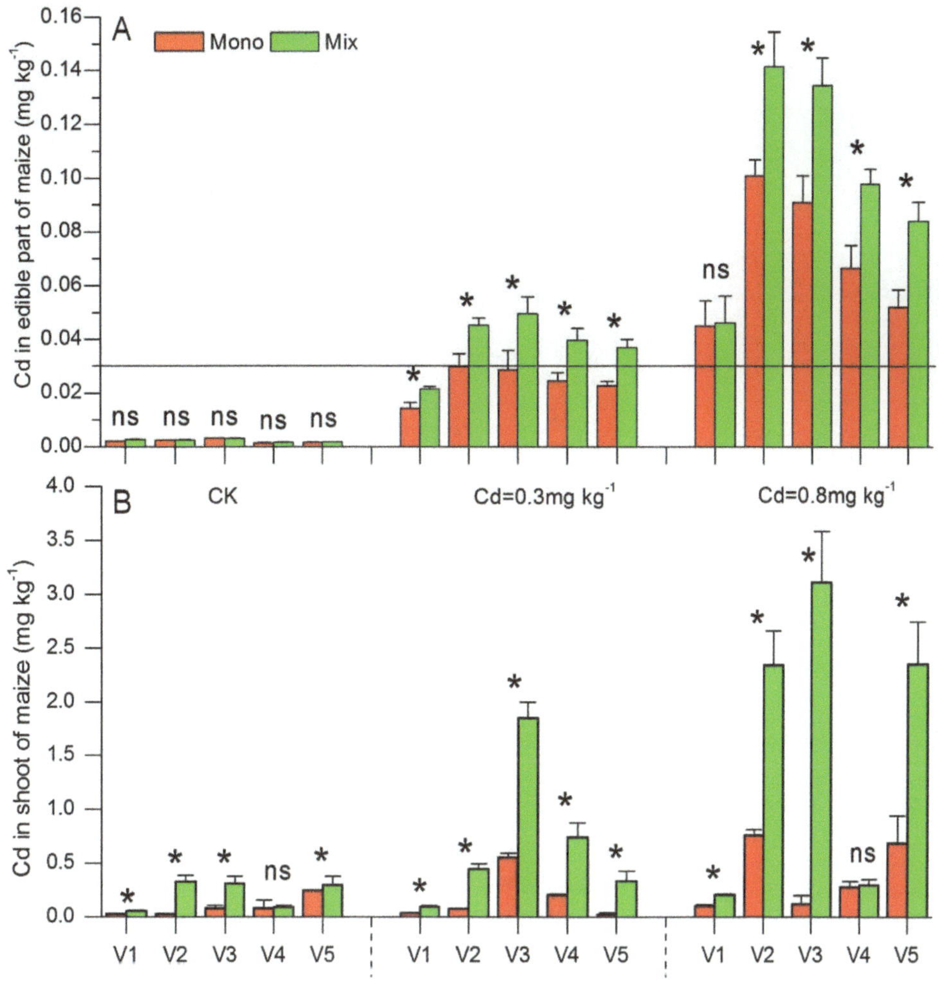

Figure 3. Cd concentrations in edible parts (A) and shoots (B) of five maize genotypes grown under monoculture and mixture. V1 (Zhengdan-958), V2 (Suyunuo-2), V3 (Huyunuo-2), V4 (Jinzhumichaotian), V5 (Shentian-1) indicate the different maize genotypes. The horizontal line indicates the Cd MPC for maize. Values are means (±SE) of four replications. Values with asterisk (*) are significantly different (LSD at P<0.05).

crops tested and affected Cd accumulation in tomato and pakchoi. Thus, the exact mechanism(s) governing Cd accumulation for each legume-crop pair needs further research.

Soil polluted by heavy metals and metalloids is an increasing problem in many developing countries, resulting from increased industrial activities and chemical fertilizer utilization [20]. In these countries, the increasing use of synthetic nitrogen fertilizers to drive higher crop yield has also resulted in soil N depletion [21]. Alternative approaches to producing more food while maintaining soil fertility are urgently needed in view of decreasing farmlands [22–26]. Legume-based intercropping has been widely promoted as one of the better alternate strategy to increase crop production and maintaining soil fertility [5,22]. However, our results do not support growing legumes in soils with elevated Cd levels even though these Cd levels are near to or below the pollution threshold as the co-cultivated legumes will biologically increase the Cd level in the adjacent crop. This new revelation suggests that the future design of cropping systems involving legumes should be carefully considered to avoid food contamination by Cd although legumes and crops vary in responding to a specific legume -crop system.

Materials and Methods

Field Study

Ethics statement. The field study was conducted in the area that corn and vegetable are main crops. It locates in Taizhou City (28°30′N, 121°22′E), Zhejiang Province, China. The field for the experiments is belong to the Long-term Experimental Station of Taizhou Academy of Agriculture Sciences, which is authorized by the government of Taizhou City, Zhejiang Province, China. The field is not privately-owned or protected. No specific permits were required for the described field studies. During the experiment, no other specific permission was required because we only carried out normal agricultural activities. Also our field study did not involve endangered or protected species.

Climate and soil. The climate in the field study area is typically of the subtropical monsoon with a mean annual air temperature of 18.6°C and a mean annual precipitation of 1693 mm. The soil was a sandy loam with a pH of 5.6. The soil contained 38.1 g kg^{-1} organic matter, 2.01 g kg^{-1} total nitrogen (N), 183.3 mg kg^{-1} extractable phosphorus (P), and 106.8 mg kg^{-1} extractable potassium (K) (extracted by 2 M HNO_3 and determined by flame atomic absorption spectrophotometry, flame-AAS). Total and exchangeable Cd concentrations in the soil was

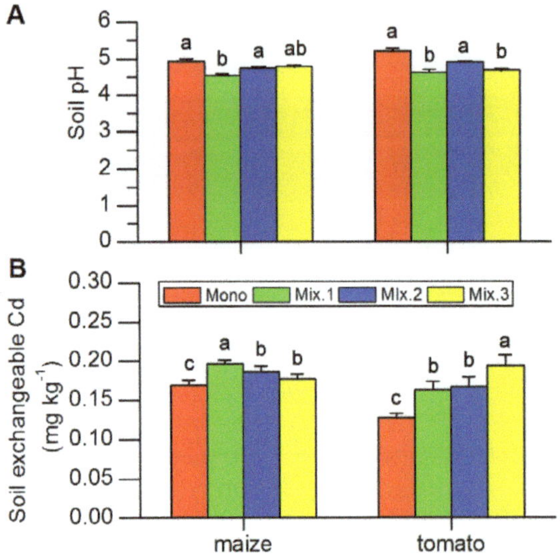

Figure 4. Soil pH (A) and Cd concentrations of exchangeable form (B) in soil grown under monoculture and intercropping. Mono.: crop monoculture. Mix.1: co-planted with Japanese clover. Mix. 2: co-planted with soybean. Mix.3: co-planted with cowpea. Values are means (±SE) of four replications. Values with different letters are significantly different (LSD at $P<0.05$).

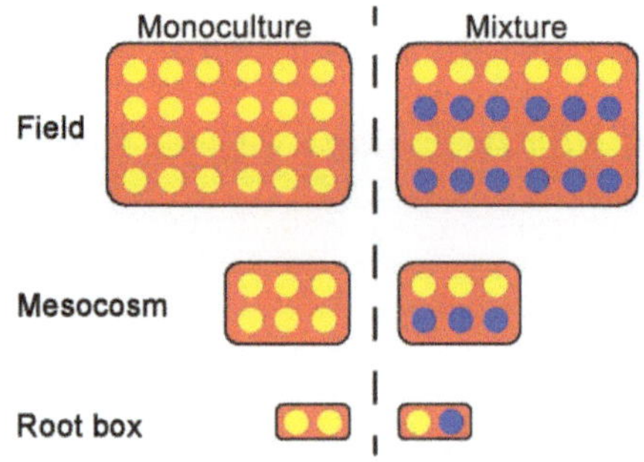

Figure 5. Experimental design. Yellow dot refers to the test crop plant; blue dot refers to the interplanted legume plant. For maize and tomato, row space and distance between plants within row were 45 cm and 30 cm respectively; for cabbage and pakchoi, row space and distance between plants within row were 35 cm and 20 cm respectively.

0.283 and 0.128 mg kg^{-1} respectively (determined by Inductively Coupled Plasma Mass Spectrometry, ICP-MS), which is considered to be slightly polluted but is below the limit for Cd as indicated by the Environmental Quality Standards set by the State Environmental Protection Administration for soils in China [15].

Legumes and crops. Four crops and three legumes that are cultivated extensively in the local region were used in the field study. The four crop species were maize (*Zea mays* L. var. Shentian-1), tomato (*Solanum lycopersicum* L. var. Zhongshu-4), cabbage (*Brassica oleracea* L. var. Jingfeng-1), and pakchoi (*Brassica chinensis* (L.) Makino. var. Youdonger-Hangzhou). The three legumes used as partner plants were Japanese clover [*Kummerowia striata* (Thunb.) Schindl.], soybean (*Glycine max*(L.)Merr.) and cowpea (*Vigna unguiculata* (Linn.) Walp.). Japanese clover is a common legume cover crop that is planted as forage, and soybean and cowpea are common legume crops in this area.

Experimental design. Three experiments were designed to test the effects of the three partner legumes on their neighboring crops at the same field (Fig. 5) in different years. In each experiment, there were nine treatments (one legume monoculture and four crop monoculture, and four interplanting). Each of the nine treatments was represented by four replicate plots (3×5 m per plot), and the 36 plots were arranged in randomized complete blocks.

The three experiments were conducted separately by using Japanese clover, or soybean, or cowpea as one of the intercrops in 2009, 2010 and 2011 respectively. In early March, the seeds of tomato, cabbage, pakchoi and legume partner were germinated and pre-cultured in the plastic mesh plate with vermiculite and peat in greenhouse under natural light and temperature. After 40 days, the pre-cultured seedlings were transplanted and the seeds of maize were sown simultaneously. Solid compound fertilizer (N:P:K = 15:15:15) was applied at 250 kg ha^{-1} in all plots for the duration of the experiment.

Measurements. Pakchoi was harvested 70 days after planting, and tomato and cabbage were harvested 95 days after transplanting, and maize was harvested 105 days after sowing. Edible parts of tomato and maize were harvested at their mature stage. Ten plants were randomly sampled in each plot for determination of crop yield and for quantification of Cd in shoots and the edible part. For pakchoi and cabbage, both the edible part and shoot are the leaves; for maize, the edible part is the ear without husk; for tomato, the edible part is the fresh fruit.

The edible parts of the crops were briefly washed with tap water and then with deionized water to remove surface dust and soil. After the fresh weights were recorded, the tissues (except the edible part of tomato) were dried at 80°C for 48 h and weighed. The fresh tomato fruit were processed in a blender and kept in clean polyethylene bags for metal analysis. Samples of edible crop parts that were used for the determination of Cd concentration were dried and ground to <0.25 mm in a stainless steel mill. Subsamples (1.0 g) of plant tissues were digested with concentrated HNO_3 and $HClO_4$ at a ratio of 4:1 (*V*/*V*), and the digest was adjusted to a final volume of 50 mL with deionized water. The concentrations of metals in the solutions extracted from the plant tissues were analyzed by ICP-MS (PerkinElmer, ELAN DRC-e, USA). The accuracy and reproducibility of the method for Cd in crop edible parts were tested by using a plant standard reference material Rice Powder GBW08502 [27]. The Cd content in the edible parts of crops was calculated on a per unit fresh weight basis by using the ratio of dry weight and fresh weight.

After each experiment, soil samples were collected from five locations in each plot. Total and exchangeable Cd concentrations in the soils were determined by Inductively Coupled Plasma Mass Spectrometry (ICP-MS).

Statistic analysis. In each legume experiment, one-way ANOVA was performed to determine the effect of planting patterns (monoculture or mixture with legume) on Cd concentration in the shoot and edible part for each crop, and on total and extractable Cd concentrations in soils after each experiment. The general linear model (GLM) in the SPSS V.10.0 (SPSS Inc., Chicago, USA) was used, and means were compared with the least significant difference at the 5% confidence level.

Mesocosm study

Mesoscosm experiments were conducted to verify that the effects of the legumes on neighboring crops to accumulate Cd were not by chance. The experimental soil was collected from the surface layer (0–20 cm) of an unpolluted vegetable field in Hangzhou City (28°54′N, 111°30′E), Zhejiang Province, in southeastern China. The basic properties of the soil were pH 5.9, organic matter content 38.6 g kg^{-1}, total N 2.49 g kg^{-1}, extractable P 98.03 mg kg^{-1}, extractable K 264.72 mg kg^{-1}, and total Cd 0.047 mg kg^{-1}. There were four replicates with five maize varieties, grown under two planting patterns (monoculture or mixture, Fig. 5). Japanese clover was used as partner legume. The five varieties were Zhengdan-958 (V1), Suyunuo-2 (V2), Huyunuo-2 (V3), Jinzhumichaotian (V4), and Shentian-1 (V5). Cadmium was added in the form of $3CdSO_4{\cdot}8H_2O$ and Cd treatments of 0.0, 0.3, or 0.8 mg Cd per kg of soil were used to simulate soil without Cd contamination, with slight Cd contamination, and moderate Cd contamination, respectively. The mesocosms were plastic containers that were 42 cm long, 32 cm wide and 20 cm deep. Each mesocosm was filled with 20 kg of well-mixed soil. Soil moisture content was adjusted to about 60% of water-holding capacity based on weight and the soil was placed in the mesocosm one month before planting to allow various sorption mechanisms and other processes to stabilize. During that month, water was added to maintain soil moisture at 60% of water-holding capacity. Mesocosms were arranged in a greenhouse in a randomized complete block design. Plants were maintained under natural light and temperature conditions with an average air temperature of 18–30°C during the course of the experiment from April to August. Plants were watered daily to maintain soil moisture at 70–90% of water-holding capacity. No additional nutrients were applied.

The edible part of each maize variety was harvested at the mature stage, and Cd concentrations were determined as described for the field experiment.

For each maize variety, two-way ANOVA was performed to determine the effects of planting patterns and soil Cd level on Cd concentration in the edible part. The general linear model (GLM) in the SPSS V.10.0 (SPSS Inc., Chicago, USA) was used, and means were compared with the least significant difference at the 5% confidence level.

Microcosm Study

To test the hypothesis that legumes enhanced Cd accumulation in their neighbors by changing the rhizosphere environment, we conducted a microcosm experiment by using the same three legumes mentioned in the field experiment. The two crops used were maize var. Shentian-1 and tomato var. Zhongshu 4, grown with three partner legumes (Japanese clover, soybean and cowpea) under two plant patterns (monoculture and mixture) (Fig. 5).

The soil used for the microcosm experiment was collected from the surface layer (0–20 cm) of field experiment location. The soil was collected, air dried, mixed, crushed, and passed through a 5-mm sieve. The basic properties and Cd contents of the soil were the same as the field experiment. Plastic root boxes of 22 cm length, 14 cm width, and 15 cm depth used in the microcosm experiment. Each root box was filled with 4 kg of well-mixed soil. Root boxes were arranged in the greenhouse in a complete randomized block design with four replicates. Plants were maintained under the conditions as described by the mesocosm experiment. No additional nutrients were applied.

Soil samples of 50 g were collected 15 days after the first flowering of the maize and tomato using a soil sampling tube from five locations in each microcosm. Soil pH was measured in a 1:5 (*W*/*V*) suspension of soil and deionized water. The chemical forms of Cd in soil were quantified by a sequential extraction procedure [27] and were analyzed by flame-AAS.

For each crop (maize or tomato), a two-way ANOVA was performed to determine the effects of legumes, planting patterns on soil pH and exchangeable forms of Cd in each rhizosphere soil of crops. The general linear model (GLM) in the SPSS V.10.0 (SPSS Inc., Chicago, USA) was used, and means were compared with the least significant difference at the 5% confidence level.

Acknowledgments

The authors thank Dr. Zhongchang Wu and Dr. Yu Liu at Zhejiang University for helping the measurement of Cd in plant tissue and soil.

Author Contributions

Conceived and designed the experiments: XC JY JT. Performed the experiments: LL QZ LH JT XY LX. Analyzed the data: LL QZ XC. Wrote the paper: XC JY QZ LH.

References

1. Siddique I, Engel VL, Parrotta JA, Lamb D, Nardoto GB, et al. (2008) Dominance of legume trees alters nutrient relations in mixed species forest restoration plantings within seven years. Biogeochemistry 88: 89–101.
2. Padilla FM, Pugnaire FI (2006) The role of nurse plants in the restoration of degraded environments. Front Ecol Environ 4: 196–202.
3. Padilla FM, Ortega R, Sanchez J, Pugnaire FI (2009) Rethinking species selection for restoration of arid shrublands. Basic Appl Ecol 10: 640–647.
4. Chaer GM, Resende AS, Campello EFC, de Faria SM, Boddey RM (2011) Nitrogen-fixing legume tree species for the reclamation of severely degraded lands in Brazil. Tree Physiol 31: 139–149.
5. Li L, Li SM, Sun JH, Zhou LL, Bao XG, et al. (2007) Diversity enhances agricultural productivity via rhizosphere phosphorus facilitation on phosphorus-deficient soils. Proc Natl Acad Sci USA 104: 11192–11196.
6. Ghosh PK, Bandyopadhyay KK, Wanjari RH, Manna MC, Misra AK, et al. (2007) Legume effect for enhancing productivity and nutrient use-efficiency in major cropping systems-an Indian perspective: a review. J Sustain Agr 30:59–86
7. Luo SM (2010) Principle and technology of agro-biodiversity utilization. Beijing: Chemical Industrial Press.
8. Vandermeer J (1989) The Ecology of Intercropping. Cambridge: Cambridge University Press.
9. Omondi EC, Ridenour M., Ridenour C, Smith R. (2010) The effect of intercropping annual ryegrass with pinto beans in mitigating iron deficiency in calcareous soils. J Sustain Agr 34: 244–257.
10. Hauggaard-Nielsen H, Gooding M, Ambus P, Corre-Hellou G, Crozat Y, et al. (2009) Pea-barley intercropping and short-term subsequent crop effects across European organic cropping conditions. Nutr Cycl Agroecosys 85: 141–155.
11. Li NY, Li ZA, Zhuang P, Zhou B, McBride M (2009) Cadmium uptake from soil by maize with intercrops. Water Air Soil Pollut 199:45–56.
12. Liu L, Li YF, Hu LL, Tang JJ, Chen X (2011) Plant coexistence can enhance phytoextraction of cadmium by tobacco (*Nicotiana tabacum* L.) in contaminated soil. J Environ Sci 23: 453–460.
13. Mench MJ (1998) Cadmium availability to plants in relation to major long-term changes in agronomy systems. Agr Ecosyst Environ 67: 175–187.
14. Ministry of Health of the People's Republic of China (2005) National Standards of the People's Republic of China: Maximum levels of contaminants in foods (GB2762–2005). Beijing: Standard Press.
15. State Environmental Protection Administration of China (1995) National Standards of the People's Republic of China: Environmental Quality Standard for Soils (GB15618–1995).Bejing: Standard Press.
16. Ahumada I, Mendoza J, Navarrete E, Ascar L (1999) Sequential extraction of heacy metals in Soils irrigated with Wastewater. Commun Soil Sci Plant Anal 30:1507–1519
17. Krishnamurti GSR, Cieslinski G, Huang PM, van Rees KCJ (1997) Kinetics of cadmium release from soils as influenced by organic acids: Implication in cadmium availability. J Environ Qual 26:271–277.
18. Yang ZF, Chen YL, Qian X, Guo L, Zhu HY (2005) A study of the effect of soil pH on chemical species of cadmium by simulated experiments. Earth Sci Front 12: 252–260.
19. Sappin-Didier V, Vansuyts G, Mench M, Briat JF (2005) Cadmium availability at different soil pH to transgenic tobacco overexpressing ferritin. Plant Soil 270: 189–197.
20. Nriagu J.O., PacynaJ.M., Quantitative assessment of worldwide contamination of air, water and soils by trace-metals. Nature, 333, 134 (1988)

21. Mulvaney RL, Khan SA, Ellsworth TR. Synthetic Nitrogen Fertilizers Deplete Soil Nitrogen: A Global Dilemma for Sustainable Cereal Production J Environ. Qual 38, 2295–2314 (2009)
22. Rao MR, Mathuva MN (2000) Legumes for improving maize yields and income in semi-arid Kenya. Agr Ecosyst Environ 78: 123–137.
23. Maeder P, Fliessbach A, Dubois D, Gunst L, Fried P, et al. (2002) Soil fertility and biodiversity in organic farming. Science 296: 1694–1697.
24. Godfray HCJ, Beddington JR, Crute IR, Haddad L, Lawrence D, et al. (2010) Food Security: The Challenge of Feeding 9 Billion People. Science 327: 812–818.
25. Bromley DW (2010) Food security: beyond technology. Science 329:169.
26. Liu L, Hu LL, Tang JJ, Li YF, Zhang Q, et al. (2012) Food safety assessment of planting patterns of four vegetable-type crops grown in soil contaminated by electronic waste activities. J Environ Manage 93: 22–30.
27. Tessier A, Campbell PGC, Bisson M (1979) Sequential extraction procedure for the speciation of particulate trace metals. Anal Chem 51:844–851.

11

Changes in N-Transforming Archaea and Bacteria in Soil during the Establishment of Bioenergy Crops

Yuejian Mao[1,2], Anthony C. Yannarell[1,2,3], Roderick I. Mackie[1,2,4*]

1 Energy Biosciences Institute, University of Illinois, Urbana, Illinois, United States of America, **2** Institute for Genomic Biology, University of Illinois, Urbana, Illinois, United States of America, **3** Department of Natural Resources and Environmental Sciences, University of Illinois, Urbana, Illinois, United States of America, **4** Department of Animal Sciences, University of Illinois, Urbana, Illinois, United States of America

Abstract

Widespread adaptation of biomass production for bioenergy may influence important biogeochemical functions in the landscape, which are mainly carried out by soil microbes. Here we explore the impact of four potential bioenergy feedstock crops (maize, switchgrass, *Miscanthus X giganteus*, and mixed tallgrass prairie) on nitrogen cycling microorganisms in the soil by monitoring the changes in the quantity (real-time PCR) and diversity (barcoded pyrosequencing) of key functional genes (*nifH*, bacterial/archaeal *amoA* and *nosZ*) and 16S rRNA genes over two years after bioenergy crop establishment. The quantities of these N-cycling genes were relatively stable in all four crops, except maize (the only fertilized crop), in which the population size of AOB doubled in less than 3 months. The nitrification rate was significantly correlated with the quantity of ammonia-oxidizing archaea (AOA) not bacteria (AOB), indicating that archaea were the major ammonia oxidizers. Deep sequencing revealed high diversity of *nifH*, archaeal *amoA*, bacterial *amoA*, *nosZ* and 16S rRNA genes, with 229, 309, 330, 331 and 8989 OTUs observed, respectively. Rarefaction analysis revealed the diversity of archaeal *amoA* in maize markedly decreased in the second year. Ordination analysis of T-RFLP and pyrosequencing results showed that the N-transforming microbial community structures in the soil under these crops gradually differentiated. Thus far, our two-year study has shown that specific N-transforming microbial communities develop in the soil in response to planting different bioenergy crops, and each functional group responded in a different way. Our results also suggest that cultivation of maize with N-fertilization increases the abundance of AOB and denitrifiers, reduces the diversity of AOA, and results in significant changes in the structure of denitrification community.

Editor: Jack Anthony Gilbert, Argonne National Laboratory, United States of America

Funding: This work was funded by the Energy Biosciences Institute, Environmental Impact and Sustainability of Feedstock Production Program at the University of Illinois, Urbana. The funders had no role in study design, data collection and analysis, decision to publish, or preparation of the manuscript.

Competing Interests: The authors have declared that no competing interests exist.

* E-mail: r-mackie@illinois.edu

Introduction

Bioenergy derived from cellulosic ethanol is a potential sustainable alternative to fossil fuel-based energy, since the energy from green plants is renewable and largely carbon neutral in comparison to fossil fuel combustion. Perennial grasses, such as switchgrass (*Panicum virgatum*) and *Miscanthus*×*giganteus*, with large annual biomass production potential, are proposed as biofuel feedstocks that can maximize ethanol production without adversely affecting the market for food crops (e.g. maize). However, our knowledge of the impacts of various bioenergy feedstock production systems on the soil microbial ecosystem is still very limited. The chemistry of perennial crop residues and plant root exudates may stimulate or inhibit the growth and activity of different fractions of the soil microbial community, and thus the planting of different crops can result in distinct microbial communities [1,2,3]. Differences in management techniques between traditional row-crop agriculture and perennial biomass feedstocks represent different soil disturbance regimes, altered water use, differing rates of fertilizer application, etc., and these should have a direct impact on soil microbial dynamics, subsequently influencing the terrestrial biogeochemical cycles. In particular, we predict that the cultivation of high nitrogen-use efficiency perennial grasses will result in altered nitrogen-transforming microbial communities in comparison to those found under N-fertilized maize.

The biological nitrogen cycle is one of the most important nutrient cycles in the terrestrial ecosystem. It includes four major processes: nitrogen fixation, mineralization (decay), nitrification and denitrification. Because many of the microorganisms responsible for these processes are recalcitrant to laboratory cultivation, previous studies of the distribution and diversity of nitrogen-transforming microorganisms have employed cultivation-independent techniques targeting functional genes: *nifH*, *amoA* and *nosZ* genes, which encode the key enzymes in nitrogen fixation, ammonia oxidization and complete denitrification, respectively [4,5,6,7,8,9].

Biological nitrogen fixation, which converts atmospheric N_2 into ammonium that is available to organisms, is an important natural input of available nitrogen in many terrestrial habitats [10]. Although nitrogen fixation in terrestrial ecosystems is thought to be mainly carried out by the symbiotic bacteria in association with plants, free-living diazotrophs in soils can play important roles in N cycling in a number of ecosystems [11,12]. In average, 2–3 kg N ha^{-1} $year^{-1}$ could be imported by free living N-fixers [13]. Various field experiments have shown that the biomass yield of one candidate biofuel feedstock crop, *Miscanthus×giganteus*, is not significantly increased by the addition of mineral N fertilizer [14]. The lack of response to nitrogen fertilization and the high biomass production suggest that biological nitrogen fixation may play an important role in supplying the nitrogen needs of *Miscanthus* [15]. Plant species have previously been shown to have a significant effect on the composition of diazotrophs in the field; for example, diazotroph diversity is higher in soil under *Acacia tortilis* ssp. *raddiana* (a leguminous tree) than *Balanites aegyptiaca* (a non-leguminous tree) [16]. Plant genotype also has a strong effect on the rhizosphere diazotrophs of rice [17]. Agronomic practices can also influence soil diazotrophs, e.g. application of N-fertilization can reduce the diversity of diazotrophs [17]. Therefore, we hypothesize that the cultivation of maize with inorganic N-fertilizer will reduce the abundance and diversity of diazotrophs in the soil ecosystem, while biofuel feedstocks receiving little or no N-fertilizer (e.g. *Miscanthus*) will encourage the development of active diazotrophic communities.

Nitrification, which converts ammonium to nitrate, includes two steps: ammonia oxidation to nitrite, and nitrite oxidation to nitrate. The production of nitrate in soil not only supplies nutrition for plants, but it can also mobilize nitrogen to groundwater through nitrate leaching. Ammonia oxidation is the first and rate-limiting step of nitrification [18]. It is typically thought to be carried out by a few groups in β- and γ- Proteobacteria, referred to as ammonia-oxidizing bacteria (AOB) [18]. However, recent environmental metagenomic analyses revealed that ammonia monooxygenase α-subunit (*amoA*) genes are also present in archaea (AOA) [19], and archaeal *amoA* has been shown to be widespread in many environments, e.g. soils, hot springs and marine water [6,19,20,21,22]. Recent work has found that AOA can be up to 3000 times more abundant in soil than AOB [22,23,24], meaning that AOA are the most abundant ammonia oxidizing organisms in soil ecosystems [25]. The soil ammonia oxidizing community is known to be influenced by plant types and management, but different segments of this community respond differently [24,26,27]. For example, the abundance and composition of AOB is significantly altered by long-term fertilization, but AOA are rarely affected [24,27]. The nitrification activity in soil ecosystems is known to be correlated with the abundances and structures of ammonia oxidizers [24,28,29]. We therefore hypothesize that different biofuel cropping systems, especially those that rely on N-fertilization, will influence the composition of ammonium oxidizers in soil, with potential consequences for nitrification rates.

Denitrification, which reduces nitrate to N_2 gas, is carried out by a diverse group of microorganisms belonging to more than 60 genera of bacteria, archaea, and some eukaryotes [30]. Complete denitrification involves four steps: $NO_3^- \rightarrow NO_2^- \rightarrow NO \rightarrow N_2O \rightarrow N_2$. The enzyme nitrous oxide reductase (encoded by *nosZ*) that reduces N_2O to N_2 is essential for complete conversion of NO_3^- to N_2. Approximately 17 Tg N is estimated to be lost from the land surface through denitrification every year [31]. It is known that the structure and activity of denitrifiers in the terrestrial ecosystem could be significantly influenced by the plant species [7,29,32]. In a study of a maize-cropped field, it was found that organic or mineral fertilizer applications could affect both the structure and activity of the denitrifying community in the long term, with changes persisting for at least 14 months [33]. The potential denitrification rate was found to be significantly correlated to the denitrifier density, as estimated by the quantification of *nosZ* gene copy numbers [34]. Denitrification releases mineralized nitrogen in the soil ecosystem to the atmosphere, and thus, the balance between denitrification and N-fixation, can determine the biologically available N for the biosphere (Arp, 2000).

It is known that plant species can change the soil microbial community [1]. However, while much previous work has examined the microbial community differences between the established crops [7,29,34,35,36], less is known about how microbial communities in the agricultural soils develop during the transition from one cropping system to another (e.g. annual row crops to perennial biofuel feedstocks). Thus, to improve our knowledge of the effects of bioenergy feedstock production on the complex N-cycling microbial communities of terrestrial ecosystems, we followed the changes in soil microbial communities during a two-year establishment period of maize, switchgrass, *Miscanthus×giganteus*, and mixed tallgrass prairie. We monitored the abundance of key genes for nitrogen fixation, ammonia oxidation and complete denitrification (*nifH*, bacterial/archaeal *amoA* and *nosZ*) as well as the structural changes of these N-cycling genes and bacterial/archaeal 16S rRNA genes using real-time PCR and barcoded pyrosequencing methods respectively.

Materials and Methods

Study site and sampling

The experiment was conducted at the Energy Biosciences Institute's Energy Farm located southwest of Urbana, Illinois, USA. Miscanthus (*Miscanthus×giganteus*, MG), switchgrass (*Panicum virgatum*, PV), maize (*Zea mays*, ZM) and restored tallgrass prairie (used as control, NP) were planted in the spring of 2008. Miscanthus was replanted in the spring of 2009 due to its poor growth in 2008. Each crop was planted in a randomized block design, with a 0.7-ha plot of each crop randomly positioned within four blocks (n = 4 for each crop). Samples were collected in April 2008, before planting of these crops, in order to characterize the background soil microbial communities. Bulk soil samples were collected at monthly intervals during the growing seasons (June–September 2008 and 2009). 10 soil cores (0–10 cm depth, 1.8 cm diameter) were collected from each plot and homogenized in an ethanol-sanitized, plastic bucket. About 60 g of the well-mixed soil was then subsampled into a 50 mL tube for each plot, and kept on ice until brought to the lab and stored in a −80°C freezer. In total, 112 soil samples were collected. Following standard agricultural practices, only maize was fertilized (17 g/m^2 in 2008, 20 g/m^2 in 2009) with a mixture of urea, ammonia and nitrate (28% UAN). Herbicides were applied in these crops except the restored tallgrass prairie: 1.56 l/ha of Roundup (only applied in 2008) and 4.13 l/ha of Lumax for maize; 1.37 l/ha of 2,4-Dichlorophenoxyacetic acid for Miscanthus; 1.37 l/ha of 2,4-Dichlorophenoxyacetic acid for switchgrass in 2008.

DNA extraction and purification

Soil samples were freeze-dried overnight until completely dry and then manually homogenized with a sterile screwdriver. DNA was extracted from 0.3 g soil using the FastDNA SPIN Kit For Soil (MP Biomedicals) according to manufacturer's protocol. Extracted DNA was then purified using CTAB. DNA concentrations were determined using the Qubit quantification platform with Quant-iT^{TM} dsDNA BR Assay Kit (Invitrogen). DNA was diluted to 10 ng/μL and stored in −80°C freezer for the following molecular applications.

Real-time PCR

The abundances of *nifH*, archaeal *amoA*, bacterial *amoA* and *nosZ* genes in all the soil samples were quantified using real-time PCR. Quantitative real-time PCR was performed according to the methods modified from previous studies: *nifH* (as a measure of N-fixing bacteria) used primers PolF (5′- TGC GAY CCS AAR GCB

GAC TC-3′) and PolR (5′-ATS GCC ATC ATY TCR CCG GA-3′) [5]; archaeal *amoA* (as a measure of ammonia-oxidizing archaea) used primers Arch-amoAF (5′-STA ATG GTC TGG CTT AGA CG-3′) and Arch-amoAR (5′-GCG GCC ATC CAT CTG TAT GT-3′) [6]; bacterial *amoA* (as a measure of ammonia-oxidizing bacteria) used primers amoA-1F (5′-GGG GTT TCT ACT GGT GGT-3′) and amoA-2R (5′-CCC CTC KGS AAA GCC TTC TTC-3′) [4]; and *nosZ* (as a measure of denitrification bacteria) used primers nosZ-F (5′-CGY TGT TCM TCG ACA GCC AG-3′) [37] and nosZ 1622R (5′-CGS ACC TTS TTG CCS TYG CG-3′) [7]. Purified PCR products from a common DNA mixture (equal amounts of DNA from all samples collected in August of 2008 and 2009) were used to prepare sample-derived quantification standards as previously described [38]. The copy number of gene in each standard was calculated by DNA concentration (ng/μL, measured by Qubit) divided by the average molecular weight (estimated based on the barcoded-pyrosequencing results) of that gene. In comparison to using a clone (plasmid) as standard, this method avoids the difference of PCR amplification efficiency between standards and samples caused by the different sequence composition in the PCR templates (single sequence in a plasmid for the standard vs. mixture of thousands of sequences in a soil sample). The 10 μL reaction mixture contained 5 μL of 2× Power SYBR Green Master Mix (Applied Biosystems), 0.5 μL of BSA (10 mg/mL, New England Biolabs), 0.4 μL of each primer (10 μM) and 5 ng of DNA template. Real-time amplification was performed in an ABI Prism 7700 Sequence Detector with MicroAmp Optical 384-Well Reaction Plate and Optical Adhesive Film (Applied Biosystems) using the following program: 94°C for 5 min; 40 cycles of 94°C for 45 s, 56°C for 1 min (54°C for *nifH* gene), 72°C for 1 min. A dissociation step was added at the end of the qPCR to assess amplification quality. The specificity of the PCR was further evaluated by running twenty randomly selected samples (for each gene) on a 1% (w/v) agarose gel. The corresponding real-time PCR efficiency for each of these genes was estimated based on a two-fold serial dilution of the common DNA mixture described above. The qPCR efficiency (E) was calculated according to the equation $E = 10^{[-1/\text{slope}]}$ [39]. Triplicate qPCR repetitions were performed for each of the gene for all the samples. The real-time PCR amplification efficiency of *nifH*, archaeal *amoA*, bacterial *amoA* and *nosZ* genes was 1.90±0.01, 1.90±0.06, 1.76±0.01 and 1.82±0.01, respectively. The R^2 of all these standard curves was higher than 0.99. The detection limit of this real-time PCR assay was 10 copies/μL.

The copy numbers of these genes per gram of dry soil was calculated by the copy numbers of the gene per ng of DNA multiplied by the amount of DNA contained in each gram of dry soil. The quantities of these genes were corrected assuming a DNA extraction efficiency of 30% [40,41].

T-RFLP

The soil samples collected from four replicated (blocks) plots of the four crops prior to planting, and then August of establishment years 1 and 2 (2008 and 2009; 48 samples in total) were analyzed by terminal restriction fragment length polymorphism (T-RFLP). The *nifH*, archaeal *amoA*, bacterial *amoA*, *nosZ* gene were amplified from these samples with FAM-labeled (on forward primer) primers PolF/R, Arch-amoAF/R, amoA-1F/2R and nosZ-F/1622R (see Table S1). The 16S rRNA gene was amplified with 8F (5′-FAM-AGAGTTTGATCMTGGCTCAG-3′) and 1492R (5′-GGTT-ACCTTGTTACGACTT-3′) [42,43]. The PCR reaction mixture (25 μl) contained 5 μl GoTaq Flexi Buffer (5×), 2 μl $MgCl_2$ (25 mM), 0.25 μl DNA Polymerase (5 U/μl, Promega, Madison, Wis.), 1.25 μl BSA (10 mg/ml), 1 μl forward primer (10 μM), 1 μl reverse primer (10 μM), 1.25 μl dNTP Mix (10 mM), 2 μl DNA template (10 ng/μl). The PCR reaction was performed in a thermo cycler (BioRad, Hercules, CA) using a 5-min heating step at 94°C followed by 30 cycles of denaturing at 94°C for 1 min, annealing at 60°C (54°C for *nifH*) for 45 s, and extension at 72°C for 1 min, with a final extension step of 5 min at 72°C. The PCR products were purified by QIAquick PCR Purification Kit (QIAGEN) and digested at 37°C overnight in a 20-μl mixture containing 2 μl NEB Buffer (10×), 0.5 μl AluI/HhaI (20 U/μl) and 5 μl PCR product. 5 μl of the digested product was sent to the Core Sequencing Facility (University of Illinois at Urbana-Champaign) for fragment analysis. ROX1000 was used as inner standard. T-RFLP profiles were analyzed by GeneMarker (v 1.85) according to the manufacturer's instruction. Fragments with sizes between 50 bp and the length of the PCR products and peak area >500 were selected for T-RFLP profile statistical analysis. The profile data were normalized by calculating the relative abundance (percentage) of each fragment (individual peak area divided by the total peak area).

Barcoded pyrosequencing

The same samples used in T-RFLP were also used in barcoded pyrosequencing. The four replicated samples collected from the same crop at the same time were combined in to one composite sample for the construction of sequence libraries. Altogether, nine samples (one sample for background soil [mixed from the soils from all the plots before planting these crops], four samples from each of the different crops at Aug 2008, and four from Aug 2009) were obtained. Furthermore, all of these N-transforming genes of the sample collected from MG at the end of the second year (MG2) were sequenced twice in two different lanes to estimate the variation of the sequencing method. The 16S rRNA gene of ZM2 was also sequenced twice. Details of primers and PCR conditions used in the study are listed in Table S1.

The *nifH*, archaeal *amoA*, bacterial *amoA*, *nosZ* and 16S rRNA genes (V4–V5 region) were amplified using the barcode primers PolF/R, Arch-amoAF/R, amoA-1F/2R, nosZ-F/1622R and U519F/U926R, respectively (primers are shown in Table S1). The primers (HPLC purified, Integrated DNA Technologies) were designed as 5′ Fusion Primer+barcode+gene specific primer-3′ (Fusion Primer A, 5′- CGTATCGCCTCCCTCGCGCCAT-CAG-3′; Fusion Primer B, 5′-CTATGCGCCTTGCCAGC-CCGCTCAG-3′). The PCR conditions were optimized and primers with appropriate barcodes (10 bp) were selected. The barcodes used for each primer are described in NCBI SRA, with accession number SRA023700. The 50 μL PCR mixture contained 10 μL of 5× Phusion HF Buffer (Phusion GC Buffer was used for bacterial *amoA* gene amplification, both buffer contains 7.5 mM $MgCl_2$), 1 μL of 10 mM dNTPs, 2.5 μL of 10 mg/mL BSA, 0.5 μL of 2 U/μL Phusion Hot Start DNA Polymerase (FINNZYMES), 4 μL of 10 μM forward/reverse primer mixture and 4 μL of 10 ng/μL DNA templates. 1 μL of 100% DMSO was supplemented into the PCR mixture in bacteria *amoA* gene (GC rich) amplification. The PCR amplification was performed in a thermal cycler (BioRad, Hercules, CA) using the program 98°C for 2 min; 30 cycles of 98°C for 10 s, 60°C (54°C for *nifH* gene and 56°C for bacterial *amoA* gene) for 30 s, 72°C for 20 s; 72°C for 5 min. The PCR product was first checked on a 1.2% w/v agarose gel, and then purified by QIAquick PCR Purification Kit (QIAGEN). The DNA concentration of the purified PCR product was measured by Qubit Fluorometer using the Quant-iT™ dsDNA BR Assay Kit (Invitrogen) according to the manual. PCR products of the same gene, to be run together in the same lane (1/16 plate) in 454 sequencing, were mixed in equal mole amounts and run on a 2% w/v agarose gel. The target bands were cut from the gel and purified by QIAquick Gel Extraction Kit (QIAGEN). The DNA concentrations of the purified PCR products were measured by Qubit Fluorometer and adjusted to 50 nM. The *nifH*, archaeal *amoA*, bacterial *amoA* and *nosZ* genes PCR products were then mixed in equal mole amounts and sequenced on a Genome Sequencer FLX Instrument (Roche) using GS FLX Titanium series reagents. The 16S rRNA gene was run in a separate lane.

Sequence analysis

Sequences were first extracted from the raw data according the Genome Sequencer Data Analysis Software Manual (Software Version 2.0.00, October 2008) by the sequencing center (Roy J. Carver Biotechnology Center, University of Illinois at Urbana-Champaign). The sequences with low quality (length <50 bp, which ambiguous base 'N', and average base quality score <20, for detail see manual) were removed. The sequences that fully matched with the barcodes were selected and distributed to separate files for each of the different genes, after removal of the barcode, using RDP Pipeline Initial Process (http://pyro.cme.msu.edu/). For each gene, the sequences that didn't match with the gene specific primers or had a read length shorter than 350 bp were removed. The sequences that matched with the reverse primer were converted to their reverse complement counterparts using BioEdit to make all the sequences forward-oriented.

The 16S rRNA gene sequences were aligned by NAST (Greengenes). The sequences with significant matched minimum length <300 and identity <75% were removed. The aligned 16S rRNA gene sequences were used for chimera check using Bellerophon method in Mothur [44]. Distance matrices were calculated by ARB using the neighbor joining method [45]. A lane mask was used in calculating the 16S rRNA gene sequences to filter out the hyper variable regions. Operational Taxonomic Units (OTUs) were then classified using a 97% nucleotide sequence similarity cutoff and rarefaction curves were constructed based on the distance matrices (both of nucleotide and amino acid sequences) using DOTUR [46]. The phylogenetic affiliation of each 16S rRNA gene sequence was analyzed by RDP CLASSIFIER (http://rdp.cme.msu.edu/) using confidence level of 80%.

The 16S rRNA gene sequences were also processed by QIIME pipeline and denoised by Denoiser V0.91 according to the manual [47,48]. The results were compared to that obtained by RDP pipeline. In total, 26,431 valid sequences were obtained after denoising using QIIME, which is 12.2% less than that obtained by RDP pipeline (without denoising). Using the 97% similarity cutoff, 8,568 OTUs were obtained, which is 4.7% lower than that observed by RDP pipeline. After random re-sampling to the same sequence depth (1789 sequences per sample) using Daisy_chopper (http://www.genomics.ceh.ac.uk/GeneSwytch/Tools.html), the number of OTUs for each sample obtained by two different processing methods (QIIME, denoised and RDP, non-denoised) was compared (Fig. S1). The estimated number of OTUs after denoising was similar to that obtained by RDP pyrosequencing pipeline (without denoising), showing that the denoising process had a very limited influence on our diversity analysis. The data reported in this paper was analyzed using RPD pipeline described in the previous paragraph.

The *nifH*, archaeal *amoA*, bacterial *amoA* and *nosZ* genes sequences were blasted against a non-redundant protein sequence database (download from NCBI) using BLASTX with an *E*-value cutoff of 0.001. The top 10 closest matches of each sequence were estimated using a custom made Perl script to remove possible chimeras and sequences with sequencing errors causing frameshifts. Sequences with different regions matching the same sequence in the database but with different frame positions were considered to be frameshifts. Sequences that matched two or more different origin sequences were classified as chimeras. The nucleotide sequences of *nifH*, archaeal *amoA*, bacterial *amoA* and *nosZ* genes were translated into amino acid by Geneious (http://www.geneious.com/) based on the frame positions obtained from BLASTX. The redundant sequences (identical sequences) were removed using CD-Hit [49], and the representatives with longest length were selected for following phylogenetic analysis. Both of the nucleotide and amino acid sequences of these N-cycling genes were aligned by MUSCLE 3.7 [50] using program default settings. Operational Taxonomic Units (OTUs) were then classified and rarefaction curves were constructed based on the distance matrices (both of nucleotide and amino acid sequences) using DOTUR [46]. Previous studies showed that the amino acid sequences of AmoA and NosZ similarity around 90% is generally relevant to 97% similarity of 16S rRNA gene [51,52]. Thus, all these N-cycling gene sequences were classified into OTUs using a 90% amino acid sequence similarity cutoff, and phylogenetic trees were built in ARB using the neighbor-joining method. Sequences of all the samples and genes were also randomly re-sampled to identical sequencing depth (the smallest sequencing effort) using Daisy_chopper (http://www.genomics.ceh.ac.uk/GeneSwytch/Tools.html) to avoid the potential bias caused by sequencing effort difference [53].

The 454-pyrosequencing data were deposited in NCBI SRA under accession number SRA023700.

Statistical analysis

ANOVA combined with post hoc Tukey B test was used to estimate the difference of archaeal/bacterial *amoA*, *nifH* and *nosZ* genes abundances under different crops based on the quantitative PCR results from the replicated plots. The T-RFLP data from the replicated plots were used to follow the structural changes of soil microbial communities by plant types, and significance tests for these changes were conducted using Analysis of Similarity (ANOSIM) based on Bray–Curtis similarity coefficients. Correspondence analysis (CA) and Canonical correspondence analysis (CCA) were also used to visualize the predominant microbial community changes of archaeal/bacterial *amoA*, *nifH*, *nosZ* and 16S rRNA genes after planting bioenergy crops based on the T-RFLP data. These statistical analyses were done using the free software PAST (http://folk.uio.no/ohammer/past/). Based on our extensive pyrosequencing library, the OTUs/genera that showed monotonic (i.e. continuously increasing or decreasing) trends for each crop treatment over the two year establishment were presumed to be particularly noteworthy in terms of crop impact. The populations with continuously increased or decreased abundance in the two-year period after planting these bioenergy crops were selected using a custom Perl script.

Results

Quantification of *nifH*, archaeal *amoA*, bacterial *amoA* and *nosZ* genes

Quantities of AOA in all of crops fluctuated over the two growing seasons in a similar pattern (Fig. 1), but the abundance of this gene was always higher in MG than NP and PV. The quantity of bacterial *amoA* genes in ZM significantly increased from $1.47\pm0.61\times10^8$ to $3.26\pm0.94\times10^8$ during the first three months of establishment and thereafter remained higher in ZM than in the other cropping systems. The nitrification rates under these crops were analyzed in the second year by estimating the accumulation of the nitrate in buried soil bags, and linear regression revealed

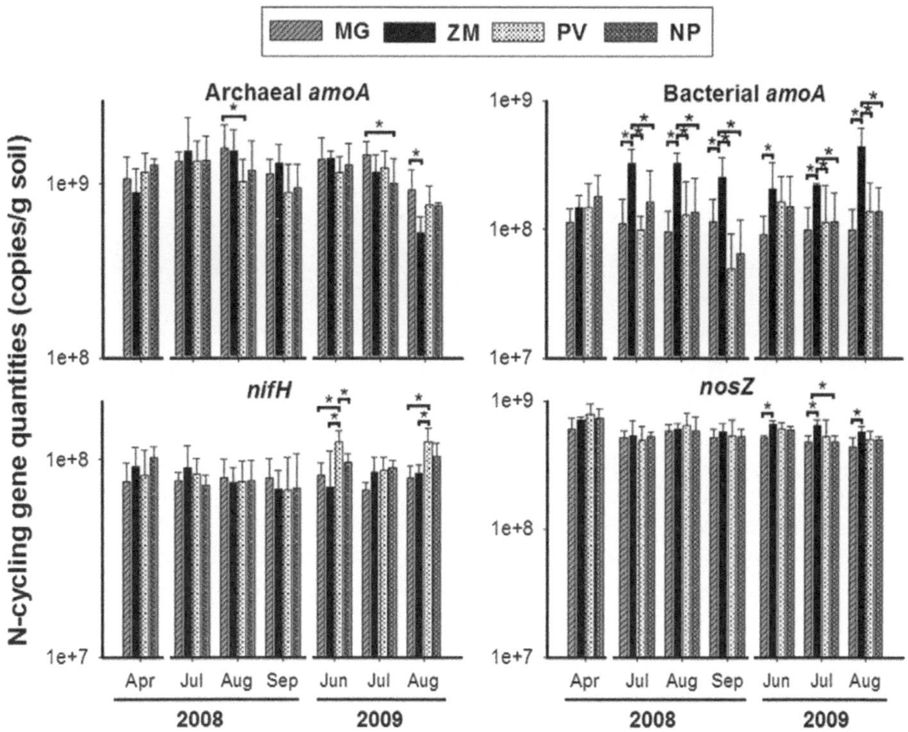

Figure 1. Changes in abundance of *nifH*, archaeal *amoA*, bacterial *amoA* and *nosZ* genes in plots after planting *Miscanthus×giganteus* (MG), *Panicum virgatum* (PV), restored prairie (NP) and *Zea mays* (ZM). The copy number of genes in each gram of dry soil was estimated based on the results of real-time PCR (copy number in each ng DNA) and the average amount of extracted DNA (6.23 µg per dry soil) and assuming DNA extraction efficiency was 30% [40]. The R^2 of the standard curve of all these genes was higher than 0.99. The real-time PCR amplification efficiency of *nifH*, archaeal *amoA*, bacterial *amoA* and *nosZ* genes was 1.90±0.01, 1.90±0.06, 1.76±0.01 and 1.82±0.01 respectively. *Represents values that are significantly different (P<0.01).

that the nitrification rates were significantly related to the quantities (log) of archaeal *amoA* genes ($R^2 = 0.61$, $P = 0.03$, $n = 12$), but not to the quantities of bacterial *amoA* genes (Fig. 2). The abundance of *nifH* genes remained stable for all the crops, ranging from 7×10^7 to 9×10^7 copies per gram of dry soil (Fig. 1). No significant differences were observed among the different crops in the first year. In the second year, the population sizes of diazotrophs in PV and NP had significantly increased in comparison to the first year ($P = 0.0001$ and 0.0002). The population size of denitrifiers was less variable in comparison to the other N-cycling populations; however, the copy number of *nosZ* increased in ZM during the second year of the study and remained higher than in MG (Fig. 1).

Structural changes of N-cycling genes and microbial communities after planting of bioenergy crops

The community structural differentiation of *nifH*, archaeal *amoA*, bacterial *amoA*, *nosZ* and 16S rRNA genes under different bioenergy crops were analyzed by T-RFLP. These analyses used the fully replicated sample set from the randomized block design.

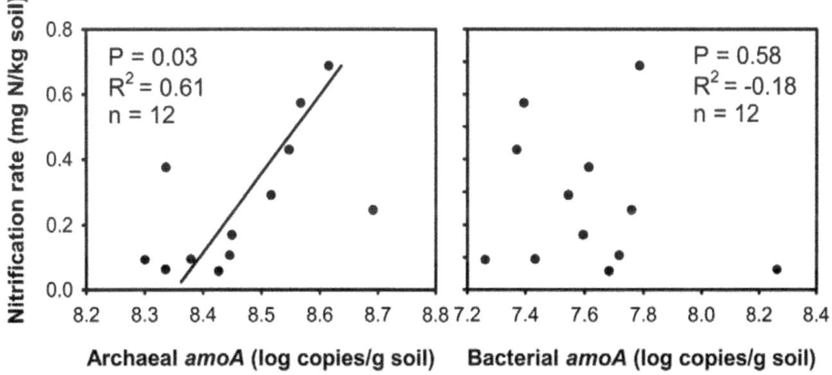

Figure 2. Relationship between the concentration of ammonia-oxidizing archaea/bacteria and nitrification rate. The nitrification rate was determined over the same time as our sample collection in 2009. Nitrification rate was calculated based on the accumulation of nitrate in soil bags incubated in the field (0–10 cm depth) for 15 to 32 days.

Correspondence analysis (Fig. 3) showed that the soil microbial communities in the initial plots did not show any relationship with the crop treatments being applied. During the establishment of these bioenergy crops, the community composition of denitrification bacteria (*nosZ*) under ZM was completely separated (ANOSIM, $P<0.05$, Table S2) from those under the other crops by the end of the second year along the first-axis, which explained 51.7% of the total variance. None of the other groups showed significant clustering by plant, although the community composition of AOB (bacteria *amoA*) under ZM appeared to be separated from that of MG (ANOSIM, $P=0.17$) along the second axis, which explained 14.2% of the total variance. In addition to plant species, the changes of soil microbial communities also could be caused by the variation of environmental conditions. To compare the magnitude of the changes of soil microbial communities related only to plant species, a constrained ordination method was also used. Canonical Correspondence analysis (CCA, Fig. S2) revealed that, at the end of the second year, the microbial communities under ZM were most different from the three cropping systems for bacterial *amoA*, *nosZ*, and 16S rRNA. Archaeal *amoA* was most distinct under MG, followed by ZM, and *nifH* was equally separated under all cropping systems (Fig. S2).

Diversity of *nifH*, archaeal *amoA*, bacterial *amoA*, *nosZ* and 16S rRNA genes

To further understand the composition of microbial community in the field, the *nifH*, archaeal *amoA*, bacterial *amoA*, *nosZ* and 16S rRNA genes were deeply sequenced using the pyrosequencing approach. In total, 143,487 reads were obtained for these genes. The numbers and qualities of these sequences are described in Table 1, Table S3 and Text S1. The reproducibility of the pyrosequencing result was estimated by comparing the observed microbial composition between repeat sequencing runs for all these genes (Fig. 4). Linear regression analysis indicated a high reproducibility ($R^2=0.95$) of our pyrosequencing.

High diversity of *nifH*, archaeal *amoA*, bacterial *amoA* and *nosZ* genes were observed with 10899, 3187, 3945 and 11242 unique nucleotide sequences and 2286, 2246, 3633 and 4208 unique deduced amino acid sequences respectively (Fig. S3). These sequences were then translated to amino acid sequence according

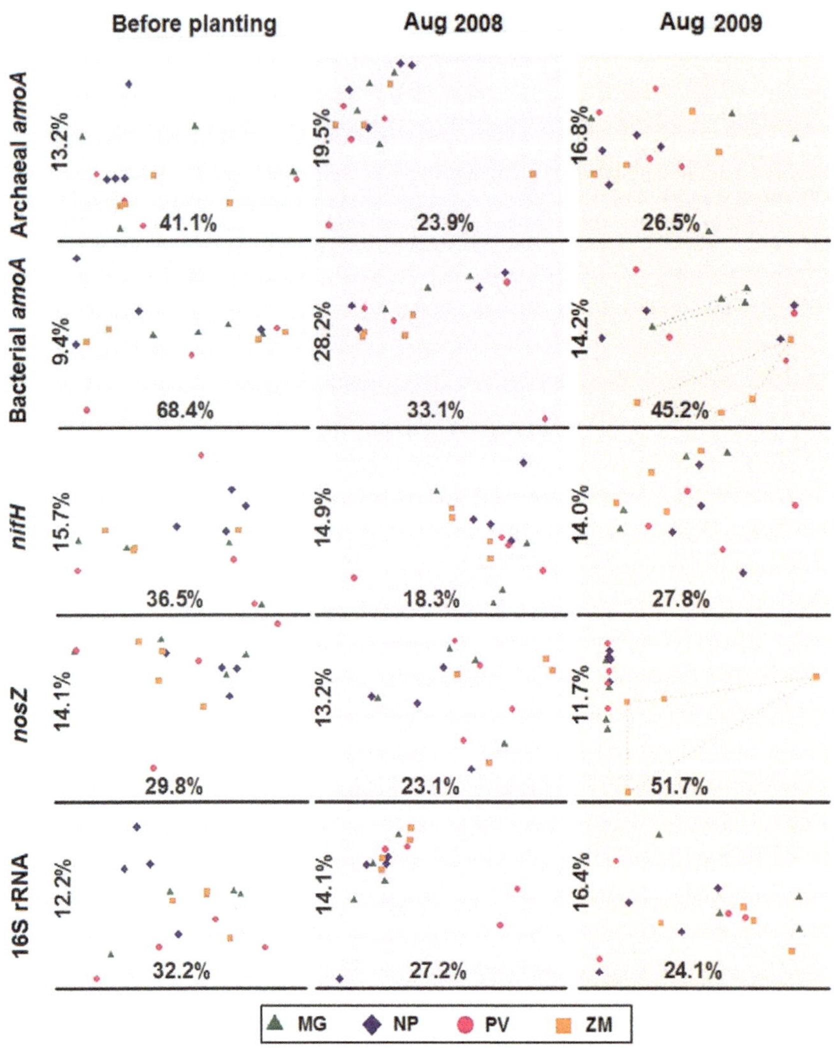

Figure 3. Structural changes of archaeal *amoA*, bacterial *amoA*, *nifH*, *nosZ* and 16S rRNA genes after planting *Miscanthus×giganteus* (MG), *Panicum virgatum* (PV), restored prairie (NP) and *Zea mays* (ZM) revealed by T-RFLP and Correspondence analysis (CA). The number on each axis shows the percentage of total variation explained. The soil samples were collected from four replicate plots for each plant at each time point.

Table 1. Quality of barcoded pyrosequencing reads.

	Number of sequences			
Genes	Correct barcode and primer	Length >350 bp	[a]Valid	Each sample (range)
nifH	28,334	27,781	21,111	1,312–2,956
Archaeal *amoA*	16,978	16,226	14,025	697–1,792
Bacterial *amoA*	28,254	27,874	21,817	1,726–2,569
nosZ	33,838	28,819	22,590	1,600–2,951
16S rRNA	30,487	30,175	30,101	2,034–3,488
Total	137,891 (96.1%)	130,875 (91.2%)	109,644 (76.4%)	697–3,488

Total number of raw reads was 143,487.

[a]Valid sequences of *nifH*, archaeal *amoA*, bacterial *amoA* and *nosZ* genes were defined as high quality sequences with correct barcode and primer (at 5′-end), length >350 bp and that did not have frameshifts and chimeric structure. The possible sequencing errors causing frameshifts and chimeras were removed based on the BLASTX result. Sequences with different regions matching the same sequence in the database but with different frame positions were considered to be frameshifts. Sequences that matched two or more different origin sequences were classified as chimeras. Valid sequences of 16S rRNA gene were sequences with correct barcode and primer, length >350 bp and passed the chimeric check program in Greengenes with the Bellerophon method. The sequence numbers for each sample are listed in Table S3.

to the BLASTX report. The amino acid sequences of *nifH*, archaeal *amoA*, bacterial *amoA* and *nosZ* genes were classified into 229, 309, 330 and 331 OTUs, respectively, with a similarity cutoff of 90%. After random re-sampling to the same sequencing depth (697 sequences for each sample), the adjusted total number of OTUs for these genes were 217, 303, 319 and 278, respectively (Table S4). Rarefaction analysis of these genes showed that the diversity of archaeal *amoA* gene in ZM2 (second year ZM) and MG1 (first year MG) was markedly lower than the others (Fig. S4). The diversity of *nifH* and *nosZ* genes slightly decreased in ZM2.

The diversity of bacterial and archaeal 16S rRNA genes was much higher than these N-cycling genes. In total, 19,824 unique 16S rRNA gene sequences and 8,989 species (OTUs classified at 97% similarity cutoff) were observed. RDP classification showed that these sequences covered 16 bacterial and 1 archaeal phyla, including 201 genera (Fig. S5). Proteobacteria and Acidobacteria were the most predominant phyla in the soil (>20%). The

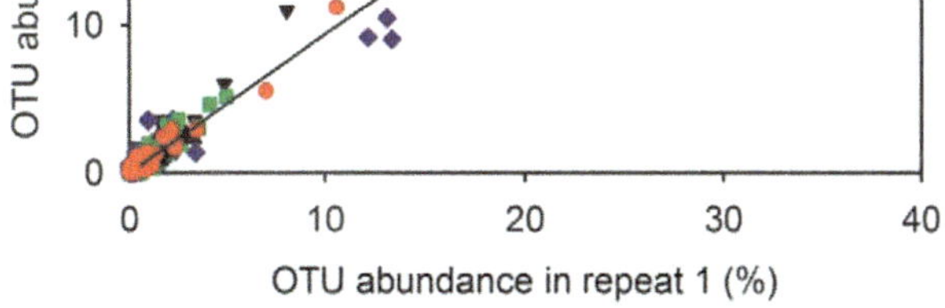

Figure 4. Reproducibility of the pyrosequencing replicates. OTUs of *nifH*, archaeal *amoA*, bacterial *amoA* and *nosZ* genes were classified at a nucleotide similarity cutoff 90%. The 16S rRNA gene sequences were classified to genus level by RDP classifier. One of the samples was duplicated for each gene.

sequences belonging to Proteobacteria were distributed over 86 different genera, while 94.1% of the sequences in Acidobacteria belonged to Family Gemmatimonadaceae, with GP1 as the most predominant genus (accounted 35.9% of the Acidobacteria).

To understand which phylotypes were impacted by vegetation type, the OTUs of *nifH*, archaeal *amoA*, bacterial *amoA*, *nosZ* and 16S rRNA (genus for 16S rRNA) genes that continuously increased or decreased over the two-year establishment of these bioenergy crops were identified (Fig. 5 and Fig. S6, S7, S8, S9). After planting of these bioenergy crops, 27.5%, 15.4%, 22.7% and 14.5% of the total archaeal *amoA*, bacterial *amoA*, *nifH*, and *nosZ* phylotypes, respectively, were found to be continuously increasing or decreasing (Table 2). Details of these continuously changed N-cycling OTUs are described in Text S1 and Figure S6, S7, S8, S9.

Pyrosequencing of 16S rRNA gene revealed 19.9% of the bacterial genera (39), spanning six phyla, continuously changed after planting of these bioenergy crops (Fig. 5). Only genus *Methylibium* was changed in all the crops, with decreased abundance in MG and increased abundance in the other crops. *Rhodanobacter* only appeared after planting of ZM (7 sequences for both of ZM1 and ZM2), and it was undetectable either in the background soil or in the soil under other crops. Consistent with the changes of *Nitrosospira*-like bacterial *amoA* OTU (see above), the abundance of genus *Nitrosospira* in the 16S rRNA library also increased in ZM. The abundance of genus *Nitrospira*, which is known as a nitrite-oxidizing bacteria, also increased in ZM. Most of the changed genera in MG decreased or even disappeared, except *Terrabacter* and *Herbaspirillum*. All of them were found at low abundance (<1%). Although many genera in Proteobacteria were changed, the total abundance of this most predominant phylum was quite stable under all of the crops (Fig. S5).

Discussion

In this study, we monitored the structural and quantitative changes of the key genes involved in N-cycling as well as the overall bacterial/archaeal community during two-year establishment of four different bioenergy feedstock crops, and analyzed the shifts of specific soil microbial populations in response to different types of crops. We were able to detect significant changes in the abundance of many of these microbial functional groups within 2 years of initial crop establishment. We also found that traditional row-crop agriculture of maize has a larger

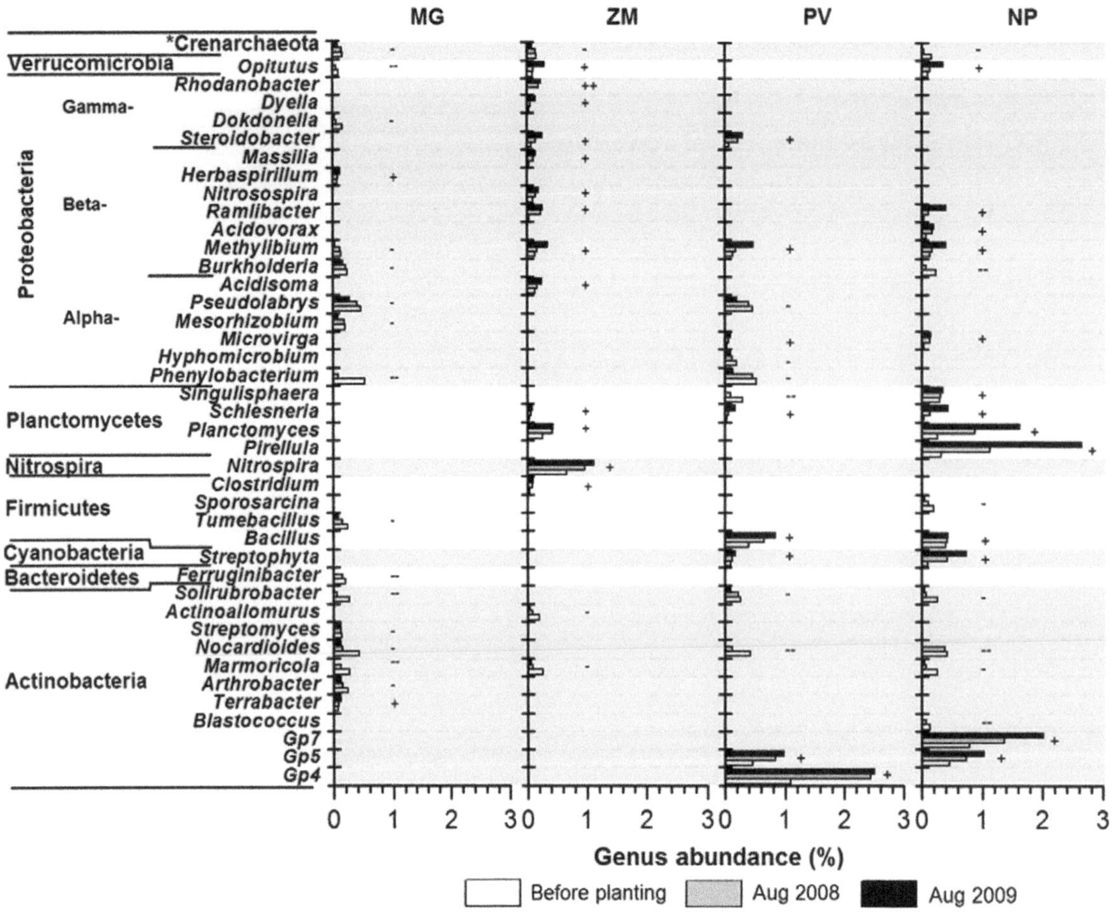

Figure 5. Microbial genera that changed after planting of *Miscanthus×giganteus* (MG), *Panicum virgatum* (PV), restored prairie (NP) and *Zea mays* (ZM). Sequences were classified by RDP Classifier project. −/+ represents the genus continuously decreased/increased after planting the crops; − −/++ represents the genus disappeared/appeared after planting the crops. *sequences belonging to Crenarchaeota, which could not be classified to genus level.

impact on the soil N-cycling community than any of the perennial bioenergy feedstock crops (Figs. 1, S2), while the perennial crops were associated with overall community shifts in the phyla Planctomyces, Firmicutes, and Actinobacteria (Fig. 5).

Table 2. Number of OTUs or genera that changed continuously after planting *Miscanthus×giganteus* (MG), *Panicum virgatum* (PV), restored prairie (NP) and *Zea mays* (ZM).

Genes	MG	PV	ZM	NP	*Total
Archaeal *amoA*	4	16	61	23	85
Bacterial *amoA*	18	18	7	20	51
nifH	8	21	14	23	52
nosZ	12	18	24	16	48
16S rRNA	17	15	14	21	40

OTUs of N-cycling genes were classified based on a cutoff of 90% amino acid sequences similarity. 16S rRNA genes were classified into genus level by RDP Classifier. Details of the abundance changes of the OTUs or genera are shown in Fig. 5 and S6, S7, S8, S9.
*Total number of changed unique OTUs or genera.

Traditional maize cultivation significantly increased the total abundance of ammonia-oxidizing bacteria and denitrifying bacteria (Fig. 1), altered the community composition of denitrifying bacteria (Fig. 3) and decreased the diversity of ammonia-oxidizing archaea, denitrifying bacteria, and diazotrophs (Fig. S4). This may be due to the application of N-fertilizer, which occurred only in ZM plots. Ammonia oxidizers are sensitive to N-fertilizer [24,27], and these responses were manifested in the increased population size of AOB and the high number of markedly changed AOA species. The nitrification rate was significantly correlated with the quantity of archaeal *amoA*, but not bacterial *amoA*, indicating AOA was the major ammonia-oxidizer.

Deep understanding of the structural shifts of key functional genes can help us to better understand changes in microbial activity in the environment. From the present database, we know that the global diversity of the *nifH*, archaeal *amoA*, bacterial *amoA* and *nosZ* genes, as well as the other functional genes of microorganisms, is high. The traditional approaches (e.g. clone library, DGGE and T-RFLP) used in previous studies may largely underestimate the diversity of microbial communities involved in soil nitrogen cycling. Mounier et al. (2004) revealed that even a large library with 713 clones was insufficient to enumerate the diversity *nosZ* gene in maize rhizosphere, showing the high complexity of N-cycling genes [54]. Thus, high-throughput deep sequencing approaches are essential to improve our knowledge of the diversity of these functional genes. In the present study, using barcoded 454-pyrosequencing approach, we found high diversity (ranging from ~3100 to ~11200 unique nucleotide sequences) of *nifH*, archaeal *amoA*, bacterial *amoA* and *nosZ* genes in the soil ecosystem, which far surpasses the diversity of the N-cycling genes observed in previous studies [54,55,56,57,58,59,60,61]. The rarefaction curves of these genes were close to saturation after sequencing ~1000 for each sample, indicating that such a sequencing effort is sufficient to elucidate the diversity and structure of the complex soil N-cycling communities. The high similarity between repeat runs of these genes (Fig. 4) demonstrates the high reproducibility and reliability of this barcoded pyrosequencing method. This result also indicates that the variation of pyrosequencing, resulting from random sampling of gene targets during emulsion PCR [62], can be greatly reduced by increasing the sequencing depth and library coverage.

During the establishment of these bioenergy crops, about 15%–30% of N-cycling genes and the detected bacterial/archaeal genera were continuously changed, indicating that a large proportion of soil microbes were affected by the transition to different bioenergy feedstock systems. Most of these phylotypes changed uniquely in one of the crops, indicating that the changes were mainly caused by the particular experimental crop treatment (specific plant species or management, such as fertilization) and not due to the environmental conditions that fluctuated in all treatments (e.g. temperature and moisture). Contrary changes of certain populations in different crops also support this conclusion; for example, the abundance of *Methylibium* (belonging to β-Proteobacteria) decreased in MG but increased in the other crops. The abundance of Bacteroidetes was previously found to be lower in the soil of Miscanthus-dominated grasslands (4%) in comparison to forest soil (6%) [63]. We found that the decrease of Bacteroidetes in MG was mainly due to the disappearance of the genus *Ferruginibacter* (Fig. 5 and S5).

The structure of *nifH* gene was completely separated according to vegetation by the end of the second year (Fig. S2), which suggests that the structure of N-fixation bacterial population was particularly sensitive to plant genotype. Tan et al. (2003) has revealed that the structure of diazotrophs was not only different among rice species, but also changed rapidly with fertilization. The diversity of the *nifH* gene was obviously reduced within 15 days after fertilization [17]. Thus, the decreased diversity of *nifH* in ZM may be also due to the application of fertilizer not the presence of maize. However, the population size of N-fixing bacteria did not change under ZM, indicating N-fertilization may not change the quantity of soil diazotroph [64]. The N-fixing activity was expected to increase in MG [14,15]. Although the population size of total free-living soil N-fixing bacteria was not significantly increased by growth of MG in the two year period, the abundance of genus *Herbaspirillum* increased. *Herbaspirillum* species are known as endophytic diazotrophs that are enriched by C4-prennial grasses including Miscanthus [65,66,67]. Thus, we speculate that the abundance increase of *Herbaspirillum* in the bulk soil was likely due to the root exudates (e.g. organic carbons) released by Miscanthus, which favored the growth of this population. Our results also suggest that Miscanthus may only selectively enhance the activity of specific diazotrophs, not the whole N-fixing microbial community.

The AOA are thought to be more stable and less responsive to environmental differences than AOB, as revealed by previous quantitative studies [6,68]. However, in the present study we found that, while the population size of AOA was relatively stable, the structure of the AOA community was sensitive to the different cropping systems. The diversity of AOA markedly decreased after planting of maize, with 41 of the AOA phylotypes disappearing (Fig. S6). In contrast, the population size of AOB significantly increased after planting of maize in both the qPCR of bacterial *amoA* and the 16S rRNA pyrosequencing results. In addition to the increase of genus *Nitrosospira* (AOB) [69], the abundance of *Nitrospira* (nitrite-oxidizing bacteria) [70] also increased in N-fertilized maize (Fig. 5). However, the number of changed bacterial *amoA* phylotypes in ZM was much less than the other crops. Therefore, these two different groups of ammonia-oxidizers respond to the N-fertilization in a very different way. The population size of AOB increased immediately in the first growing season, thus, we hypothesize that the increased AOB abundance in maize was likely due not to the growth of maize plants, but to the application of N-fertilizer, which increased the ammonia content in the soil [24,27]. It has been found that AOB population size increased in seven days after applying of N-fertilizer, and it was still significantly greater than unfertilized soil 8 months after the last application of ammonia [41]. Consistently, we found that the population size of AOB doubled in less than three months and maintained a relatively high level over the two-year study even though measurable NH_4^+ in the soil declined over this time period to levels close to that of the unfertilized plots (Fig. 1 and S10). These results indicate the AOB population size can be quickly increased by N-fertilization and can remain for a long period even after the measurable ammonia has been consumed.

The nitrification rate was found to be significantly correlated with the quantity of archaeal *amoA* rather than bacterial *amoA*, indicating AOA rather than AOB may be the major active ammonia-oxidizer in these soils. Contradictory conclusions on the relative importance of AOA and AOB in soil nitrification have been previously reported where nitrification was found to be associated with the changes of archaeal *amoA* abundance or higher archaeal transcriptional activity in some of the soils [58,71,72]. In contrast, the nitrification kinetics in the other soils were correlated with the growth of AOB [73,74]. It has been found that the ammonia affinity of "*Candidatus* Nitrosopumilus maritimus" (a marine AOA) is much higher than AOBs, and its growth may be enhanced by relatively low ammonia concentration [75]. Thus, the contradictory conclusions from these studies may be due to the different soil ammonia concentration used in these experiments [58,73]. These results hint that AOB may be more active in soils amended with ammonia, while AOA are more active in soils with low ammonia concentration [58]. The nitrification rate outlier (ZM Aug; Fig. 2) had highest population size of AOB, suggesting that the activity of AOB was enhanced by N-fertilization and also supported the above speculation. In support of this speculation, a recent publication shows that recovery of nitrification potential after disruption was dominated by AOB in cropped soils while AOA were responsible RNP in pasture soils [76].

It is known that nitrogen fertilization can change the structure and activity of denitrifying community, and subsequently affect the N_2O emission [33,34,77,78]. Large amounts (1.3%) of the applied N-fertilizer in maize fields (north Colorado) are converted to N_2O by the combination of nitrification and denitrification [79]. However, it is still unclear how fertilization changes the microbial community, since most of the previous studies are based on the already established fields [7,29,33]. Our study revealed that the structure of denitrifying bacteria in maize soil was significantly differentiated from the other crops at the second year (Fig. 3).

However, the population size of denitrifiers was relatively stable in all the crops in comparison to other N-cycling microbial communities, which only slightly increased in maize at the second year. The high stability of denitrifying population abundance could be explained by the high diversity and functional redundancy of denitrification community [30,80].

In conclusion, our two-year study of transitional agriculture shows that specific N-transforming microbial communities develop in the soil in response to different bioenergy crops. Each N-cycling microbial group responded in a different way after planting with different bioenergy crops. In general, planting of maize has a larger impact on the soil N-cycling community than the other bioenergy crops. Our results also indicate that application of N-fertilizer may not only cause short-term environmental problems, e.g. water contamination, but also can have long-term influence on the global biogeochemical cycles through changing the soil microbial community structure and abundance. Since soil types and other environmental factors may also impact the N-cycling microbial community, the universality of our findings needs to be confirmed by additional study at different sites.

Supporting Information

Figure S1 Number of OTUs obtained by two different processing methods: QIIME (denoised) and RDP pyrosequencing pipeline (non-denoised).

Figure S2 Structural changes of archaeal *amoA*, bacterial *amoA*, *nifH*, *nosZ* and 16S rRNA genes after planting *Miscanthus*×*giganteus* (MG), *Panicum virgatum* (PV), restored prairie (NP) and *Zea mays* (ZM) revealed by T-RFLP and Canonical correspondence analysis (CCA). The number on each axis shows the explained total variation. The soil samples were collected from four replicated plots for each plant at each time point. * Correspondence analysis was used for the samples collected before planting bioenergy crops.

Figure S3 OTU classification of valid sequences at different distance levels based on nucleotide and deduced amino acid sequences.

Figure S4 Rarefaction analysis of the diversities of *nifH*, archaeal *amoA*, bacterial *amoA*, *nosZ* and 16S rRNA genes in the soil underneath different bioenergy crops. The OTUs of *nifH*, archaeal *amoA*, bacterial *amoA* and *nosZ* genes were classified at 90% similarity cutoff based on amino acid sequences, and 16S rRNA gene was classified at 97% similarity cutoff on nucleotide sequences. BG0 represents the samples collected before planting bioenergy crops. MG, PV, NP, and ZM represent *Miscanthus*×*giganteus*, *Panicum virgatum*, restored prairie and *Zea mays* respectively. 1 and 2 represent samples collected in the first and second growing seasons.

Figure S5 Phylum level microbial community composition in the soil under different plants before and for two years after transition to bioenergy cropping. * represent significantly changed phylum. MG, *Miscanthus*×*giganteus*; PV, *Panicum virgatum*; NP, restored prairie; ZM, *Zea mays*.

Figure S6 (a) Phylogenetic tree of and (b) abundance of archaeal *amoA* OTUs that continuously changed after planting *Miscanthus*×*giganteus* (MG), *Panicum virgatum* (PV), restored prairie (NP) and *Zea mays* (ZM). OTUs were classified based on a cutoff of 90% amino acid sequence similarity.

Figure S7 (a) Phylogenetic tree of and (b) abundance of bacterial *amoA* OTUs that continuously changed after planting *Miscanthus*×*giganteus* (MG), *Panicum virgatum* (PV), restored prairie (NP) and *Zea mays* (ZM). OTUs were classified based on a cutoff of 90% amino acid sequence similarity.

Figure S8 (a) Phylogenetic tree of and (b) abundance of *nifH* OTUs that continuously changed after planting *Miscanthus*×*giganteus* (MG), *Panicum virgatum* (PV), restored prairie (NP) and *Zea mays* (ZM). OTUs were classified based on a cutoff of 90% amino acid sequence similarity.

Figure S9 (a) Phylogenetic tree of and (b) abundance of *nosZ* OTUs that continuously changed after planting *Miscanthus*×*giganteus* (MG), *Panicum virgatum* (PV), restored prairie (NP) and *Zea mays* (ZM). OTUs of were classified based on a cutoff of 90% amino acid sequence similarity.

Figure S10 Nitrate and ammonia concentration in bulk soil. Soil samples for chemical and microbiological analysis were collected in the same week for each time point, except Sep 2008 when the nitrate and ammonia concentrations were not measured. MG, *Miscanthus*×*giganteus*; PV, *Panicum virgatum*; NP, restored prairie; ZM, *Zea mays*. These data were measured over the same period of our sample collection by the Biogeochemistry laboratory (C. Smith and M. David).

Figure S11 Pyrosequencing read length based on the raw sequence reads.

Table S1 Primers and annealing temperature for *nifH*, archaeal *amoA*, bacterial *amoA*, *nosZ* and 16S rRNA genes.

Table S2 Comparsion of the microbial community structures between different crops.

Table S3 Number of valid sequences for each gene in each sample.

Table S4 Number of OTUs observed after random re-sampling to the identical sequencing depth (697 sequences/sample.

Acknowledgments

The authors gratefully acknowledge help from A. Duong, D. Fieckert, A. Groll, N. Peld, and M. Masters with field sampling and laboratory work. C. Smith and M. David kindly provided denitrification rate estimates for sampled plots. P.Y. Hong provided helpful comments on this manuscript.

Author Contributions

Conceived and designed the experiments: YM ACY RIM. Performed the experiments: YM ACY. Analyzed the data: YM. Contributed reagents/materials/analysis tools: YM. Wrote the paper: YM ACY RIM.

References

1. Garbeva P, van Veen JA, van Elsas JD (2004) Microbial diversity in soil: selection microbial populations by plant and soil type and implications for disease suppressiveness. Annu Rev Phytopathol 42: 243–270.
2. Kowalchuk GA, Buma DS, de Boer W, Klinkhamer PGL, van Veen JA (2002) Effects of above-ground plant species composition and diversity on the diversity of soil-borne microorganisms. Antonie Van Leeuwenhoek International Journal of General and Molecular Microbiology 81: 509–520.
3. Wardle DA, Bardgett RD, Klironomos JN, Setala H, van der Putten WH, et al. (2004) Ecological linkages between aboveground and belowground biota. Science 304: 1629–1633.
4. Rotthauwe JH, Witzel KP, Liesack W (1997) The ammonia monooxygenase structural gene amoA as a functional marker: molecular fine-scale analysis of natural ammonia-oxidizing populations. Appl Environ Microbiol 63: 4704–4712.
5. Poly F, Ranjard L, Nazaret S, Gourbiere F, Monrozier LJ (2001) Comparison of *nifH* gene pools in soils and soil microenvironments with contrasting properties. Appl Environ Microbiol 67: 2255–2262.
6. Francis CA, Roberts KJ, Beman JM, Santoro AE, Oakley BB (2005) Ubiquity and diversity of ammonia-oxidizing archaea in water columns and sediments of the ocean. Proc Natl Acad Sci U S A 102: 14683–14688.
7. Ruiz-Rueda O, Hallin S, Baneras L (2009) Structure and function of denitrifying and nitrifying bacterial communities in relation to the plant species in a constructed wetland. FEMS Microbiol Ecol 67: 308–319.
8. Bru D, Ramette A, Saby NP, Dequiedt S, Ranjard L, et al. (2011) Determinants of the distribution of nitrogen-cycling microbial communities at the landscape scale. ISME J 5: 532–542.
9. Orr CH, James A, Leifert C, Cooper JM, Cummings SP (2011) Diversity and activity of free-living nitrogen-fixing bacteria and total bacteria in organic and conventionally managed soils. Appl Environ Microbiol 77: 911–919.
10. Zehr JP, Jenkins BD, Short SM, Steward GF (2003) Nitrogenase gene diversity and microbial community structure: a cross-system comparison. Environ Microbiol 5: 539–554.
11. Cleveland CC, Townsend AR, Schimel DS, Fisher H, Howarth RW, et al. (1999) Global patterns of terrestrial biological nitrogen (N-2) fixation in natural ecosystems. Global Biogeochemical Cycles 13: 623–645.
12. Hsu SF, Buckley DH (2009) Evidence for the functional significance of diazotroph community structure in soil. Isme Journal 3: 124–136.
13. Son Y (2001) Non-symbiotic nitrogen fixation in forest ecosystems. Ecological Research 16: 183–196.
14. Schwarz H, Liebhard P, Ehrendorfer K, Ruckenbauer P (1994) The effect of fertilization on yield and quality of *Miscanthus sinensis* 'Giganteus'. Ind Crop Prod 2: 153–159.
15. Davis S, Parton W, Dohleman F, Smith C, Grosso S, et al. (2010) Comparative biogeochemical cycles of bioenergy crops reveal nitrogen-fixation and low greenhouse gas emissions in a *Miscanthus×giganteus* agro-ecosystem ecosystems DOI 10.1007/s10021-009-9306-9.
16. Demba Diallo M, Willems A, Vloemans N, Cousin S, Vandekerckhove TT, et al. (2004) Polymerase chain reaction denaturing gradient gel electrophoresis analysis of the N_2-fixing bacterial diversity in soil under *Acacia tortilis* ssp. raddiana and *Balanites aegyptiaca* in the dryland part of Senegal. Environ Microbiol 6: 400–415.
17. Tan Z, Hurek T, Reinhold-Hurek B (2003) Effect of N-fertilization, plant genotype and environmental conditions on *nifH* gene pools in roots of rice. Environ Microbiol 5: 1009–1015.
18. Kowalchuk GA, Stephen JR (2001) Ammonia-oxidizing bacteria: a model for molecular microbial ecology. Annu Rev Microbiol 55: 485–529.
19. Schleper C, Jurgens G, Jonuscheit M (2005) Genomic studies of uncultivated archaea. Nat Rev Microbiol 3: 479–488.
20. Konneke M, Bernhard AE, de la Torre JR, Walker CB, Waterbury JB, et al. (2005) Isolation of an autotrophic ammonia-oxidizing marine archaeon. Nature 437: 543–546.
21. Hatzenpichler R, Lebedeva EV, Spieck E, Stoecker K, Richter A, et al. (2008) A moderately thermophilic ammonia-oxidizing crenarchaeote from a hot spring. Proceedings of the National Academy of Sciences of the United States of America 105: 2134–2139.
22. Leininger S, Urich T, Schloter M, Schwark L, Qi J, et al. (2006) Archaea predominate among ammonia-oxidizing prokaryotes in soils. Nature 442: 806–809.
23. He JZ, Shen JP, Zhang LM, Zhu YG, Zheng YM, et al. (2007) Quantitative analyses of the abundance and composition of ammonia-oxidizing bacteria and ammonia-oxidizing archaea of a Chinese upland red soil under long-term fertilization practices. Environ Microbiol 9: 2364–2374.
24. Shen JP, Zhang LM, Zhu YG, Zhang JB, He JZ (2008) Abundance and composition of ammonia-oxidizing bacteria and ammonia-oxidizing archaea communities of an alkaline sandy loam. Environ Microbiol 10: 1601–1611.
25. Wessen E, Soderstrom M, Stenberg M, Bru D, Hellman M, et al. (2011) Spatial distribution of ammonia-oxidizing bacteria and archaea across a 44-hectare farm related to ecosystem functioning. ISME J.
26. Briones AM, Okabe S, Umemiya Y, Ramsing NB, Reichardt W, et al. (2002) Influence of different cultivars on populations of ammonia-oxidizing bacteria in the root environment of rice. Appl Environ Microbiol 68: 3067–3075.
27. Wang Y, Ke X, Wu L, Lu Y (2009) Community composition of ammonia-oxidizing bacteria and archaea in rice field soil as affected by nitrogen fertilization. Syst Appl Microbiol 32: 27–36.
28. Le Roux X, Poly F, Currey P, Commeaux C, Hai B, et al. (2008) Effects of aboveground grazing on coupling among nitrifier activity, abundance and community structure. Isme J 2: 221–232.
29. Patra AK, Abbadie L, Clays-Josserand A, Degrange V, Grayston SJ, et al. (2006) Effects of management regime and plant species on the enzyme activity and genetic structure of N-fixing, denitrifying and nitrifying bacterial communities in grassland soils. Environ Microbiol 8: 1005–1016.
30. Philippot L, Hallin S, Schloter M (2007) Ecology of denitrifying prokaryotes in agricultural soil. Advances in Agronomy, Vol 96 96: 249–305.
31. Schlesinger WH (2009) On the fate of anthropogenic nitrogen. Proc Natl Acad Sci U S A 106: 203–208.
32. Rich JJ, Heichen RS, Bottomley PJ, Cromack K, Jr., Myrold DD (2003) Community composition and functioning of denitrifying bacteria from adjacent meadow and forest soils. Appl Environ Microbiol 69: 5974–5982.
33. Dambreville C, Hallet S, Nguyen C, Morvan T, Germon JC, et al. (2006) Structure and activity of the denitrifying community in a maize-cropped field fertilized with composted pig manure or ammonium nitrate. FEMS Microbiol Ecol 56: 119–131.
34. Hallin S, Jones CM, Schloter M, Philippot L (2009) Relationship between N-cycling communities and ecosystem functioning in a 50-year-old fertilization experiment. Isme J 3: 597–605.
35. Knops JMH, Bradley KL, Wedin DA (2002) Mechanisms of plant species impacts on ecosystem nitrogen cycling. Ecology Letters 5: 454–466.
36. Priha O, Grayston SJ, Pennanen T, Smolander A (1999) Microbial activities related to C and N cycling and microbial community structure in the rhizospheres of *Pinus sylvestris*, *Picea abies* and *Betula pendula* seedlings in an organic and mineral soil. FEMS Microbiol Ecol 30: 187–199.
37. Kloos K, Mergel A, Rosch C, Bothe H (2001) Denitrification within the genus *Azospirillum* and other associative bacteria. Australian Journal of Plant Physiology 28: 991–998.
38. Chen J, Yu ZT, Michel FC, Wittum T, Morrison M (2007) Development and application of real-time PCR assays for quantification of erm genes conferring resistance to macrolides-lincosamides-streptogramin B in livestock manure and manure management systems. Applied and Environmental Microbiology 73: 4407–4416.
39. Bustin SA (2000) Absolute quantification of mRNA using real-time reverse transcription polymerase chain reaction assays. Journal of Molecular Endocrinology 25: 169–193.
40. Mumy KL, Findlay RH (2004) Convenient determination of DNA extraction efficiency using an external DNA recovery standard and quantitative-competitive PCR. Journal of Microbiological Methods 57: 259–268.
41. Okano Y, Hristova KR, Leutenegger CM, Jackson LE, Denison RF, et al. (2004) Application of real-time PCR to study effects of ammonium on population size of ammonia-oxidizing bacteria in soil. Appl Environ Microbiol 70: 1008–1016.
42. Heuer H, Krsek M, Baker P, Smalla K, Wellington EM (1997) Analysis of actinomycete communities by specific amplification of genes encoding 16S rRNA and gel-electrophoretic separation in denaturing gradients. Appl Environ Microbiol 63: 3233–3241.
43. Lane DJ (1991) 16S/23S rRNA sequencing. InIn E. Stackebrandt, M. Goodfellow, eds. Nucleic acid techniques in bacterial systematics. pp 115–175.
44. Schloss PD, Westcott SL, Ryabin T, Hall JR, Hartmann M, et al. (2009) Introducing mothur: open-source, platform-independent, community-supported software for describing and comparing microbial communities. Appl Environ Microbiol 75: 7537–7541.
45. Ludwig W, Strunk O, Westram R, Richter L, Meier H, et al. (2004) ARB: a software environment for sequence data. Nucleic Acids Research 32: 1363–1371.
46. Schloss PD, Handelsman J (2005) Introducing DOTUR, a computer program for defining operational taxonomic units and estimating species richness. Applied and Environmental Microbiology 71: 1501–1506.
47. Caporaso JG, Kuczynski J, Stombaugh J, Bittinger K, Bushman FD, et al. (2010) QIIME allows analysis of high-throughput community sequencing data. Nat Methods 7: 335–336.
48. Reeder J, Knight R (2010) Rapidly denoising pyrosequencing amplicon reads by exploiting rank-abundance distributions. Nat Methods 7: 668–669.
49. Li W, Godzik A (2006) Cd-hit: a fast program for clustering and comparing large sets of protein or nucleotide sequences. Bioinformatics 22: 1658–1659.
50. Edgar RC (2004) MUSCLE: multiple sequence alignment with high accuracy and high throughput. Nucleic Acids Research 32: 1792–1797.
51. Palmer K, Drake HL, Horn MA (2009) Genome-derived criteria for assigning environmental *narG* and *nosZ* sequences to operational taxonomic units of nitrate reducers. Appl Environ Microbiol 75: 5170–5174.
52. Koops H-P, Purkhold U, Pommerening-Röser A, Timmermann G, Wagner M (2006) The lithoautotrophic ammonia-oxidizing bacteria. Prokaryotes 5: 788–811.
53. Gilbert JA, Field D, Swift P, Thomas S, Cummings D, et al. (2010) The taxonomic and functional diversity of microbes at a temperate coastal site: a 'multi-omic' study of seasonal and diel temporal variation. PLoS One 5: e15545.

54. Mounier E, Hallet S, Cheneby D, Benizri E, Gruet Y, et al. (2004) Influence of maize mucilage on the diversity and activity of the denitrifying community. Environ Microbiol 6: 301–312.
55. Teng Q, Sun B, Fu X, Li S, Cui Z, et al. (2009) Analysis of nifH gene diversity in red soil amended with manure in Jiangxi, South China. J Microbiol 47: 135–141.
56. Duc L, Noll M, Meier BE, Burgmann H, Zeyer J (2009) High diversity of diazotrophs in the forefield of a receding alpine glacier. Microb Ecol 57: 179–190.
57. Palmer K, Drake HL, Horn MA (2010) Association of novel and highly diverse acid-tolerant denitrifiers with N_2O fluxes of an acidic fen. Appl Environ Microbiol 76: 1125–1134.
58. Zhang LM, Offre PR, He JZ, Verhamme DT, Nicol GW, et al. (2010) Autotrophic ammonia oxidation by soil thaumarchaea. Proc Natl Acad Sci U S A 107: 17240–17245.
59. Reed DW, Smith JM, Francis CA, Fujita Y (2010) Responses of ammonia-oxidizing bacterial and archaeal populations to organic nitrogen amendments in low-nutrient groundwater. Appl Environ Microbiol 76: 2517–2523.
60. Moin NS, Nelson KA, Bush A, Bernhard AE (2009) Distribution and diversity of archaeal and bacterial ammonia oxidizers in salt marsh sediments. Appl Environ Microbiol 75: 7461–7468.
61. Nicol GW, Leininger S, Schleper C, Prosser JI (2008) The influence of soil pH on the diversity, abundance and transcriptional activity of ammonia oxidizing archaea and bacteria. Environ Microbiol 10: 2966–2978.
62. Zhou J, Wu L, Deng Y, Zhi X, Jiang YH, et al. (2011) Reproducibility and quantitation of amplicon sequencing-based detection. ISME J.
63. Lin YT, Lin CP, Chaw SM, Whitman WB, Coleman DC, et al. (2010) Bacterial community of very wet and acidic subalpine forest and fire-induced grassland soils Plant and Soil DOI:10.1007/s11104-010-0308-3.
64. Wakelin SA, Colloff MJ, Harvey PR, Marschner P, Gregg AL, et al. (2007) The effects of stubble retention and nitrogen application on soil microbial community structure and functional gene abundance under irrigated maize. FEMS Microbiol Ecol 59: 661–670.
65. Rothballer M, Eckert B, Schmid M, Fekete A, Schloter M, et al. (2008) Endophytic root colonization of gramineous plants by *Herbaspirillum frisingense*. FEMS Microbiol Ecol 66: 85–95.
66. Miyamoto T, Kawahara M, Minamisawa K (2004) Novel endophytic nitrogen-fixing clostridia from the grass *Miscanthus sinensis* as revealed by terminal restriction fragment length polymorphism analysis. Appl Environ Microbiol 70: 6580–6586.
67. Kirchhof G, Eckert B, Stoffels M, Baldani JI, Reis VM, et al. (2001) *Herbaspirillum frisingense* sp. nov., a new nitrogen-fixing bacterial species that occurs in C4-fibre plants. Int J Syst Evol Microbiol 51: 157–168.
68. Hai B, Diallo NH, Sall S, Haesler F, Schauss K, et al. (2009) Quantification of Key Genes Steering the microbial nitrogen cycle in the rhizosphere of sorghum cultivars in tropical agroecosystems. Applied and Environmental Microbiology 75: 4993–5000.
69. Kowalchuk GA, Stephen JR, De Boer W, Prosser JI, Embley TM, et al. (1997) Analysis of ammonia-oxidizing bacteria of the beta subdivision of the class Proteobacteria in coastal sand dunes by denaturing gradient gel electrophoresis and sequencing of PCR-amplified 16S ribosomal DNA fragments. Appl Environ Microbiol 63: 1489–1497.
70. Ehrich S, Behrens D, Lebedeva E, Ludwig W, Bock E (1995) A new obligately chemolithoautotrophic, nitrite-oxidizing bacterium, *Nitrospira moscoviensis* sp. nov. and its phylogenetic relationship. Arch Microbiol 164: 16–23.
71. Offre P, Prosser JI, Nicol GW (2009) Growth of ammonia-oxidizing archaea in soil microcosms is inhibited by acetylene. Fems Microbiology Ecology 70: 99–108.
72. Tourna M, Freitag TE, Nicol GW, Prosser JI (2008) Growth, activity and temperature responses of ammonia-oxidizing archaea and bacteria in soil microcosms. Environmental Microbiology 10: 1357–1364.
73. Jia Z, Conrad R (2009) Bacteria rather than Archaea dominate microbial ammonia oxidation in an agricultural soil. Environ Microbiol.
74. Di HJ, Cameron KC, Shen JP, Winefield CS, O'Callaghan M, et al. (2009) Nitrification driven by bacteria and not archaea in nitrogen-rich grassland soils. Nature Geoscience 2: 621–624.
75. Martens-Habbena W, Berube PM, Urakawa H, de la Torre JR, Stahl DA (2009) Ammonia oxidation kinetics determine niche separation of nitrifying Archaea and Bacteria. Nature 461: 976–979.
76. Taylor AE, Zeglin LH, Dooley S, Myrold DD, Bottomley PJ (2010) Evidence for different contributions of archaea and bacteria to the ammonia-oxidizing potential of diverse Oregon soils. Appl Environ Microbiol 76: 7691–7698.
77. Kramer SB, Reganold JP, Glover JD, Bohannan BJ, Mooney HA (2006) Reduced nitrate leaching and enhanced denitrifier activity and efficiency in organically fertilized soils. Proc Natl Acad Sci U S A 103: 4522–4527.
78. Wallenstein MD, Peterjohn WT, Schlesinger WH (2006) N fertilization effects on denitrification and N cycling in an aggrading forest. Ecol Appl 16: 2168–2176.
79. Hutchinson GL, Mosier AR (1979) Nitrous oxide emissions from an irrigated cornfield. Science 205: 1125–1127.
80. Wertz S, Degrange V, Prosser JI, Poly F, Commeaux C, et al. (2006) Maintenance of soil functioning following erosion of microbial diversity. Environ Microbiol 8: 2162–2169.

12

Assessment of the AquaCrop Model for Use in Simulation of Irrigated Winter Wheat Canopy Cover, Biomass, and Grain Yield in the North China Plain

Xiu-liang Jin[1,2,3], Hai-kuan Feng[2,3], Xin-kai Zhu[1], Zhen-hai Li[2,3], Sen-nan Song[1,2,3], Xiao-yu Song[2,3], Gui-jun Yang[2,3], Xin-gang Xu[2,3]*, Wen-shan Guo[1]*

1 Key Laboratory of Crop Genetics and Physiology of Jiangsu Province, Yangzhou University, Yangzhou, China, **2** Beijing Research Center for Information Technology in Agriculture, Beijing Academy of Agriculture and Forestry Sciences, Beijing, China, **3** National Engineering Research Center for Information Technology in Agriculture, Beijing, China

Abstract

Improving winter wheat water use efficiency in the North China Plain (NCP), China is essential in light of current irrigation water shortages. In this study, the AquaCrop model was used to calibrate, and validate winter wheat crop performance under various planting dates and irrigation application rates. All experiments were conducted at the Xiaotangshan experimental site in Beijing, China, during seasons of 2008/2009, 2009/2010, 2010/2011 and 2011/2012. This model was first calibrated using data from 2008/2009 and 2009/2010, and subsequently validated using data from 2010/2011 and 2011/2012. The results showed that the simulated canopy cover (CC), biomass yield (BY) and grain yield (GY) were consistent with the measured CC, BY and GY, with corresponding coefficients of determination (R^2) of 0.93, 0.91 and 0.93, respectively. In addition, relationships between BY, GY and transpiration (T), ($R^2 = 0.57$ and 0.71, respectively) was observed. These results suggest that frequent irrigation with a small amount of water significantly improved BY and GY. Collectively, these results indicate that the AquaCrop model can be used in the evaluation of various winter wheat irrigation strategies. The AquaCrop model predicted winter wheat CC, BY and GY with acceptable accuracy. Therefore, we concluded that AquaCrop is a useful decision-making tool for use in efforts to optimize wheat winter planting dates, and irrigation strategies.

Editor: Dafeng Hui, Tennessee State University, United States of America

Funding: The National Natural Science Foundation of China (Grant No. 41001244, 41271415), Beijing Nova Program (Grant No.2011036), the 8th Six Talents Peak Project of Jiangsu Province (2011-NY039) and the Key Projects in the National Science & Technology Pillar Program during the Twelfth Five-Year Plan Period (Grant No.2012BAH27B04). The authors are grateful to Mr. Weiguo Li, Mrs. Zhihong Ma and Hong Chang for data collection. The funders had no role in study design, data collection and analysis, decision to publish, or preparation of the manuscript.

Competing Interests: The authors have declared that no competing interests exist.

* E-mail: xxgpaper@126.com (XX); guows@yzu.edu.cn (WG)

Introduction

Winter wheat (*Triticum aestivum* L.) is an important staple food crop for the majority of the North China Plain (NCP) population [1]. However, increasing industrial, and domestic water use has resulted in a reduction in water available for irrigation of these crops. Thus there is a growing need for improvement to this region's agriculture water resources management, especially given increasing food demands of the region's increasing population.

It is widely known that well-timed irrigation can substantially increase water use efficiency (WUE) [2,3], providing an optimal growth environment throughout the season [4,5]. In fact, various studies have described several such irrigation strategies for use by farmers [6–16]. Since the mid-1960s, the relationship between water and crop yield has been described with both empirical and mechanistic models [17–20]. For example, De Wit [21] proposed that a linear relationship between yield and water consumption exists. In contrast, Downey [22], via deficit irrigation studies, suggested that there exists a nonlinear relationship between water and yield. Based on the above studies, the Minhas model [23], Rao model [18], Blank model [24], and the Stewart model [25] were developed. More recently, Wang and Sun [26] showed that a quadratic relationship between crop yield and crop water consumption did in fact exist. Their work was followed by Kang et al. [27] in which a multiple and synergistic model (developed under deficit irrigation conditions) was proposed. At present, the simulation of the soil-plant-atmosphere continuum remains an important part of such research, especially with regard to expansion of the application range of resulting models to a wider array of cropping systems.

Therefore, the Food and Agricultural Organization (FAO) developed the AquaCrop model in an effort to meet this need in 2009. This model was originated from the "yield response to water" data of Doorenbos and Kassam [28], and evolved to a normalized crop water productivity (NCWP) concept [29]. Compared with other models, AquaCrop is relatively simple to operate by those with little, or no research experience, and allows for simulation of crop performance in multiple scenarios. In addition to a high level of accuracy, this robust model requires a limited set of input parameters, most of which are relatively easy to acquire [29,30]. The AquaCrop model is also capable of predicting crop productivity, water requirements, and water use efficiency under water-limiting conditions [31]. To date, this

Table 1. Winter wheat cultivars and planting dates were selected in the 2008, 2009, 2010 and 2011.

Winter wheat cultivars	Planting dates
Nongda195, Jingdong8, Jing9428	Sep. 28th, Oct. 7th, and Oct. 20th, 2008
Nongda195, Jingdong13, Jing9428	Sep. 25th, Oct. 5th, and Oct. 15th, 2009
Nongda195, Yannong19, Jing9428	Sep. 25th, Oct. 5th, and Oct. 15th, 2010
Nongda211, Zhongmai175, Jingdong8, Jing9843	Sep. 25th, 2011

Note: There are three winter wheat cultivars and each has three planting dates in 2008, 2009 and 2010. In 2011, four cultivars are planted on the same date.

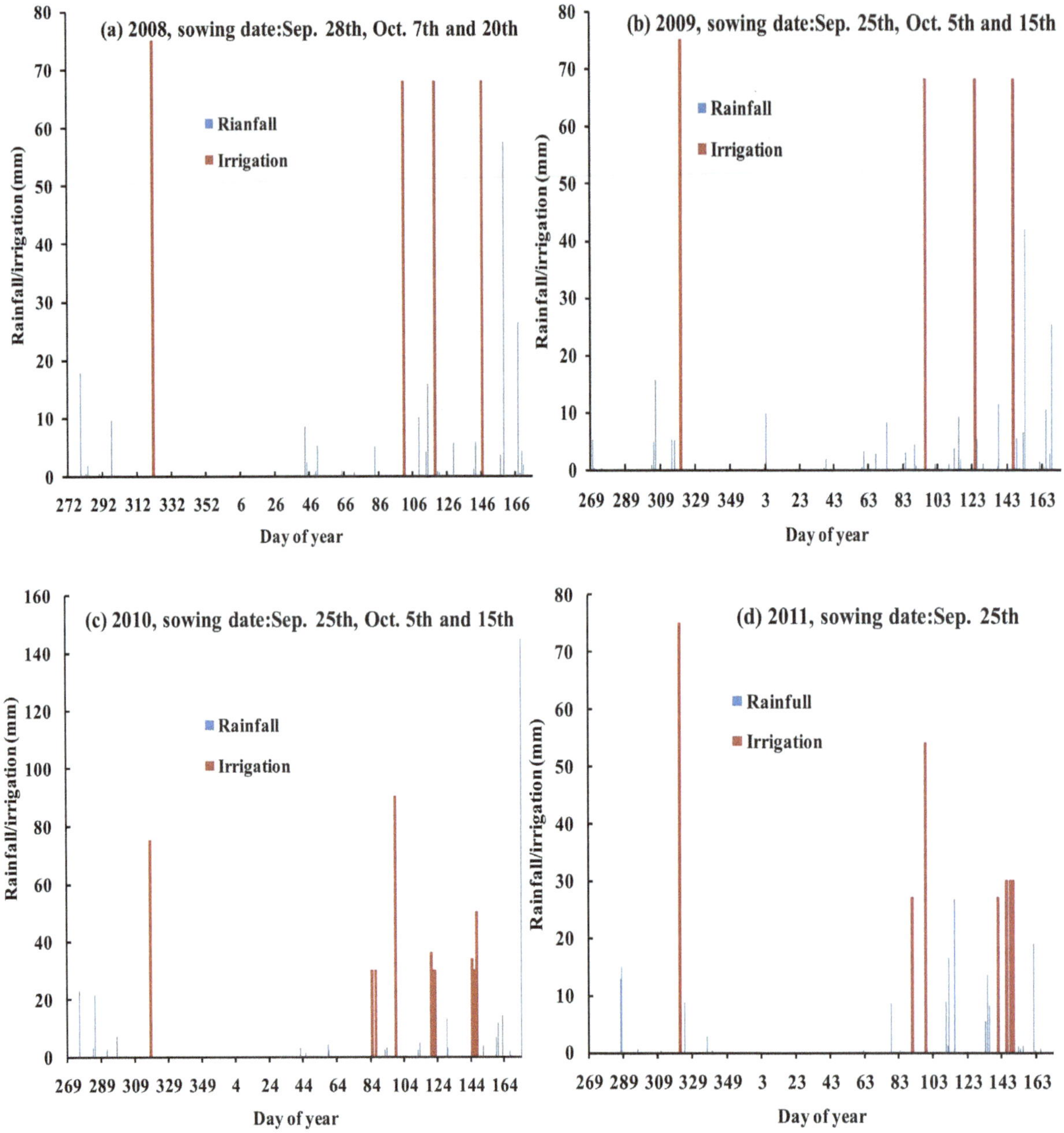

Figure 1. Daily rainfall, and supplemental irrigation for the Xiaotangshan site during the cropping seasons 2008/2009 (a), 2009/2010(b), 2010/2011 (c) and 2011/2012(d).

Table 2. Irrigation schedule during experimental period (2008/2009, 2009/2010, 2010/2011 and 2011/2012).

Day of year	Year	Month	Day	Irrigation amount (mm)
321	2008	11	16	75
100	2009	4	10	68
118	2009	4	28	68
146	2009	5	26	68
320	2009	11	15	75
95	2010	4	5	68
124	2010	5	4	68
146	2010	5	26	68
318	2010	11	13	75
84	2011	3	24	30
86	2011	3	26	30
98	2011	4	7	90
120	2011	4	29	36
121	2011	4	30	30
122	2011	5	2	30
144	2011	5	23	34
145	2011	5	24	30
146	2011	5	25	30
147	2011	5	26	54
321	2011	11	16	75
90	2012	3	30	27
98	2012	4	7	54
140	2012	5	19	27
145	2012	5	24	30
147	2012	5	26	30
149	2012	5	28	30

model has been successfully tested for cotton [32,33], maize [30,34–38], wheat [39–42], sugar beet [37], sunflower [37,43], groundnut [44], potato [45,46], quinoa [16], Teff [47], barley [48,49], green onion [50] and tomato [51] under a wide-range of environments. Previous studies have demonstrated that the AquaCrop model accurately simulates crop canopy cover (CC), biomass yield (BY) and grain yield (GY) under both regular, and deficit irrigation, and in low soil fertility conditions. In such unreliable water-limited environments as the NCP, the AquaCrop model is a potentially valuable tool for use in efforts to maximize this region's winter wheat yield. Therefore, the objective of this study was to validate this model in simulating the effects of planting date, and multiple irrigation scenarios on: (1) canopy cover, (2) biomass yield, (3) grain yield, and (4) water use efficiency of winter wheat in the NCP. These data will provide some guidelines for efforts to optimize irrigation management for winter wheat crops in this region.

Materials and Methods

Study Site

This field experiments were conducted in the 2008/2009, 2009/2010, 2010/2011 and 2011/2012 growing seasons at the Xiaotangshan experimental site (44.17° N, 116.433°E), Beijing, PR China. This area is representative of the overall soil and crop management practices in this region. The soil is fine-loamy, with a nitrate Nitrogen (NO_3–N) content of 3.16–14.82 mg kg^{-1}, an ammonium Nitrogen (NH_3–N) content of 10.20–12.32 mg kg^{-1}, an Olsen P of 3.14–21.18 mg kg^{-1}, an exchangeable K of 86.83–120.62 mg kg^{-1}, and an organic matter content of 15.84–20.24g kg^{-1} with in the uppermost 0–30 cm layer. Beijing is characterized by a typical continental climate, with maximum temperatures of 26.1°C in the summer and minimum temperatures of −4.7°C in the winter. Throughout all seasons, the temperature fluctuated daily with significant differences between night and day. During this experimental period, the average annual precipitation was 650 mm, and the frost-free period was on average 180 days.

Local winter wheat cultivars and planting dates are shown in Table 1. Each plot area is 100 m^2 in 2008, 2009 and 2010, and 300 m^2 in 2011. The experiment was designed as a 2-way factorial arrangement of treatments in a randomized complete block design, with three replications for each treatment. Plot management followed local standard practices (weed control, pest management and fertilizer application) for wheat production in this region. The Xiaotangshan Experimental Site belongs to the National Engineering Research Center for Information Technology in Agriculture. It gives some permission for us to study relative agriculture research within this area. We confirm that the field studies did not involve endangered or protected species.

Climate Data Collection and Analysis

Climate data for the experimental site was obtained from the local Xiaotangshan meteorological station. The daily reference evapotranspiration (ET_o) for the growing season from 2008 to 2012 was calculated based on the FAO Penman-Monteith method as described in Allen et al. [52], and the ET_o calculator (FAO, 2009) [53]. Daily maximum and minimum temperature, relative humidity, wind speed, rainfall, and total sunshine hours were recorded directly at the Xiaotangshan experimental site. The total rainfall, from sowing to harvest was 199, 208, 145 and 168 mm in

Table 3. Physical soil characteristics of the Xiaotangshan experimental site.

Site	Soil texture	Groundwater table	Depth (m)	Moisture content (vol%)			Ksat (mm day^{-1})	CN
				Sat	FC	WP		
Xiaotangshan	fine-loamy	3.5 m	0.0–0.1	51.1	27.3	8.8	240	75
			0.1–0.2	51.3	27.3	8.7	240	
			0.2–0.3	54.7	34.8	13.2	224	

Note: FC, field capacity; WP, wilting point; Sat, water content at saturation; CN, curve number; and K_{sat}, saturation hydraulic conductivity describes water movement through saturated media. The values of saturated hydraulic conductivity in soils vary within a wide range of several orders of magnitude, depending on the soil material.

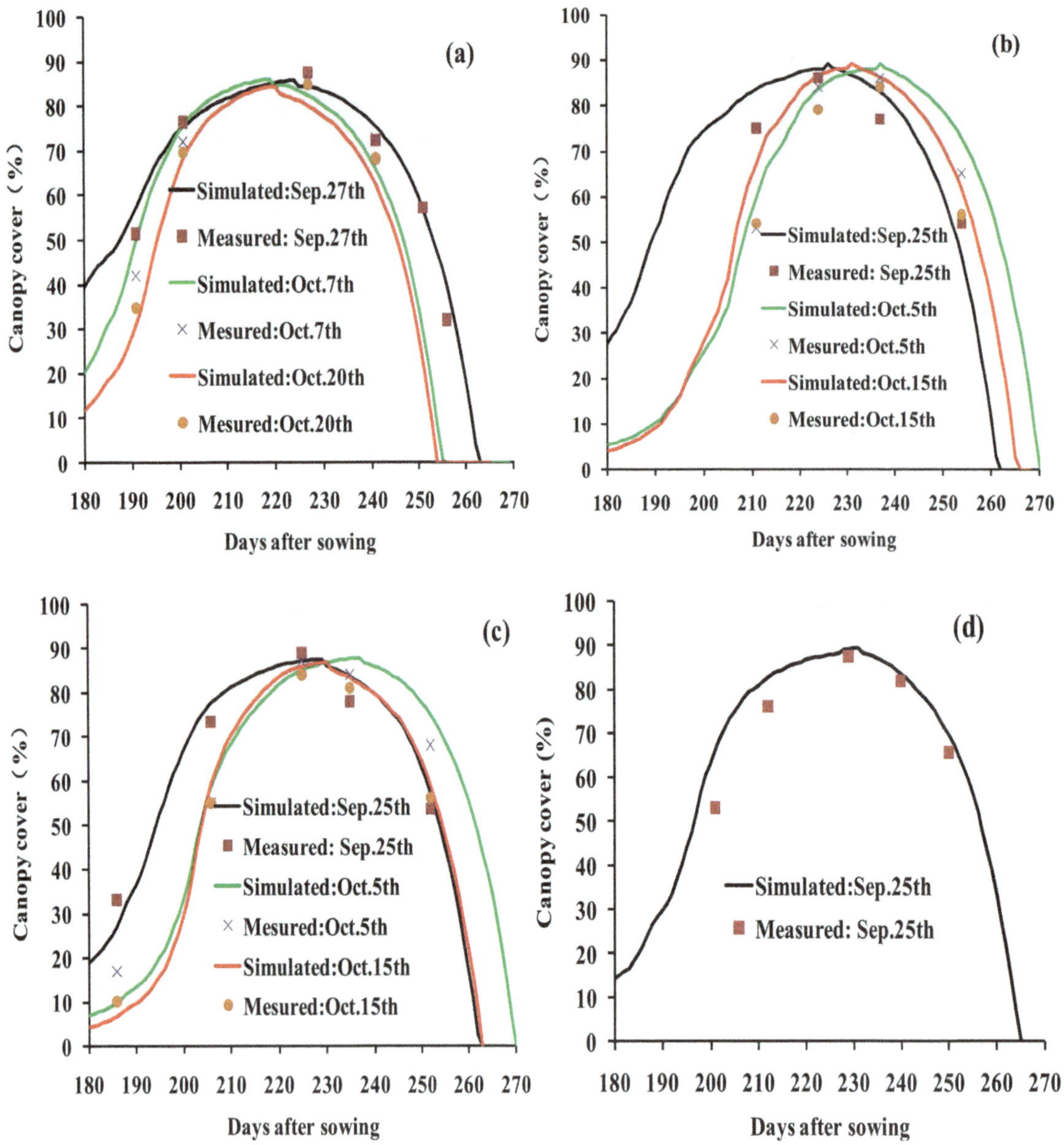

Figure 2. Simulated and measured canopy cover (CC) for winter wheat under different planting dates and irrigation strategies in the cropping season 2008/2009 (a), 2009/2010 (b), 2010/2011 (c) and 2011/2012 (d).

2008/2009, 2009/2010, 2010/2011 and 2011/2012, respectively. Supplemental irrigation was applied to treatments following cessation of any rain events (Fig. 1a–d and Table 2).

Soil Data of the Experimental Site

The soil at the Xiaotangshan experimental site represents the major soil type (fine-loamy) on which winter wheat is grown in NCP. The soil was at maximum field capacity (27.3% at 0.0–0.1 m, 27.3% at 0.1–0.2 m and 34.8% at 0.2–0.3 m) during sowing and early establishment. The physical soil characteristics were measured directly in the field and used for input into AquaCrop (Table 3).

Field Experiments and Crop Data Collection

In 2008/2009, 2009/2010, 2010/2011, and 2011/2012 aboveground biomass was determined 5–6 times from a 0.25 m^2 area by randomly cutting four representative plants from each plot. All plant samples were heated to 105°C, oven dried at 70°C to a constant weight, and final dry weight (DW) recorded.

Leaf area index (LAI) was estimated using the following two methods:

(1) By multiplying the plant population by the leaf area per plant as described in Kar et al. [54]. Area of the leaf was measured manually from 20 plants using a straightedge. Counting of plant populations was conducted manually from a 0.1 m^2 area. The LAI equation is as follows:

Table 4. Input data of crop parameters used in AquaCrop model.

Description	Value	Unit
Base temperature	0	°C
Upper temperature	26	°C
Canopy growth coefficient (CGC): Increase in CC per day	0.03	%/day
Canopy decline coefficient (CDC): Decrease in CC per day at senescence	0.09	%/day
Maximum canopy cover (CC_x)	90	%
Water productivity (NCWP)	15	g/cm^2
Reference harvest index (HI)	46	%
Upper threshold for canopy expansion (P_{upper})	0.20	% of TAW
Lower threshold for canopy expansion (P_{lower})	0.65	% of TAW
Leaf expansion stress coefficient curve shape	3.0	-
Upper threshold for stomatal closure (P_{upper})	0.65	% of TAW
Minimum effective rooting depth	0.3	m
Maximum effective rooting depth	1.2	m
Canopy senescence stress coefficient (P_{upper})	0.70	% of TAW
Shape factor describing root zone expansion	1.5	-
Crop coefficient when canopy is complete but prior to senescence	1.1	-
Senescence stress coefficient curve shape	3.0	-
Allowable maximum increase of specified HI	15	%
Minimum air temperature below which pollination starts to fail	5	°C
Maximum air temperature above which pollination starts to fail	35	°C
Water Productivity normalized for ET_o and CO_2 during yield formation	100	%
Time from sowing to emergence	7	Days
Time from sowing to flowering	232	Days
Time from sowing to start senescence	236	Days
Length of the flowering stage (days)	10	Days

$$LAI = 0.75 \times \rho \times \left(\frac{\sum_{i=1}^{m}\sum_{i=1}^{n}(L_{ij} \times B_{ij})}{m}\right) \quad (1)$$

Where ρ is plant density, m is the number of measured plants, L_{ij} is leaf length, B_{ij} is the maximum leaf width, and n is the number of leaves of the nth plant.

(2) The LAI-2000 Plant Canopy Analyzer (LI-COR Inc., Lincoln, NE, USA) was used in measuring for determination of LAI. The resulting values were similar to those obtained via the manual LAI calculations, thus the LAI-2000 data was used as the model input data.

Canopy cover was estimated from different irrigation treatments in 2008/2009, 2009/2010, 2010/2011, and 2011/2012 based on Hsiao et al. [30] using the following:

$$CC = 1 - \exp^{(-0.65 \times LAI)} \quad (2)$$

where CC is canopy cover, and LAI is the leaf area index.

Grain yield was measured following maturation from samples obtained from a 1.5 m^2 area in each plot, with three replications for each treatment. Collected grain was dried and weighed on an electronic scale (±0.01 g). As there were no significant differences

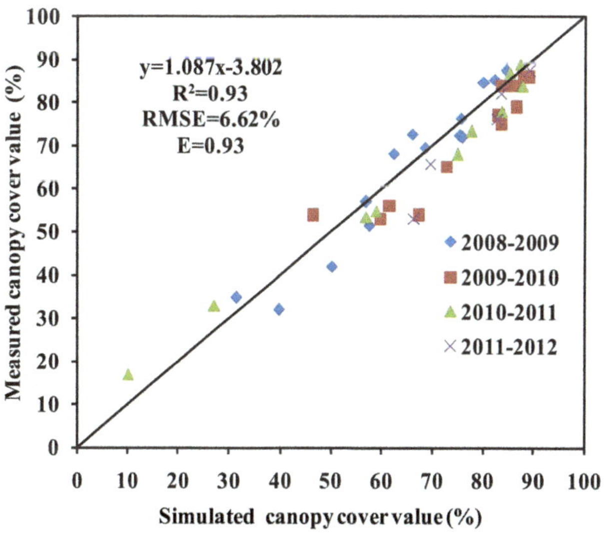

Figure 3. Relationship between the measured and simulated canopy cover (CC) in winter across 4 years. Note: x represents the simulated CC, y represents the measured CC. The intercept represents the relative error between the simulated CC and the measured CC. The slope represents the consistency between the simulated CC and the measured CC.

Table 5. Simulated and measured canopy cover (CC) for winter wheat under different planting dates in 2008/2009, 2009/2010, 2010/2011 and 2011/2012.

Year	Planting date	Slope	Intercept	R^2	RMSE (%)	E
2008–2009	28/9/2008	1.036	1.575	0.89	4.39	0.90
2008–2009	7/10/2008	1.251	−19.64	0.97	5.84	0.94
2008–2009	20/10/2008	1.018	2.032	0.98	4.25	0.96
2008–2009		1.060	−2.093	0.91	6.62	0.91
2009–2010	25/9/2009	1.201	−11.98	0.91	5.93	0.90
2009–2010	5/10/2009	1.082	−6.316	0.95	4.35	0.94
2009–2010	15/10/2009	0.930	8.443	0.92	7.02	0.91
2009–2010		1.120	−8.039	0.93	4.94	0.93
2010–2011[a]	25/9/2010	1.157	−12.31	0.97	5.73	0.94
2010–2011[a]	5/10/2010	1.180	−13.21	0.92	3.18	0.92
2010–2011[a]	15/10/2010	1.183	−7.109	0.96	3.78	0.95
2010–2011[a]		1.134	−7.996	0.96	7.19	0.94
2011–2012[a]	25/9/2011	0.743	24.35	0.96	7.15	0.93

Note: [a]Validation data set: R^2, determination coefficient; E, model efficiency; RMSE, root mean square of error.
The intercept represents the relative error between the simulated CC and the measured CC, The slope represents the consistency between the simulated CC and the measured CC.

between winter wheat varieties in many of the measured characteristics (e.g. phenological development, canopy cover, etc.), the average grain yield of the different varieties was considered in model simulations.

Water Use Efficiency (WUE)

Water use efficiency (WUE) is defined as the grain yield per unit amount of water consumed [55]. In this study grain water use efficiency (Grain-WUE), and biomass water use efficiency (Biomass-WUE) were calculated using Eq. (3) and Eq. (4), respectively, as in Araya et al. (2010b) [48]:

$$Grain\text{-WUE} = \left[\frac{\text{GY}}{\sum \text{T}}\right] \tag{3}$$

$$Biomass\text{-WUE} = \left[\frac{BY}{\sum T}\right] \tag{4}$$

Where GY is the grain yield kg ha^{-1} (measured), T is the transpiration as determined using AquaCrop model, and BY is the total final aboveground biomass yield in kg ha^{-1} (measured).

Description of AquaCrop Model

The AquaCrop model was proposed by the FAO in 2009, with a detailed description presented in Steduto et al. [29], and Raes et al. [31]. The model computes a daily water balance, and separates evapotranspiration into evaporation and transpiration components. Transpiration is correlated with canopy cover, which is proportional to the degree of soil cover, and evaporation is proportional to the area of soil not covered by vegetation. The crop's stomata conductance, canopy senescence, leaf growth, and yield response to water stress are modeled using four stress coefficients (stomata closure, leaf expansion, canopy senescence, and change in harvest index (HI)). The model subsequently estimates yield from the daily crop transpiration values.

In general, the normalized crop water productivity (NCWP) is considered constant for a given climate condition and crop (For crops not nutrient-limited, the model provides categories ranging from slight to severe deficiencies corresponding to lower water productivity (WP).) is applicable for using in different locations, seasons, and even future climates [29]. Depending on the crop, NCWP increases slightly with an increase in atmospheric CO_2 concentration [29]. NCWP is set between 13 and 20 g m^{-2} for C_3 crops. For example, NCWP is set at 15 g m^{-2} for the winter wheat according to the AquaCrop Manual (Annex I Section I.10 Wheat, Pages A39–A42) [56,57]. In our current study, we have not included any of the water stress study data; therefore NCWP remained at 15 g m^{-2} for the winter wheat. The crop's daily aboveground biomass is calculated using NCWP from the AquaCrop model [29,30]. Biomass yield (BY) is calculated by multiplying NCWP by the ratio of crop transpiration (T), and evapotranspiration (ET_o), following calculation of BY (its harvestable portion), and the grain yield (GY) is determined via harvest index (HI).

These changes are described by the following Eqs. (5) and (6):

$$BY = \sum (T/ET_O) \times NCWP \tag{5}$$

$$GY = BY \times HI \tag{6}$$

Where BY is biomass yield in kg ha^{-1}, T is crop transpiration in mm, ET_o is evapotranspiration in mm, $NCWP$ is the normalized crop water productivity in g m^{-2}, HI is harvest index, and GY is grain yield in kg ha^{-1}.

Data Analysis

Winter wheat canopy cover (CC), biomass yield (BY) and grain yield (GY) in AquaCrop were calibrated using the measured data sets of 2008/2009 and 2009/2010, and validated using the 2010/2011 and 2011/2012 measured data sets. The good fit regression equation between the simulated and observed values was corroborated using prediction error statistics. The coefficient of determination (R^2), root mean square error (RMSE), and model efficiency (E) were used as the error statistics to evaluate both calibration and validation results. The E and R^2 were used to access the predictive power of the model, and the RMSE indicated the error in model prediction. In this study, the prediction model output for CC, GY and BY during harvest was used for model evaluation. These statistical indices were used to compare measured and simulated values. Model performance was assessed using E (Nash and Sutcliffe, 1970) as follows:

$$E = 1 - \frac{\sum_{i=1}^{n} (S_i - O_i)^2}{\sum_{i=1}^{n} (O_i - \bar{O}_i)^2} \tag{7}$$

where S_i and O_i are predicted, and observed data, respectively. $\bar{O}_i$ is the mean value of O_i, and n is the number of observations.

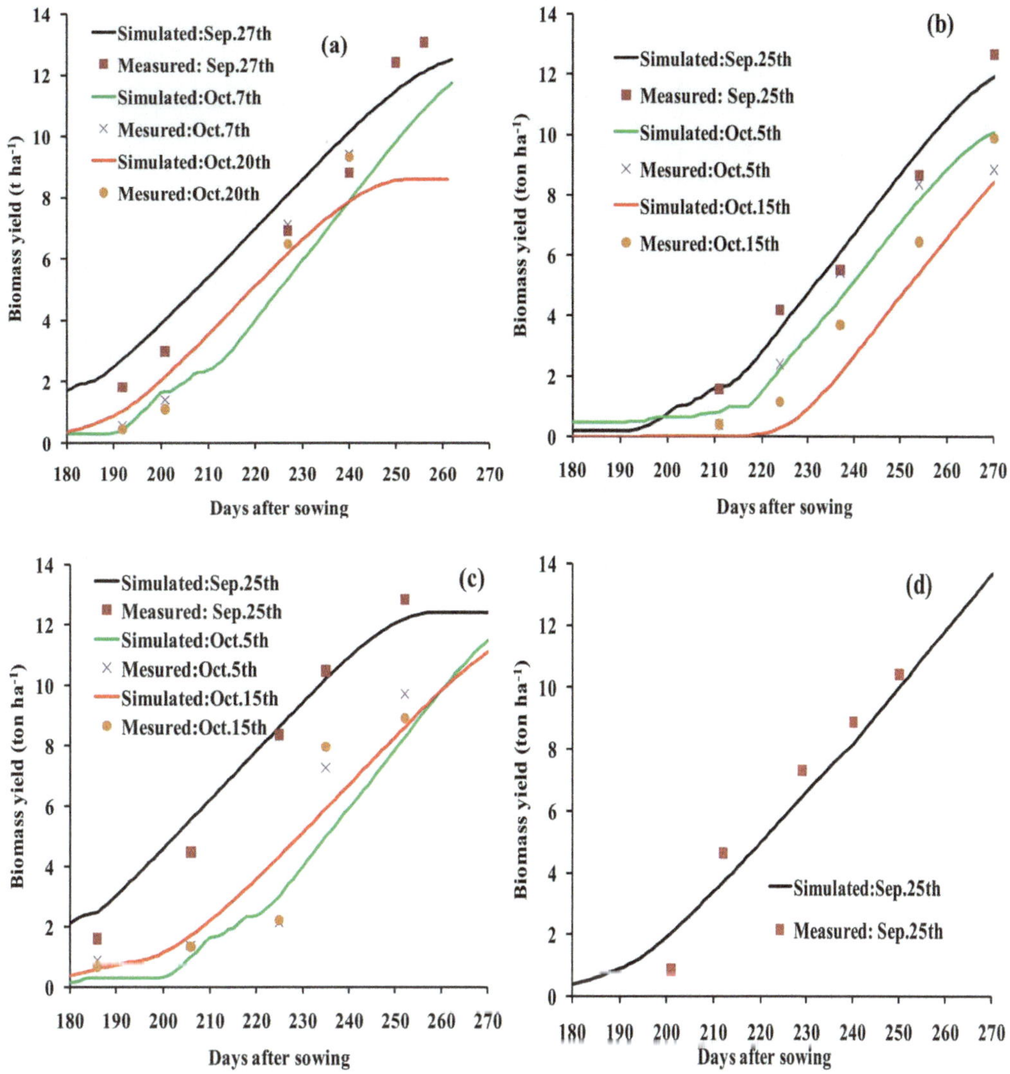

Figure 4. The simulated as compared with the measured aboveground biomass accumulation at different growth stages for winter wheat with different planting dates and irrigation strategies in the cropping season 2008/2009 (a), 2009/2010 (b), 2010/2011 (c) and 201/2012 (d).

$$RMSE = \sqrt{\frac{\sum_{i=1}^{n} (S_i - O_i)^2}{n}} \tag{8}$$

E and R^2 approaching one, and a RMSE near zero were indicators of improved model performance. Following model calibration, and validation, Grain-WUE and Biomass-WUE were calculated using Eqs. 3 and 4.

Result and Analysis

AquaCrop Model Calibration and Validation Results

The crop parameters used to calibrate the AquaCrop model are presented in Table 4. Key stress parameters (e.g. canopy growth, canopy senescence stress coefficient) (P_{upper}) were adjusted as needed to simulate CC. There was a strong linear relationship between the simulated and the measured CC ($R^2 = 0.93$, RMSE = 6.62% and E = 0.93) for winter wheat under different planting dates, and irrigation strategies in the cropping season 2008/2009, 2009/2010, 2010/2011 and 2011/2012 (Figs. 2a, 2b, 2c, 2d and 3). The R^2, RMSE and E of the simulated and measured CC were 0.91, 6.62% and 0.91, respectively in 2008/2009. And the R^2, RMSE and E values of CC were 0.93, 4.94% and 0.93 in 2009/2010, 0.96, 7.19%, 0.94 in 2010/2011, and 0.96, 7.15%, and 0.93 in 2011/2012, respectively (Table 5).

The simulated aboveground BY was similar to that measured (Figs. 4a, 4b, 4c and 4d). The stress coefficients were also adjusted, and readjusted as needed to simulated aboveground biomass. There was a strong relationship between measured and simulated BY across the four years (Fig. 5 and Table 6). The GY was also similar to the measured GY across all four years (R^2, RMSE and E values of 0.93, 0.52 ton ha^{-1} and 0.92, respectively) (Fig. 6).

The R^2, RMSE and E also showed good performance between the simulated and the measured values for CC ($R^2 = 0.89$–0.98, RMSE = 3.18–7.19% and E = 0.90–0.96) and BY ($R^2 = 0.92$–0.98, RMSE = 1.12–1.84 ton ha^{-1} and E = 0.92–0.96) (Tables 5 and 6). Higher R^2 and E values and lower RMSE values indicated good

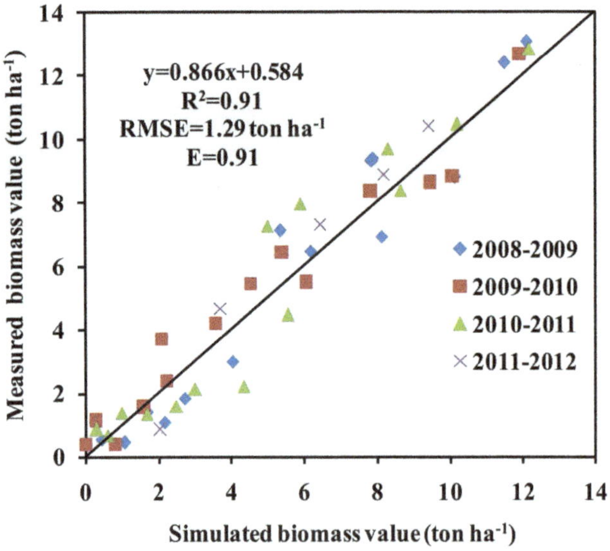

Figure 5. Relationship between the measured and simulated biomass yield (BY) in winter wheat across 4 years. Note: x represents the simulated BY, y represents the measured BY. The intercept represents the relative error between the simulated biomass yield and the measured BY. The slope represents the consistency between the simulated BY and the measured BY.

Table 6. Simulated and measured biomass yield (BY) for winter wheat under different planting dates in 2008/2009, 2009/2010, 2010/2011 and 2011–2012.

Year	Planting date	Slope	Intercept	R^2	RMSE (ton ha^{-1})	E
2008–2009	28/9/2008	0.772	1.286	0.98	1.34	0.95
2008–2009	7/10/2008	0.787	−0.270	0.98	1.54	0.96
2008–2009	20/10/2008	0.802	−0.094	0.98	1.12	0.96
2008–2009		0.778	0.361	0.94	1.21	0.93
2009–2010	25/9/2009	0.993	1.315	0.92	1.39	0.92
2009–2010	5/10/2009	0.960	1.163	0.95	1.22	0.94
2009–2010	15/10/2009	0.954	1.331	0.97	1.27	0.95
2009–2010		0.945	1.284	0.95	1.29	0.94
2010–2011[a]	25/9/2010	0.809	0.617	0.97	1.21	0.95
2010–2011[a]	5/10/2010	1.043	0.378	0.97	1.37	0.95
2010–2011[a]	15/10/2010	0.797	0.019	0.97	1.84	0.94
2010–2011[a]		0.820	0.162	0.97	1.25	0.96
2011–2012[a]	25/9/2011	0.953	0.158	0.96	0.93	0.95

Note: [a]Validation data set: R^2, determination coefficient; E, model efficiency; RMSE, root mean square of error.
The intercept represents the relative error between the simulated BY and the measured BY, The slope represents the consistency between the simulated BY and the measured BY.

model performance. The calibrated results were also consistent with the validated results for CC, BY and GY (Tables 5 and 6). These results suggest that the AquaCrop model is useful for simulating winter wheat CC, BY and GY under different planting dates, and irrigation strategies.

Biomass, Grain Yield and Water Use Efficiency

Winter wheat was planted on Sep. 25th (normal sowing), Sep. 28th (normal sowing), Oct. 5th (late sowing), Oct. 7th (late sowing), Oct. 15th (late sowing) and Oct. 20th (late sowing) during 2008/2011. Winter wheat that was planted on Sep. 25th and 28th had greater biomass and grain yield than did those planted on Oct. 5th, 7th, 15th, and 20th (Table 6). There was relatively more transpiration and biomass yield, yet lower grain yield in 2008 than in 2010, but there was relatively more transpiration (T), biomass yield, and even grain yield in 2008 than in 2009 (Table 7). The highest biomass and grain yield was obtained in crops planted on Sep. 28th 2008 and on Sep. 25th 2011, respectively; the lowest biomass and grain yield was obtained in crops planted on Oct. 15th 2009. A relatively higher grain and biomass yield per m^3 of water was obtained from crops planted in 2010 than was in 2008 or 2009. Therefore, the biomass and grain yield water use efficiency (biomass-WUE and yield-WUE) was higher in 2010 than both 2008 and 2009. The biomass yield WUE was higher in 2008 than in 2009, yet the grain yield WUE was lower in 2008 than in 2009. Frequent irrigation with a small amount of water obviously improved grain yield in 2010/2011 and 2011/2012 and increased the biomass-WUE and grain-WUE (Table 5). A good relationship did exist between GY, BY, and T (R^2 values of 0.57 and 0.71, respectively) (Fig. 7), thus suggesting that T might be used in estimating biomass and grain yield.

Discussion

In this study, the AquaCrop model successfully predicted CC, BY GY in winter wheat. The crop parameters are adjusted to simulate CC, BY and GY for winter wheat under different planting dates and irrigation strategies. These adjustments were made to obtain more stable and closer relationships between the simulated values and the measured values. The results showed that the model calibration data sets from 2008/2009 and 2009/2010 were very consistent with the model validation data sets from 2010/2011 and 2011/2012. Good relationships were obtained

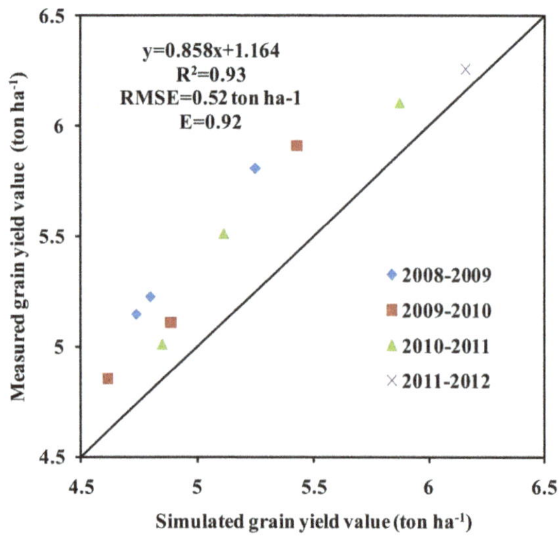

Figure 6. Relationship between the measured and simulated grain yield (GY) in winter wheat across 4 years. Note: x represents the simulated GY, y represents the measured GY. The intercept represents the relative error between the simulated GY and the measured GY. The slope represents the consistency between the simulated GY and the measured GY.

Table 7. Biomass-WUE and Grain-WUE in response to the seasonal transpiration over the different planting dates across years at Xiaotangshan experimental site.

Year	Planting date	Biomass (kg ha^{-1})	Grain (kg ha^{-1})	T (m^3 ha^{-1})	Biomass-WUE (kg m^{-3})	Grain-WUE (kg m^{-3})
2008–2009	28/9/2008	13072	5808	341.2	3.83	1.70
2008–2009	7/10/2008	11612	5227	259.1	4.48	2.02
2008–2009	20/10/2008	11144	5146	255.7	4.36	2.01
2009–2010	25/9/2009	12666	5909	322.3	3.93	1.83
2009–2010	5/10/2009	10212	5106	296.3	3.45	1.72
2009–2010	15/10/2009	9861	4852	207	4.76	2.34
2010–2011	25/9/2010	11812	6103	292.9	4.03	2.08
2010–2011	5/10/2010	11024	5512	251.1	4.39	2.20
2010–2011	15/10/2010	10062	5013	220.2	4.57	2.28
2011–2012	25/9/2011	12514	6257	306	4.09	2.04

Note: T, transpiration; WUE, water use efficiency.

between the simulated CC, BY and GY, and the measured CC, BY and GY across four years (Figs. 2,3,4,5 and 6, Tables 5 and 6). The best fitting model was obtained in 2009/2010, and the poorest fitting model was obtained in 2011/2012. These differences are very likely due to differences in rainfall and irrigation application (Fig. 1). Compared to 2008/2009, 2009/2010 and 2010/2011, the severe drought and inadequate irrigation occurred in 2011/2012. In the same year, CC, BY and GY were significantly different under different planting dates, likely due to accumulated temperature difference (Figs 3 and 4, Table 7). The results indicate that the AquaCrop model can be used to simulate CC, BY and GY for winter wheat under different planting dates and irrigation strategies. Our results (along with those in the Salemi et al. [39]), suggested that climate conditions, variety planted, and irrigation strategy could induce some differences in model simulations under different years. Heng et al. [34] demonstrated that the AquaCrop model is a good predictor of biomass and yield when irrigation is adequate; and this was corroborated by the results of this present study. In addition, the simulated BY was also consistent with the measured BY under different planting dates (Table 6 and Figs.4a–d). The CC results in this study were also similar to that observed in Salemi et al. [39], and Du et al. [40]. The results suggested that AquaCrop model could be used to simulate winter wheat CC, BY and GY under different planting dates and irrigation strategies. Winter wheat obtained higher biomass and grain yield when planted on Sep. 25th and Sep. 28th than on Oct. 5th, Oct. 7th, Oct. 15th and Oct. 20th (Table 7). This was likely due to the higher growing degree days (accumulated temperature) promoting CC growth, and BY, GY accumulation at the earlier planting date. The degree of CC affects the rate of transpiration and consequently BY and GY accumulation [32]. Therefore, the relatively higher BY and GY required relatively more temperature accumulation.

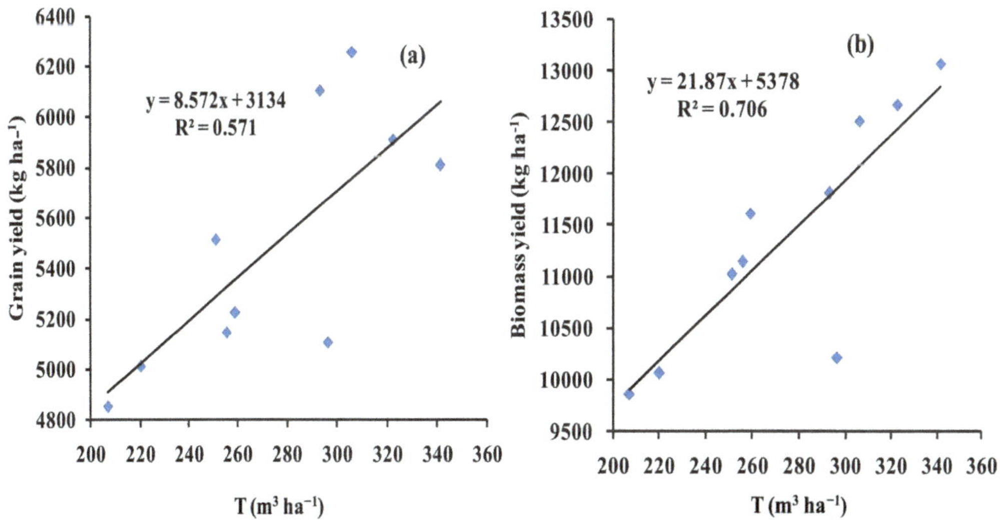

Figure 7. Relationships between the measured grain yield (GY), biomass yield (BY) and transpiration (T) in winter wheat. Note: (a) x represents transpiration, y represents the measured GY. The intercept represents the relative estimation error between transpiration and the measured GY. The slope represents the estimation consistency between transpiration and the measured GY. (b) x represents transpiration, y represents the measured BY. The intercept represents the relative estimation error between transpiration and the measured BY. The slope represents the estimation consistency between transpiration and the measured BY.

Biomass, and grain yield WUE decreased with increasing transpiration amount for all four years. This is consistent with that presented in Farahani et al. [32], but is not with that presented in Hedge [58]. In this present study, grain yield WUE ranged from 1.70 to 2.34 kg m^{-3}, reaching its maximum value on Oct. 15th 2009. Wang et al. [21] reported that grain yield WUE for winter wheat was between 0.7 and 1.3 kg m^{-3}, and Li et al. [50] reported it to be between 0.93 and 1.51 kg m^{-3} in WUE. These results are; however, not consistent with Wang et al. [26], and Li et al. [50], in which grain yield WUE was reported to be much greater. It indicated that winter wheat varieties developed faster, resulting in greater yield, leading to improvement in WUE. However, our results are consistent with that found in Fang et al. [59], in which the grain yield WUE ranged from 1.71 to 2.21 kg m^{-3}. The BY and GY in 2010/2011 and 2011/2012 were higher than in 2008/2009 and 2009/2010. This suggested that the frequent irrigation could be used to increase BY, GY, and promote biomass yield WUE and grain yield WUE for winter wheat at drought stages. This might be one of the reasons that the effects of drought on BY and GY were reduced by the frequent irrigation, thereby improving WUE.

This study demonstrated that the AquaCrop model could be used to evaluate different planting date and irrigation strategies for winter wheat in NCP. Many others have conducted AquaCrop model application studies under different crops and environment conditions [32,35,37,38,43,49]. These results indicated that AquaCrop model is stable and usable for different crops and environmental conditions. It is therefore plausible to use the AquaCrop model to improve irrigation management strategies that would maximize grain and biomass yield It is important to note that this study was limited to winter wheat in Xiaotangshan experimental site, Beijing, China. A subsequent study is focused on validating this model under deficit irrigation, thereby expanding the extent of this model's application in the future.

Conclusion

This paper demonstrated that the AquaCrop model adequately simulated the CC, BY, and GY of winter wheat under different planting dates and irrigation strategies. The simulated CC agreed well with the measured CC across all 4 years. The R^2, RMSE, E of CC winter wheat ranged from 0.89 to 0.98, 3.18% to 7.19% and 0.90 to 0.96, respectively. The measured and simulated BY were also closely related. The AquaCrop model calibrated the BY with the prediction error statistics of $0.92< R^2<0.98$, 1.12< RMSE <1.84 ton ha^{-1} and 0.92< E<0.96. The simulated GY was also consistent with the measured GY with the R^2, RMSE and E values of 0.93, 0.52 ton ha^{-1} and 0.92, respectively. The results demonstrated that frequent irrigation obviously improved BY, GY, biomass WUE and grain WUE for winter wheat in 2010/2011. These results suggest that the AquaCrop model could be used to predict CC, BY and GY of winter wheat with a high degree of reliability under various planting dates and irrigation strategies situations in the North China Plain (NCP).

Acknowledgments

We are very grateful to the reviewers for their comments, Mr. Weiguo Li and Mrs. Hong Chang for data collection.

Author Contributions

Conceived and designed the experiments: XJ HF ZL. Performed the experiments: XJ HF ZL SS. Analyzed the data: XJ XZ XS GY XX WG. Contributed reagents/materials/analysis tools: XZ XS GY XX WG. Wrote the paper: XJ. Measure data: XJ HF ZL SS.

References

1. Zhang HC (2000) Thoughts on cultivation techniques for high quality of wheat in China and its processing. Jiangsu Agricultural Sciences. 5: 2–6.(in Chinese).
2. Molden D (2003) A water-productivity framework for understanding and action. In: Kijne, J.W., Barker, R., Molden, D. (Eds.), Water Productivity in Agriculture: Limits and Opportunities for Improvement. International Water Management Institute, Colombo, Sri Lanka, pp. 1–18.
3. Zwart SJ, Bastiaanssen WGM (2004) Review of measured crop water productivity values for irrigated wheat, rice, cotton and maize. Agriculture Water Management. 69: 115–133.
4. Anac MS, Ali Ul M, Tuzel IH, Anac D, Okur B, et al. (1999) Optimum irrigation schedules for cotton under deficit irrigation conditions. In: Kirda, C.,Moutonnet, P., Hera, C., Nielsen, D.R. (Eds.), Crop Yield Response to Deficit Irrigation. Kluwer Academic Publishers, Dordrecht, Boston, London, pp. 196–212.
5. Raes D, Geerts S, Kipkorir E, Wellens J, Sahli A (2006) Simulation of yield decline as a result of water stress with a robust soil water balance model. Agriculture Water Management. 81: 335–357.
6. Hill RW, Allen RG (1996) Simple irrigation calendars: a foundation for water management. In: Food and Agricultural Organization of the United Nations (FAO) (Ed.), Irrigation Scheduling: From Theory to Practice. Rome, Italy, pp. 69–74.
7. Raes D, Sahli A, Van Looij J, Ben Mechlia N, Persoons E (2000) Charts for guiding irrigation in real time. Irrigation Drainage System.14: 343–352.
8. De Nys E, Kipkorir EC, Sahli A, Vaes R, Raes D (2001) Design of farmer oriented irrigation charts by using a soil-water balance model. In: Proceedings of the 4th Inter Regional Conference on Environment-Water, ICID, pp. 235–243.
9. Legesse A (2006) Sustainable irrigation in the Tigray Highlands of Northern Ethiopia. Master dissertation of the Master in Water Resources engineering. K.U. Leuven University, Leuven, Belgium.
10. Farre F, Faci JM (2009) Deficit irrigation in maize for reducing agricultural water use in a Mediterranean environment. Agriculture Water Management. 96: 384–394.
11. Kipkorir EC (2002) Optimal planning of deficit irrigation for multiple crop systems according to user specified strategy. Dissertations de Agricultura No. 514. Faculty of Agricultural Sciences. K.U. Leuven University, Belgium.
12. Debaek P, Aboudrare A (2004) Adaptations of crop manage to water-limited environments. European Journal of Agronomy. 21: 433–446.
13. Ali MH, Talukder MSU (2008) Increasing water productivity in crop production. A synthesis. Agricultural Water Management. 95: 1201–1213.
14. Behera SK, Panda RK (2009) Integrated management of irrigation water and fertilizers for wheat crop using field experiments and simulation modeling. Agricultural Water Management. 96, 1532–1540.
15. Blum FA (2009) Effective use of water (EUW) and not water-use efficiency (WUE) is the target of crop yield improvement under drought stress. Field Crops Research. 112: 119–123.
16. Geerts S, Raes D, Gracia M, Miranda R, Cusicanqui JA, et al. (2009) Simulating yield response of Quinoa to water availability with AquaCrop. Agronomy Journal. 101: 499–508.
17. Jensen ME (1968) Water consumption by agricultural plants. In:Water Deficits in Plant Growth (T T Koslowski, ed.). Academic Press, New York, 1: 1–22.
18. Rao NH, Sarma PBS, Chander S (1988) Irrigation scheduling under limited water supply. Agriculture Water Management. 15: 165–175.
19. Rao NH, Sarma PBS, Chander S (1992) Real time adaptive irigation scheduling under a limited water supply. Agriculture Water Management. 20: 267–279.
20. Penning de Vries FWT, Jansen DM, Berge ten HFM, Bakema A (1989) Simulation of Ecophysiological process of growth in several annual crops. Simulation Monographys, Pudoc, Wagenningen, pp. 82–96.
21. De Wit CT (1970) Dynamic concepts in biology. In: Prediction and Management of Photosynthetic Productivity. Proceedings International Biological Program/Plant Production Technical Meeting. Wageningen, Netherlands: PUDOC, pp. 17–23.
22. Downey LA (1972) Water-yield relations for non-forage crops. Journal of Irrigation and Drainage Division, American Society of Civil Engineers. 98: 107–115.
23. Minhas BS, Parikh KS, Srinivasan TN (1974) Towards the structure of a production function for wheat yields with dated in Puts of irrigation water. Water Resource Research. 10: 383–393.
24. Blank H (1975) Optimal irrigation decisions with limited water (D). Ph. D. Thesis, Department of Civil Engineering, Colorado State University, Fort Collins, CO.
25. Stewart JI, Hagan RM, Pruitt WO (1976) Production function sand predicted irrigation programs for Principal crops as required for water resources planning and increased water use efficiency (R). Final Report, U.S. Department of Interior, Washington DC, 1976, pp. 80.

26. Wang JP, Sun XF (2001) A Combining Water-Crop Relation Model. Irrigation and Drainage. 2: 76–78.
27. Kang SZ, Cai HJ, Feng SY (2004) Technique innovation and research fields of modern agricultural and ecological water-saving in the future. Transactions of the Chinese Society of Agricultural Engineering. 20: 1–6. (in Chinese).
28. Doorenbos J, Kassam AH (1979) Yield response to water. FAO irrigation and drainage paper no. 33. FAO, Rome.
29. Steduto P, Hsiao TC, Raes D, Fereres E (2009) AquaCrop-The FAO crop model to simulate yield response to water. I. Concepts. Agronomy Journal. 101: 426–437.
30. Hsiao TC, Heng L, Steduto P, Roja-Lara B, Raes D, et al. (2009) AquaCrop-The FAO model to simulate yield response to water: parametrization and testing for maize. Agronomy Journal. 101: 448–459.
31. Raes D, Steduto P, Hsiao TC, Fereres E (2009a) AquaCrop-The FAO Crop Model to Simulate Yield Response to Water: Reference Manual Annexes., www.fao.org/nr/water/aquacrop.html.
32. Farahani HJ, Izzi G, Oweis TY (2009) Parameterization and evaluation of the AquaCrop model for full and deficit irrigated cotton. Agronomy journal.101: 469–476.
33. Garcia-Vila M, Fereres E, Mateos L, Orgaz F, Steduto P (2009) Deficit irrigation optimization of cotton with AquaCrop. Agronomy journal. 101: 477–487.
34. Heng LK, Hsiao TC, Evett S, Howell T, Steduto P (2009) Validating the FAO AquaCrop model for irrigated and water deficient field maize. Agronomy Journal 101: 488–498.
35. Vanuytrecht E, Raes D, Willems P (2011) Considering sink strength to model crop production under elevated atmospheric CO_2. Agricultural and Forest Meteorology 15: 1753–1762.
36. Zinyengere N, Mhizha T, Mashonjowa E, Chipindu B, Geerts S, et al. (2011) Using seasonal climate forecasts to improve maize production decision support in Zimbabwe. Agricultural and Forest Meteorology. 151: 1792–1799.
37. Stricevic R, Cosic M, Djurovic N, Pejic B, Maksimovic L (2011) Assessment of the FAO AquaCrop model in the simulation of rainfed and supplementally irrigated maize, sugar beet and sunflower. Agriculture Water Management. 98: 1615–1621.
38. Abedinpour M, Sarangi A, Rajput TBS, Singh M, Pathak H, et al. (2012) Performance evaluation of AquaCrop model for maize crop in a semi-arid environment. Agricultural Water Management. 110: 55–66.
39. Salemi H, Mohd-Soom MA, Lee TS, Mousavi SF, Ganji A, et al. (2011) Application of AquaCrop model in deficit irrigation management of Winter wheat in arid region. African Journal of Agricultural Research. 610: 2204–2215.
40. Du WY, He XK, Shamaila Z, Hu ZF, Zeng AJ, et al. (2011) Yield and biomass prediction testing of aquaCrop model for winter wheat. Transactions of the Chinese Society for Agricultural Machinery.42: 174–183. (in Chinese).
41. Wang XX, Wang QJ, Fan J, Fu QP (2013) Evaluation of the AquaCrop model for simulating the impact of water deficits and different irrigation regimes on the biomass and yield of winter wheat grown on China's Loess Plateau. Agricultural Water Management 129: 95–104.
42. Soddu A, Deidda R, Marrocu M, Meloni R, Paniconi C, et al. (2013) Climate variability and durum wheat adaptation using the AquaCrop model in southern Sardinia. Procedia Environmental Sciences 19: 830–835.
43. Todorovic M, Albrizio R, Zivotic L, Abi saab M, Stwckle C, et al. (2009) Assessment of AquaCrop, CropSyst, and WOFOST models in the simulation of sunflower growth under different water regimes. Agronomy Journal. 101: 509–521.
44. Karunaratne AS, Azam-ali SN, Izzi G, Steduto P (2011) Calibraion and validation of FAO-AquaCrop model for irrigated and water deficient bambara groundnut. Experimental Agriculture 47: 509–527.
45. Garcia-Vila M, Fereres E (2012) Combining the simulation crop model AquaCrop with an economic model for the optimization of irrigation management at farm level. European Journal Agronomy. 36: 21–31.
46. Yuan M, Zhang L, Gou F, Su Z, Spiertz JHJ, et al. (2013) Assessment of crop growth and water productivity for five C_3 species in semi-arid Inner Mongolia. Agricultural Water Management 122: 28–38.
47. Araya A, Keesstra SD, Stroosnijder L (2010a) Simulating yield response to water of Teff (Eragrostis tef) with FAO's AquaCrop model. Field Crops Research. 116: 196–204.
48. Araya A, Habtu S, Hadgu KM, Kebede A, Dejene T (2010b) Test of AquaCrop model in simulating biomass and yield of water deficient and irrigated barley (Hordeum vulgare). Agricultural Water Management. 97: 1838–1846.
49. Abrha B, Delbecque N, Raes D, Tsegay A, Todorvovic M, et al. (2012) Sowing strategies for barley (Hordeum vulgare L.) based on modeled yield response to water with AquaCrop. Experimental Agriculture. 48: 252-271.
50. Li ZZ, Xu Y, Lu XJ, Hu KL, Jiang LH, et al. (2011) Evaluation of the AquaCrop model for simulating biomass for Chinese green onion and soil water storage. Journal o f China Agricultural University.16: 59–66. (in Chinese).
51. Katerji N, Campi P, Mastrorilli M (2013) Productivity, evapotranspiration, and water use efficiency of corn and tomato crops simulated by AquaCrop under contrasting water stressconditions in the Mediterranean region. Agricultural Water Management 130: 14–26.
52. Allen RG, Periera LS, Raes D, Smith M (1998) Crop evapotranspiration. Guidelines for computing crop water requirement. FAO Irrigation and Drainage Paper No.56. FAO, Rome.
53. FAO (2009) ETo calculator version 3.1. In: Evapotranspiration from reference surface, FAO, Land and Water Division, Rome, Italy.
54. Kar G, Verma HN, Singh R (2006) Effect of winter crop and supplementary irrigation on crop yield, water us efficiency and profitability in rainfed rice based on cropping system of eastern India. Agriculture Water Management. 79: 280–292.
55. Oweis T, Hachum A (2006) Water harvesting and supplementary irrigation for improved water productivity of dry farming systems in West Asia and North Africa. Agriculture Water Management. 80: 57–73.
56. Raes D, Steduto P, Hsiao TC, Fereres E (2009b) Crop Water Productivity. Calculation Procedures and Calibration Guidance. AquaCrop version 3.0. FAO, Land and Water Development Division, Rome.
57. Raer D, Steduto P, Hsiao TC, Fereres E (2010) Reference Manual. Annexes-AquaCrop, Crop parameters. Annex I Section I.10 Wheat, A39–A42. FAO, Land and Water Development Division, Rome.
58. Hedge DM (1987) The effect of soil water potential, method of irrigation, canopy temperature, yield and water use of radish. Horticultural Science. 62: 507–511.
59. Fang QX, Chen YH, Li QQ, Yu SZ, Luo Y, et al. (2004) Effect of irrigation on water use efficiency of winter wheat. Transactions of the Chinese Society of Agricultural Engineering. 20: 34–39. (in Chinese).

13

Convergent Adaptations: Bitter Manioc Cultivation Systems in Fertile Anthropogenic Dark Earths and Floodplain Soils in Central Amazonia

James Angus Fraser[1,2*], Alessandro Alves-Pereira[3,4], André Braga Junqueira[5,6], Nivaldo Peroni[7], Charles Roland Clement[4,5]

1 Departamentos de Antropología e Historia, Universidad de los Andes, Bogotá, Cundinamarca, Colombia, **2** Department of Anthropology, University of Sussex, East Sussex, United Kingdom, **3** Post-graduate Program in Genetics and Plant Breeding, Universidade de São Paulo, Piracicaba, Estado de São Paulo, Brazil, **4** Laboratório de Evolução Aplicada, Universidade Federal do Amazonas, Manaus, Amazonas, Brazil, **5** Coordenação de Tecnologia e Inovação, Instituto Nacional de Pesquisas da Amazônia, Manaus, Amazonas, Brazil, **6** Centre for Crop Systems Analysis/Technology and Agrarian Development Group Wageningen University, Wageningen, Gelderland, The Netherlands, **7** Departamento de Ecologia e Zoologia, Universidade Federal de Santa Catarina, Florianópolis, Santa Catarina, Brazil

Abstract

Shifting cultivation in the humid tropics is incredibly diverse, yet research tends to focus on one type: long-fallow shifting cultivation. While it is a typical adaptation to the highly-weathered nutrient-poor soils of the Amazonian *terra firme*, fertile environments in the region offer opportunities for agricultural intensification. We hypothesized that Amazonian people have developed divergent bitter manioc cultivation systems as adaptations to the properties of different soils. We compared bitter manioc cultivation in two nutrient-rich and two nutrient-poor soils, along the middle Madeira River in Central Amazonia. We interviewed 249 farmers in 6 localities, sampled their manioc fields, and carried out genetic analysis of bitter manioc landraces. While cultivation in the two richer soils at different localities was characterized by fast-maturing, low-starch manioc landraces, with shorter cropping periods and shorter fallows, the predominant manioc landraces in these soils were generally not genetically similar. Rather, predominant landraces in each of these two fertile soils have emerged from separate selective trajectories which produced landraces that converged for fast-maturing low-starch traits adapted to intensified swidden systems in fertile soils. This contrasts with the more extensive cultivation systems found in the two poorer soils at different localities, characterized by the prevalence of slow-maturing high-starch landraces, longer cropping periods and longer fallows, typical of previous studies. Farmers plant different assemblages of bitter manioc landraces in different soils and the most popular landraces were shown to exhibit significantly different yields when planted in different soils. Farmers have selected different sets of landraces with different perceived agronomic characteristics, along with different fallow lengths, as adaptations to the specific properties of each agroecological micro-environment. These findings open up new avenues for research and debate concerning the origins, evolution, history and contemporary cultivation of bitter manioc in Amazonia and beyond.

Editor: Lee A. Newsom, The Pennsylvania State University, United States of America

Funding: The research was funded by Leverhulme Trust (Grant F/00 230/W) and the CNPq (Brazilian National Research Council) (CT Amazonia 575588/08-0). The funders had no role in study design, data collection and analysis, decision to publish, or preparation of the manuscript.

Competing Interests: The authors have declared that no competing interests exist.

* E-mail: james.angus.fraser@gmail.com

Introduction

Shifting cultivation has been a predominant mode of traditional agriculture in the rainforests of the lowland Neotropics [1], humid Africa [2], the Indian subcontinent [3] and parts of South East Asia [4] and Oceania [5] for thousands of years. Shifting cultivation in the humid tropics is incredibly diverse [6], yet both empirical studies and theoretical discussions tend to focus on one type: long-fallow shifting cultivation (LFSC). LFSC is an *extensive* form of crop production, which entails a relatively short cropping period of 1–3 years, followed by a longer fallow period (normally 10–20 years), with land cleared by slashing and burning fallow vegetation (or sometimes mature forest) [7]. The long fallow period is necessary because soils of the *terra firme* (non-flooded upland plateaus) in the tropical world tend to be highly-weathered infertile Oxisols and Ultisols (US classification system). Hence, this form of land-use is spatially (fields must shift frequently) and temporally (fields must be fallowed for years before they can be cultivated again) *extensive*. Extensive systems are widely held to be well adapted to such soils because a nutrient flush provided by the burning of secondary or mature forest is used for a single cropping period, before swidden fields are left to fallow in order to restore fertility via the growth of secondary forest [8]. It has long been assumed that any intensification of LFSC (i.e., a reduction of fallow periods) in the humid tropics will cause crop yields to decline, as infertile Oxisols and Ultisols are not capable of withstanding a sustained reduction of fallow periods, leading to the eventual collapse of the system. These assumptions about the prevalence and precariousness of extensive shifting cultivation have influenced theories of cultural development in Amazonia, where Steward [9] and Meggers [10] held that the small scattered

Amerindian settlements practicing LFSC found in the interfluvial *terra firme* today represent an optimal adaptation to a low productivity environment. While their theories have long since gone out of fashion [11], studies of the cultivation of the Amazonian staple bitter manioc (*Manihot esculenta* Crantz) continue to focus almost exclusively on LFSC practiced by current Native Amazonian populations [12] and other traditional populations [13] in the poor soils of marginal interfluvial *terra firme* environments.

In Amazonia, however, the most populated areas – both today and in the late pre-Columbian period – are along major white-water rivers, such as the Solimões, Amazon and Madeira, all of which have nutrient-rich floodplain soils used for agriculture and adjacent paleo-floodplains that are somewhat more nutrient-rich than the typical Oxisols and Ultisols [14], as well as areas of Anthropogenic Dark Earths [15]. These regions are among the most suitable for human habitation and population growth, since they are characterized by an abundance of fish as well as fertile soils. Students of pre-Columbian Amazonia have long emphasized the agricultural potential of these regions. Lathrap [16] argued that the productivity of the white-water floodplains underwrote the growth of large and dense human settlements, sparking competition and warfare between groups for these rich but spatially and temporally limited areas. However, although the flood pulse provides a yearly flush of nutrients, extreme floods once every decade or so can destroy crops, settlements and even the high floodplain itself, which makes these environments relatively unpredictable. Denevan [17] argued that agriculture began as multi-cropping in floodplain environments and developed later in forested uplands as intensive multi-cropping systems with short fallow. For him, long fallow, extensive cultivation is a late development associated with the introduction of efficient forest clearance tools, especially metal axes and machetes.

Denevan recognized that large scale settlements could not have been supported exclusively by seasonal floodplain cultivation, given the riskiness of producing food in areas subjected to floods. He proposed that settlements would preferentially be located on bluffs along the river margins, allowing the exploitation of both the fertile floodplain zones and the *terra firme* areas that were safe from flooding but located on less fertile soils [17,18]. This occupation pattern would ultimately result in the transformation of upland soils: many bluffs in central Amazonia feature fertile anthropogenic soils known as Anthropogenic Dark Earths (ADE) [19–21]. These soils are associated with Native Amazonian settlements of the late pre-Columbian period (2000–500 years before present) [22,23], and are most abundant and largest in whitewater regions, because these are where the largest populations were located in pre-Columbian times [24]. Amazonian ADE form through human inputs of organic and inorganic matter [e.g., biomass wastes, manure, bones, ash, charcoal and ceramics] [15]. Amazonian ADE are enduringly fertile in part due to their historical enrichment in highly stable black carbon, which has a half-life of 1000 years [25]. ADE exhibit much higher levels of chemical elements essential for plant growth, such as phosphorous, calcium, magnesium, zinc and manganese, than the weathered and infertile soils in which they were formed. ADE sites exhibit a highly fertile 'core area', which grades into more subtly modified soils, with a continuum of fertility between them and surrounding soils [26,27]. It is likely that ADE were appreciated for plant cultivation by pre-Columbian Native Amazonian peoples, just as they are by Amazonian people today [28].

These historical arguments point to the possibility that local people may also intensify shifting cultivation when circumstances permit today. Fertile soils, such as ADE and floodplain soils, provide farmers with the opportunity to intensify shifting cultivation, increasing both crop productivity and frequency of cultivation [29], raising the carrying capacity of the landscape [30]. Therefore, the study of swidden systems on fertile soils allows us to examine the extent to which fallow lengths, bitter manioc landraces, and other aspects of swidden systems vary on different soils. While numerous studies have addressed many dimensions of shifting cultivation, few have compared swidden systems on different soils and in different ecological contexts within a circumscribed geographical area [31].

The Amazonian Staple: Bitter Manioc

Manioc landraces are classified as either bitter or sweet depending on cyanogenic glucoside (CG) content. In Central Amazonia, bitter manioc is the staple crop today, whilst sweet manioc is only of secondary importance [32]. Sweet manioc has low CG content in their tuberous roots (<100 ppm fresh weight), while bitter manioc has larger amounts of cyanogenic glycosides (>100 ppm fresh weight) [33]. There are an estimated 7000 landraces of manioc worldwide [34], but this is surely an underestimate given farmers' continual selection and propagation of new landraces. Differences in the color, form and size of leaves and stems, in the size, number, color and cyanide content of roots, and in the rate of growth of tuberous roots are determined by genetic differences, although the environmental conditions in which plants grow also affect manioc phenotypes, including CG content [33]. Farmers comprehend a "landrace" as a set of individuals sharing particular morphological characteristics that differentiate them from other landraces; they also distinguish them by giving them a particular name [35,36]. These morphological characteristics are shaped by genotype-environment interactions that can result in different phenotypic expressions of the same genotype, and create morphotypes that are identified as distinct landraces [37]. Landraces are the result of generations of farmer selection in local environments, and are therefore well adapted to local growing conditions, which has been demonstrated in South America [13,38,39], in humid Africa [40] and Asia [41]. Several students of manioc have asserted that bitter manioc does not yield well in fertile soils [42–44], perhaps because it has been so thoroughly studied on nutrient-poor soils. However, about 30% of Central Amazonian manioc, both bitter and sweet, is produced in the floodplain [45]. It was observed that on highly fertile soils manioc plants tend to invest more in aboveground biomass than in the tubers [29,43]. However, there are bitter manioc landraces that yield well in the floodplain and in ADE, possibly even better than in Oxisols and Ultisols [46].

Manioc is vegetatively propagated via stem cuttings, which grow into plants that are genetically identical to the mature plants from which they were cut. However, manioc retains its ability to reproduce sexually and produces seeds that lie dormant in fallow vegetation [47]. When fallows are cleared for cultivation, seeds are stimulated to germinate by increasing temperatures caused by both soil exposure and the heat of the burn phase in swidden-fallow systems [48], and seedlings appear [49]. When cuttings from seedlings are incorporated into a landrace, its genetic diversity is increased because it becomes a polyclonal landrace [36,48,50–52]. This increased genetic diversity provides the raw material for adaptation to new conditions, such as ADE and the floodplain, as different clones in the landrace have different possibilities for adaptation to different agro-ecosystems. The extent to which volunteer seedlings are incorporated is variable in modern bitter manioc cultivation in the Neotropics. Studies in Guyana [53], French Guiana [54] and the Atlantic Forest in Brazil [55] found a relatively high level of incorporation. However,

Stocker found only a small amount of seedling incorporation amongst farmers in Pará, in eastern Brazilian Amazonia ([56]:162–63). Nonetheless, this is a recurring pattern in many traditional farming systems in Amazonia and beyond [13].

This article examines the hypothesis that contemporary Amazonian people have developed divergent bitter manioc cultivation systems as adaptations to the properties of different soils in the landscapes that they inhabit. We test this hypothesis by comparing swidden systems in fertile (ADE and floodplain) and infertile (Oxisols and Ultisols) soils with respect to their assemblages of manioc landraces, the performance (from the farmer's point of view) and productivity of the most common landraces in different soils, along with the genetic relationships between the most common landraces, and length of fallow periods. Hence, we will examine both the adaptations of manioc landraces and of swidden-fallow systems. In earlier publications [46,57,58], we hypothesized, based on initial interpretations of ethnobotanical data, that "weak" bitter manioc landraces grown in ADE originated in the floodplain, given farmers' knowledge and similar adaptations (fast maturation and low starch content).

Results and Discussion

Farmer theory and practice

In the preliminary stage of research we conducted multi-sited participant observation in numerous communities of the study region (Figure 1). Subsequent quantitative work was driven by hypotheses generated during this stage. While qualitative data have been presented at length elsewhere [57], here we summarize major findings and their interplay with quantitative data. We found that, contrary to what was expected from the literature [29,31,43,59–61], bitter manioc was widely cultivated in ADE. More specifically, farmers stated that not only did they *i*) perceive that certain landraces yield better in certain soils (the basis for our *pri* index and yield observations, see below), but that this informed their *ii*) selection of different landraces for planting in different soils (the basis for the *la* index), and *iii*) that fertile soils (ADE/floodplain) could be farmed with much shorter fallow periods than infertile soils (the basis for our fallow length data).

Most intriguingly, however, was the finding that farmers expressed these perceptions and behaviors in the form of a local theory of strength and weakness (the categories strong and weak are present in traditional agriculture in other areas of the world ([62]:148)). Along the middle Madeira River, the categories weak and strong express the suitability of different landraces for planting in different soil-successional scenarios. Briefly, weak landraces are fast yielding (5 months–1 year), but rot if left too long in the ground, and produce less starch than stronger landraces. Strong landraces, on the other hand, are slow yielding (1–3 years), but produce more starch than weaker landraces. Farmers claim that weak landraces are best suited to planting in "weak" land (land with young fallow), whilst strong landraces are said to be suited to "strong" land (land with older fallow). Elsewhere, we reported that landraces described as weak were planted more frequently in ADE and the floodplain, whilst those described as strong were planted more frequently in Oxisols and Ultisols ([63]:400). The specifics of this local theory have been elaborated at length [46,57,58], and here we would just emphasize two points. This local theory would appear to provide evidence of the cognitive aspects of adaptation – local people have theorized the emergent properties of their adaptive knowledge and practices – and express these using the simple yet powerful metaphors of strength and weakness. Secondly, this theory is not limited to the middle Madeira; an independent study recently found an identical theory of weak and strong manioc amongst floodplain and *terra firme* manioc farmers in and around the Brazilian Sustainable Development Reserves of Mamirauá and Amanã, close to the town of Tefé on the middle Solimões River [64]. The fact that the same local theory exists in two localities hundreds of kilometers apart suggests that this theory is likely to inform adaptive aspects of manioc farming in various locations throughout Amazonia.

Diversity of manioc landraces

A total of 50 landraces were found, with 29 in ADE, 20 in Oxisols, 20 in Ultisols and 23 in the floodplain. Most of the landraces cultivated on the *terra firme* are planted in more than one type of soil: among the 29 landraces cultivated in ADE, 18 (62%) are also planted in Ultisols and 19 (65.5%) in Oxisols. Ultisols and Oxisols shared 75% (15) of their landraces. On the other hand, most of the landraces that occurred in the floodplains (19, or 82.6%) only occurred in this environment, with only four landraces shared with other types of soil (one shared with Ultisols and ADE, and three that occurred in all four types of soil). The number of landraces cultivated in each village varied between 5 (Verdum) and 21 (Vista Alegre), and tended to be lower in floodplain villages ($N = 5$; mean ± standard deviation: 8.2 ± 3.1) than villages located on the *terra firme* ($N = 4$; 15.7 ± 5.0), principally because villages on the *terra firme* have access to more than one type of soil (and, in the case of Água Azul, even cultivate in the floodplain). The average number of landraces cultivated in each manioc field was similar among the different types of soil (ADE: 2.8 ± 1.9; Oxisols: 3.4 ± 1.5; Ultisols; 3.2 ± 2.0; Floodplain: 3.1 ± 1.3) (Dataset S1).

Along the Upper Negro River, communities commonly have 60 to 89 landraces, somewhat higher when compared to other local communities in Amazonia and in the Atlantic Forest in Brazil (53–58; [13]). The diversity of bitter manioc landraces on the *terra firme* along the middle Madeira is quite low, although higher than what Grenand reported along the Cuieiras River near Manaus (six) [65]. Perhaps this is due to a greater market orientation in the Madeira River communities, and higher population pressure. By contrast, the number of landraces we registered on the floodplain is much higher than reports for Careiro Island (three; [65]) at the confluence of the Negro and Solimões Rivers, and for communities between Santarém and Óbidos (maximum of four), along the middle Amazon River [66], although Pereira [67] reported 43 bitter manioc landraces at the confluence of the Solimões and Japurá Rivers. The generally low numbers in the literature most likely reflect the lack of attention to manioc in the floodplain compared to the *terra firme*, as pointed out by Denevan [68] for studies of agriculture in the floodplain in general. Various studies mention fast maturing manioc in the floodplain, without reporting the number of landraces [69–71].

Landrace composition of manioc fields in different soils

The composition of manioc landraces in fields on different types of soil is significantly different, which is shown both in the general model (NPMANOVA; $F = 16.46$, $p = 0.01$) and also in all possible pairwise comparisons (six) between the four different soil types (all with p values lower than 0.05, corrected for multiple comparisons) (Figure 2). In general, manioc fields in Oxisols and Ultisols cluster together, while in Floodplains they form an almost completely separate group, with fields in ADE occupying an intermediate area in the ordination space (Figure 2). Also, fields in Oxisols and Ultisols are more homogeneous, while in Floodplains and ADE they are more heterogeneous, especially in the latter case. The areas of overlap and the dispersion of some points through the figure indicate that the composition of manioc landraces is very

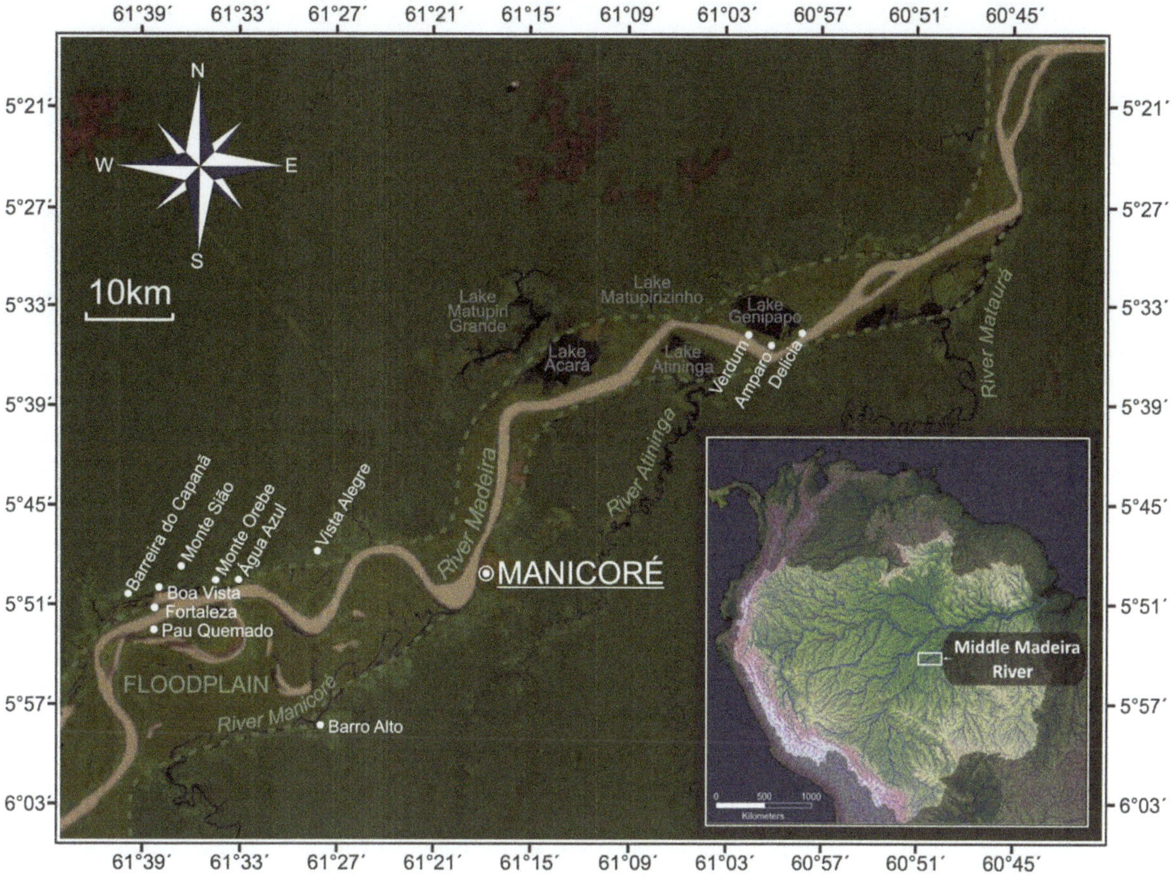

Figure 1. Map of the middle Madeira River region, Amazonas State, Brazil, showing communities where research was carried out. The inset map shows the location of the middle Madeira River in Northern South America. Map by Victoria Frausin.

heterogeneous and highlights the practice of landrace exchange between soils. Within-field heterogeneity is commonly reported [12,13,38,72], as is exchange among farmers [35,36,38–40,48,73], but exchange among farmers on different soil types has not been reported previously, perhaps because the heterogeneity of soils has been less carefully studied than that of bitter manioc landraces.

Farmer perceptions of landrace performance

A Performance Ranking Index (*pri*) for each landrace in each type of soil on the *terra firme* was calculated. Most of the landraces mentioned by the farmers in the interviews in all three soils on the *terra firme* (ADE, Oxisol and Ultisol) have a very low *pri* score, either because they are rarely cited in that specific soil or because farmers do not rank them among the best performers in that soil (Figure 3) (Dataset S1). *Roxinha*, *Tartaruga*, *Arroz* and *Aruari* are among the five landraces with higher *pri* in all three soils. However, while the landraces *Jabuti* and *Arroz* are considered the best performers in Oxisols and Ultisols, in ADE this place is occupied by *Tartaruga*, *Pirarucu Branco* and *Roxinha Branca* (Figure 3).

There were strong positive correlations between the area occupied by each landrace in the fields (*la*) and their performance (*pri*) in the soils (Pearson's Correlation - Ultisol: $\rho = 0.901$; Oxisol: $\rho = 0.912$; ADE: $\rho = 0.898$). This high degree of correlation between *pri* and *la* supports the notion that farmers' planting behavior is shaped by their perceptions of the performance of different landraces in different types of soil. While this pattern is common and has been mentioned elsewhere (e.g. [74]:50), the correlations between this pair of indices permits a quantitative demonstration of this behavior.

Farmer estimates of manioc yields

Our farmer yield observations reveal that there are significant differences between yields in five of the six most popular landraces when planted in different soils along the middle Madeira River (Table 1). These findings support our inference of a high degree of adaptation in bitter manioc systems found on different soils, since landraces that perform better in particular soils tend to be more predominant in those soils, as farmers respond positively to good yields by planting more of specific landraces. This is reflected in the *la* and *pri* indices: *Jabuti* and *Roxinha* are significantly more predominant in Oxisols and Ultisols than in ADE, while *Tartaruga* and *Pirarucu Branco* are more predominant in ADE than in Oxisols or Ultisols.

Fallow length and the intensification of swidden systems

Intensification in swidden-fallow systems refers, among other factors, to a shortening of fallow length [75]. Intensification on infertile soils can result in exhaustion of fertility and breakdown of the system. Fertile soils provide opportunities for sustainable reduction of fallow lengths. The average fallow length for 55 fields on Ultisols was 13.2±8.1 years and for 64 Oxisol fields the average length was 21.3±15.4 years, while for 71 ADE fields it was 6.5±6.8 years and for 59 floodplain fields it was 1.6±4.0. There were significant differences between fallow lengths in different soils

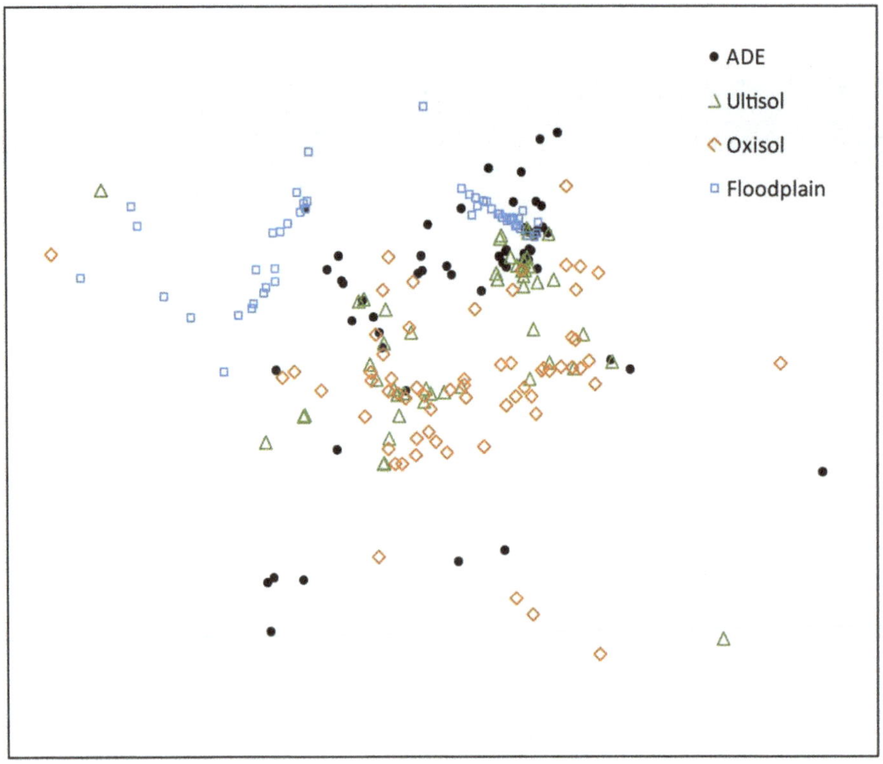

Figure 2. Non-metric multidimensional scaling (NMDS) showing the composition of bitter manioc landraces sampled in fields on Ultisols, Oxisols, Anthropogenic Dark Earths (ADE) and on floodplains along the middle Madeira River (percentage of explanation of the bidimensional model: 35.8%). Each point represents a bitter manioc field, and its position in the graph is a bidimensional representation of the Bray-Curtis dissimilarity between the fields.

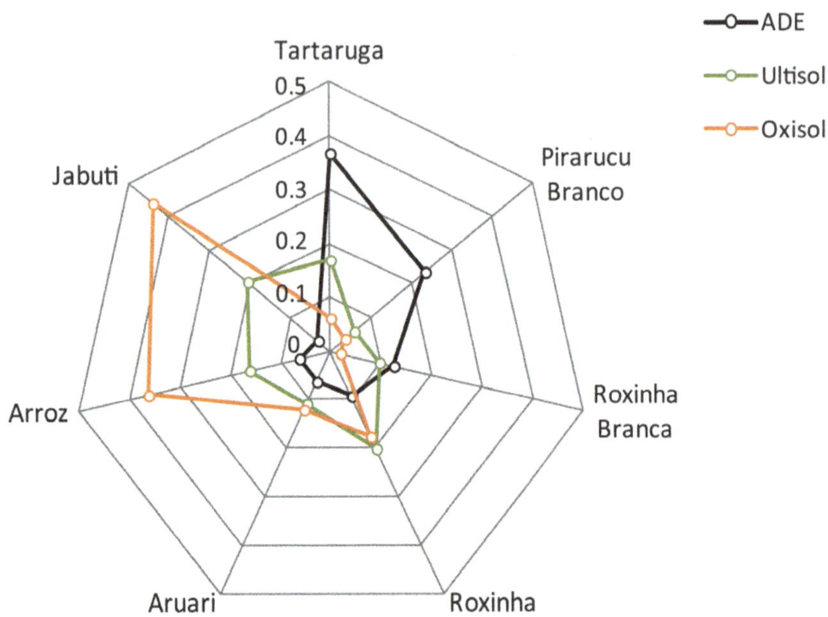

Figure 3. Farmer perceptions of performance for the six most frequent manioc landraces cultivated in Oxisols, Ultisols and Anthropogenic Dark Earths (ADE) along the middle Madeira River. Numbers in the axis indicate the Performance-Rank Index (*pri*) of each landrace in each type of soil, calculated based on the perception of 162 farmers interviewed on the Middle Madeira. Dots indicate the value of *pri* for each variety in the three types of soil.

Table 1. Farmer-recorded production of 50 kg sacks of manioc flour that would be obtained from approximately 0.5 ha of swidden plot at various localities along the Middle Madeira River, Central Amazonia, Brazil, in 2007–2008.

Manioc landrace	ADE		Oxisol/Ultisol		Floodplain		ANOVA	
	n	(avg ± SD)	n	(avg ± SD)	n	(avg ± SD)	F	p
Jabuti	5	24.2±5.1	6	44.2±8.6			20.6	0.001
Arroz	4	31±2.6	4	30.5±4.2			0.04	ns
Tartaruga	6	44.4±6.1	7	26.7±3.5			43.0	<0.001
Roxinha	5	25.2±3.11	7	40.1±4.3			43.5	<0.001
Pirarucu Branco	4	53.2±7.0[a]	4	25.7±4.6[b]	5	55±11.2[a]	15.8	0.001
Pirarucu Amarelo	4	66.7±24.9[a]	4	30.5±4.4[b]	4	70.5±11.3[a]	7.6	0.012

Superscript characters indicate significant differences in pairwise comparisons using Tukey's HSD test ($p<0.05$).

(ANOVA, $F=44.51$, $p<0.001$, while Tukey's post hoc test revealed significant differences between all possible pairwise comparisons except ADE vs Floodplain). It is important to emphasize that local factors, such as population density, also play important roles in shaping fallow lengths, and these are discussed at length for each locality in the Supporting Information, along with other factors (Text S1).

Generation, selection and exchange of landrace diversity

Local farmers identify and exchange genetic diversity of bitter manioc landraces [38,72]. Along the middle Madeira River, we observed that certain people that we categorized as "key individuals" play an important role in the exchange of new landraces originating from seedlings, conservation of existing landraces and knowledge associated with them. Key individuals were identified as being the individuals who most informants mentioned during open interviews as being responsible for introducing new landraces and keeping a stock of all landraces present in each community. Interviews with these key individuals revealed that they consciously recognize new phenotypes and are always intentionally experimenting with new landraces from seedlings and with clonal material from their kin in other communities [13]. These key individuals perform a vital role in the identification and distribution of new manioc genetic diversity, because they try out new landraces in local micro-environments and, if they prove to be exceptional, are responsible for their distribution within communities, and to kin in other communities. The fact that each locality only has a few key individuals points to the critical role that they play in the circulation of exceptional landraces throughout the region, but some new genotypes can circulate unconsciously mixed with other individuals that are morphologically similar.

At the six localities examined in this study, both the incorporation of seedlings into pre-existing landraces and the creation of new landraces from seedlings are relatively common practices. Farmers recognize that seedlings often appear in the newly burnt fields *before* cuttings have been planted and are morphologically different from the planted landraces, as they tend to grow taller and only have a single tuberous root growing straight downwards. Most farmers (53–66% across communities) interviewed simply ignore or weed out the seedlings, while others (11–32%, including, but not restricted to key individuals) take cuttings from mature seedlings and plant them separately (Table 2). When mature, these volunteers are either: i) incorporated into an existing landrace, e.g., *Jabuti*; ii) incorporated as a sub-landrace of an existing landrace (such as *Jabuti-Preto*); iii) named as a combination of two landraces (such as *Jabuti-Arroz*); or iv) established as a new landrace with a different name.

All the landraces analyzed with microsatellite markers had observed heterozygosities (H_O) higher than expected heterozygosities (Table 3), which appears to be related to both the incorporation of seedlings and the selection of heterozygous individuals [76]. For example, the landrace *Arroz* from ADE at Barreira do Capanã had an observed heterozygosity of 0.495 and 5 different multi-locus genotypes (MLGs) in the 20 individuals examined. The landraces *Jabuti* from Oxisols and *Tartaruga* from floodplain had 7 and 6 different MLGs, and H_O of 0.567 and 0.707, respectively. The number of MLGs reflects either the incorporation of seedlings or the unintentional "mixing" of landraces (which may happen, for example, when they are morphologically similar), both of which contribute to increasing heterozygosity. Incorporation of seedlings into recognized landraces increases intra-landrace heterogeneity, because farmers tend to select the largest and healthiest volunteers, which tend to be the most heterozygous [53,54]. These practices are critical in maintaining and amplifying the genetic diversity of manioc landraces, and in adapting landraces to new environmental conditions, such as different soils.

Along the middle Madeira River the performance in different soils is a major factor in varietal selection by farmers (Figure 4). The selection of distinct traits in different environments where the landraces are cultivated may be correlated to the genetic differentiation found among the bitter manioc landraces grown in different soil types (Figure 5; [77]). Evidence of genetic divergence among landraces cultivated in floodplain from those cultivated in ADE and infertile soils was also found when analyzing intra-varietal genetic diversity. In all examples the landraces grown in ADE and infertile soils had the same most common MLGs. On the other hand, landraces grown in the floodplain, except *Pirarucu Amarelo*, had distinct MLGs from those with the same landrace name but grown in ADE or infertile soils. Additionally, in the floodplain some farmers observed that volunteers occur more frequently in higher areas of the floodplain. These areas are not flooded every year and the seeds have time to be acted upon by agro-ecological management. The planting of manioc in different zones, the flood regime, farmer perceptions and the combination of landraces they select drive the generation of new landraces in the floodplain. This is consistent with the somewhat higher number of landraces from the floodplain observed in this study, and may be related to the genetic differentiation among the landraces grown in the floodplain from those grown in upland soils (Figure 4).

Table 2. Number of informants, both total and key individuals for manioc management, number of landraces cultivated, and the ways that farmers manage seedlings at six communities along the middle Madeira River, municipality of Manicoré, Amazonas, Brazil.

	Informants		Number of landraces	Seedling Management		
	Total	Key		Intentionally incorporate	Incorporate at random	Remove
Barro Alto	37	3	12	0.32	0.14	0.54
Barreira do Capanã/Boa Vista	29	6	19	0.17	0.14	0.66
Água Azul	13	4	9	0.23	0.31	0.46
Vista Alegre	11	4	21	0.27	0.09	0.55
Água Azul floodplain	9	3	13	0.11	0.33	0.55
Genipapo floodplain	15	4	14	0.26	0.20	0.53

Farmer selection for agronomic performance? Comparative perspectives

Our findings call for re-evaluation of the widespread notion that farmer identification and selection of manioc landraces is weakly associated with agronomic performance. This notion is based on Boster's research on the perceptual distinctiveness of manioc amongst the Aguaruna Jivaro of the Peruvian Amazon. Boster concluded that "there are significant differences between the [manioc] cultivars in their responses to environmental factors…", but "the Aguaruna are generally unaware of or unconcerned with these differences" ([72]:40). Boster's own garden experiments did, however, show that "some cultivars grow well in sandy alluvial soils on river islands and do poorly on other sites, while other cultivars yield about the same regardless of soil conditions." Boster also noted that the Cashinahua, another Native Amazonian group that resides in Peru and Brazil, do appear to select particular landraces for planting in certain kinds of soil, according to their well-known ethnographer, Ken Kensinger [78]. The Aguaruna Jivaro reside in hill country, where there is certainly a lot of soil variation and there will be some high quality soils, and the Cashinauhua live along the fertile whitewater Juruá River (Tarauacá, Jordão, Breu, Muru, Envira) and in the Purus basin (Upper Purus and its affluent the Curanja) [79,80]. Each of these areas offers a variety of very different soils to the manioc farmer. However, even in the Upper Negro Region, an area notorious for infertile soils, according to Wilson [81], there appears to be non-random planting of sweet manioc in more fertile soils by Tukanoan manioc farmers. Heckler and Zent, on the other hand, describing research amongst the Piaroa in Venezuela, found that "specific genotype-micro-condition interactions were demonstrably not a significant factor in determining planting patterns" ([82]:684). Similarly, Salick and colleagues, reporting on a study amongst the Amuesha in Peruvian Amazonia, found that "soil seemed to have little predictive power for cassava phenotype occurrence." ([38]:11).

Table 3. Indices of intra-varietal genetic diversity [Observed (H_O) and expected (H_E) heterozygosities, and number of multi-locus genotypes (No. MLGs)] for five bitter manioc landraces cultivated in different soil types in the middle Madeira River region, based on variation detected with 10 microsatellite markers.

Landrace	Soil	Locality	H_O	H_E	No. MLGs	MLGs
Pirarucu Branco	ADE	Barreira do Capanã	0.495	0.250	2	**A**, g
	FP	Pau Queimado	0.505	0.255	2	**B**, h
Tartaruga	ADE	Barro Alto	0.503	0.268	2	**B**,F
	OX	Barro Alto	0.507	0.276	4	**B**, F, i, j
	FP	Verdum	0.707	0.361	5	**C**, k, l, m, n
Arroz	ADE	Barreira do Capanã	0.495	0.460	5	A, **D**, o, p, q
	OX	Água Azul	0.503	0.257	2	**D**, r
Jabuti	OX	Barreira do Capanã	0.567	0.445	6	D, **E**, s, t, u, v
Pirarucu Amarelo	ADE	Água Azul	0.607	0.306	2	**F**, x
	FP	Água Azul	0.607	0.307	1	**F**

Capital letters indicate the MLGs that were present in more than one landrace. Boldfaced letters indicate the most common MLG for a given landrace, while small letters indicate MLGs present at low frequencies.

The findings of Boster, Heckler and Zent, and Salick et al. explain the persistence of the notion that farmer identification and selection of manioc landraces is weakly associated with agronomic performance. Our findings, however, when combined with those of Kensinger and Wilson, support the conclusion that the way in which *at least some* Amazonian farmers consciously identify and select landraces *is* shaped by agronomic performance. Indeed, the perceptual weak-strong theory of manioc detailed above and elsewhere appears to be built upon and indeed express in schematic/metaphoric form local knowledge of selection for agronomic performance. This raises questions regarding the distribution of practices of selection for agronomic performance, and which factors (such as the presence/absence of fertile soils, indigenous/non-indigenous farmers etc.) contribute to the presence of these practices and associated knowledge in different parts of Amazonia. We conjecture that selection for agronomic performance would be more likely to emerge amongst people inhabiting areas with greater heterogeneity of soil types.

Conclusions

We examined manioc cultivation in four different soil types at six localities, and found that farmers plant different assemblages of bitter manioc landraces in different soils and that shorter fallow

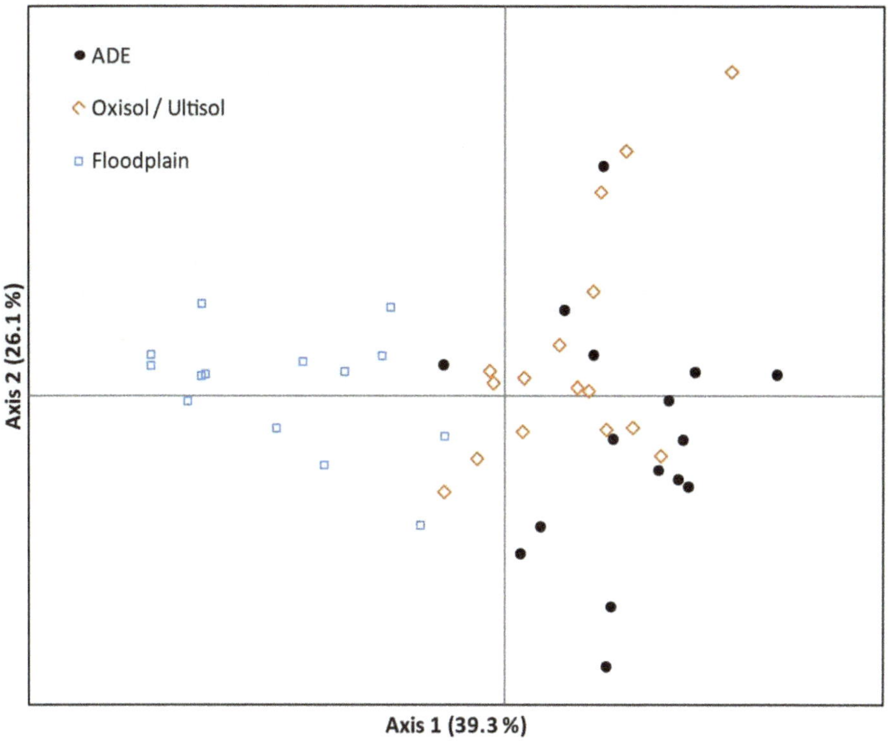

Figure 4. Principal Coordinates Analysis based on diversity revealed by 10 microsatellites markers showing the dispersion of 48 swiddens from three soil types (17 in ADE soils, 14 in floodplain soils and 17 in Oxisols/Utilsols) in six communities along the middle Madeira River. The two coordinates together explain 65.4% of the variation in the matrix.

lengths were strongly associated with more fertile soils. Most popular landraces were shown to exhibit significantly different yields when planted in different soils. From this we can infer that farmers have selected different sets of landraces with different perceived agronomic characteristics, along with different fallow lengths, as adaptations to the specific properties of each agroecological micro-environment. On the *terra firme*, intentional selection of certain landraces for cultivation in certain soils was demonstrated by the close relationship between *Performance Ranking Index* and *Landrace Area* for landraces in different soil types. These findings suggest that the local theory revolving around metaphors of strength and weakness may represent a cognitive manifestation of adaptive farming behavior: a way to express in simple terms local knowledge of the relationship between different sets of bitter manioc landrace traits and soil properties and fallow stages. Although landraces grown in ADE and the floodplain share similar phenotypical characteristics, specifically fast maturation and low starch content, and are both described as "weak" by

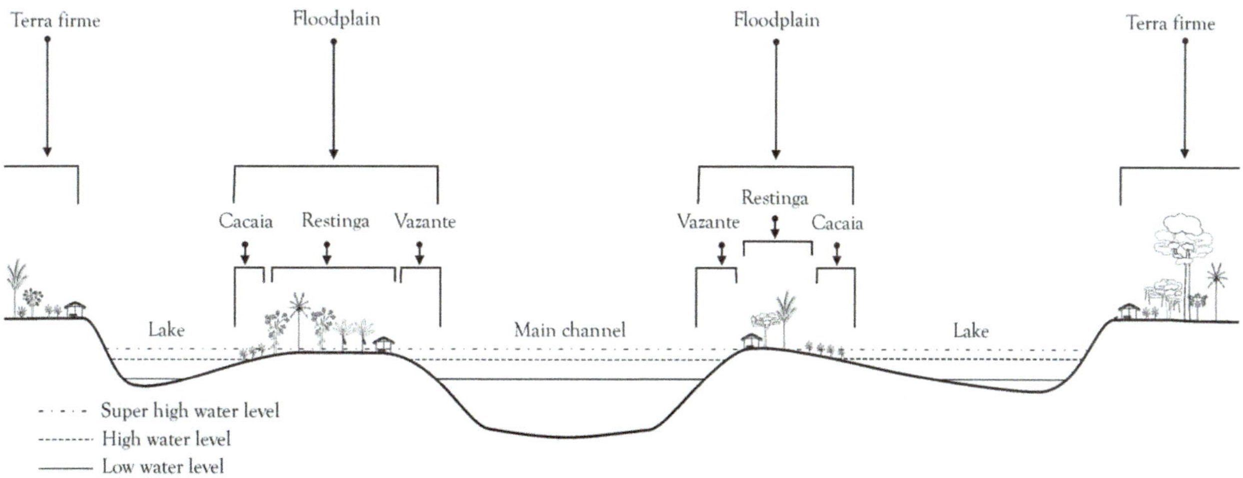

Figure 5. Schematic representation of floodplain zones relative to the main channel, lakes and *terra firme* (non-flooded upland plateaus). *Vazante* is the local term on the middle Madeira for the banks of the main channel, *Restinga* for the high levee floodplain and *Cacaia* for the back-swamp area. Drawn by Victoria Frausin.

farmers, these groups of landraces were not genetically more related to each other, as we had predicted from ethnobotanical observation. Rather, the landraces grown in upland soils (ADE and Oxisols/Ultisols) are more related to each other and genetically differentiated from floodplain landraces. For landraces cultivated in the floodplain and in ADE, the selection for convergent adaptive traits appears to be associated with the similar ecological adaptations to nutrient-rich soils and short periods for rapid growth and yield before stress. The stresses are quite different, however: floods in the floodplain and enhanced weed pressure in ADE. Nonetheless, they yield convergent adaptive results.

If the similar ecological adaptations observed for ADE and floodplain landraces are outcomes of adaptive convergence of traits directed by farmer's selection, what drives that selection? On ADE, farmer selection for fast maturing landraces coupled with a reduction in fallow periods is likely to be driven by population pressure and permitted by ADE's high nutrient availability. When ADE is cultivated at Barro Alto and Água Azul, where population pressure exerts more of an influence, the most intensive swidden-fallow systems were found (Text S1). These intensive systems were associated with a reduction in the diversity of landraces in general, suggesting that when farmers reduce fallow lengths they may select only the fastest maturing landrace that is widely available for planting. In the floodplain, selection for fast maturing landraces is certainly a strategy to cope with the high seasonality of the floodplain determined by the flood pulse. Hence two different swidden-fallow systems have developed as convergent adaptations to intensification in nutrient-rich anthropogenic and floodplain soils. This intensification may be driven by population pressure (*terra firme*) and the flood pulse (floodplain); alternatively, owing to their greater fertility, it may be that the floodplain and ADE are the only places that offer the opportunity to produce food and/or money in a shorter time period – and this could also be a driver for the selection of these common traits.

Our findings broaden our understanding of the diversity of bitter manioc swidden-fallow systems in contemporary Amazonia, and allow some ethnographic projection. Firstly, the intensive short-fallow shifting cultivation in the floodplain and on bluffs with ADE that we reported along the middle Madeira are clearly modern analogs of Denevan's [17,18] hypothesis that pre-conquest food production systems were much more intensive than the extensive long-fallow shifting cultivation common today, which he argues is a post-conquest adaptation. Secondly, Arroyo-Kalin [32] recently hypothesized that during the pre-Columbian period manioc was selected for high toxicity in swiddens in the infertile agricultural hinterland further from settlements, but farmers were selecting for lower toxicity in fertile ADE soils forming in and around homegardens. The phenomenon of farmer selection being driven by landrace agronomic performance in different soils would support the possibility that farmer selection in the pre-Columbian period for planting in different environments was a key driver in the emergence of "bitter" and "sweet" manioc and the continuum of toxicity that underlie them. Given the importance of bitter manioc cultivation to our understanding of pre-Columbian and contemporary populations in Central Amazonia, we hope that this study will encourage further research into the under-investigated area of intensified bitter manioc cultivation in fertile soils.

Materials and Methods

Study area and sampling design

We focus on the middle Madeira River, Amazonas State, Brazil (Figure 1), where traditional farmers plant their staple crop, bitter manioc, in infertile Oxisols and Ultisols of the *terra firme*, and in fertile ADE and floodplain soils. Amazonian Dark Earths are easily recognizable by their very dark brown or black coloring, high fertility, and pottery shards. The other *terra firme* soils were classified in the field according to local ethnopedological knowledge and physical properties (color, presence of potsherds, etc.). It was found that the local ethnopedological category "*barro*" (clay), recognized by its red/yellow coloring and high clay content in the A horizon, is broadly coterminous with Oxisols, while the local category "*areia*" (sand), recognized by dark brown, grey or black coloring and highly friable and "sandy" (although probably pseudosands) A horizon is broadly coterminous with Ultisols [26]. Along the Madeira River, farmers recognize three major zones in the floodplain: i) the highly fertile sides of the main channel, locally known as the *vazante*; ii) the high floodplain, known as the *restinga*; iii) and the back-swamp area, known as the *cacaia* (Figure 5). These terms are local and identify fewer categories than Denevan [68] observed along the floodplain of the Ucayali River, in eastern Peru, between the towns of Pucallpa and Panaillo.

Both the collection of plant material and the interviews took place after prior informed consent was obtained at each community. Anthropological fieldwork was authorized by a scientific expedition (EXC 022/05) granted by the Brazilian National Research Council (CNPq). Four *terra firme* localities (the second and third comprised of several communities) were selected for semi-structured quantitative interviews on bitter manioc cultivation on ADE, Oxisols and Ultisols: Barro Alto, Barreira do Capanã/Boa Vista, the Água Azul Coast, and Vista Alegre (Figure 1). These localities were selected because they had the greatest numbers of farmers cultivating this crop on ADE [83]. All farmers cultivating bitter manioc on ADE in these localities at the time of research were interviewed. We selected an equivalent number of families farming Oxisols and Ultisols using the snowball method whereby new families were enrolled through those who had already been interviewed ([84]:184–85). Two floodplain localities, each comprising several communities, were selected. Upstream from Manicoré families resident in the floodplain communities Fortaleza and Pau Queimado, and a few families at Água Azul and Monte Sião who had fields in the floodplain were selected. Downstream from Manicoré are the communities Verdum, Amparo and Delícia. These communities are among the longest established floodplain communities on the middle Madeira River. All farmers present at these floodplain communities during visits were interviewed. Key individuals who maintained the highest number of bitter manioc landraces were identified through open-interviews with different community members. In order to minimize gender/age biases, and to verify information from multiple sources, interviews were conducted in the household, normally with all of the family present, as family members would often intervene to correct the orator or provide additional information.

Participant observation and open interviews

JAF conducted extensive participant observation in all communities of the study area prior to and during subsequent quantitative work (from September 2006 to March 2008), in order to build rapport and open up areas of interest that cannot be gleaned from close reading of the literature [85]. This entailed engaging in daily activities, including planting, harvesting and processing manioc, and conducting open, unstructured interviews with local people. The advantage of this type of initial qualitative approach is that it places no limitations on and is non-reductive in relation to reality. This open-ended approach led to the generation of the hypothesis that this paper addresses (that there are different adaptive manioc systems in different soils) and facilitated the discovery of the local theory of weakness and strength. The disadvantage of such an unstructured qualitative approach is that it cannot test the hypotheses it generates, and this is why we generated various novel quantitative indicators to measure aspects of bitter manioc cultivation systems from which adaptability could be inferred (planting behaviour, perceptions of landrace performance in different soils, yields and fallow lengths).

Ethnobotanical Data, Indices and Analyses

In total, we conducted 249 semi-structured interviews at the six localities between May 2007 and March 2008. Quantitative ethnobotanical data were gathered on a) the area occupied by each manioc landrace (landrace area – *la* – see below) in fields on four soil types (ADE, Ultisols, Oxisols and Floodplain), b) perceptions of relative performance (performance ranking index – *pri* – see below) of the same landraces in the three different *terra firme* soils, c) fallow lengths, and d) incorporation of seedlings. In order to test if fields on different soils exhibit different compositions of landraces, data on *la* were ordered through a Non-Metric Multidimensional Scaling (NMDS) based on the Bray-Curtis dissimilarity, followed by non-parametric multivariate analyses of variance (MANOVA; [86]). Non-parametric MANOVAs were performed for the general model (i.e., including all soil types simultaneously) and also for the six possible pairwise comparisons between the soil types. Critical p-values were adjusted by the Holm correction for multiple comparisons [87]. The relationship between *la* and *pri* was investigated using simple correlations. Composite soil samples were collected in each field and the results from chemical analysis were used for an *a posteriori* classification of soil types [26].

Landrace Area Index and Fallow Lengths

In order to identify the different landraces cultivated and quantify their abundance in the fields, we relied on the local nomenclature for the landraces and on numerical estimates provided by each farmer, using a local unit commonly used for quantifying manioc stem cuttings. In general, farmers along the middle Madeira recall the amount of each landrace they plant in terms of the number of bundles ("*feixes*") of each landrace they planted in their fields. We asked each farmer (a) which landraces he/she cultivated, (b) how many bundles of each landrace he had planted in his/her current manioc field and (c) how big was his/her field. Using this information we calculated the area occupied by each landrace in each field (*la*) through the formula:

$$la = \frac{nbv}{nbT} \times arf$$

where *la* is the *landrace area* index, *nbv* is the number of bundles of a given landrace planted in a given field, *nbT* is the total number of bundles planted in that field, and *arf* is the area of the field. Since we did not observe differences in the spacing used to plant the landraces, we regard the area occupied by each landrace (*la*) as a good estimator of the abundance of each landrace in the field. When all landraces are taken into account, we interpret this data as the varietal composition of the fields. We collected data on fallow lengths by asking each farmer how old the fallow was that he/she had cleared in order to establish the field. Data for the calculation of *la* and on fallow lengths were collected with 249 farmers at all six localities, of which 190 farmers on the *terra firme* and 59 on the floodplain.

Performance Ranking Index and seedling management

During participant observation it emerged that farmers claimed that certain landraces performed better in particular soils [83]. In order to quantify this, we asked all 190 *terra firme* farmers (a) to name all landraces that they cultivated in each type of soil and then (b) to rank the landraces mentioned according to their performance in that specific soil. Based on Sutrop's "Cognitive Salience Index" [88], we combined data on the frequency with which a given landrace was cited, its mean position in the performance ranking and calculated a Performance Ranking Index (*pri*) of each landrace in each type of soil, using the formula:

$$pri = \frac{F}{(N \times mP)}$$

where *pri* is the Performance Ranking index, *F* is the frequency of citations of the landrace, *N* is the number of farmers interviewed who cultivate in this specific type of soil, and *mP* is the mean position of the landrace in the rankings.

Participant observation also revealed that some farmers purposefully separate cuttings from volunteer seedlings for later planting, others simply harvest them together with other landraces (resulting in their random incorporation into pre-existing landraces), and others purposefully exclude volunteers [83]. Once these three categories were established, the incorporation of seedlings was measured by asking 114 farming families if they: a) *separated* seedlings, that is took cuttings from seedlings and planted them apart to see how they did; b) *randomly* incorporated seedlings, where they do plant cuttings of seedlings volunteers, but randomly mixed up with the other landraces; or c) *removed* them, cutting the volunteers out as weeds when they appeared in the field.

Farmer landrace yield estimates

Thirty one farmers were selected to record the production of different bitter manioc landraces in mono-varietal patches within 43 fields on the four different soil types. They were selected on the basis of literacy, their expressed interest in participating in the study (i.e., willingness to commit to recording production), and, crucially, the fact that they had mono-varietal patches within manioc fields during the study period. Bitter manioc fields are harvested bit by bit depending on labor availability and demand for manioc flour for subsistence consumption and sale. Amazonian farmers also do not tend to have accurate scales. They take a great interest in the amount of manioc flour produced, however, since it is equivalent to, as one woman put it, their "daily bread" (e.g., carbohydrate staple) and additionally an important product for sale in markets. In order to measure production therefore we asked farmers to record how many sacks of manioc flour (each sack is 50 kg) they produced from each mono-varietal-landrace patch. Fields were visited and we did not record significant differences in spacing, so we consider them to be all planted at similar density (around 1 meter between plants – c. 5000 plants per 0.5 ha). The size of mono-varietal patches varied (mean 0.286 ha±0.123 SD), so we corrected each one up to 0.5 ha in order to make the data commensurate and comparable using ANOVA (Table 1).

Collection of genetic data

The collection of plant material for genetic analyses was carried out in 2009. Authorization for interviewing farmers was obtained from *Instituto Nacional de Pesquisas da Amazônia*'s Committee for Research Ethics (protocol 235/09) and our collecting was registered in the System for Authorization and Information on Biodiversity, coordinated by the Chico Mendes Institute for

Biodiversity, of the Ministry of the Environment (number of register: 10020-5). No proprietary traditional knowledge was accessed, allowing us to meet Resolution 21 requirements for basic research that does not require authorization from Brazil's Council for Genetic Patrimony (CGEN in the Brazilian acronym), which was consulted before field work.

To investigate how the genetic diversity of bitter manioc was organized in different environments of cultivation we collected leaf samples of each landrace present in each of a total of 48 swiddens in all the localities except Vista Alegre: 17 in ADE soils, 14 in floodplain soils and 17 in Oxisols/Utilsols. A total of 184 individuals were sampled, representing 43 different bitter manioc landraces. The genetic variation of landraces was accessed using ten microsatellite loci [89,90], which are short sequences of repetitive DNA. Seven of them (GA21, GA126, GA131, GA134, GA136, GA140, GAGG5) were described by Chavarriaga-Aguirre *et al.* [89] and three (SSRY13, SSRY89, SSRY 164) by Mba *et al.* [90]. Each microsatellite locus refers to a unique genomic region and the genetic variation results from differences in the number of repetitive units among individuals.

To investigate the extent of genetic variation within landraces and the genetic identity of landraces with the same name but grown in different soil types, some of the most commonly cultivated bitter manioc landraces in the region (*Pirarucu Branco*, *Tartaruga*, *Arroz*, *Jabuti* and *Pirarucu Amarelo*) had 20 or 30 individuals sampled (Dataset S2). To evaluate the dispersion of bitter manioc swiddens along the genetic diversity revealed by microsatellites, a Principal Coordinate Analysis (PCoA), based on Euclidian distances, was carried out with GenAlEx v.6.4 [91]. Among the parameters used to describe intra-varietal genetic diversity, we estimated the observed (H_O) and expected (H_E) heterozygotes with GenAlEx v.6.4, while the number of multi-locus genotypes (MLGs) was analysed with GenClone v.2.0 [92]. H_O varies from 0 to 1, and corresponds to the probability that a given microsatellite locus has two different numbers of repetitions in a given individual. H_E also varies from 0 to 1, and corresponds to the proportion of heterozygote individuals that were expected under the Hardy-Weinberg Equilibrium based on the number of different microsatellite forms found for a given landrace. The MLGs describe the number of different microsatellite combinations present in a given group of individuals (in this case, the landraces). The detailed sampling strategy, material and methods and other genetic results are described elsewhere [76,77].

Acknowledgments

We extend our heartfelt thanks to the people of the Middle Madeira River for participating in our research. JAF thanks the CNPq (Brazilian National Research Council) for granting research permission (EXC 022/05). AAP thanks the *Coordenação de Aperfeiçoamento de Pessoal de Nível Superior*—CAPES for a master's scholarship. CRC thanks CNPq for a research fellowship (*bolsas de produtividade*). NP thanks CAPES for a research fellowship (PRODOC).

Author Contributions

Conceived and designed the experiments: JAF AAP CRC NP. Performed the experiments: JAF AAP NP ABJ. Analyzed the data: JAF AAP NP ABJ CRC. Wrote the paper: JAF AAP NP ABJ CRC.

References

1. Arroyo-Kalin M (2011) Slash-burn-and-churn: Landscape history and crop cultivation in pre-Columbian Amazonia. Quaternary International 249: 4–18.
2. Neumann K, Bostoen K, Höhn A, Kahlheber S, Ngomanda A, et al. (2011) First farmers in the Central African rainforest: A view from southern Cameroon. Quaternary International 249: 53–62.
3. Kingwell-Banham E, Fuller DQ (2011) Shifting cultivators in South Asia: Expansion, marginalisation and specialisation over the long term. Quaternary International 249: 84–95.
4. Barker G, Gilbertson D, Reynolds T, editors (2012) Rainforest Foraging and Farming in Island Southeast Asia: The Archaeology and Environmental History of the Niah Caves, Sarawak. Cambridge: Cambridge University Press.
5. Denham T, Haberle S (2008) Agricultural emergence and transformation in the Upper Wahgi valley, Papua New Guinea, during the Holocene: theory, method and practice. Holocene 18: 481–496.
6. Brookfield H, Padoch C (1994) Appreciating agrodiversity: A look at the dynamism and diversity of indigenous farming practices. Environment 36: 6–21.
7. Conklin H (1961) The study of shifting cultivation. Current Anthropology 2: 27–61.
8. Nye PH, Greenland DJ (1960) The Soil under Shifting Cultivation. Harpenden, UK: Technical Communication 51, Commonwealth Bureaux of Soils. 156 p.
9. Steward JH, editor (1948) Handbook of South American Indians. Volume 3: The Tropical Forest Tribes. Washington, D.C: Bulletin 143, Bureau of American Ethnology, Smithsonian Institution. 567 p.
10. Meggers BJ (1971) Amazonia: Man and Culture in a Counterfeit Paradise. Washington: Smithsonian Institution Press. 194 p.
11. Stahl PW (2002) Paradigms in paradise: revising standard Amazonian prehistory. The Review of Archaeology 23: 39–51.
12. Rival L, McKey D (2008) Domestication and diversity in manioc (Manihot esculenta Crantz ssp. esculenta, Euphorbiaceae). Current Anthropology 49: 1119–1128.
13. Emperaire L, Peroni N (2007) Traditional management of agrobiodiversity in Brazil: A case study of manioc. Human Ecology 35: 761–768.
14. Irion G (1984) Sedimentation and sediments of Amazon rivers and evolution of the Amazon landscape since Pliocene times. In: Sioli H, editor. The Amazon - Limnology and Landscape Ecology of a Mighty Tropical River and its Basin. The Hague: Springer. pp. 201–204.
15. Glaser B, Birk JJ (2012) State of the scientific knowledge on properties and genesis of Anthropogenic Dark Earths in Central Amazonia (terra preta de Indio). Geochimica et Cosmochimica Acta 82: 39–51.
16. Lathrap DW (1970) The Upper Amazon. Southampton: Thames & Hudson. 256 p.
17. Denevan WM (2001) Cultivated Landscapes of Native Amazonia and the Andes; Clark G, Goudie A, Peach C, editors. Oxford: Oxford University Press. 400 p.
18. Denevan WM (1996) A bluff model of riverine settlement in prehistoric Amazonia. Annals of the Association of American Geographers 86: 654–681.
19. Glaser B, Woods WI, editors (2004) Amazonian Dark Earths: Explorations in Space and Time. Berlin: Springer. 216 p.
20. Lehmann J, Kern DC, Glaser B, Woods WI, editors (2003) Amazonian Dark Earths. Origins, Properties and Management. Dordrecht: Kluwer Press. 523 p.
21. Woods WI, Teixeira WG, Lehmann J, Steiner C, WinklerPrins AMGA, et al. (2009) Amazonian Dark Earths: Wim Sombroek's Vision. Berlin: Springer. 504 p.
22. Neves EG (2008) Ecology, ceramic chronology and distribution, long-term history, and political change in the Amazonian floodplain. In: Silverman H, Isbell WH, editors. Handbook of South American Archaeology: Springer. pp. 359–403.
23. Heckenberger M, Neves EG (2009) Amazonian archaeology. Annual Review of Anthropology 38: 251–266.
24. Petersen JB, Neves EG, Heckenberger MJ (2001) Gift from the past: *terra preta* and prehistoric occupation in Amazonia. In: McEwan C, Barreto C, Neves EG, editors. Unknown Amazon Culture in Nature in Ancient Brazil. 1st ed. London: The British Museum Press. pp. 86–107.
25. Kuzyakov Y, Subbotina I, Bogomolova I, Xu XL (2009) Black carbon decomposition and incorporation into soil microbial biomass estimated by ^{14}C labeling. Soil Biology & Biochemistry 41: 210–219.
26. Fraser J, Teixeira W, Falcão N, Woods W, Lehmann J, et al. (2011) Anthropogenic soils in the Central Amazon: From categories to a continuum. Area 43: 264–273.

27. Schmidt MJ (2010) Reconstructing tropical nature: Prehistoric and modern anthrosols (*terra preta*) in the Amazon rainforest, upper Xingu river, Brazil [PhD Dissertation]: Department of Geography, University of Florida. 1594 p.
28. Erickson CL (2008) Amazonia: The Historical Ecology of a Domesticated Landscape. In: Silverman H, Isbell WH, editors. The Handbook of South American Archaeology. New York: Springer. pp. 157–183.
29. German LA (2003) Historical contingencies in the coevolution of environment and livelihood: contributions to the debate on Amazonian Black Earth. Geoderma 111: 307–331.
30. Smith NJH (1980) Anthrosols and human carrying capacity in Amazonia. Annals of the Association of American Geographers 70: 553–566.
31. German LA (2004) Ecological praxis and blackwater ecosystems: A case study from the Brazilian Amazon. Human Ecology 32: 653–683.
32. Arroyo-Kalin M (2010) The Amazonian formative: Crop domestication and anthropogenic soils. Diversity 2: 473–504.
33. McKey D, Cavagnaro TR, Cliff J, Gleadow R (2010) Chemical ecology in coupled human and natural systems: people, manioc, multitrophic interactions and global change. Chemoecology 20: 109–133.
34. Hershey CH (1994) Manihot Genetic Diversity: International network for cassava genetic resources. International Crop Network Series, IPGRI 10.
35. Boster JS (1985) Selection for perceptual distinctiveness: evidence from Aguaruna cultivars of Manihot esculenta. Economic Botany 39: 310–325.
36. Elias M, McKey D (2000) The unmanaged reproductive ecology of domesticated plants in traditional agroecosystems: An example involving cassava and a call for data. Acta Oecologica-International Journal of Ecology 21: 223–230.
37. Emperaire L, Pinton F, Second G (1998) Gestion dynamique de la diversité varie tale du manioc en Amazonie du nord-ouest. Nature, Science et Societé 6: 27–42.
38. Salick J, Cellinese N, Knapp S (1997) Indigenous diversity of cassava: Generation, maintenance, use and loss among the Amuesha, Peruvian Upper Amazon. Economic Botany 51: 6–19.
39. Elias M, Rival L, McKey D (2000) Perception and management of cassava (Manihot esculenta Crantz) diversity among Makushi Amerindians of Guyana (South America). Journal of Ethnobiology 20: 239–265.
40. Delêtre M, McKey D, Hodkinson T (2011) Marriage exchanges, seed exchanges, and the dynamics of manioc diversity. Proceedings of the National Academy of Sciences of the United States of America 108: 18249–18254.
41. Ellen RF, Soselisa HL (2012) A Comparative Study of the Socio-ecological Concomitants of Cassava (Manihot esculenta Crantz) Diversity, Local Knowledge and Management in Eastern Indonesia. Ethnobotany Research & Applications 10: 15–35.
42. Jones W (1959) Manioc in Africa: Stanford University Press. 315 p.
43. Roosevelt AC (1980) Parmana: Prehistoric Maize and Manioc Susbistence along the Amazon and Orinoco. New York: Academic Press. 320 p.
44. Ohly J, Junk W (1999) Multiple Use of Central Amazon Floodplains: Reconciling Ecological Conditions, Requirements for Environmental Protection, and Socioeconomic Needs. In: Padoch C, Ayres JM, Pinedo-Vasquez M, Henderson A, editors. Varzea: Diversity, Development and Conservation of Amazonias's whitewater floodplains. New York: New York Botanical Garden. pp. 283–299.
45. Gutjahr E (2000) Prospects for arable farming in the floodplains of the Central Amazon. In: W. J . Junk, J. J . Ohly, Piedade MTF, Soares MGM, editors. The Central Amazon Floodplain: Actual Use and Options for a Sustainable Management. Leiden: Backhuys. pp. 141–170.
46. Fraser JA (2010) The diversity of bitter manioc (Manihot esculenta Crantz) cultivation in a whitewater Amazonian landscape. Diversity 2: 586–609.
47. McKey D, Emperaire L, Elias M, Pinton F, Robert T, et al. (2001) Local management and regional dynamics of varietal diversity of cassava in Amazonia. Genetics Selection Evolution 33: S465–S490.
48. Pujol B, Gigot G, Laurent G, Pinheiro-Kluppel M, Elias M, et al. (2002) Germination ecology of cassava, Manihot esculenta Crantz, Euphorbiaceae, in traditional agroecosystems: Seed and seedling biology of a vegetatively propagated domesticated plant. Economic Botany 56: 366–379.
49. Elias M, Penet L, Vindry P, McKey D, Panaud O, et al. (2001) Unmanaged sexual reproduction and the dynamics of genetic diversity of a vegetatively propagated crop plant, cassava (Manihot esculenta Crantz), in a traditional farming system. Molecular Ecology 10: 1895–1907.
50. Elias M, Muhlen GS, McKey D, Roa AC, Tohme J (2004) Genetic diversity of traditional south american landraces of cassava (Manihot esculeatta Crantz): An analysis using microsatellites. Economic Botany 58: 242–256.
51. Pujol B, Renoux F, Elias M, Rival L, McKey D (2007) The unappreciated ecology of landrace populations: Conservation consequences of soil seed banks in Cassava. Biological Conservation 136: 541–551.
52. Peroni N (1998) Taxonomia folk e diversidade intraespecífica de mandioca (Manihot esculenta Crantz) em roças de agricultura tradicional em áreas de Mata Atlântica, São Paulo. [Masters Dissertation]. São Paulo: University of São Paulo.
53. Elias M, McKey D, Panaud O, Anstett MC, Robert T (2001) Traditional management of cassava morphological and genetic diversity by the Makushi Amerindians (Guyana, South America): Perspectives for on-farm conservation of crop genetic resources. Euphytica 120: 143–157.
54. Pujol B, David P, McKey D (2005) Microevolution in agricultural environments: how a traditional Amerindian farming practice favours heterozygosity in cassava (Manihot esculenta Crantz, Euphorbiaceae). Ecology Letters 8: 138–147.
55. Peroni N, Kageyama PY, Begossi A (2007) Molecular differentiation, diversity, and folk classification of "sweet" and "bitter" cassava (Manihot esculenta) in Caicara and Caboclo management systems (Brazil). Genetic Resources and Crop Evolution 54: 1333–1349.
56. Stocker P (2006) Family farmers and manioc in contemporary Brazil: The management of agrobiodiversity and change [PhD Dissertation]: University of Edinburgh. 359 p.
57. Fraser JA (2010) Caboclo horticulture and Amazonian dark earths along the middle Madeira River, Brazil. Human Ecology 38: 651–662.
58. Fraser JA, Clement CR (2008) Dark earths and manioc cultivation in Central Amazonia: A window on pre-Colombian agricultural systems? Boletim do Museu Paraense Emílio Goeldi Ciências Humanas 3: 175–194.
59. Hiraoka M, Yamamoto S, Matsumoto E, Nakamura S, Falesi I, et al. (2003) Contemporary use and management of Amazonian Dark Earths. In: Lehmann J, Kern D, Glaser B, Woods W, editors. Amazonian Dark Earths: Origin, properties, and management. Dordrecht: Kluwer Academic Publishers. pp. 387–406.
60. German LA (2001) The dynamics of *Terra Preta*: an integrated study of human-environmental interaction in a nutrient-poor Amazonian ecosystem [Unpublished Ph.D Dissertation]: University of Georgia. 312 p.
61. German LA (2003) Ethnoscientific understandings of Amazonian Dark Earths. In: Lehmann J, Kern D, Glaser B, Woods WI, editors. Amazonian Dark Earths: Origin, Properties and Management. Dordrecht: Kluwer. pp. 179–201.
62. Harlan J (1992) Crops and Man. Madison, Wisconsin.: American Society of Agronomy Inc. Crop Science Society of America Inc. 284 p.
63. Fraser JA, Junqueira AB, Kawa NC, Moraes CP, Clement CR (2011) Crop Diversity on Anthropogenic Dark Earths in Central Amazonia. Human Ecology 39: 395–406.
64. Lima D, Steward A, Richers BT (2012) Trocas, experimentações e preferências: um estudo sobre a dinâmica da diversidade da mandioca no Médio Solimões, Amazonas. Boletim do Museu Paraense Emílio Goeldi: Ciências Humanas.
65. Grenand F (1993) Bitter manioc in the lowlands of Tropical America: from myth to commercialization. In: C. M. Hladik, A. Hladik, O. F. Linares, H. Pagezy, Semple A et al., editors. Tropical Forests and Food: Biocultural Interactions and Applications to Development. Paris: UNESCO and Parthenon.
66. Castro F (2009) Patterns of Resource Use by Caboclo Communities in the Middle-Lower Amazon. In: C. R. Adams, R. Murrieta, W. Neves, Harris M, editors. Amazon peasant societies in a changing environment: Political ecology, invisibility and modernity in the rainforest. New York: Springer. pp. 157–177.
67. Pereira KJC (2008) Agricultura tradicional e manejo da agrobiodiversidade na Amazônia Central: Um estudo de caso nos roçados de mandioca das Reservas de Desenvolvimento Sustentável Amanã e Mamirauá, Amazonas [PhD Dissertation]. São Paulo: University of São Paulo. 223 p.
68. Denevan W (1984) Ecological heterogeneity and horizontal zonation of agriculture in the Amazon floodplain. In: M . Schmink and C . Wood, editor. Frontier Expansion in Amazonia. Gainsville: University of Florida Press. pp 311–336.
69. Smith NJH (1999) The Amazon River Forest. A Natural History of Plants, Animals, and People. Oxford: Oxford University Press. 208 p.
70. Noda SN, Noda H, Santos HP (2000) Family farming systems in the floodplains of the State of Amazonas. In The Central Amazon Floodplain: Actual Use and Options for a Sustainable Management. In: W. J. Junk, J. J. Ohly, Piedade MTF, Soares MGM, editors. The Central Amazon Floodplain: Actual Use and Options for a Sustainable Management. Leiden: Backhuys. pp. 215–242.
71. Lima RMB, Saragoussi M (2000) Floodplain home gardens on the Central Amazon in Brazil. In: Junk WJ, J. J. Ohly, Piedade MTF, Soares MGM, editors. The Central Amazon Floodplain: Actual Use and Options for a Sustainable Management. Leiden: Backhuys. pp. 243–268.
72. Boster JS (1984) Classification, cultivation and selection of Aguaruna cultivars of Manihot esculenta (Euphorbiaceae). Advances in Economic Botany 1: 34–47.
73. Pujol B, Gigot G, Laurent G, Pinheiro-Kluppel M, Elias M, et al. (2002) Germination ecology of cassava (Manihot esculenta Crantz, Euphorriaceae) in traditional agroecosystems: Seed and seedling biology of a vegetatively propagated domesticated plant. Economic Botany 56: 366–379.
74. Lebot V (2009) Tropical Root and Tuber Crops: Cassava, Sweet Potato, Yams and Aroids. Oxford, UK: CAB International.
75. Boserup E (1993 (1965)) The conditions of agricultural growth. The economics of agrarian change under population pressure. London: Earthscan Publications Ltd.
76. Alves Pereira A (2011) Dinâmica evolutiva de mandioca (Manihot esculenta Crantz) em três tipos de solos manejados por Caboclos na região do Médio Rio Madiera, Amazonas. Manaus, Brazil: Instituto Nacional de Pesquisas da Amazônia. 87 p.
77. Alves-Pereira A, Peroni N, Abreu AG, Gribel R, Clement CR (2012) Genetic structure of traditional varieties of bitter manioc in three soils in Central Amazonia. Genetica 139: 1259–1271.
78. Kensinger KM (1975) Studying the Cashinahua. In: Kensinger KM, editor. The Cashinahua of Eastern Peru Studies in Anthropology and Material Culture, vol 1. Providence, Rhode Island: Brown University. pp. 9–85.
79. Aquino TVd, Iglesias MP (2002) Habitantes: Os Kaxinawá. In: Carneiro da Cunha M, Almeida MBd, editors. Enciclopédia da floresta O alto juruá: práticas e conhecimentos das populações. São Paulo: Companhia das Letras. pp. 147–160.

80. Ricardo CA, editor (2000) Povos Indígenas no Brasil, 1996–2000. São Paulo: Instituto Socioambiental.
81. Wilson W (2002) Soils Utilized for Gardens by Tukanoans in Northwestern Amazonia and their Impact on Cassava (Manihot esculenta Crantz) Cultivar Selection. Culture & Agriculture 24: 20–30.
82. Heckler S, Zent S (2008) Piaroa Manioc Varietals: Hyperdiversity or Social Currency? Human Ecology 36: 679–697.
83. Fraser JA (2009) Amazonian dark earths and *Caboclo* subsistence on the Middle Madeira River, Brazil [PhD Dissertation]: University of Sussex.
84. Bryman A (2008) Social Research Methods. Oxford, UK: Oxford University Press. 800 p.
85. Bernard HR (2006) Research methods in anthropology: qualitative and quantitative approaches. Oxford: Altamira Press. 803 p.
86. McCune B, Grace JB (2002) Analysis of Ecological Communities: MjM Software Design. 304 p.
87. Wright SP (1992) Adjusted p-values for simultaneous inference. Biometrics 48: 1005–1013.
88. Sutrop U (2001) List task and a cognitive salience index. Field Methods 13: 263–276.
89. Chavarriaga-Aguirre P, Maya MM, Bonierbale MV, Kresovich S, Fregene MA, et al. (1998) Microsatellites in cassava (Manihot esculenta Crantz): discovery, inheritance and variability. Theoretical and Applied Genetics 97: 493–501.
90. Mba REC, Stephenson P, Edwards K, Melzer S, Nkumbira J, et al. (2001) Simple sequence repeat (SSR) markers survey of the cassava (Manihot esculenta Crantz) genome: towards an SSR-based molecular genetic map of cassava. Theoretical and Applied Genetics 102: 21–31.
91. Peakall R, Smouse PE (2006) Genalex 6: Genetic analysis in Excel. Population genetic software for teaching and research. Molecular Ecology Notes 6: 288–295.
92. Arnaud-Haond S, Belkhir K (2007) GENCLONE: a computer program to analyse genotypic data, test for clonality and describe spatial clonal organization. Molecular Ecology Notes 7: 15–17.

14

Arbuscular Mycorrhizas Reduce Nitrogen Loss via Leaching

Hamid R. Asghari[1], Timothy R. Cavagnaro[2,3]*

1 Faculty of Agriculture, Shahrood University of Technology, Shahrood, Iran, 2 School of Biological Sciences, Monash University, Clayton, Victoria, Australia, 3 Australian Centre for Biodiversity, Monash University, Clayton, Victoria, Australia

Abstract

The capacity of mycorrhizal and non-mycorrhizal root systems to reduce nitrate (NO_3^-) and ammonium (NH_4^+) loss from soils via leaching was investigated in a microcosm-based study. A mycorrhiza defective tomato mutant and its mycorrhizal wildtype progenitor were used in this experiment in order to avoid the indirect effects of establishing non-mycorrhizal control treatments on soil nitrogen cycling and the wider soil biota. Mycorrhizal root systems dramatically reduced nitrate loss (almost 40 times less) via leaching, compared to their non-mycorrhizal counterparts, following a pulse application of ammonium nitrate to experimental microcosms. The capacity of AM to reduce nutrient loss via leaching has received relatively little attention, but as demonstrated here, can be significant. Taken together, these data highlight the need to consider the potential benefits of AM beyond improvements in plant nutrition alone.

Editor: Mari Moora, University of Tartu, Estonia

Funding: Murray Darling Basin Authority provided financial support for plant analysis. HRA was supported by Shahrood University of Technology of Iran and with an Endeavor award from the Australian Government. TRC also wishes to acknowledge the Monash Research Accelerator Program for supporting this work. The funders had no role in study design, data collection and analysis, decision to publish, or preparation of the manuscript.

Competing Interests: The authors have declared that no competing interests exist.

* E-mail: timothy.cavagnaro@monash.edu

Introduction

Agricultural systems are being intensified to meet the world's increasing demand for food and fiber [1]. To meet these demands fertilizer use is expected to increase dramatically [2]. However, excess application and inefficient use of fertilizers can have considerable negative economic and environmental consequences. For example, nitrate (NO_3^-), a highly mobile form of nitrogen, is readily lost from agricultural lands via leaching [3]. This can lead to contamination of drinking water supplies and eutrophication of water bodies [4,5,6]. Thus, interception of nitrate before it leaches below the root zone of plants is a high priority, both in terms of improving fertilizer use efficiency, and reducing the risk of environmental degradation [7].

Most terrestrial plant species, including the majority of crops, form arbuscular mycorrhizas (AM) [8]. These associations, between plant roots and a specialized group of soil fungi, play an important role in plant acquisition of nutrients, including P, N, Zn and others [9,10]. While most research has focused on P, there is increasing evidence of an important role for AM in acquisition of N from both inorganic [11,12,13,14] and organic [15,16] sources in the soil. Whereas AM are typically considered in terms of their potential to improve plant nutrition, they have also been found to have an important, but often overlooked, role to play in reducing the loss of nutrients (both P and N) via leaching [17,18,19]. Thus, maintaining and enhancing levels of AM in ecosystems where the risk of nutrient leaching is high may be important.

To study mycorrhizal functioning, plants that are colonized by arbuscular mycorrhizal fungi (AMF) are normally compared to plants that are not colonized by AMF. These non-mycorrhizal treatments are typically established by sterilizing soil to eliminate the fungi. However, soil sterilization changes soil chemistry, and eliminates other soil microbes involved in nutrient cycling [20].One option to overcome this issue is the use of mycorrhiza defective plant mutants, and their mycorrhizal wildtype progenitors, as a means of establishing non-mycorrhizal controls [13,20,21]. The advantage of this approach is that it allows for the direct investigation of mycorrhizal effects on soil and plant processes with the wider soil biota intact. This is particularly important with respect to the role of AM in increasing plant nitrogen acquisition, as the cycling of nitrogenin the soil, which is extremely rapid and dynamic, is in large part driven by microbial processes [7].

Here we present results of a microcosm-based study investigating the capacity of mycorrhizal and non-mycorrhizal root systems to reduce nitrate (NO_3^-) and ammonium (NH_4^+) loss from soils via leaching. A mycorrhiza defective tomato mutant, and its mycorrhizal wild-type progenitor [22] were used to establish mycorrhizal and non-mycorrhizal treatments. The results of this study are considered in the context of the potential for mycorrhizal root systems to reduce nutrient loss (especially nitrate) from soils via leaching.

Materials and Methods

Experimental design

A glasshouse experiment was carried out to investigate the effects of forming AM on the capacity of root systems to reduce soil nitrate and ammonium leaching. A mycorrhiza defective tomato mutant with reduced mycorrhizal colonization (named *rmc*) and its mycorrhizal wildtype progenitor (named 76R) [22] were used to establish mycorrhizal and non-mycorrhizal treatments in experimental microcosms, to which either ammonium nitrate or water (control) were added. This approach is that it avoids the need to sterilize soils to establish non-mycorrhizal controls [13,20]. Thus, mycorrhizal and non-mycorrhizal plants can be compared with the wider soil biota, including those involved in soil N cycling, in-tact [23]; this is a key point of novelty of this experiment. Importantly, the growth of these genotypes is match when grown under a wide range of experimental conditions [13,24,25,26],with one exception [27]. Furthermore, when grown in the absence of AMF the growth of the genotypes is matched [24], indicating that the mutation affecting the formation of AM by the *rmc* genotype has no pleiotropic effects on other plant processes.

Soil, plants and nutrient addition

Microcosms, as described previously [19], were established as follows: a 30 mm layer ofdried washed sand (140 g) was placed on a layer of cotton mesh at the base of PVC columns (90 mm diameter×400 mm deep) with a PVC cap (with a central hole, 15 mm in diameter) on their base. To each column 2.5 kg of a soil:sand mixture (40:60% W/W) was added to a final bulk density of 1.4 g.cm^{-3}. A soil:sand mix was used in this experiment as it provides a very even mixture, uniform leaching conditions, and ready extraction of roots and hyphae at the time of harvest [17,19,28]. The soil, which was air-dried and passed through a 2 mm sieve, was collected from the 0–15 cm layer of restored riparian zone adjacent to Faithfuls Creek in the southern Murray-Darling Basin in southeastern Australia (see www.mdba.gov.au/riparian-restoration-experiment/). The soil at this site is a quaternary red chromosol, fluvial silt-sand, with a pH of 6.0, plant available (Olsen) P of 4.5 mg kg^{-1}, total C of 1.9% and total N of 0.12% (T.R. Cavagnaro, unpublished). This soil was selected as riparian zones commonly experience large nutrient inputs in rapid "pulse–based" events (as simulated here, see below), for example, following large rainfall events or at the break of seasons or droughts [29]. The sand used in the mix was a coarse grained and washed river sand, as in our earlier work on soil leaching [19].

Seeds of a mycorrhiza defective tomato (*Solanum lycopersicum* L. mutant (*rmc*), and its wildtype progenitor (76R) [22] were surface sterilized and imbibed prior to planting [19]. Seeds of either genotype were planted in each column, kept moist with RO water, and after 2 weeks, seedlings were thinned to one per column. Columns were then irrigated (to weight) with RO water every second day, to 80% of the field capacity, thereby ensuring that no water leached out of the columns during the plant growth phase of the experiment [17]. Plants were grown in a glasshouse with supplemental lighting: mean day time temperature was 22.1°C, min 18.3°C, max 25.9°C; night time temperature was mean 20.1°C, min 17.3°C, max, 22.6°C; and mean daily photon load of 495.1±108.6 mol quanta m^{-2}. Four weeks after planting, all plants were supplied with a 20 ml of a modified Long Ashton nutrient solution minus P [30], once a week for 3 weeks. Each treatment was replicated four times; however, one replicate was lost from the *rmc* treatments during the course of the experiment.

Nutrient addition treatments

Nine weeks after planting half of the pots were supplied with a pulse of nitrogen as 143 mg of ammonium nitrate (NH_4NO_3) dissolved in 10 ml RO water; this nitrogen addition treatment was equivalent to an input of 280 kg N ha^{-1} [31]. This amount of N, which was used in earlier leaching experiments using this soil [19], is within the range applied to commercial tomato crops. To the remaining control columns, 10 ml of RO water was added. Following the addition of the N pulse (or RO water for the controls), the cores were watered to 80% of field capacity, in order to water in the added nutrients, and to maintain constant soil moisture content for the remainder of the experiment.

Harvesting and leachate collection

Ten weeks after planting (i.e. 7 days after nutrient addition treatments were applied) all cores were destructively harvested. The shoots of plants were removed (to eliminate water loss via transpiration), and the columns immediately flushed with 700 ml of RO water to leach soil nutrients from the columns. This approach was taken in an effort to simulate a large rainfall event as typically occurs at the site from where the soil used in this experiment was collected. The leachate was collected from the columns until the cessation of leaching (24 hrs) and the concentration of NH_4^+-N and NO_3^--N determined colourimetrically, as for soil extracts (see below). Soil samples were collected from three layers (0–5, 10–15, 20–25 cm) for analysis of soil NO_3^--N and NH_4^+-N concentrations (in duplicate). Briefly, the soils were extracted using a 2 M KCl solution and inorganic nitrogen content determined colorimetrically using a modification of the methods of Miranda et al. [32] for NO_3^- (plus NO_2) and Forster [33] for NH_4^+. Mycorrhizal hyphal length was determined [34] on the middle soil layer;preliminary analyses revealed that hyphal length densities did not differ between soil layers (data not shown). The roots were then carefully washed from all of the remaining soil with RO water. Mycorrhizal colonization of a sub-sample of roots was determined using the gridline intersect method [35], following clearing and staining of roots with Trypan Blue (omitting phenol from all reagents) [36]. All remaining plant material was dried at 60°C, and shoot dry weights (SDW) and root dry weights (RDW) determined. Plant material was then ground to a fine powder and the concentration of nitrogen determined by dry combustion (CHN 2000 analyzer, LECO Corporation, USA). Here we present plant nutrient data on a whole plant nutrient content basis (i.e. N per plant), rather than on a tissue concentration basis, as our emphasis is on nutrient interception, rather than nutrient concentrations in plant tissues (see results).

Statistical analysis

Data from mycorrhizal treatments and nutrient addition treatments were analyzed using two-way analysis of variance (ANOVA) using JMP® (version 8.0.2. SAS Institute). Where significant differences between treatments were found ($P<0.05$), differences between individual treatment means were determined using Tukeys HSD tests. Additional targeted data analysis was undertaken to further explore genotypic differences (using t-tests) within the different nutrient addition treatments; these additional targeted analyses are indicated in the relevant sections of the results.

Results

Mycorrhizal colonization, plant growth and nutrition

Roots of the 76R genotype were well colonized, with levels of colonization not significantly different between the nitrogen (52±8% root length colonized: values are mean ± SE) and control (39±6% root length colonized) nutrient addition treatments. Conversely, colonization of *rmc* roots was less than 1%, and was restricted to the root epidermis. The length of external hyphae of arbuscular mycorrhizal fungi (AMF) was significantly higher in columns containing 76R plants (10.9±2.2 and 16.0±1.2 m.g^{-1} dry soil, for the nitrogen and control nutrient addition treatments, respectively) compared to those containing *rmc* plants (4.2±1.1 and 5.0±3.1 m.g^{-1} dry soil, for thenitrogenand control nutrient addition treatments, respectively). There was no significant difference in hyphal length densities between nutrient addition treatments.

The RDW of plants did not differ between any of the treatments (Fig. 1a). Whereas, the SDW (Fig. 1a) of the 76R genotype was greater than that of the *rmc* genotype ($F_{(1,3)} = 5.06$, $P<0.05$), irrespective of nutrient addition treatment, total plant dry weight (SDW+RDW) did not differ between any of the treatments (see Fig. 1a). Shoot nitrogen content ($mg.plant^{-1}$) of 76R plants was higher ($F_{(1,3)} = 4.95$, $P = 0.05$) than that of *rmc* plants (Fig. 1b).Similarly, the nitrogen content of plants grown in the nitrogen addition treatment (irrespective of genotype) was significantly ($F_{(1,3)} = 17.32$, $P = 0.0019$) higher than that of plants in the control (water addition) treatment. There were no significant differences in root nitrogen contents (Fig. 1b) between any of the experimental treatments. Importantly, whole plant nitrogen content (shoots+roots) was significantly higher in the 76R than *rmc* genotype, irrespective of nutrient addition treatment, and in the nitrogen addition treatment, irrespective of genotype (see Figure 1b).

Soil nitrogen pools and interception

The volume of leachate collected from the experimental columns did not differ between any of the experimental treatments (477 ml±48 ml per column). The concentration of nitrate in the leachate collected from columns (Fig. 2a) containing *rmc* root systems was significantly higher in the nitrogen addition treatment, when compared to all other treatments (Fig. 2a) ($F_{(1,3)} = 160.75$, $P<0.0001$).When the control treatment was considered separately from the nitrogen addition treatment (targeted t-test), the concentration of NO_3^- leached from columns containing *rmc* plants was significantly higher than that from columns containing 76R plants. The concentration of ammonium in the leachate was low, and did not differ between any of the experimental treatments (Fig. 2b).

At the end of the experiment, the amount of $N\text{-}NO_3^-$ remaining in the soil was relatively low (Fig. 3a); the concentration of $N\text{-}NO_3^-$ was higher in the nitrogen addition treatment, in the surface (and to a lesser extent the middle and lower) soil layer, irrespective of genotype. Similarly, the concentration of $N\text{-}NO_3^-$ was lower in the surface and middle soil layers, where plants were mycorrhizal, irrespective of nutrient addition treatment. The concentration of $N\text{-}NH_4^+$ in the soil (Fig. 3b) at the end of the experiment was higher in the surface soil layers, than at depth. Ammonium was especially high in the nitrogen addition treatment, irrespective of genotype.

Discussion

Mycorrhizal root systems dramatically reduced nitrate loss (almost 40 times less) via leaching, compared to their non-mycorrhizal counterparts, following a pulse application of ammonium nitrate to experimental microcosms. The capacity of

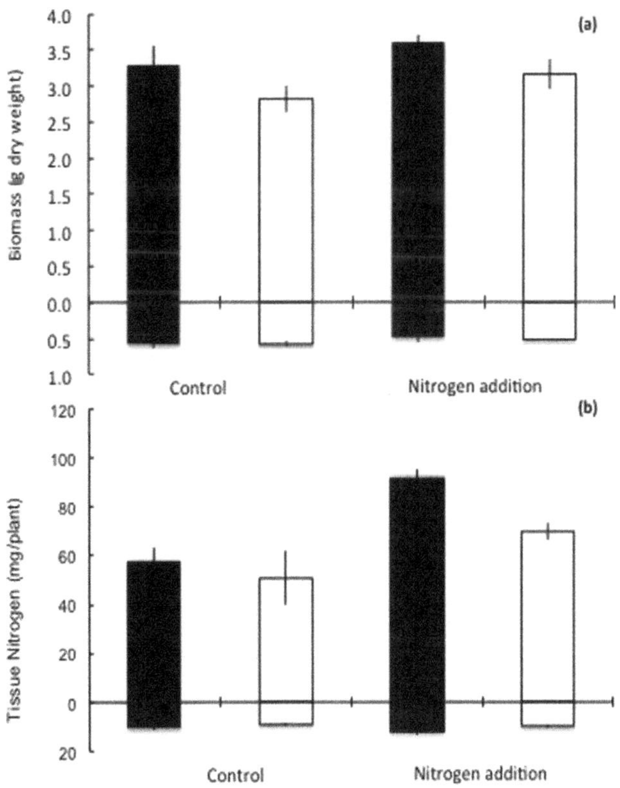

Figure 1. Mean shoot (above *X*-axis) and root (below *X*-axis). (a) dry weights, and (b) plant nitrogen contents, of the 76R (black bars) and *rmc* (white bars) genotypes of tomato, following the application of nutrient addition treatments (nitrogen or control). Values are mean ± S.E. The shoot dry weight of 76R plants was significantly greater than of *rmc* plants, irrespective of nutrient addition treatments; the nitrogen content differed significantly between genotypes irrespective of nutrient addition treatments, and vice versa; see text for details.

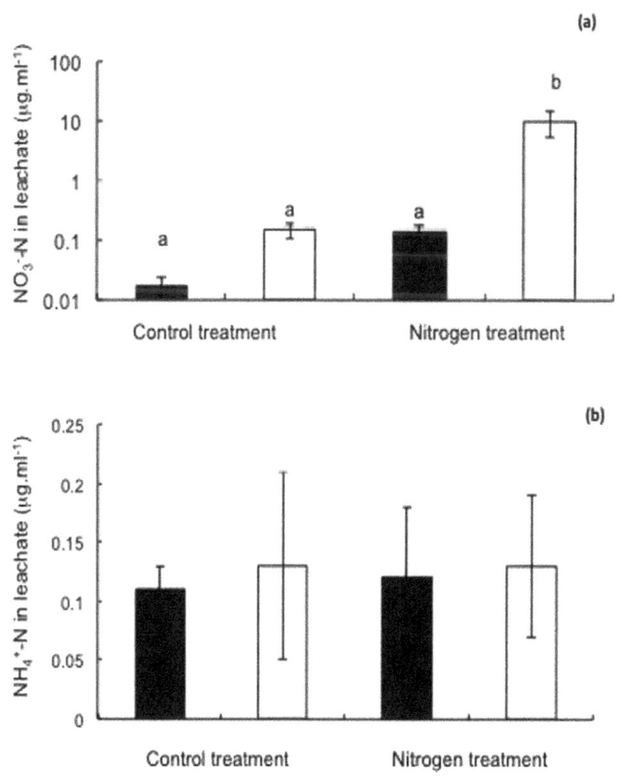

Figure 2. Concentration of (a) NO_3^--N and (b) NH_4^+-N in leachate, collected from columns containing 76R (black bars) and *rmc* (white bars) genotypes of tomato, following the application of nutrient addition treatments (nitrogen or control). Values are mean ± S.E. Means followed by the same letter are not significantly different at the $P<0.05$ level. See text for results of targeted statistical analyses comparing genotypes within specific nutrient addition treatments.

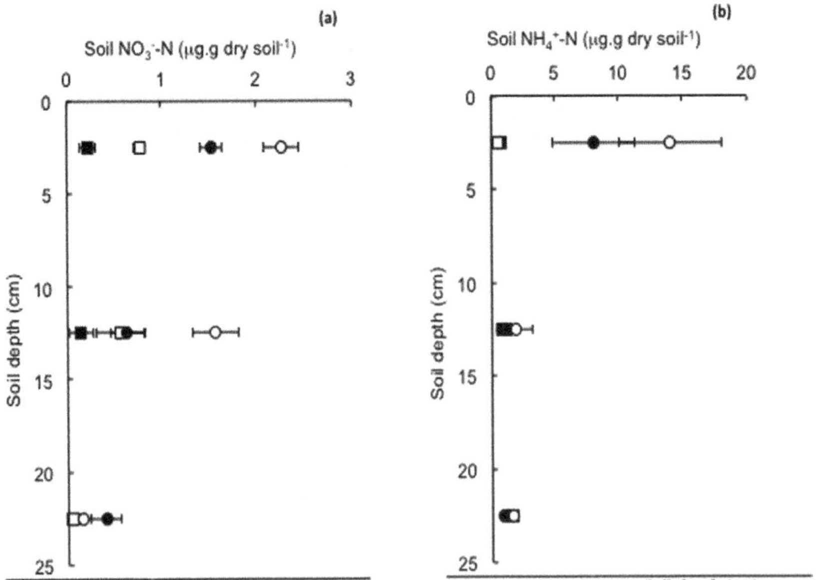

Factor	Soil depth (cm) 0-5	10-15	20-25
AM	***	***	ns
Nutrient	***	***	*
AM × Nutrient	ns	ns	ns

Factor	Soil depth (cm) 0-5	10-15	20-25
AM	ns	ns	ns
Nutrient	***	ns	ns
AM × Nutrient	ns	ns	ns

Figure 3. Soil (a) NO_3^--N and (b) NH_4^+-N, concentrations with depth (data plotted at mid-point of sampling depth) in columns containing mycorrhizal (closed symbols) and non mycorrhizal (open symbols) tomato plants in the nitrogen addition (circular symbols) and water control nutrient addition (square symbols) treatments. Values are means ± standard error. ANOVA Tables are given below Figures, see text for additional details of statistical analysis. *, **, *** = significant at $P<0.05$, 0.01, 0.001 levels respectively. ns = not significant, $P>0.05$.

AM to reduce nutrient loss (both N and P) via leaching has received relatively little attention, but is potentially very significant [17,18,19]. A decrease in nitrate loss via leaching on the scale seen here may be especially important in ecosystems where the risk of leaching is high, such as at the interface of agricultural and natural lands [4,5,6]. Taken together, these data highlight the need to consider the potential benefits of AM beyond improvements in plant nutrition alone.

The decrease in nitrate loss from columns containing mycorrhizal root systems seen here cannot solely be attributed to size asymmetry between mycorrhizal and non-mycorrhizal plants. Importantly, the mycorrhizal tomato genotype was well colonized, whereas the mycorrhiza defective tomato mutant was not colonized. By effectively increasing the absorptive surface of root system, AMF allow exploration of a larger soil volume, and hence, access to more nutrients [8]. The reduction in nitrate leachate from columns containing mycorrhizal plants was paralleled by a lower concentration of nitrate in the soil and higher plant nitrogen content. This is consistent with earlier work demonstrating that AM enhance the capacity of plants to acquire N from inorganic sources, under both laboratory and field conditions [11,12,13,14,37]. Further, using the same genotypes as in the present study Ruzicka et al. [14] found that the regulation of key genes involved in the N transport and assimilation indicating a shift towards N uptake via the mycorrhizal pathway in the mycorrhizal genotype. Together, this suggests that the reduction of nitrogen lost via leaching of nitrate was due to enhanced nitrogen interception and immobilization by mycorrhizal root systems [12,38].

Given its relatively low mobility in soils, it is not surprising that there was little ammonium was lost via leaching [39]. Although most of the added ammonium was retained in the upper soil layer, mycorrhizal interception of ammonium can be important [12], which may indirectly reduce the risk of nitrogen loss as nitrate. That is, ammonium not taken up by roots or AMF can be readily transformed into nitrate under aerobic conditions (via nitrification), such as those in the columns during the week following the application of ammonium nitrate to the soil [40]. This example serves to highlight the need for carefully controlled studies of soil-plant-AMF-nitrogen dynamics.

The use of a mycorrhiza defective tomato mutant and its mycorrhizal wildtype progenitor to establish mycorrhizal treatments avoided the use of non-specific sterilization techniques to establish non-mycorrhizal controls, which can eliminate soil biota and alter soil nitrogen cycling [13,20]. This allows us to attribute the differences seen here to the mycorrhizal status of the root systems. It may also explain why the reduction in N lost via leaching in the present study is substantially greater than in our earlier work using soil sterilization for the establishment of non-mycorrhizal controls [19]. While the very large differences in nitrate loss via leaching were likely due to plant/AMF immobilization, it will be important in future studies to begin to consider potential changes in microbial communities in the rhizosphere of these genotypes, as have been reported in earlier studies [27]. For example, possible differences in rhizodeposition between the genotypes, which may in turn affect denitrification and other nitrogen cycling processes, may need to be considered.

To further explore the role of AMF in reducing nitrate loss via leaching, the use of nitrogen isotopes will be especially informative. For example, the capacity to trace the applied nitrogen through the soil, plant, leachate and atmospheric (i.e. N_2O and N_2 efflux) poolswould permit calculation of full nitrogen loss budgets, and the impacts of AM on them. Similarly, studies with in-tact soil cores (rather than a soil-sand mix) and field based experiments are also likely to be especially informative. Nevertheless, it is clear from the data presented here, and in our earlier studies [19], that mycorrhizal root systems (i.e. roots plus fungi) can have a role to play in reducing nitrate loss via leaching.

Conclusions

The importance of AM in improving plant nutrition has been long known [8]. The results of this study highlight the potential of AM to reduce nitrate leaching from soil on a previously unrecognized scale. This suggests that managing farming systems to maximize mycorrhizal colonization of roots, especially where the risk of nitrate loss via leaching is high, should be of high priority [7]. Similarly, maintaining and enhancing mycorrhizas in vegetated buffer strips between potential sources of nitrogen pollution, such as farms and urban areas, and potential sinks for nitrogen, such as natural lands and water bodies, is also important. Effective management of AM in such systems will have the benefits of reducing the risk of nitrogen loss via leaching, as well as improving plant nitrogen nutrition.

Acknowledgments

The authors wish to thank members of the "Cav-Lab" for valuable discussions.

Author Contributions

Conceived and designed the experiments: HRA TRC. Performed the experiments: HRA. Analyzed the data: HRA TRC. Contributed reagents/materials/analysis tools: TRC. Wrote the paper: HRA TRC.

References

1. Ainsworth EA, Rogers A, Leakey ADB (2008) Targets for crop biotechnology in a future high-CO_2 and high-O_3 world. Plant Physiol 147: 13–19.
2. Cordell D, Drangert J-O, White S (2009) The story of phosphorus: global food security and food for thought. Glob Environ Change 19: 292–305.
3. Puckett LJ, Cowdery TK, Lorenz DL, Stoner JD (1999) Estimation of nitrate contamination of an agro-ecosystem outwash aquifer using a nitrogen mass-balance budget. J Appl Ecol 28: 2015–2025.
4. Likens GE, Bormann FH, Johnson NM, Fisher DW, Pierce RS (1970) Effects of forest cutting and herbicide treatment on nutrient budgets in the Hubbard Brook watershed-ecosystem. Ecol Monogr 40: 23–47.
5. Lowrance R, Altier LS, Newbold JD, Schnabel RR, Groffman PM, et al. (1997) Water quality functions of riparian forest buffers in Chesapeake Bay watersheds. Environ Manage 21: 687–712.
6. Lowrance R, Todd R, Frail JJ, Hendrickson OJ, Leonard R, et al. (1984) Riparian forests as nutrients filters in agricultural watersheds. Biosci 34: 374–377.
7. Jackson LE, Burger M, Cavagnaro TR (2008) Roots, nitrogen transformations, and ecosystem services. Ann Rev Plant Biol 59: 341–363.
8. Smith SE, Read DJ (2008) Mycorrhizal Symbiosis. Cambridge, UK: Academic Press Ltd.
9. Hoeksema JD, Chaudhary VB, Gehring CA, Collins-Johnson N, Karst J, et al. (2010) A meta-analysis of context dependency in plant response to inoculation with mycorrhizal fungi. Ecol Let 13: 394–407.
10. Cavagnaro TR (2008) The role of arbuscular mycorrhizas in improving plant zinc nutrition under low soil zinc concentrations: a review. Plant Soil 304: 315–325.
11. Ames RN, Reid CPP, Porter L, Cambardella C (1983) Hyphal uptake and transport of nitrogen from two 15N-labelled sources by Glomus mosseae, a vesicular arbuscular mycorrhizal fungus. New Phytol 95: 381–396.
12. Tanaka Y, Yano K (2005) Nitrogen delivery to maize via mycorrhizal hyphae depends on the form of N supplied. Plant Cell Environ 28: 1247–1254.
13. Cavagnaro TR, Jackson LE, Six J, Ferris H, Goyal S, et al. (2006) Arbuscular mycorrhizas, microbial communities, nutrient availability, and soil aggregates in organic tomato production. Plant Soil 282: 209–225.
14. Ruzicka DR, Barrios-Masias FH, Hausmann NT, Jackson LE, Schachtman DP (2010) Tomato root transcriptome response to a nitrogen-enriched soil patch. BMC Plant Biol 10: 75–94.
15. Hodge A, Campbell CD, Fitter AH (2001) An arbuscular mycorrhizal fungus accelerates decomposition and acquires nitrogen directly from organic material. Nature 413: 297–299.
16. Hodge A, Fitter AH (2010) Substantial nitrogen acquisition by arbuscular mycorrhizal fungi from organic material has implications for N cycling. PNAS 107: 13754–13759.
17. Asghari HR, Chittleborough DJ, Smith FA, Smith SE (2005) Influence of arbuscular mycorrhizal (AM) symbiosis on phosphorus leaching through soil cores. Plant Soil 275: 181–193.
18. van der Heijden MGA (2010) Mycorrhizal fungi reduce nutrient loss from model grassland ecosystems. Ecology 91: 1163–1171.
19. Asghari HR, Cavagnaro TR (2011) Arbuscular mycorrhizas enhance plant interception of leached nutrients. Funct Plant Biol 38: 219–226.
20. Rillig MC (2004) Arbuscular mycorrhizae and terrestrial ecosystem processes. Ecol Lett 7: 740–754.
21. David-Schwartz R, Gadkar V, Wininger S, Bendov R, Galili G, et al. (2003) Isolation of a premycorrhizal infection (pmi2) mutant of tomato, resistant to arbuscular mycorrhizal fungal colonization. MPMI 16: 382–388.
22. Barker SJ, Stummer B, Gao L, Dispain I, O'Connor PJ, et al. (1998) A mutant in *Lycopersicon esculentum* Mill. with highly reduced VA mycorrhizal colonization, isolation and preliminary characterisation. Plant J 15: 791–797.
23. Cavagnaro TR, Jackson LE, Scow KM, Hristova KR (2007) Effects of arbuscular mycorrhizas on ammonia oxidizing bacteria in an organic farm soil. Microb Ecol 54: 618–626.
24. Cavagnaro TR, Smith FA, Hay G, Carne-Cavagnaro VL, Smith SE (2004) Inoculum type does not affect overall resistance of an arbuscular mycorrhiza-defective tomato mutant to colonisation but inoculation does change competitive interactions with wild-type tomato. New Phytol 161: 485–494.
25. Poulsen KH, Nagy R, Gao L-L, Smith SE, Bucher M, et al. (2005) Physiological and molecular evidence for Pi uptake via the symbiotic pathway in a reduced mycorrhizal colonization mutant in tomato associated with a compatible fungus. New Phytol 168: 445–453.
26. Cavagnaro TR, Martin AW (2011) Arbuscular mycorrhizas in southeastern Australian processing tomato farm soils. Plant Soil 340: 327–336.
27. Marschner P, Timonen S (2004) Interactions between plant species and mycorrhizal colonization on the bacterial community composition in the rhizosphere. Appl Soil Ecol 28: 23–36.
28. Gao L, Smith FA, Smith SE (2006) The *rmc* locus does not affect plant interactions of defence-related gene expression when tomato (*Solanum lycopersicum*) is infected with the root fungal parasite, *Rhizoctonia*. Funct Plant Biol 33: 289–296.
29. Burger B, Reich P, Cavagnaro TR (2010) Trajectories of change: riparian vegetation and soil conditions following livestock removal and replanting. Austral Ecol 35: 980–987.
30. Cavagnaro TR, Smith FA, Lorimer MF, Haskard KA, Ayling SM, et al. (2001) Quantitative development of *Paris*-type arbuscular mycorrhizas formed between *Asphodelus fistulosus* and *Glomus coronatum*. New Phytol 149: 105–113.
31. Entry J, Sojka R (2007) Matrix Based fertilizers reduce nitrogen and phosphorus leaching in greenhouse column studies. Water Air Soil Pollut 180: 283–292.
32. Miranda KM, Espey MG, Wink DA (2001) A rapid, simple spectrophotometric method for simultaneous detection of nitrate and nitrite. Nitric Oxide 5: 62–71.
33. Forster JC (1995) Soil nitrogen. In: Alef K, Nannipiero P, eds. Methods in applied soil microbiology and biochemistry. San DiegoCA: Academic Press.
34. Jakobsen I, Abbott LK, Robson AD (1992) External hyphae of vesicular-arbuscular mycorrhizal fungi associated with *Trifolium subterraneum* L. 1. Spread of hyphae and phosphorus inflow into roots. New Phytol 120: 371–380.
35. Giovannetti M, Mosse B (1980) An evaluation of techniques for measuring vesicular-arbuscular mycorrhizal infection in roots. New Phytol 84: 489–500.
36. Phillips JM, Hayman DS (1970) Improved procedures for clearing roots and staining parasitic and vesicular-arbuscular mycorrhizal fungi for rapid assessment of infection. Trans Br Mycol Soc 55: 158–161.
37. Cavagnaro TR, Barrios-Masias FH, Jackson LE. Arbuscular mycorrhizas and their role in plant growth, nitrogen interception and soil gas efflux in an organic production system. Plant Soil, in press.
38. Johansen A, Jakobsen I, Jensen ES (1993) Hyphal transport by a vesicular-arbuscular mycorrhizal fungus of N applied to the soil as ammonium or nitrate. Biol Fert Soil 16: 66–70.
39. Tinker PB, Nye PH (2000) Solute Movement in the Rhizosphere. Oxford, Great Britain: Oxford University Press.
40. Bateman EJ, Baggs EM (2005) Contributions of nitrification and denitrification to N_2O emissions from soils at different water-filled pore space. Biol Fert Soil 41: 379–388.

15

Remotely Sensed Rice Yield Prediction Using Multi-Temporal NDVI Data Derived from NOAA's-AVHRR

Jingfeng Huang[1,2,3], Xiuzhen Wang[4*], Xinxing Li[1,2,3], Hanqin Tian[5,6], Zhuokun Pan[1,3]

1 Institute of Agricultural Remote Sensing & Information Application, Zijingang Campus, Zhejiang University, Hangzhou, China, **2** China Ministry of Education Key Laboratory of Environmental Remediation and Ecological Health, Zhejiang University, Hangzhou, China, **3** Key Laboratory of Agricultural Remote Sensing and Information System, Zhejiang Province, China, **4** Institute of Remote Sensing and Earth Sciences, Hangzhou Normal University, Hangzhou, China, **5** International Center for Climate and Global Change Research, Auburn University, Auburn, Alabama, United States of America, **6** Ecosystem Dynamics and Global Ecology (EDGE) Laboratory, School of Forestry and Wildlife Sciences, Auburn University, Auburn, Alabama, United States of America

Abstract

Grain-yield prediction using remotely sensed data have been intensively studied in wheat and maize, but such information is limited in rice, barley, oats and soybeans. The present study proposes a new framework for rice-yield prediction, which eliminates the influence of the technology development, fertilizer application, and management improvement and can be used for the development and implementation of provincial rice-yield predictions. The technique requires the collection of remotely sensed data over an adequate time frame and a corresponding record of the region's crop yields. Longer normalized-difference-vegetation-index (NDVI) time series are preferable to shorter ones for the purposes of rice-yield prediction because the well-contrasted seasons in a longer time series provide the opportunity to build regression models with a wide application range. A regression analysis of the yield versus the year indicated an annual gain in the rice yield of 50 to 128 kg ha^{-1}. Stepwise regression models for the remotely sensed rice-yield predictions have been developed for five typical rice-growing provinces in China. The prediction models for the remotely sensed rice yield indicated that the influences of the NDVIs on the rice yield were always positive. The association between the predicted and observed rice yields was highly significant without obvious outliers from 1982 to 2004. Independent validation found that the overall relative error is approximately 5.82%, and a majority of the relative errors were less than 5% in 2005 and 2006, depending on the study area. The proposed models can be used in an operational context to predict rice yields at the provincial level in China. The methodologies described in the present paper can be applied to any crop for which a sufficient time series of NDVI data and the corresponding historical yield information are available, as long as the historical yield increases significantly.

Editor: Wengui Yan, National Rice Research Center, United States of America

Funding: The authors' work was supported from National Key Technology R&D Program of China Grant(2011BAD32B01), the National Natural Science Foundation of China (NSFC) grant (40875070 and 40871158) and Zhejiang Provincial Natural Science Foundation of China grant (Y5100021). The funders had no role in study design, data collection and analysis, decision to publish, or preparation of the manuscript.

Competing Interests: The authors have declared that no competing interests exist.

* E-mail: wxz05160516@126.com

Introduction

Paddy rice is one of the most important and widely grown crops in China. The total paddy-rice production in 2009 reached 195.1 million tons, and it accounted for 40.5% of the total grain production in China (481.563 million tons) [1]. Timely, objective and quantitative information regarding to paddy-rice yield can provide important information for government agencies and producers that can be used for planning harvest, storage and marketing activities. Therefore, paddy-rice-yield prediction is important for the food security of China and is considered to be one of the most challenging tasks in agricultural research [2]. The traditional approach of crop-yield forecasting, the use of ground-based data collection is expensive, time-consuming, labor-intensive, and often difficult [3]. Crop-yield prediction using remotely sensed data has already represented a very active field of research and application [3–5]. Notable advances in remote-sensing technology over the last several decades are now providing scientists with valuable information for yield and production forecast. Time series of normalized-difference-vegetation-index (NDVI), derived from the satellite data, have been used for crop-yield predictions since the 1980's. Most of the studies that related NDVI measurements to crop yield have been concentrated on staple crops such as wheat [4,6–38] and maize [3,13,18,20,21,24,29,39–49] and rice [2,15,37,44,50–52]. Many researchers have also found that NDVI variables are very good at grain yield predictors of millet [53–57], sorghum [24,56,58,59], barley [19,24,29,60,61], soybean [3,24,62,63], ground nut [54,59], sugar beet [29], alfalfa [29], rye [29], pea [19,29], and canola [19] (Literature review was summarized in Table 1). However, remotely sensed yield prediction appears limited in rice.

Different methods have been developed to predict crop yields using remotely sensed data, and the most common approach is, by generating regression model, to develop direct empirical relationships between the NDVI measurements and the crop yield [15,19,45,57]. These approaches assume that measures of the photosynthetic capacity from spectral-vegetation indices are directly related to crop yield. This assumption is used because many of the conditions that affect crop growth, development and ultimately yield could be captured through spectra measurements such as the NDVI [64]. By using long-term historical-yield data as a dependent variable and remotely sensed data as an independent variable, a statistical regression function was generated to perform crop-yield predictions, whereas the actual crop yields depend on many more factors than the presence of spectral-vegetation indices [37]. Tilman et al. [65] noted that increased yields in cereal are mainly the result of greater inputs of fertilizer, water and pesticides, new crop species, and the improvement of management over the last decades. For all developing countries, modern varieties accounted for 21% of the growth in crop yields during the early Green Revolution period [66]. In Asia, rice production has more than doubled as a result of the expansion of cultivated area, the adoption of modern cultivars, increased investments in irrigation, and an increased use of fertilizer over the past 4 decades [67]. Hafner [68] found that linear growth has been the most common trend in maize, rice, and wheat yields for 188 nations over the past 40 years. This scenario also occurs in China. Although the inter-annual variability of NDVI (probably due to unexpected weather conditions or disasters) can reveal crop yield fluctuations [19,59]; however, remotely sensed-NDVI cannot detect those human-induced factors that resulted in increase of rice yield. Therefore, to monitor and predict crop-yield cannot use NDVI measurements solely.

For unit-yield estimation, using one simple regression function (usually known as: $Y = a+b * NDVI$) would be incompatible as the advance of years, because simple regression would be likely neglect those man-induced factors in yield increase. However, few studies have analyzed the time trends of crop yields, which reflect the influence of technology development, fertilizer application, and management improvement. Moreover, the regression model between statistical data and NDVI cannot be extendable [19,45] because cropping system and rice yield level is natural condition-dependent in China.

In consideration of social factors and regional differences for remotely sensed crop yield estimation in China, the objective of the present paper was to develop a methodological framework that may be adopted for the regional-, national- and international-scale prediction of crop yields. This methodology was based on a time series analysis of historical-yield information. Paddy rice was chosen to test the proposed methodology. To accomplish this objective, we needed to: (1) geographically regionalize rice cultivation area for remotely sensed monitoring; (2) analyze the historical trends in the grain yield of rice; (3) decompose the remotely sensed yield of rice from the long-term historical data; (4) select the optimal predictors, based on a correlation analysis between the remotely sensed yield and the AVHRR-derived NDVIs; (5) construct prediction models for rice yield; and (6) evaluate the potential for rice-grain-yield prediction in China using AVHRR NDVI data as predictors.

Materials and Methodology

2.1. The Remote-Sensing dataset

The research presented in this paper relies on a time series of AVHRR NDVI composite imagery from July 1981 to December 2006, derived from the National Oceanic and Atmospheric Administration's (NOAA) series of Advanced Very High Resolution Radiometer (AVHRR) instruments, with a spatial resolution of 8 km, by the NASA Global Inventory Monitoring and Modeling Systems (GIMMS) group at the Laboratory for Terrestrial Physics. There are two 15-day composites per month: the first (15a) is a maximum value composite from the first day to $15^{th}$of the month; and the 15b composite is from days 16 till the end of the month. All data are available from the University of Maryland Global Land Cover Facility (http://glcf.umiacs.umd.edu/data/gimms/).

Table 1. Relevant literatures that linked with crop yield forecast using remotely sensed data literatures are sorted according to the crop types.

Crop	reference
wheat	MacDonald et al., 1980; Rudorff et al., 1991; , Bullock, 1992; Benedetti et al., 1993; Gupta et al., 1993; Benedetti et al., 1993; Cheng, 1994; Dubey et al., 1994; Sridhar et al., 1994; Doraiswamy et al., 1995, 2003; Smith et al., 1995; Hochheim et al., 1998; Huang et al., 1999; Maselli et al., 2001; Boken et al., 2002; Labus et al., 2002; Manjunath et al., 2002; Mika et al., 2002; Bastiaanssen, et al., 2003; Kalubarme et al., 2003; Ferencz et al., 2004; Zhang et al., 2004; Kastensa et al., 2005; Mo et al., 2005; Wang et al., 2005; Patel et al., 2006; Ren et al., 2006; Moriondo et al., 2007; Prasad et al., 2007; Balaghi et al., 2008; Ren et al., 2008; Wall et al., 2008; Schut et al., 2009; Becker-Reshef et al., 2010; Mkhabela et al., 2011
maize	Quarmby et al., 1993; Hayes et al., 1996; Unganai et al., 1998; Lewis et al., 1998; Lee et al., 1999; Reynolds et al., 2000; Seiler et al., 2000; Maselli et al.,2001; Mika et al., 2002; Wannebo et al., 2003; Ferencz et al., 2004; Kastensa et al., 2005; Mkhabela et al., 2005; Mo et al., 2005; Prasad et al., 2006; Rojas, 2007; Ren, et al., 2008; Funk et al., 2009
millet	Rasmussen, 1992, 1997, 1998; Groten, 1993; Maselli et al.,2000
sorghum	Potdar, 1993; Fuller, 1998; Maselli et al., 2000; Kastensa et al., 2005
barley	Wendroth et al., 2003; Ferencz et al., 2004; Kastensa et al., 2005; Weissteiner et al., 2005; Mkhabela et al., 2011
soybean	Liu et al., 2002; Kastensa et al., 2005; Prasad et al., 2006; Esquerdo et al., 2011
ground nut	Rasmussen, 1997; Fuller, 1998
sugar beet	*Ferencz et al., 2004*
alfalfa	*Ferencz et al., 2004*
rye	Ferencz et al., 2004
pea	Ferencz et al., 2004; Mkhabela et al., 2011
canola	Mkhabela et al., 2011
rice	Tennakoon et al., 1992; Quarmby et al., 1993; Huang et al., 2002; Wang et al., 2002; Bastiaanssen, et al., 2003; Prasad et al., 2007; Huang et al., 2010

Table 2. NVDI variables and their calculation formulas.

	NDVIs	Description of formulas
1	$NDVI_{maxb1}$	the first biweekly NDVI before $NDVI_{max}$
2	NDVImaxb2	the second biweekly NDVI before $NDVI_{max}$
3	$NDVI_{maxb3}$	the third biweekly NDVI before $NDVI_{max}$
4	$NDVI_{maxb4}$	the fourth biweekly NDVI before $NDVI_{max}$
5	$NDVI_{max}$	the maximum NDVI during the growth period
6	$NDVI_{maxa1}$	the first biweekly NDVI after $NDVI_{max}$
7	$NDVI_{maxa2}$	the second biweekly NDVI after $NDVI_{max}$
8	$mNDVI_{maxb4\text{-}b3}$	$(NDVI_{maxb4}+ NDVI_{maxb3})/2$
9	$mNDVI_{maxb4\text{-}b2}$	$(NDVI_{maxb4}+ NDVI_{maxb3}+ NDVI_{maxb2})/3$
10	$MNDVI_{maxb4\text{-}b1}$	$(NDVI_{maxb4}+ NDVI_{maxb3}+ NDVI_{maxb2}+ NDVI_{maxb1})/4$
11	$mNDVI_{maxb4\text{-}max}$	$(NDVI_{maxb4}+ NDVI_{maxb3}+ NDVI_{maxb2}+ NDVI_{maxb1}+ NDVI_{max})/5$
12	$mNDVI_{maxb4\text{-}a1}$	$(NDVI_{maxb4}+ NDVI_{maxb3}+ NDVI_{maxb2}+ NDVI_{maxb1}+ NDVI_{max}+ NDVI_{maxa1})/6$
13	$mNDVI_{maxb4\text{-}a2}$	$(NDVI_{maxb4}+ NDVI_{maxb3}+ NDVI_{maxb2}+ NDVI_{maxb1}+ NDVI_{max}+ NDVI_{maxa1}+ NDVI_{maxa2})/7$
14	$mNDVI_{maxb3\text{-}b2}$	$(NDVI_{maxb3}+ NDVI_{maxb2})/2$
15	$mNDVI_{maxb3\text{-}b1}$	$(NDVI_{maxb3}+ NDVI_{maxb2}+ NDVI_{maxb1})/3$
16	$mNDVI_{maxb3\text{-}max}$	$(NDVI_{maxb3}+ NDVI_{maxb2}+ NDVI_{maxb1}+ NDVI_{max})/4$
17	$mNDVI_{maxb3\text{-}a1}$	$(NDVI_{maxb3}+ NDVI_{maxb2}+ NDVI_{maxb1}+ NDVI_{max}+ NDVI_{maxa1})/5$
18	$mNDVI_{maxb3\text{-}a2}$	$(NDVI_{maxb3}+ NDVI_{maxb2}+ NDVI_{maxb1}+ NDVI_{max}+ NDVI_{maxa1}+ NDVI_{maxa2})/6$
19	$mNDVI_{maxb2\text{-}b1}$	$(NDVI_{maxb2}+ NDVI_{maxb1})/2$
20	$mNDVI_{maxb2\text{-}max}$	$(NDVI_{maxb2}+ NDVI_{maxb1}+ NDVI_{max})/3$
21	$mNDVI_{maxb2\text{-}a1}$	$(NDVI_{maxb2}+ NDVI_{maxb1}+ NDVI_{max}+ NDVI_{maxa1})/4$
22	$mNDVI_{maxb2\text{-}a2}$	$(NDVI_{maxb2}+ NDVI_{maxb1}+ NDVI_{max}+ NDVI_{maxa1}+ NDVI_{maxa2})/5$
23	$mNDVI_{maxb1\text{-}max}$	$(NDVI_{maxb1}+ NDVI_{max})/2$
24	$mNDVI_{maxb1\text{-}a1}$	$(NDVI_{maxb1}+ NDVI_{max}+ NDVI_{maxa1})/3$
25	$mNDVI_{maxb1\text{-}a2}$	$(NDVI_{maxb1}+ NDVI_{max}+ NDVI_{maxa1}+ NDVI_{maxa2})/4$
26	$mNDVI_{max\text{-}a1}$	$(NDVI_{max}+ NDVI_{maxa1})/2$
27	$mNDVI_{max\text{-}a2}$	$(NDVI_{max}+ NDVI_{maxa1}+ NDVI_{maxa2})/3$
28	$mNDVI_{maxa1\text{-}a2}$	$(NDVI_{maxa1}+ NDVI_{maxa2})/2$

Pinzon et al. [69] and Tucker et al. [70] described in detail how the GIMMS data set was developed. A number of improvements have been made on the GIMMS NDVI database, with respect to previous NDVI data sets, including corrections for: (1) sensor degradation; (2) inter-sensor differences; (3) solar-illumination angle and sensor-view angle effects due to satellite drift; (4) volcanic stratospheric aerosol corrections for 1982–1984 and 1991–1994; (5) missing data in the Northern Hemisphere during winter, using interpolation; and (6) short-term atmospheric aerosol effects, atmospheric water-vapor effects, and cloud-cover physics [69,70]. This data set is considered to be the most accurate, long-term AVHRR data record [71]. By comparing these data to new, improved coarse-resolution remotely sensed data from SPOT Vegetation instrument and MODIS instruments, recent study confirmed its suitability for long-term vegetation studies [72].

2.2. NDVI Variables

A large number of studies found a close relationship between crop yields and NDVI variables. The theory is: the NDVI value presents the yield level corresponding to every single pixel. Therefore, a simple regression function can be explained the yield: yield = a*NDVI + b; then the total yield can be obtained by multiplying planting area. By literature review, previous studies suggest three types of NDVI variables: original NDVI [13,23,42,63], cumulative NDVI [8,23,38,42,45,63,73,74], and average NDVI [34,45,63]. The cumulative NDVI and the corresponding average NDVI for the same period were highly correlated because of the linear nature of the operations involved. Only the original NDVIs and the average NDVIs were selected as input data for the prediction models in the present paper.

NDVI variables around the time of the maximum are strongly correlated with final yields [31,35,75]. Specifically, the rice yield is most determined by crop conditions during the heading (i.e. peak phenological phase of growth); and yield-reflectance relationships are typically the strongest after mid-season. In contrast, NDVI value changes that occur outside of the rice-growing period maybe not positively related to yield [52]. These relationships within changes of NDVI value suggests that the NDVIs during the mid-to-late growing period should be a good indicator of rice yield; meanwhile this phenomenon provides an approach to discriminate rice planting area from remote sensing image. Therefore, the first step of this study was to extract the maximum NDVI during the rice-growth period ($NDVI_{max}$) for each studied province from the remote sensing dataset from the year 1982 to 2006. The maximum NDVI is equal to the peak value of the seasonal NDVI profile. Then, six other original NDVIs were calculated: the first, second,

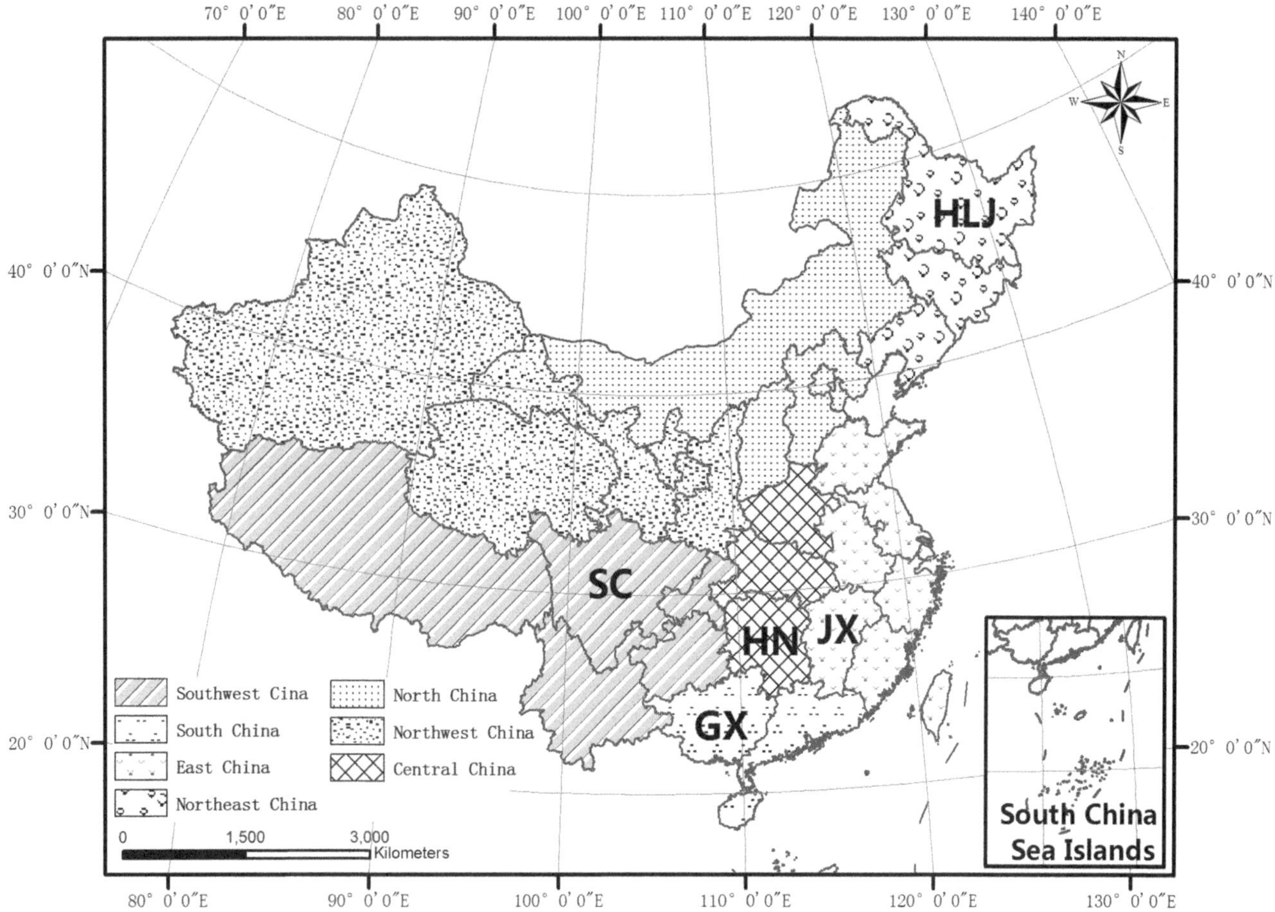

Figure 1. The locations of the study areas within Mainland China. Heilongjiang is designated by HLJ, Jiangxi by JX, Guangxi by GX, Sichuan by SC, and Hunan by HN.

third and fourth biweekly NDVIs prior to the $NDVI_{max}$ ($NDVI_{maxb4}$) and the first and second biweekly NDVIs after the $NDVI_{max}$ ($NDVI_{maxa2}$). These seven biweekly composites span 3 months of raw AVHRR imagery, corresponding to the rice-growth period. Focusing on the NDVI response during the rice-growth period helps to identify rice-specific vegetation changes.

Hochheim and Barber [27] also found that NDVI estimators with longer integration periods minimized variability in yield prediction. Therefore, based on the seven original NDVIs, twenty-one average NDVIs, clustered around the time of the peak NDVI, were calculated using a rigorous arithmetic mean framework (Table 2). In total, 28 NDVI variables were generated. They include all of the possible combinations of the original seven NDVIs.

2.3. Official Statistical Data of Rice Yield

Historical rice-yield data were acquired from the China Statistical Year Book by the National Bureau of Statistics of China (NBSC) from the years 1979 to 2009 [1]. The NBSC is the agency responsible for collecting and publishing agricultural statistics at the national and provincial levels. The NBSC crop statistics are based on data obtained from sub-province sample surveys and released in official documents. Customarily, Chinese provinces have been geographically grouped into 7 regions to present a spatial pattern for paddy rice planting area: Northeastern China (Heilongjiang, Jilin, and Liaoning), Northern China (Inner Mongolia, Hebei, Shanxi, Beijing, and Tianjin), Northwestern China (Ningxia, Shaanxi, Gansu, Qinghai, and Xinjiang), Central China (Henan, Hunan, and Hubei), Eastern China (Shandong, Jiangsu, Shanghai, Zhejiang, Anhui, and Jiangxi), Southwestern China (Chongqing, Sichuan, Guizhou, Yunnan, and Xizang), Southern China (Guangdong, Guangxi, Hainan) (see Figure 1). Unfortunately, rice planting area and yield information for Hong Kong- Macao-Taiwan areas was not available. According to NBSC crop statistical data (see Table 3), Eastern China was the region with the highest rice acreage and production levels (9808.60 kha and 64984.00 kt, respectively) in 2009. Central China ranked second in both rice acreage and production (6703.60 kha and 46215.00 kt, respectively). The third-largest rice cultivation and production area was Southwestern China (4448.10 kha and 31214.00 kt, respectively). Southern China and Northeastern China ranked fourth and fifth, respectively, in both rice acreage and production (4402.40 kha and 23499.00 kt; 3777.90 kha and 25855.00 kt, respectively). The total rice cultivation area in Eastern China, Central China, Southwestern China, Southern China, and Northeastern China is 29140.60 kha and accounts for 98.36% of the total rice cultivation area in the conterminous China. The total rice production in Eastern China, Central China, Southwestern China, Southern China, and Northeastern China was 191767.00 kt and accounted for 98.29% of the total rice production in the conterminous China in 2009. Northern China and Northwestern China constitute less

Table 3. Planted area and production changes for rice between 1979 and 2009 for different regions in the conterminous China.

Regions	Area (Kha)				Production (Kt)			
	1979	% of China	2009	% of China	1979	% of China	2009	% of China
Northeastern China	841.73	2.49	3777.90	12.75	3860.00	2.69	25855.00	13.25
Northern China	264.07	0.78	204.40	0.69	1165.00	0.81	1343.00	0.69
Northwestern China	315.27	0.93	281.70	0.95	1305.00	0.91	1993.00	1.02
Central China	7639.13	22.55	6703.60	22.63	34260.00	23.83	46215.00	23.69
Eastern China	12926.33	38.16	9808.60	33.11	56230.00	39.12	64984.00	33.31
Southwestern China	4803.73	14.18	4448.10	15.01	21440.00	14.91	31214.00	16.00
Southern China	7082.40	20.91	4402.40	14.86	25490.00	17.73	23499.00	12.04
Total	33872.67	100.00	29626.70	100.00	143750.00	100.00	195103.00	100.00

than 2% of the national rice harvested area and production and were less important on a national scale in 2009.

2.4. Description of Study Area

We divided China into 7 regions together with 5 representative provinces selected to convey the information of paddy rice planting area: Heilongjiang (HLJ) in Northeastern China, Hunan (HN) in Central China, Jiangxi (JX) in Eastern China, Sichuan (SC) in Southwestern China, and Guangxi (GX) in Southern China. These provinces were selected as the study areas for the present research because these locations: (1) represented the typical cropping system in China, (2) are located in primary rice-production regions, and (3) are geographically and climatologically different (see Figure 1 and Table 4). The life span, cropping system, and planting schedule are all depend on regional hydro-thermal condition. The general information on life span, cropping system, total annual rainfall (mm), annual accumulated temperature (°C), area (kha), and production levels (kt) for the selected provinces is shown in Table 4. The total combined rice-cultivation area in Heilongjiang (HLJ), Hunan (HN), Jiangxi (JX), Sichuan (SC), and Guangxi (GX) is 13942.2 kha, and these regions accounted for 47.06% of the total rice-cultivation area in China in 2009. The total combined rice production in Heilongjiang (HLJ), Hunan (HN), Jiangxi (JX), Sichuan (SC), and Guangxi (GX) was 87251 kt and accounted for 44.72% of the total rice production in China in 2009. The time series of the NBSC province-level rice yields were used to train and develop the prediction models for these five provinces.

2.5. Calibration of Rice-Yield Prediction Models

The gradual trend in yields is due to the influence of technological development, fertilizer application, and improved management on the rice cultivation. The results of this analysis suggest that the most common trend of rice yield is a linear growth. The province-specific intercepts account for spatial variations in rice management and soil quality; province-specific time trends account for yield growth due to technology gains. This indicates us the yield is composed from the intrinsic and extrinsic factors. Therefore, we decomposed the historical rice yield Y into the trend yield Y_t and the remotely sensed yield Y_{RS}, using the following equation:

$$Y = Y_t + Y_{RS} \quad (1)$$

Y_t, represents the component that is regulated by agricultural technology, including (1) the usual biological-chemical technologies (new varieties, fertilizers, herbicides, insecticides, etc.) and the mechanical technologies (machinery, equipment, etc.); (2) the management practices, which involve changes such as the timing of field operations and other practices which may or may not be involved in the purchase of new inputs. Y_{RS} is defined as the component regulated by natural environmental conditions, such as temperature, precipitation, pests and disease; these environment factors can be detected by a remote sensor.

To quantify past trends in yields, many different yield de-trend methods have been reported, including: least-squares regressions [76,77], moving averages [78,79], exponential algorithms [80], and polynomial regressions [81]. For rice-yield predictions in the present investigation, a linear regression model and a moving average are both generated to fit each separated provincial rice dataset (also see in Figure 2):

$$Y_t = \alpha + \beta t \quad (2)$$

where Y_t is the trend yield in a given province during a given year (kg ha^{-1}), t represents the year of harvest (the year 1979 was numeral 1979, 1980 was numeral 1980, etc., until 2009 was numeral 2009), α and β are the province-specific linear regression coefficients.

In our study, a moving average is used with historical crop-yield data to smooth out short-term fluctuations and highlight longer-term trends. Rice yields were de-trended using their deviations from the 5-year moving average. The mean changes in provincial historical rice yield (Y), the trend yield (Y_t) and the remotely sensed yield (Y_{RS}) were calculated for each period as an average of the changes from each single preceding year to the next by using a moving average method. Generally, the moving average method is used to calculate arithmetic mean of each five of the entire dataset: $y_{i-n}, y_{i-n+1}, \ldots, y_i, \ldots, y_{i+n}$. Such method has been usually employed in meteorological data analysis to remove the stochastic errors from long-time series of data. Hence, an algorithm for a 5-year moving average is as follows:

Table 4. General information on Rice cropping system, Life span, Total annual rainfall (mm), Annual accumulated temperature (≥10°C), Area (kha) and Production (kt) for the study areas.

Provinces	Climate region	Rice cropping system	Life span	Total annual rainfall (mm)	Annual accumulated temperature (≥10 °C)	Planting Area in 2009(kha)	Percent age of China (%)	Production in 2009 (kt)	Percenta ge of China (%)
Heilongjiang (HLJ)	Temperate continental monsoon climate	Single cropping	May - Oct	450–650	2000–3700	2460.80	8.31	15745.00	8.07
Hunan (HN)	Subtropical monsoon climate	Double cropping	Mar - Aug, Jun - Nov	1200–1700	4500–6500	4047.20	13.66	25786.00	13.22
Jiangxi (JX)	Subtropical monsoon climate	Double cropping	Mar - Aug, Jun - Nov	1300–2000	4500–6500	3282.10	11.08	19059.00	9.77
Sichuan (SC)	Subtropical humid climate	Single cropping	Mar - Aug	950–1200 (Sichuan Basin)	4000–6000 (Sichuan Basin)	2027.10	6.84	15202.00	7.79
Guangxi (GX)	Subtropical monsoon climate	Double cropping	Mar - Aug, Jun - Nov	1300–2000	5800–9300	2125.00	7.17	11459.00	5.87

$$\begin{cases} Y_{t,i}=\frac{Y_{t,i-2}+Y_{t,i-1}+Y_{t,i}+Y_{t,i+1}+Y_{t,i+2}}{5}(3\le i\le n-2) \\ Y_{t,1}=\frac{3Y_{t,1}+2Y_{t,2}+Y_{t,3}-Y_{t,5}}{5} \\ Y_{t,2}=\frac{4Y_{t,1}+3Y_{t,2}+2Y_{t,3}+Y_{t,4}}{5} \\ Y_{t,n-1}=\frac{Y_{t,n-3}+2Y_{t,n-2}+3Y_{t,n-1}+4Y_{t,n}}{5} \\ Y_{t,n}=\frac{-Y_{t,n-4}+Y_{t,n-2}+2Y_{t,n-1}+3Y_{t,n}}{5} \end{cases} \quad (3)$$

Where $Y_{t,\ i}$ is the trend yield in a given province during a given year (kg ha^{-1}); n represents the number of data points; i represents the year of the harvest (e.g. the year 1979 was numeral 1, 1980 was numeral 2, etc., until 2006 it should be numeral 31); $Y_{t,1}$ and $Y_{t,2}$ are the trend yields for the first two harvested years; then $Y_{t,\ n-1}$ and $Y_{t,\ n}$ are the trend yields for the last two harvested years within the 5 years.

Consequently, the trend yield Y_t was obtained. To remove the technological influences, it is necessary to remove the yield trend to produce a new time series that is directly related to the NDVIs. We defined this new time series as the remotely sensed yield. According to Eq. (1), the remotely sensed yield can be calculated by the following equation:

$$Y_{RS}=Y-Y_t \quad (4)$$

Next, correlation analysis was performed between the remotely sensed yield and the NDVI variables. The correlation coefficient is a measure of the strength and the direction of a linear relationship between two variables. The symbol r in Eq. (5) represents the samples' correlation coefficient; x and y represent the remotely sensed yield and the NDVI variables respectively; n is the number of data pairs.

$$r=\frac{n\sum xy-(\sum x\sum y)}{\sqrt{n\sum x^2-(\sum x)^2}\sqrt{n\sum y^2-(\sum y)^2}} \quad (5)$$

Statistical regression models are the most commonly used method for crop-yield prediction based on remotely sensed data [8,36]. They do not require numerous inputs and can be performed directly; also because it requires little computing power and the selected variables are distinctive and non-overlapping. Therefore, each of the provincial Y_{RS} and NDVI dataset was analyzed separately by means of stepwise regression techniques. These models were constructed via the 'STEPWISE' regression process which was available in software Statistical Product and Service Solutions (SPSS) 17.0 [82]. The probability significance thresholds for the entry and retention of candidate independent variables in the model were both set to $\alpha=5\%$.

2.6. Evaluation of Rice-Yield Prediction Models

The rice-yield prediction models were evaluated using the following indicators:

Root mean square error (RMSE):

$$RMSE=\sqrt{\frac{\sum_{i=1}^{n}(Y_i-Y_i')^2}{n}} \quad (6)$$

Coefficient of determination (R^2):

$$R^2=\frac{\sum_{i=1}^{n}(Y_i'-\bar{Y})^2}{\sum_{i=1}^{n}(Y_i-\bar{Y})^2} \quad (7)$$

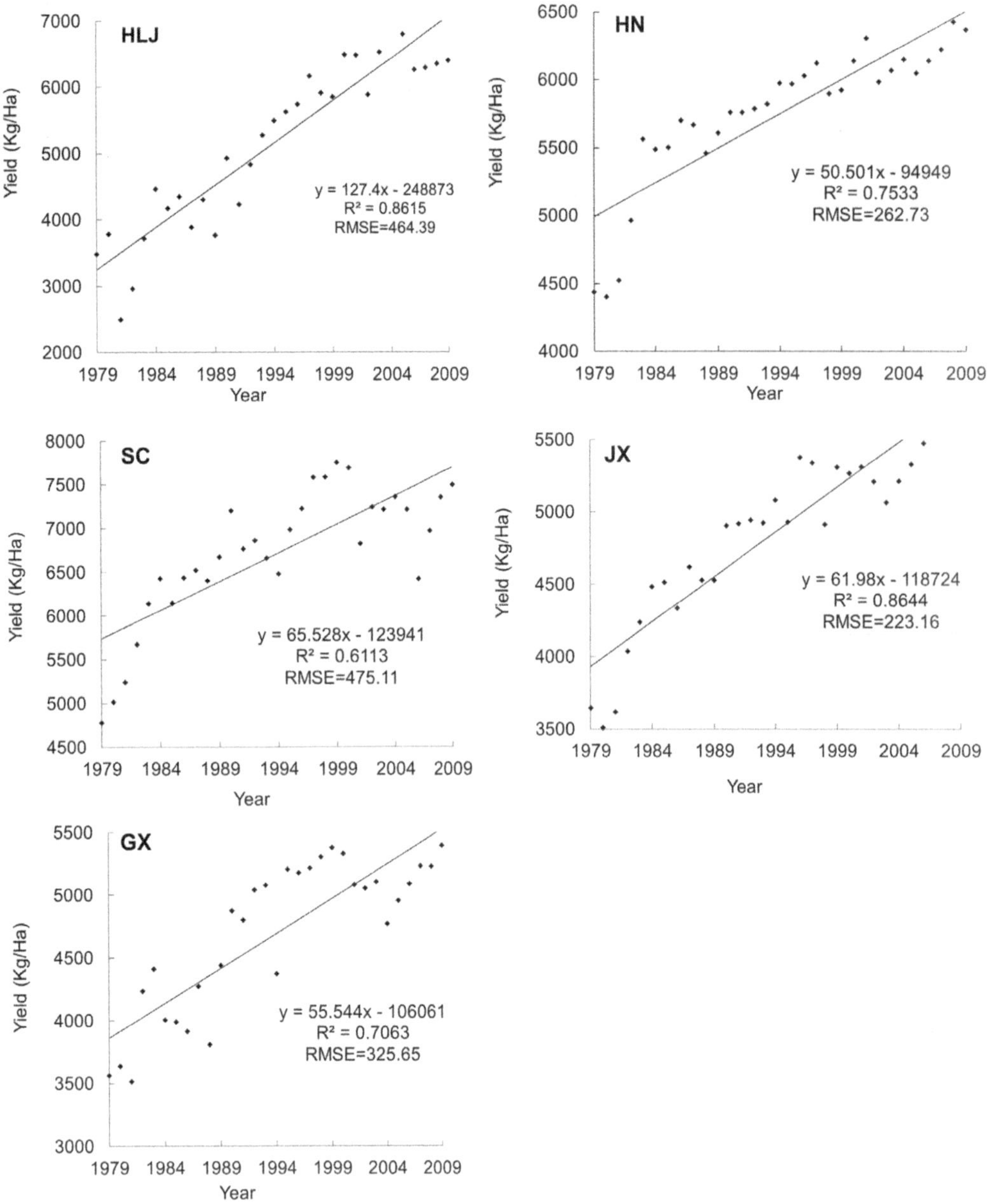

Figure 2. Rice yield trends for the provinces' of Heilongjiang (HLJ), Hunan (HN), Jiangxi (JX), Sichuan (SC) and Guangxi (GX) from 1979 to 2006.

F-value (F):

$$F = \frac{\sum_{i=1}^{n} (Y_i' - \bar{Y})^2 / k}{\sum_{i=1}^{n} (Y_i - Y_i')^2 / (n-k-1)} \quad (8)$$

and relative error (RE):

$$RE = \frac{Y_i - Y_i'}{Y_i} \quad (9)$$

Together with the above, where n is the number of comparisons; k is the number of predictors; Y_i is the statistical rice yield; $\bar{Y}$ is the average rice yield, and Y_i' is the predicted yield.

Results and Discussion

3.1. Rice Yield Trend Analysis

Figure 2 presents the evolution of the average rice-grain yield in Heilongjiang (HLJ), Jiangxi (JX), Guangxi (GX), Sichuan (SC), and Hunan (HN) from 1979 to 2009; according to their R-square and RMSE, all rice yields showed a visible and significant growth trend over time. Understanding the past rice-yield trends can help us to gauge the importance of the preprocessing procedure for rice-yield prediction using remotely sensed data. The statistical data of rice yield together with average yield growth trend from 1979 to 2009 in five provinces of China is summarized in Table 5.

As analysis above (see Figure 2), the social input and advance of technology account for the linear trend of the rice-yield growth, whereas such human-induced factors could not be detected using remotely sensed data. To overcome this problem and make rice-yield prediction methods more robust and easily exportable, one possible strategy is to integrate remote-sensing data with the rice yield time series analysis. De-trending is necessary to properly identify the remote-sensible effects in these panel datasets. Therefore, before the rice-yield predicting models are established using remotely sensed variables as predictors, we suggest that the statistical yield should be decomposed into the trend yield and the remotely sensed yield, methodology was described in Section 2.5.

3.2. Correlation Coefficients between the Remotely Sensed Yield and NDVI Variables

The correlation coefficients between Y_{RS} and the NDVI variables for the rice-growth period from the fourth 15-day period before $NDVI_{max}$ ($NDVI_{maxb4}$) to the second 15-day period after $NDVI_{max}$ ($NDVI_{maxa2}$) for each of the studied provinces are summarized in Table 6. By comparing the correlation coefficients (Column 2 and 3 in Table 6), the Y_{RS} that was de-trended by linear regression performed better than the Y_{RS} that was de-trended by a 5-year moving average against the NDVI variables.

The correlation coefficients between the Y_{RS} that were de-trended by linear regression and the NDVI variables were generally high in HN and SC. According to Table 6, for HN, the correlation coefficients were significant at the 0.01 level between the Y_{RS} that was de-trended by linear regression and $NDVI_{maxb1}$, $NDVI_{max}$, $NDVI_{maxa1}$, $mNDVI_{maxb4\text{-}a2}$,$mNDVI_{maxb3\text{-}a1}$, $mNDVI_{maxb3\text{-}a2}$, $mNDVI_{maxb2\text{-}b1}$, $mNDVI_{maxb2\text{-}max}$, $mNDVI_{maxb2\text{-}a1}$, $mNDVI_{maxb2\text{-}a2}$, $mNDVI_{maxb1\text{-}max}$, $mNDVI_{maxb1\text{-}a1}$, $mNDVI_{maxb1\text{-}a2}$, $mNDVI_{max\text{-}a1}$, $mNDVI_{max\text{-}a2}$, and $mNDVI_{maxa1\text{-}a2}$; the correlation coefficients were significant at the 0.05 level between the Y_{RS} that was de-trended by linear regression and $NDVI_{maxa2}$, $mNDVI_{maxb4\text{-}a1}$, and $mNDVI_{maxb3\text{-}max}$. For SC, the correlation coefficients were significant at the 0.01 level between the Y_{RS} that was de-trended by linear regression and $NDVI_{maxb4}$, $NDVI_{maxb3}$, $NDVI_{maxb2}$,$mNDVI_{maxb4\text{-}b3}$, $mNDVI_{maxb4\text{-}b2}$, $mNDVI_{maxb4\text{-}b1}$, $mNDVI_{maxb4\text{-}max}$, $mNDVI_{maxb4\text{-}a1}$, $mNDVI_{maxb4\text{-}a2}$, $mNDVI_{maxb3\text{-}b2}$, $mNDVI_{maxb3\text{-}b1}$, and $mNDVI_{maxb3\text{-}max}$, $mNDVI_{maxb3\text{-}a1}$; the correlation coefficients were significant at the 0.05 level between the Y_{RS} that was de-trended by linear regression and $mNDVI_{maxb3\text{-}a2}$, $mNDVI_{maxb2\text{-}b1}$, $mNDVI_{maxb2\text{-}max}$, and $mNDVI_{maxb2\text{-}a1}$. The highest correlation coefficient between the Y_{RS} that was de-trended by linear regression and the NDVI variables occurred in the second 15-day period after $NDVI_{max}$ ($NDVI_{maxa2}$) and was significant at the 0.05 level for HLJ. The correlation coefficients between the Y_{RS} that was de-trended by linear regression and $NDVI_{maxb1}$, $mNDVI_{maxb1\text{-}max}$, $mNDVI_{maxb1\text{-}a1}$, and $mNDVI_{maxb1\text{-}a2}$ were significant at the 0.05 level in JX. The correlation coefficients between the Y_{RS} that was de-trended by linear regression and the NDVI variables ranged from – 0.14 to 0.38 in GX.

The correlation coefficients between the Y_{RS} that were de-trended by a 5-year moving average and the NDVI variables were generally low in HLJ, HN, and JX. For SC, the correlation coefficients were significant at the 0.01 level between the Y_{RS} that was de-trended by a 5-year moving average and $NDVI_{maxb4}$, and the correlation coefficients were significant at the 0.05 level between the Y_{RS} that was de-trended by a 5-year moving average and $NDVI_{maxb3}$, $mNDVI_{maxb4\text{-}b3}$, $mNDVI_{maxb4\text{-}b2}$, $mNDVI_{maxb4\text{-}b1}$, $mNDVI_{maxb4\text{-}max}$ and $mNDVI_{maxb3\text{-}b2}$. The correlation coefficients were significant at the 0.01 level between the Y_{RS} that was de-trended by a 5-year moving average and $NDVI_{maxb4}$, $NDVI_{maxb3}$, and $NDVI_{maxb4\text{-}b3}$.

3.3. Remotely Sensed Yield-Prediction Models

Conclusions drawn in the yield-trend analysis and the correlation analysis between Y_{RS} and the NDVI variables encouraged us to attempt to build a simple remotely sensed yield-prediction model for rice based on the NDVI variables. According to the correlation coefficient result summarized in Table 6, the Y_{RS} values that were de-trended by linear regression were used as dependent variables in HLJ, HN, JX, and SC. The Y_{RS} values that were de-trended by a 5-year moving average were used as dependent variables in GX. The NDVIs were used as independent variables. These models were constructed through the 'STEPWISE' regression process in SPSS software. Each model contains variables using the data period from 1982 to 2004. The correlation coefficients of the selected models ranged from 0.42 to 0.92, and all models were significant at the 0.01 level, except for HLJ which is significant at the 0.05 level (see Table 7). This means that increases in NDVI during the rice-growth period are generally related to the final rice-grain yield. The influence of NDVI always had a positive impact on yield. These results are consistent with numerous previous studies [31,36,12,75]. Data from 2005 to 2006 were used for model validation.

3.4. Validation of Rice Yield Prediction Models

The remotely sensed yield (Y_{RS}) of rice was calculated using the NDVI variables required by each model described in Table 7. The final rice yield (Y) was the sum of the trend yield (Y_t) and the remotely sensed yield (Y_{RS}). Figure 3 shows a scatter plot of the predicted and observed final rice yields for HLJ, HN, JX, SC, and GX from 1982 to 2004, expressed in units of kilogram per hectare. The models performed well, showing a good similarity between the predicted values and the official statistical values in HLJ, HN, JX, SC, and GX from 1982 to 2004 and capturing the fluctuations of rice yields over time. The regression line between the predicted

Table 5. Trends in rice yield for five selected-provinces in China from 1979 to 2009.

Province	Yield in 1979 ($kgha^{-1}$)	Yield in 2009($kgha^{-1}$)	Annual increase, 1979–2009 ($kgha^{-1}yr^{-1}$)
Heilongjiang (HLJ)	3480	6398.3	94.14
Hunan (HN)	4440	6371.3	62.30
Jiangxi (JX)	3645	5807	69.74
Sichuan (SC)	4777.5	7499.4	87.80
Guangxi (GX)	3562.5	5392.5	59.03

Table 6. Correlation coefficient (R) between the remotely sensed yields and NDVI variables during the rice growth period.

Variables	the remotely sensed yields de-trended by linear models					the remotely sensed yields de-trended by 5-year moving average				
	HLJ	HN	JX	SC	GX	HLJ	HN	JX	SC	GX
$NDVI_{maxb4}$	−0.02	−0.08	0.05	0.68**	0.24	−0.12	0.14	0.04	0.51**	0.54**
$NDVI_{maxb3}$	−0.16	−0.02	0.14	0.73**	0.36	−0.21	0.13	0.10	0.46*	0.52**
$NDVI_{maxb2}$	−0.08	0.38	0.21	0.57**	−0.14	−0.06	0.34	0.14	0.39	−0.30
$NDVI_{maxb1}$	−0.06	0.56**	0.42*	0.32	0.19	−0.03	0.22	−0.04	0.16	0.09
$NDVI_{max}$	0.13	0.60**	0.39	−0.06	−0.04	0.20	0.20	0.10	0.10	−0.26
$NDVI_{maxa1}$	0.42*	0.62**	0.28	0.29	−0.01	0.35	0.27	−0.05	0.08	−0.28
$NDVI_{maxa2}$	0.20	0.49*	0.32	−0.11	0.38	0.28	0.18	0.01	−0.22	0.39
$mNDVI_{maxb4\text{-}b3}$	−0.08	−0.05	0.10	0.73**	0.31	−0.16	0.14	0.08	0.50*	0.57**
$mNDVI_{maxb4\text{-}b2}$	−0.09	0.12	0.16	0.73**	0.19	−0.14	0.25	0.11	0.50*	0.32
$mNDVI_{maxb4\text{-}b1}$	−0.08	0.25	0.26	0.66**	0.22	−0.13	0.28	0.08	0.43*	0.30
$mNDVI_{maxb4\text{-}max}$	−0.07	0.33	0.29	0.64**	0.22	−0.11	0.30	0.09	0.44*	0.26
$mNDVI_{maxb4\text{-}a1}$	0.09	0.47*	0.31	0.61**	0.18	0.03	0.33	0.07	0.39	0.12
$mNDVI_{maxb4\text{-}a2}$	0.15	0.56**	0.33	0.54**	0.25	0.12	0.35	0.06	0.31	0.20
$mNDVI_{maxb3\text{-}b2}$	−0.14	0.23	0.20	0.70**	0.10	−0.15	0.28	0.14	0.46*	0.06
$mNDVI_{maxb3\text{-}b1}$	−0.12	0.37	0.32	0.60**	0.15	−0.13	0.30	0.09	0.38	0.08
$mNDVI_{maxb3\text{-}max}$	−0.10	0.45*	0.35	0.58**	0.14	−0.09	0.31	0.10	0.38	0.02
$mNDVI_{maxb3\text{-}a1}$	0.13	0.57**	0.35	0.55**	0.09	0.10	0.34	0.07	0.33	−0.10
$mNDVI_{maxb3\text{-}a2}$	0.19	0.64**	0.37	0.46*	0.18	0.19	0.35	0.06	0.24	0.03
$mNDVI_{maxb2\text{-}b1}$	−0.08	0.51**	0.35	0.47*	0.00	−0.05	0.33	0.07	0.29	−0.15
$mNDVI_{maxb2\text{-}max}$	−0.03	0.59**	0.38	0.44*	−0.01	0.01	0.34	0.08	0.29	−0.21
$mNDVI_{maxb2\text{-}a1}$	0.25	0.66**	0.37	0.43*	−0.01	0.23	0.34	0.04	0.25	−0.25
$mNDVI_{maxb2\text{-}a2}$	0.27	0.69**	0.38	0.33	0.10	0.29	0.33	0.04	0.15	−0.09
$mNDVI_{maxb1\text{-}max}$	0.02	0.64**	0.46*	0.25	0.13	0.07	0.24	0.01	0.17	−0.05
$mNDVI_{maxb1\text{-}a1}$	0.34	0.69**	0.42*	0.30	0.06	0.31	0.28	−0.01	0.15	−0.19
$mNDVI_{maxb1\text{-}a2}$	0.30	0.69**	0.41*	0.18	0.19	0.32	0.27	−0.01	0.02	0.01
$mNDVI_{max\text{-}a1}$	0.40*	0.66**	0.35	0.23	−0.02	0.36	0.27	0.00	0.12	−0.31
$mNDVI_{max\text{-}a2}$	0.33	0.66**	0.37	0.07	0.16	0.34	0.26	0.01	−0.06	−0.02
$mNDVI_{maxa1\text{-}a2}$	0.32	0.62**	0.32	0.09	0.19	0.33	0.25	−0.02	−0.09	0.04

*significant at 0.05 level; ** significant at 0.01 level, n = 23.

values and the observed values was close to the diagonal (intercept = 0, slope = 1), and the coefficients of determination for the five study areas ranged from 0.84 to 0.98, indicating that the reliability of the forecasts are very high.

The yield data for 2005 and 2006 were not included in the model construction and instead were used to evaluate the prediction models independently. These data provide independent estimates of the predictive power of the selected models (Table 8).

Table 7. Results of the stepwise regression models for remotely sensed rice yield using AVHRR-derived NDVI measures as independent variables.

Study areas	Model	R	F-test value	RMSE
HLJ	$Y_{RS} = -849.158 + 0.137 NDVI_{maxa1}$	0.42*	4.508	361.99
HN	$Y_{RS} = -1240.690 + 0.229\ mNDVI_{maxb1\text{-}a2}$	0.69**	19.342	114.57
JX	$Y_{RS} = -1553.145 + 0.261\ mNDVI_{maxb1\text{-}max}$	0.46**	5.689	166.38
SC	$Y_{RS} = -1495.515 + 0.403\ mNDVI_{maxb4\text{-}b3}$	0.73**	24.238	207.07
GX	$Y_{RS} = -1832.285 + 1.138\ mNDVI_{maxb4\text{-}b3} + 0.214 NDVI_{maxa2} - 1.315\ mNDVI_{maxb4\text{-}b2} + 0.307\ mNDVI_{maxb2\text{-}b1}$	0.92**	25.103	87.70

R: multiple correlation coefficient.
*significant at 0.05 level; ** significant at 0.01 level.

The differences between the predicted values and the official statistical values were 5% or less in seven out of ten years. These results demonstrate the potential of a NDVI rice-yield estimate that is based on model calibration with historical data at the provincial level. However, it is noticeable that the predicted relative errors were greater than 10%, but less than 19% in both 2005 and 2006 for SC and in 2006 in HLJ when compared with the official statistical data. These error rates are likely due to a number of contamination sources that can confound the potential relationship between NDVIs and rice yield. For instance, cloud and atmospheric-moisture contamination can influence the NDVI signal. Vegetation signals from before or after the selected NDVIs can impact the final yield of rice.

Conclusion

This study focused on the obvious and important role that advance of technology plays in rice yields increase. The results of this analysis suggest that the most common trend of rice yields in China during the years 1979–2009 is a linear growth. In the light of rice-yield trend could not be detected directly by a satellite remote sensor therefore, yield de-trended analysis was necessary to properly identify the remote-sensible effects and obtain an accurate prediction for rice yield. Only with de-trending analysis could we interpret the NDVI's evolution as being mainly due to variations in the photosynthetic activity and growth conditions of rice and then predict the rice yield using NDVI variables.

The AVHRR-based indices explored in the present research were useful for the remotely sensed rice yield-prediction in major

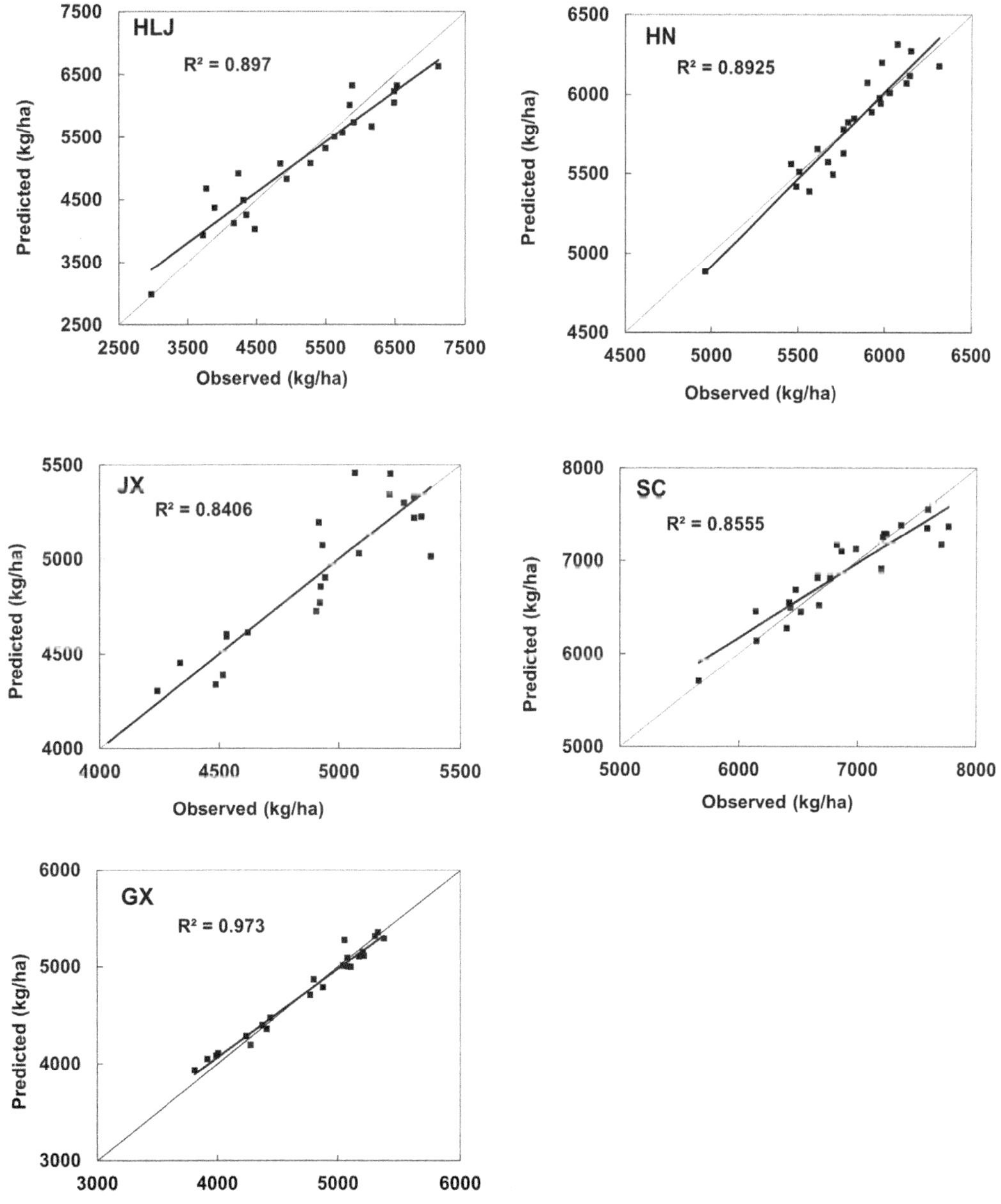

Figure 3. Observed versus predicted yields of rice (kg/ha) for the provinces of Heilongjiang (HLJ), Hunan (HN), Jiangxi (JX), Sichuan (SC) and Guangxi (GX) over the period 1982–2004.

Table 8. Observed and predicted rice yields (independent test).

Provinces	Year	Observed(kg/ha)	Predicted(kg/ha)	Relative Error (%)
Heilongjiang (HLJ)	2005	6795.7	6780.7	−0.22
	2006	6261.3	6897.8	10.17
Hunan (HN)	2005	6050.3	6337.5	4.75
	2006	6141.3	6441.2	4.88
Jiangxi (JX)	2005	5328.2	5545.9	4.09
	2006	5475.1	5634.9	2.92
Sichuan (SC)	2005	7213.0	8018.4	11.17
	2006	6420.7	7680.3	19.62
Guangxi (GX)	2005	4953.0	5028.98	1.53
	2006	5088.0	5053.44	−0.68

rice cultivation areas of China. This method allowed us to have a fine provincial estimate which satellite image could be difficult to obtain, or else a similar cost and a similar time frame data is easily available. However, it is cautious to restrict these analysis to those areas where the common trend of the crop yield is linear growth for the period considered.

The two steps for de-trending the statistical yield to obtain new time series, that are the trend yield (Y_t) and the remotely sensed yield (Y_{RS}); And by constructing the prediction models of Y_{RS} using NDVI variables enabled the development of a robust, simple, remotely sensed data-based model that was applicable at the provincial level in China. We believe the approach introduced here has a wide applicability to other rice-producing countries as well as other crops, such as wheat and corn.

More empirical studies should be performed on the use of AVHRR-derived NDVI time series as predictors for crop yield to enhance the understanding its forecasting capacity and limitations, and to validate the methods of remotely sensed yield estimation further. A future study should also include the application of a longer AVHRR NDVI time series in combination with other data sets such as SPOT-VEG, MODIS and SeaWiFS, especially in the event of one of these dataset's unexpected absence.

Acknowledgments

We thank the NASA/GSFC Global Inventory Modeling and Mapping Studies (GIMMS) group for providing access to the AVHRR NDVI data and the National Bureau of Statistics of China for providing rice yield data. We also appreciate professor Jiwei (jiwei@umkc.edu) and anonymous reviewers who provided very valuable comments.

Author Contributions

Conceived and designed the experiments: JH XW. Performed the experiments: JH XL. Analyzed the data: JH XW XL. Wrote the paper: JH XW XL ZP. Reviewed and revised this manuscript: XL HT ZP.

References

1. National Bureau of Statistics of China (2010) China Statistical Year Book. Beijing: China Statistical Press.
2. Wang RC, Huang JF (2002) Rice yield estimation using remote sensing data. Beijing: China Agriculture Press. 287 p. (in Chinese with English abstract).
3. Prasad AK, Chai L, Singh RP, Kafatos M (2006) Crop yield estimation model for Iowa using remote sensing and surface parameters. International Journal of Applied Earth Observation and Geoinformation 8: 26–33.
4. Manjunath KR, Potdar MB, Purohit NL (2002) Large area operational wheat yield model development and validation based on spectral and meteorological data. International Journal of Remote Sensing 23: 3023–3038.
5. Salazar L, Kogan F, Roytman L (2007) Use of remote sensing data for estimation of winter wheat yield in the United States. International Journal of Remote Sensing 28: 3795–3811.
6. Zhang F, Wu BF, Luo ZM (2004) Winter wheat yield predicting for America using remote sensing data. Journal of remote sensing 8: 611–617. (in Chinese with English abstract).
7. Wang CY, Lin WP (2005) Winter wheat yield estimation based on MODIS EVI. Transactions of the Chinese Society of Agricultural Engineering 21: 90–94. (in Chinese with English abstract).
8. Wall L, Larocque D, Leger PM (2008) The early explanatory power of NDVI in crop yield modelling. International Journal of Remote Sensing 29: 2211–2225.
9. Sridhar VN, Dadhwal VK, Chaudhari KN, Sharma R, Bairagi GD, et al. (1994) Wheat production forecasting for a predominantly unirrigated region in Madhya-Pradesh (India). International Journal of Remote Sensing 15: 1307–1316.
10. Smith RCG, Adams J, Stephens DJ, Hick PT (1995) Forecasting wheat yield in a Mediterranean-type environment from the NOAA satellite. Australian Journal of Agricultural Research 46: 113–125.
11. Schut AGT, Stephens DJ, Stovold RGH, Adams M, Craig RL (2009) Improved wheat yield and production forecasting with a moisture stress index, AVHRR and MODIS data. Crop & Pasture Science 60: 60–70.
12. Rudorff BFT, Batista GT (1991) Wheat yield estimation at the farm level using TM-Landsat and agrometeorological data. International Journal of Remote Sensing 12: 2477–2484.
13. Ren JQ, Chen ZX, Zhou QB, Tang HJ (2008) Regional yield estimation for winter wheat with MODIS-NDVI data in Shandong, China. International Journal of Applied Earth Observation and Geoinformation 10: 403–413.
14. Ren JQ, Chen ZX, Tang HJ (2006) Regional scale remote sensing-based yield estimation of winter wheat by using MODIS-NDVI data: A case study of Jining City in Shandong Province. Chinese Journal of Applied Ecology 17: 2371–2375. (in Chinese with English abstract).
15. Prasad AK, Singh RP, Tare V, Kafatos M (2007) Use of vegetation index and meteorological parameters for the prediction of crop yield in India. International Journal of Remote Sensing 28: 5207–5235.
16. Patel NR, Bhattacharjee B, Mohammed AJ, Tanupriya B, Saha SK (2006) Remote sensing of regional yield assessment of wheat in Haryana, India. International Journal of Remote Sensing 27: 4071–4090.
17. Moriondo M, Maselli F, Bindi M (2007) A simple model of regional wheat yield based on NDVI data. European Journal of Agronomy 26: 266–274.
18. Mo X, Liu S, Lin Z, Xu Y, Xiang Y, et al. (2005) Prediction of crop yield, water consumption and water use efficiency with a SVAT-crop growth model using remotely sensed data on the North China Plain. Ecological Modelling 183: 301–322.
19. Mkhabela MS, Bullock P, Raj S, Wang S, Yang Y (2011) Crop yield forecasting on the Canadian Prairies using MODIS NDVI data. Agricultural and Forest Meteorology 151: 385–393.
20. Mika J, Kerenyi J, Rimoczi-Paal A, Merza A, Szinell C, et al. (2002) On correlation of maize and wheat yield with NDVI: Example of Hungary (1985–1998). In: Fellous JL, LeMarshall JF, Choudhury BJ, Menenti M, Paxton LJ et al., editors. Earth's Atmosphere, Ocean and Surface Studies. 2399–2404.
21. Maselli F, Rembold F (2001) Analysis of GAC NDVI data for cropland identification and yield forecasting in Mediterranean African countries. Photogrammetric Engineering and Remote Sensing 67: 593–602.

22. MacDonald R, Hall F (1980) Global crop forecasting. Science 208: 670.
23. Labus MP, Nielsen GA, Lawrence RL, Engel R, Long DS (2002) Wheat yield estimates using multi-temporal NDVI satellite imagery. International Journal of Remote Sensing 23: 4169–4180.
24. Kastens JH, Kastens TL, Kastens DLA, Price KP, Martinko EA, et al. (2005) Image masking for crop yield forecasting using AVHRR NDVI time series imagery. Remote Sensing of Environment 99: 341–356.
25. Kalubarme AH, Potdar MB, Manjunath KR, Mahey RK, Siddhu SS (2003) Growth profile based crop yield models: a case study of large area wheat yield modelling and its extendibility using atmospheric corrected NOAA AVHRR data. International Journal of Remote Sensing 24: 2037–2054.
26. Huang JF, Wang RC, Wang XZ, Liu SM, Zhang JH (1999) Study on multiple yield estimation models of winter wheat using remote sensing data. Journal of Zhejiang University (Agric & Life Sci) 25: 519–523. (in Chinese with English abstract).
27. Hochheim KP, Barber DG (1998) Spring wheat yield estimation for Western Canada using NOAA NDVI data. Canadian Journal of Remote Sensing 24: 17–27.
28. Gupta R, Prasad S, Rao G, Nadham T (1993) District level wheat yield estimation using NOAA/AVHRR NDVI temporal profile. Advances in Space Research 13: 253–256.
29. Ferencz C, Bognar P, Lichtenberger J, Hamar D, Tarscai G, et al. (2004) Crop yield estimation by satellite remote sensing. International Journal of Remote Sensing 25: 4113–4149.
30. Dubey RP, Ajwani N, Kalubarme MH, Sridhar VN, Navalgund RR, et al. (1994) Preharvest wheat yield and production estimation for the Punjab, India. International Journal of Remote Sensing 15: 2137–2144.
31. Doraiswamy PC, Cook PW (1995) Spring wheat yield assessment using NOAA AVHRR data. Canadian Journal of Remote Sensing 21: 43–51.
32. Cheng Q (1994) The use of vegetation index for monitoring drought and winter wheat yield estimation. Remote sensing technology and application 9: 12–18. (in Chinese with English abstract).
33. Bullock PR (1992) Operational estimates of western Canadian grain production using NOAA AVHRR LAC data. Canadian Journal of Remote Sensing 18: 23–29.
34. Boken VK, Shaykewich CF (2002) Improving an operational wheat yield model using phenological phase-based Normalized Difference Vegetation Index. International Journal of Remote Sensing 23: 4155–4168.
35. Benedetti R, Rossini P (1993) On the use of NDVI profiles as a tool for agricultural statistics: the case study of wheat yield estimate and forecast in Emilia Romagna. Remote Sensing of Environment 45: 311–326.
36. Becker-Reshef I, Vermote E, Lindeman M, Justice C (2010) A generalized regression-based model for forecasting winter wheat yields in Kansas and Ukraine using MODIS data. Remote Sensing of Environment 114: 1312–1323.
37. Bastiaanssen WGM, Ali S (2003) A new crop yield forecasting model based on satellite measurements applied across the Indus Basin, Pakistan. Agriculture Ecosystems & Environment 94: 321–340.
38. Balaghi R, Tychon B, Eerens H, Jlibene M (2008) Empirical regression models using NDVI, rainfall and temperature data for the early prediction of wheat grain yields in Morocco. International Journal of Applied Earth Observation and Geoinformation 10: 438–452.
39. Wannebo A, Rosenzweig C (2003) Remote sensing of US cornbelt areas sensitive to the El Nino-Southern Oscillation. International Journal of Remote Sensing 24: 2055–2067.
40. Unganai LS, Kogan FN (1998) Drought monitoring and corn yield estimation in Southern Africa from AVHRR data. Remote Sensing of Environment 63: 219–232.
41. Seiler RA, Kogan F, Wei G (2000) Monitoring weather impact and crop yield from NOAA AVHRR data in Argentina. In: Gupta RK, editor. Remote Sensing for Land Surface Characterisation. 1177–1185.
42. Rojas O (2007) Operational maize yield model development and validation based on remote sensing and agro-meteorological data in Kenya. International Journal of Remote Sensing 28: 3775–3793.
43. Reynolds CA, Yitayew M, Slack DC, Hutchinson CF, Huete A, et al. (2000) Estimating crop yields and production by integrating the FAO Crop specific Water Balance model with real-time satellite data and ground-based ancillary data. International Journal of Remote Sensing 21: 3487–3508.
44. Quarmby N, Milnes M, Hindle T, Silleos N (1993) The use of multi-temporal NDVI measurements from AVHRR data for crop yield estimation and prediction. International Journal of Remote Sensing 14: 199–210.
45. Mkhabela MS, Mashinini NN (2005) Early maize yield forecasting in the four agro-ecological regions of Swaziland using NDVI data derived from NOAAs-AVHRR. Agricultural and Forest Meteorology 129: 1–9.
46. Lewis JE, Rowland J, Nadeau A (1998) Estimating maize production in Kenya using NDVI: some statistical considerations. International Journal of Remote Sensing 19: 2609–2617.
47. Lee R, Kastens D, Price K, Martinko E. Forecasting corn yield in Iowa using remotely sensed data and vegetation phenology information; 2000 January 10–12; Lake Buena Vista, Florida. 460–467.
48. Hayes M, Decker W (1996) Using NOAA AVHRR data to estimate maize production in the United States Corn Belt. International Journal of Remote Sensing 17: 3189–3200.
49. Funk C, Budde ME (2009) Phenologically-tuned MODIS NDVI-based production anomaly estimates for Zimbabwe. Remote Sensing of Environment 113: 115–125.
50. Tennakoon SB, Murty VVN, Eiumnoh A (1992) Estimation of cropped area and grain-yield of rice using remote-sensing data. International Journal of Remote Sensing 13: 427–439.
51. Huang JF, Yang ZE, Wang RC, Xu HW, Jiang HX (2002) The rice production forecasting models using NOAA/AVHRR data based on GIS. Remote sensing technology and application 17: 125–128. (in Chinese with English abstract).
52. Huang JF, Wang FM, Wang XZ (2010) Hyperspectral experiment for paddy rice remote sensing; Huang JQ, Chen JY, editors. Hangzhou: Zhejiang University Press. 315 p. (in Chinese with English abstract).
53. Rasmussen MS (1998) Developing simple, operational, consistent NDVI-vegetation models by applying environmental and climatic information: Part I. Assessment of net primary production. International Journal of Remote Sensing 19: 97–117.
54. Rasmussen MS (1997) Operational yield forecast using AVHRR NDVI data: Reduction of environmental and inter-annual variability. International Journal of Remote Sensing 18: 1059–1077.
55. Rasmussen MS (1992) Assessment of millet yields and production in northern Burkina Faso using integrated NDVI from the AVHRR. International Journal of Remote Sensing 13: 3431–3442.
56. Maselli F, Romanelli S, Bottai L, Maracchi G (2000) Processing of GAC NDVI data for yield forecasting in the Sahelian region. International Journal of Remote Sensing 21: 3509–3523.
57. Groten SME (1993) NDVI – Crop Monitoring and Early Yield Assessment of Burkina-Faso. International Journal of Remote Sensing 14: 1495–1515.
58. Potdar MB (1993) Sorghum yield modelling based on crop growth parameters determined from visible and near-IR channel NOAA AVHRR data. International Journal of Remote Sensing 14: 895–905.
59. Fuller DO (1998) Trends in NDVI time series and their relation to rangeland and crop production in Senegal, 1987–1993. International Journal of Remote Sensing 19: 2013–2018.
60. Wendroth O, Reuter HI, Kersebaum KC (2003) Predicting yield of barley across a landscape: a state-space modeling approach. Journal of Hydrology 272: 250–263.
61. Weissteiner C, Kühbauch W (2005) Regional Yield Forecasts of Malting Barley (Hordeum vulgare L.) by NOAA-AVHRR Remote Sensing Data and Ancillary Data. Journal of agronomy and crop science 191: 308–320.
62. Liu WT, Kogan F (2002) Monitoring Brazilian soybean production using NOAA/AVHRR based vegetation condition indices. International Journal of Remote Sensing 23: 1161–1179.
63. Esquerdo J, Zullo J, Antunes JFG (2011) Use of NDVI/AVHRR time-series profiles for soybean crop monitoring in Brazil. International Journal of Remote Sensing 32: 3711–3727.
64. Tucker CJ (1979) Red and photographic infrared linear combinations for monitoring vegetation. Remote Sensing of Environment 8: 127–150.
65. Tilman D, Cassman KG, Matson PA, Naylor R, Polasky S (2002) Agricultural sustainability and intensive production practices. Nature 418: 671–677.
66. Evenson RE, Gollin D (2003) Assessing the impact of the Green Revolution, 1960 to 2000. Science 300: 758.
67. Peng S, Laza R, Visperas R, Sanico A, Cassman KG, et al. (2000) Grain yield of rice cultivars and lines developed in the Philippines since 1966. Crop Science 40: 307–314.
68. Hafner S (2003) Trends in maize, rice, and wheat yields for 188 nations over the past 40 years: a prevalence of linear growth. Agriculture, ecosystems & environment 97: 275–283.
69. Pinzon J, Brown ME, Tucker CJ (2004) Satellite time series correction of orbital drift artifacts using empirical mode decomposition. Hilbert-Huang Transform: Introduction and Applications 10: 285–295.
70. Tucker CJ, Pinzon JE, Brown ME, Slayback DA, Pak EW, et al. (2005) An extended AVHRR 8-km NDVI dataset compatible with MODIS and SPOT vegetation NDVI data. International Journal of Remote Sensing 26: 4485–4498.
71. Fensholt R, Nielsen TT, Stisen S (2006) Evaluation of AVHRR PAL and GIMMS 10-day composite NDVI time series products using SPOT-4 vegetation data for the African continent. International Journal of Remote Sensing 27: 2719–2733.
72. Fensholt R, Rasmussen K, Nielsen TT, Mbow C (2009) Evaluation of earth observation based long term vegetation trends – Intercomparing NDVI time series trend analysis consistency of Sahel from AVHRR GIMMS, Terra MODIS and SPOT VGT data. Remote Sensing of Environment 113: 1886–1898.
73. Genovese G, Vignolles C, Negre T, Passera G (2001) A methodology for a combined use of normalised difference vegetation index and CORINE land cover data for crop yield monitoring and forecasting. A case study on Spain. Agronomie 21: 91–111.
74. Freund JT (2005) Estimating Crop Production in Kenya: A Multi-Temporal Remote Sensing Approach [Masters]. Santa Barbara: University of California. 59 p.
75. Tucker CJ, Holben BN, Elgin JH, McMurtrey JE (1980) Relationship of spectral data to grain yield variation. Photogrammetric Engineering and Remote Sensing 46: 657–666.
76. Dyson T (1999) World food trends and prospects to 2025. Proceedings of the National Academy of Sciences 96: 5929.

77. Clarke FR, Baker RJ, Depauw RM (1994) Moving mean and least-squares smoothing for analysis of grain-yield data. Crop Science 34: 1479–1483.
78. Peltonen-Sainio P, Jauhiainen L, Laurila IP (2009) Cereal yield trends in northern European conditions: Changes in yield potential and its realisation. Field Crops Research 110: 85–90.
79. Epplin FM, Peeper TF (1998) Influence of planting date and environment on Oklahoma wheat grain yield trend from 1963 to 1995. Canadian Journal of Plant Science 78: 71–77.
80. Stergiou KI, Christou ED (1996) Modelling and forecasting annual fisheries catches: Comparison of regression, univariate and multivariate time series methods. Fisheries Research 25: 105–138.
81. Foody GM, Boyd DS, Cutler MEJ (2003) Predictive relations of tropical forest biomass from Landsat TM data and their transferability between regions. Remote Sensing of Environment 85: 463–474.
82. Sirkin RM (2006) Statistics for the social sciences. Thousand OaksCalifornia: Sage Publications, Inc. 244 p.

Not All Kinds of Revegetation Are Created Equal: Revegetation Type Influences Bird Assemblages in Threatened Australian Woodland Ecosystems

David B. Lindenmayer*, Amanda R. Northrop-Mackie, Rebecca Montague-Drake, Mason Crane, Damian Michael, Sachiko Okada, Philip Gibbons

Fenner School of Environment and Society, ARC Centre of Excellence for Environmental Decisions, and National Environment Research Program, The Australian National University, Canberra, Australia

Abstract

The value for biodiversity of large intact areas of native vegetation is well established. The biodiversity value of regrowth vegetation is also increasingly recognised worldwide. However, there can be different kinds of revegetation that have different origins. Are there differences in the richness and composition of biotic communities in different kinds of revegetation? The answer remains unknown or poorly known in many ecosystems. We examined the conservation value of different kinds of revegetation through a comparative study of birds in 193 sites surveyed over ten years in four growth types located in semi-cleared agricultural areas of south-eastern Australia. These growth types were resprout regrowth, seedling regrowth, plantings, and old growth. Our investigation produced several key findings: **(1)** Marked differences in the bird assemblages of plantings, resprout regrowth, seedling regrowth, and old growth. **(2)** Differences in the number of species detected significantly more often in the different growth types; 29 species for plantings, 25 for seedling regrowth, 20 for resprout regrowth, and 15 for old growth. **(3)** Many bird species of conservation concern were significantly more often recorded in resprout regrowth, seedling regrowth or plantings but no species of conservation concern were recorded most often in old growth. We suggest that differences in bird occurrence among different growth types are likely to be strongly associated with growth-type differences in stand structural complexity. Our findings suggest a range of vegetation growth types are likely to be required in a given farmland area to support the diverse array of bird species that have the potential to occur in Australian temperate woodland ecosystems. Our results also highlight the inherent conservation value of regrowth woodland and suggest that current policies which allow it to be cleared or thinned need to be carefully re-examined.

Editor: Matt Hayward, Australian Wildlife Conservancy, Australia

Funding: This work was supported by major grants from the Murray Catchment Management Authority, the Australian Research Council and the Australian Government's Caring for our Country Program. The funders had no role in study design, data collection and analysis, decision to publish, or preparation of the manuscript.

Competing Interests: The authors have declared that no competing interests exist.

* E-mail: david.lindenmayer@anu.edu.au

Introduction

Much has been written in many parts of the world about the value of regrowth vegetation, including its importance for biodiversity conservation [1–5]. This includes regrowth after logging, after vegetation clearing, and after agricultural land abandonment. Indeed, ecological studies of regrowth vegetation have been a classic part of ecology for a long time [6] and a wide range of studies from around the world have demonstrated that regrowth vegetation can be important for a range of species. This is true both in tropical and temperate ecosystems (e.g. [4,7–13]), although such kinds of vegetation can often support different suites of species compared with, for example, old growth vegetation [4,14–16].

While the value of regrowth vegetation is increasingly recognised, there can be different kinds of revegetation that have different origins. That is, different starting conditions and/or disturbance regimes can give rise to structurally different kinds of revegetation. For example, in the threatened temperate woodlands of southern Australia (where the study we report here is taking place), different kinds of regrowth vegetation can include: **(1)** resprout regrowth vegetation which develops following logging, fire or partial clearing, e.g. [17], and **(2)** seedling regrowth which develops after a reduction in grazing pressure by domestic livestock [18] or after some kinds of disturbance like ploughing after droughts. In addition, throughout threatened Australian temperate woodlands, there are extensive efforts to replant native vegetation [19–22]. A key question is: Are there differences in the richness and composition of biotic communities in different kinds of regrowth and how does this compare to old growth and plantings? The answer to this question remains either unknown or poorly known in many ecosystems worldwide. Yet it is critical to know whether different kinds of restoration efforts like the deliberate planting of vegetation or the natural regeneration of vegetation lead to the development of different suites of species that are associated with them. This is, in part, because the costs of deliberately planting vegetation can be very high but those associated with passive revegetation (i.e. resprout and seedling regrowth) can be comparatively much lower [21]. In addition, some authors, e.g. [23] have argued that biodiversity has been negatively affected in some replanted areas because stem density has been too high relative to that typical of passively regenerating areas. For example, this may impede foraging by bats and slow the rate of development of key structures like large cavity trees [24].

We addressed key knowledge gaps about the conservation value of different kinds of revegetation in the temperate woodlands ecosystems of southern New South Wales, south-eastern Australia. Temperate woodlands are some of the most heavily cleared, extensively degraded, and highly threatened ecosystems on the Australian continent [25] and there is an urgent need for vegetation restoration in many areas [18]. However, to the best of our collective knowledge, no-one has previously compared the biodiversity value of different kinds of revegetation. We focused on birds in our comparative study of growth types. This was because: **(1)** there is a wide range of bird species of conservation concern in Australian temperate woodlands [26,27], **(2)** past work has indicated that some elements of the temperate woodland bird biota respond strongly to key attributes of stand structure (e.g. [11,28]) which are likely, in turn, to vary substantially between different kinds of revegetation, and **(3)** birds are widely considered to play important roles in maintaining some ecosystem processes [29].

Using an extensive dataset gathered at a large number of sites (N = 193) that have been surveyed repeatedly over the past ten years, we posed the following series of broad and inter-connected questions:

- Is there a difference in the bird species richness and the composition of bird assemblages between old growth temperate woodland and different kinds of revegetation, including plantings? We postulated at the outset of this study that there would be marked differences in bird species richness and assemblage composition between old growth, resprout regrowth, seedling regrowth and planted areas. This was because of likely major differences in vegetation structure between growth types and previously well documented relationships between vegetation structure and bird responses as reflected through ecological theories like the structural complexity hypothesis [30], the intermediate disturbance hypothesis [31], and the biological legacies concept [32]. In addition, there are marked differences in starting conditions between types of revegetation and this also can influence biotic responses [33].
- Are there bird species associated with particular growth types? Guided by theory like the landscape texture hypothesis [34,35], at the onset of this investigation we predicted that small-bodied birds would be closely associated with densely structured revegetated areas [36], particularly plantings. Based on succession theory [37], we predicted that particular bird taxa like cavity-dependent species would be strongly associated with old growth where vegetation attributes like trees with hollows are likely to be most abundant.

Notably, we elected to make our investigation a comparative study of broad categories of growth types, rather than focus on relationships between birds and an array of covariates corresponding to measurements of vegetation structure and plant species composition, e.g. [11,28,38]. We made this decision because the vast majority of on-ground practitioners in south-eastern Australia charged with managing native vegetation readily recognise broad growth type categories (i.e. old growth, resprout regrowth, seedling regrowth, plantings). Moreover, such broad categories are a fundamental part of government legislation such as in Queensland and New South Wales [39]. Conversely, few practitioners have the time or expertise to complete detailed measurements of stand structure and composition and then relate them to response variables like bird species richness or the presence of individual bird species.

Addressing questions about the conservation value of different kinds of revegetation is important for several key reasons. First, approximately 40% of the planet's terrestrial land surface is used being used for agriculture [40,41] but 16–40% of that area is lightly to severely degraded and in need of some form of restoration [42]. Second, because there can be substantial differences in the costs and labour requirements of different forms of revegetation (e.g. replanting versus passive regeneration) [21], it is critical to better understand the value of different areas as habitat for wildlife. Third, there has been extensive clearing of regrowth vegetation in some parts of the world (e.g. eastern Australia) [17,43]. Fourth, there are well advanced proposals to undertake management interventions like thinning in large areas of regrowth vegetation, including in many areas of threatened temperate woodland in Australia [44]. This activity is hypothesized to increase pasture growth for livestock grazing or to increase the rate of diameter increment of overstorey trees and thereby accelerate the pace at which large trees (and associated key attributes like cavities and large pieces of coarse woody debris) will develop. However, practices like thinning need to be guided by an understanding of the value of different kinds of regrowth for biodiversity.

Methods

1.1 Study area

We conducted this study within the South-west Slopes region of New South Wales in an area spanning the towns of Junee (0552952 E 6140128 N) in the north and Albury (0494981 E 6008873 N) in the south (a distance of ~150 km), and Gundagai (600532 E 6119073 N) and Howlong (467090 E 6017897 N) in the east and west respectively (a distance of ~120 km) (Figure 1). The predominant form of native vegetation in the region is temperate woodland [25,45] dominated by White Box *Eucalyptus albens*, Grey Box *E. microcarpa*, or other eucalypt tree species such as Yellow Box *E. melliodora*, Blakely's Red Gum *E. blakelyi*, Red Stringybark *E. macrorhyncha* and Red Ironbark *E. sideroxylon*. A range of broad vegetation types has been recognised in our study area [46] and these are Floodplain Transition Woodlands, Inland Riverine Forests, Upper Riverina Dry Sclerophyll Forests and Western Slopes Grassy Woodlands.

1.2 Study design

Our study comprised 193 sites on 46 individual farms that we assigned to one of four vegetation growth types plantings (63 sites), old growth woodland (71 sites), resprout regrowth woodland (27 sites), and seedling regrowth woodland (32 sites). Plantings were areas of planted native vegetation characterised by a mix of local endemic and exotic Australian ground cover, understorey and overstorey plant species. Most plants were typically spaced 2 m apart, but there was not a standard set of spacing and plant species composition protocols applied in revegetation efforts. All plantings were at least 7 years old at the start of this investigation in 2000 and many were 10–20 years old and had been established to mitigate problems associated with soil erosion and/or salinity. Resprout regrowth refers to (multi-stemmed) regrowth from existing living trees recovering after disturbance by fire, clearing or both. Seedling regrowth originates from seeds germinating after

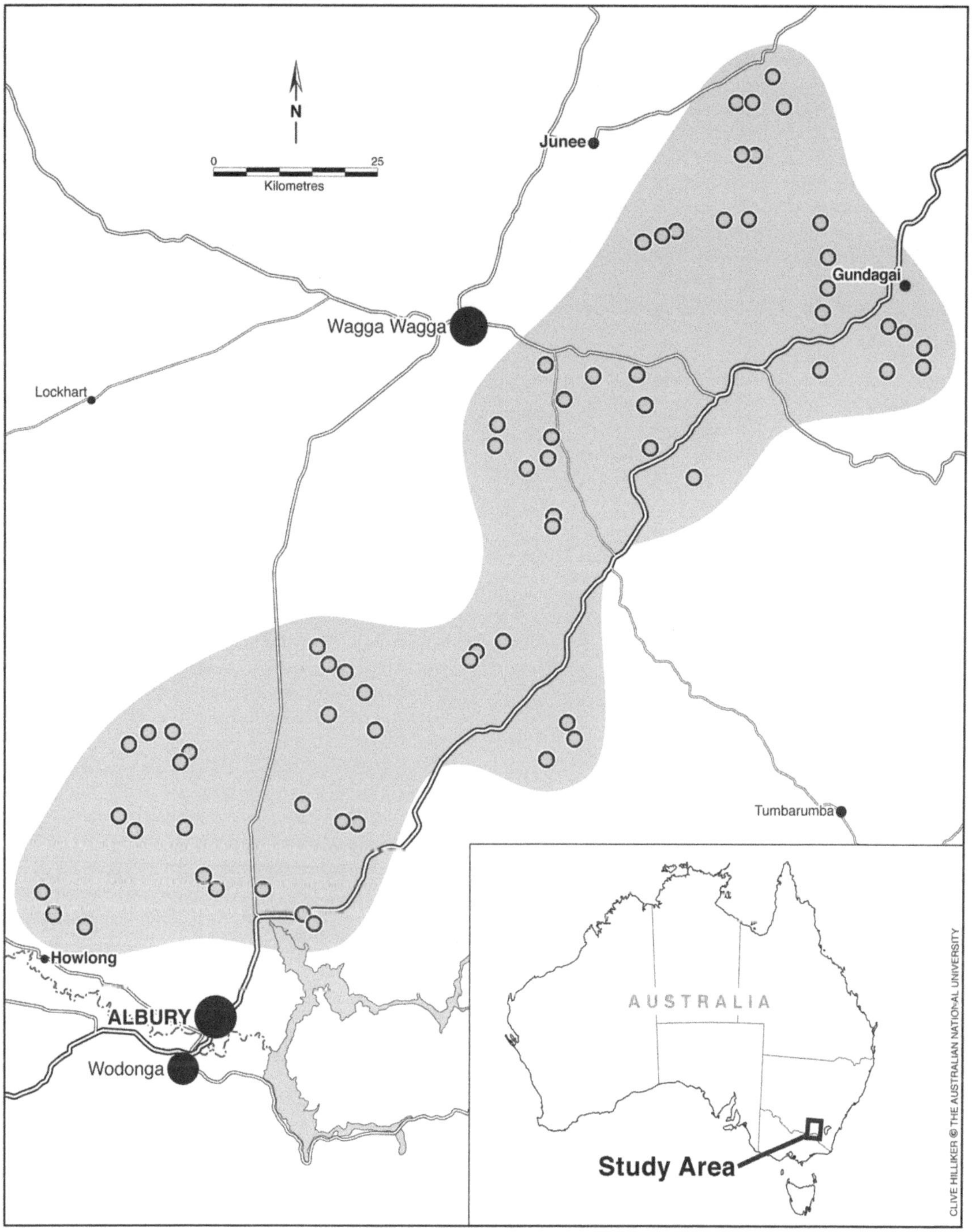

Figure 1. The South-west Slopes study region of southern New South Wales, south-eastern Australia.

being dropped by overstorey trees. As in the case of our plantings sites, the stands of resprout regrowth and seedling regrowth that we selected exceeded 7–10 years old (when we commenced our work in 2000) but many were 15–20 years old. Old growth woodland was dominated by large old scattered trees, typically 200 or more years old. Livestock grazing occurred in many of our study sites and some were in a degraded condition as a result, particularly where high-intensity set stock grazing regimes were employed. However, there also were sites in good condition in all growth types, for example, those on farms subject to cell grazing and/or short-rotation rotational grazing.

Critically for our study design, our resprout regrowth, seedling regrowth and replanting sites were approximately the same age, whereas the old growth stands were clearly much older. In addition, we worked hard to ensure that the size of patches of old growth, resprout regrowth, seedling regrowth and replanting were broadly similar to avoid the potential for confounding between patch size and growth type; see [9].

1.3 Site establishment

We established a permanent fixed 200 m long transect at each of our 193 sites. The permanent transect was where we completed counts of birds (see below) as well as completed detailed measurements of vegetation composition (see Appendix S1).

No specific permits were required for the described field studies. The relevant permissions to enter the private land involved in the study were given by the Mr Emmo Willinck, Catchment Officer, on behalf of the Murray Catchment Management Authority (MCMA). No specific permits were required for the field study locations as the owners of the private land involved had established access relationships with the MCMA. The researchers were acting as agents for the MCMA under the terms of a collaborative research partnership. However, prior to all surveys, telephone contact was made with all the relevant private landowners to indicate researcher access to their land. All native animal species and native woodland vegetation are protected in Australia, including endangered birds and plants. Our studies were observational investigations and no plants or animals were harmed in any way.

1.4 Bird counting protocols

We gathered data on temperate woodland birds at all 193 sites in our investigation between 2002 and 2009. These data comprised five spring counts and three winter counts. Thus, we surveyed each of the 193 sites eight times for birds between 2002 and 2009. Our bird counting protocols entailed repeated 5 minute point interval counts [47] at the 0 m, 100 m and 200 m points along the permanent transect at each site. Six highly experienced ornithologists participated in the surveys although they varied to some extent in their ability to detect some groups of birds. Lindenmayer et al. [48] showed that pooling counts of two or more observers at the same site could compensate for extra variability due to observer heterogeneity. Field et al. [49] showed that weather and other conditions on any given day can influence bird detectability. Thus, in each of our surveys, each permanent field site was surveyed by two different observers on different days by repeated point interval counts. We completed counts between 5.30–9.30am and did not undertake surveys on days of poor weather (rain, high wind, fog or heavy cloud cover). These protocols were identical to those employed in other long-term major studies in woodlands [11,46,50]. In summary, between 2002 and 2009, we conducted 48 individual point counts at each of our 193 permanent field sites.

1.5 Data aggregation

Our approach provided high quality presence-absence data. There were six opportunities for a given bird species to be detected within a particular site in any given survey period (i.e. 3 point counts completed by 2 observers in a given survey). This was then summarised as a single presence/absence value for a particular bird species at a site, in a particular season, in a particular year.

The species we targeted for investigation were readily recognisable and had distinctive calls. Moreover, we were extremely familiar with them from many previous studies in a range of environments in south-eastern Australia [20,51,52]. Tyre et al. [53] found that six repeated visits at a site improved the precision of estimates to levels comparable to that achieved with conventional statistics in the absence of false-negative errors. As noted above, our dataset comprised 48 point counts completed during eight surveys between 2002 and 2009. Furthermore, our preliminary data analyses revealed very few newly detected species (<2%) by our spring (2006) and winter (2007) surveys. Moreover, for our study, "plots" were sampling units rather than specific territories for which it would be appropriate to determine true occupancy. We therefore assumed that non-detection was low for the bird species included in this study. Moreover, although some individual birds may have gone undetected, this would not invalidate a comparative investigation like ours.

In summary, the work we report in this paper was a comparative study in which we were interested in quantifying differences in the bird biota of different growth types. Therefore, a key statistical underpinning is that we have used an identical field counting methodology with the same observers surveying the same sites consistently in successive surveys. The currency that we have applied in our comparative study is presence/absence.

1.6 Vegetation measurements

We completed detailed measurements of the vegetation at each of our 193 field sites to enable us to compare the vegetation structure of the different growth types (see below). First, we established a 20×20 m plot around the 0 m, 100 m and 200 m post at each site. Second, at each corner of each 20×20 m plot, we established a 1×1 m plot. This gave 3 large (20 m×20 m plots) and 12 1 a 1 m plots at each site. The array of vegetation attributes measured at each site is described in Appendix S1.

1.7 Statistical analysis

First, we analysed relationships between bird species richness and growth type using repeated measures MANOVA. For the analysis, a particular site was the subject, the between-subject effect was growth type, and the within- subject effect was time.

Second, we explored relationships between the composition of the bird assemblage and the growth type using both partial Canonical Correspondence Analysis (CCA) and partial redundancy analysis. We conducted these analyses using the package "vegan" in program R. As the results from both were similar, we elected to present only those results from CCA, which is apposite when analysing occurrence data as it detects patterns within ecological datasets that can be explained by environmental variables [54]. We used the CCA algorithm developed by Legendre and Legendre [55] to analyse relationships between bird species occurrences and growth type. Constraining variables were subjected to weighted linear regression and a correspondence analysis was conducted on the fitted values via singular value decomposition. We controlled for the effects of season and year by including them as covariates in the partial CCA [55]. We used Monte Carlo simulations, with 1000 steps, to establish the significance of all canonical axes. Tests of significance in CCA do not rely on parametric assumptions [56,57]. CCA allows a visual interpretation (a biplot) of species-environment relationships. We used species conditional scaling [58] to centre species within the sites in which they occurred. Distances between species and a particular growth type approximate their relative frequency of occurrence; see [59]. We restricted the data used in the CCA to bird species that were detected in five or more sites on average over the ten year period. This was because rare species can obscure community patterns [57,60].

Third, we examined the response of individual bird species to growth type. Some of these birds were taxa of conservation concern [11,38] and others, like the hyper-aggressive native honeyeater Noisy Miner (*Manorina melanocephala*), are known to exclude a range of other species from temperate woodland [61]. To quantify relationships between growth types and individual bird species, we conducted Cochran-Mantel-Haenszel tests [62] which were stratified by time. We then used Analysis of Means (ANOM) of proportions [63] to determine if occurrences of bird species in a particular growth type were significantly higher or lower than the average occurrences in all growth types.

Finally, we postulated that the underlying drivers of growth type differences in bird assemblages may have been a result of differences in key attributes of vegetation structure. On this basis, we explored differences in vegetation structure of the growth types to better understand the potential reasons for species' preferences. We employed a nonparametric Steel-Dwass multiple comparison procedure [64–66] to test for differences in the structure of different growth types.

Results

We recorded 178 species of birds in our dataset that was comprised of 9264 survey points. Of these, we detected 57 species of birds at fewer than four sites on average per year, and excluded them from subsequent data analyses. Our focus was on woodland birds, and we therefore removed 31 waterbirds from our analysis. This left a total 90 species for detailed investigation. We list the common and scientific names of these species in Appendix S2.

2.1 Growth type and bird species richness

Our repeated measures MANOVA showed insufficient evidence overall to conclude there were significant differences in species richness between the various growth types ($F\text{-test}_{3,159} = 0.824$, p-value = 0.4822) (Figure 2).

2.2 Growth type and the composition of the bird assemblage

We found that the first three axes of the CCA were statistically significant using 1000 permutations ($P<0.05$). The first axis accounted for 72% of the constrained variation in our data and it contrasted occurrences of bird species in plantings from the other three growth types (Figure 3). The second axis accounted for 22% of the constrained variation, and contrasted old growth from the two types of regrowth (Figure 3). The third axis, which accounted for the least amount of variation, contrasted resprout regrowth and seedling regrowth (Figure 4).

2.3 Individual species responses

We found that 29 species occurred significantly more often in plantings, 25 significantly more often seedling regrowth, 20 significantly more often in resprout regrowth, and 15 significantly more often in old growth (Appendix S2). Of the 90 species for which we completed detailed statistical analyses, 67 exhibited a significant ($P<0.05$) relationship with growth type (Appendix S2). A number of these were species of conservation concern. For example, the Grey-crowned Babbler and White-browed Babbler were significantly more likely to occur in seedling regrowth than other growth types (Appendix S2). The Black-chinned Honeyeater was most often found in seedling regrowth and resprout regrowth. In the case of the Diamond Firetail, the species occurred significantly less often in old growth relative to other growth types (Appendix S2). The Hooded Robin was significantly more likely to occur in resprout regrowth than other growth types, particularly plantings. The Brown Treecreeper, Crested Shrike-tit, Dusky Woodswallow and Jacky Winter were least likely to be recorded in plantings and most likely to be recorded in resprout regrowth and seedling regrowth. Birds of conservation concern that were most often recorded in plantings included the Red-capped Robin, Rufous Whistler, Speckled Warbler, and Flame Robin (Appendix S2). Notably, no birds of conservation concern were found the majority of the time in old growth (Appendix S2).

2.4 Vegetation structure and growth type

We completed analyses of the structure of vegetation of the four growth types using Steel-Dwass multiple comparison tests (Appendix S3). These analyses clearly indicated marked growth type differences in vegetation structure (Figure 5; Appendix S3). For example, we found that plantings had significantly higher stem density than old growth and resprout regrowth. It also had significantly higher midstorey cover and significantly lower overstorey cover than other growth types (Figure 4; Appendix S3). In contrast, old growth was characterised by significantly more trees with hollows, large logs and mistletoe than plantings and both seedling regrowth and resprout regrowth (Appendix S3).

Discussion

We have compared the bird assemblages of different growth types in the temperate woodlands of south-eastern Australia using a dataset comprised of a large number of sites that have been surveyed numerous times over the past decade. Many past studies have examined the biodiversity values of particular kinds of temperate vegetation such as old growth woodland [11], regrowth [17] or plantings [19,20–22]. However, to the best of our collective knowledge, and to the knowledge of a range of colleagues whom we contacted about this paper (see Acknowledgments), no-one in Australia has completed a comparative study of the bird biota of different growth types of temperate woodland, including different kinds of passively regenerated regrowth as well as plantings. For example, while the biodiversity values of resprout regrowth vegetation have been comparatively well studied in Europe [2,67] and North America [68], a similar level of understanding is generally lacking in Australia (although see [1,13,17]).

Our investigation produced a number of key findings. Two key ones were: **(1)** Marked differences in the bird assemblages of resprout regrowth, seedling regrowth, old growth and plantings. **(2)** Many bird species of conservation concern being significantly ($P<0.05$) more often recorded in resprout regrowth, seedling regrowth or plantings but a paucity of species of conservation concern being recorded most often in old growth. We further discuss these findings in further detail in the remainder of the paper and conclude with some of the key implications for biodiversity conservation and woodland management.

3.1 Differences in bird responses to growth types

We found highly significant differences in the bird assemblages of old growth and regrowth temperate woodland and plantings. We also found that many species of conservation concern were associated with resprout regrowth and/or seedling regrowth. A good example was the Hooded Robin which was strongly associated with resprout regrowth woodland – a result consistent with another recent study [69]. We suggest that differences in bird occurrence among different growth types are likely to be strongly associated with growth-type differences in stand structural complexity [24,70] such as stem density and the prevalence of tree hollows and logs (see Appendix S3). For example, we found that old growth woodland supported significantly higher numbers of trees with hollows than the other growth types that we examined (Appendix S3). This may explain the significantly higher number of recordings of cavity-dependent birds like the Galah, Sulphur-crested Cockatoo, Eastern Rosella and Laughing Kookaburra in this growth type (Appendix S2). Similarly, the paucity of mistletoe in plantings (Appendix S3) may explain the rarity of species closely associated with this resource like the Mistletoebird (Appendix S2). High levels of stem density in plantings and seedling regrowth may explain the prevalence of species such as

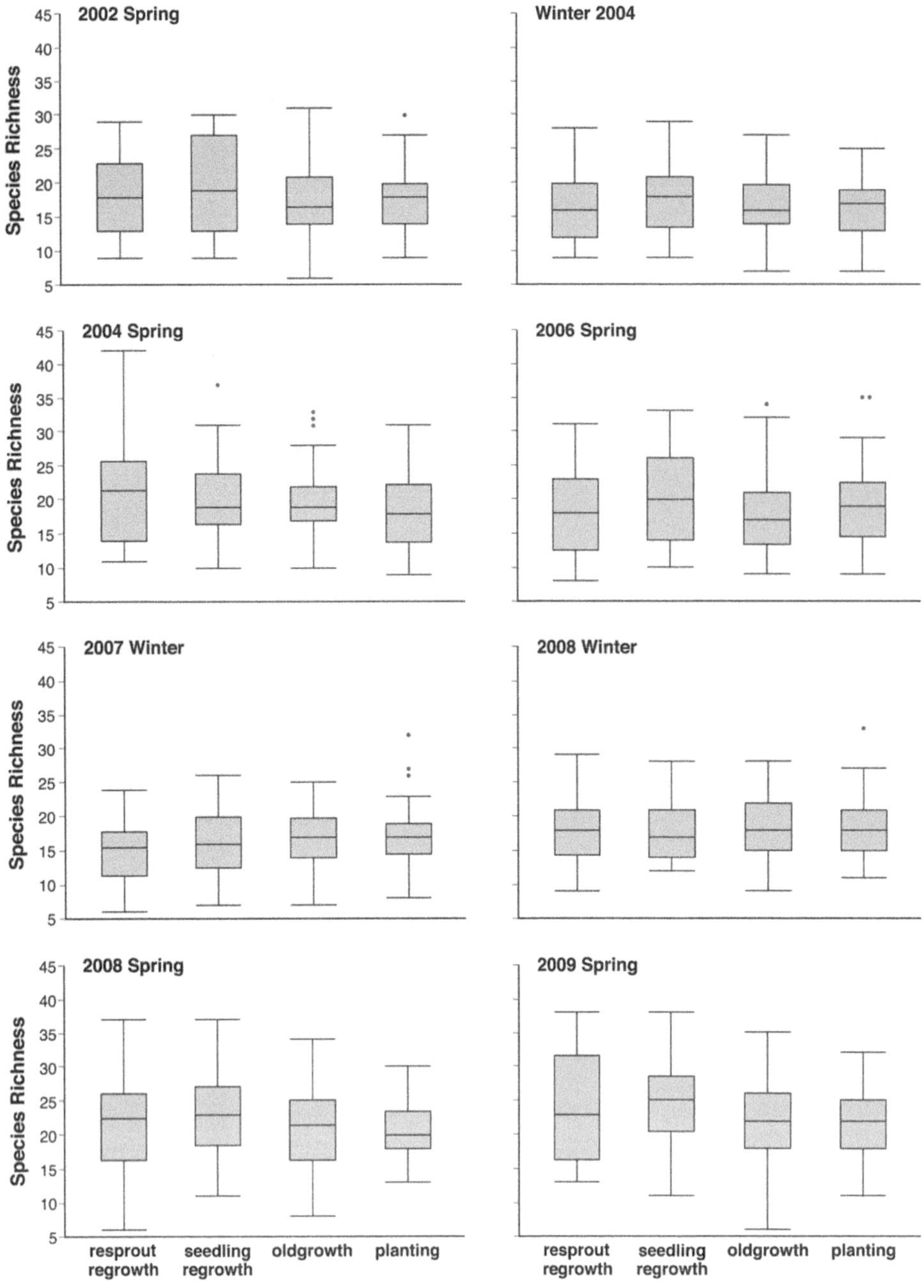

Figure 2. Average counts for growth types in a particular year and season.

the Speckled Warbler and the Eastern Yellow Robin, which are of conservation concern. However, multiple factors are likely to influence the occurrence of some bird species in particular growth types. For example, although logs were most abundant within old growth stands (Appendix S3), species such as the Brown Treecreeper, which are often closely associated with this resource [11,71,72], were more often recorded in resprout regrowth and seedling regrowth (Appendix S2). The Brown Treecreeper is an insectivore and differences in invertebrate assemblages between growth types [73] are also likely to influence many species of birds, particularly those that are insectivores or partially insectivorous.

One of our unexpected findings was that no species of conservation concern were found significantly more often in old growth (Appendix S2). The reasons for this result are unclear, but it is possible that this result is associated with the fact that old growth temperate woodland in these landscapes was often in a degraded condition [74,75] as a result of high-intensity grazing by livestock and weed invasion, thereby making these areas unsuitable for a range of species. In addition, the paucity of birds of conservation concern in old growth might be associated with the presence of the hyper-aggressive Noisy Miner (*Manorina melanocephala*) in these areas. This species often excludes smaller species of birds in temperate woodlands [61] and be common in degraded woodland, especially on high productivity areas [76] such as those subject to extensive clearing and over-grazing by domestic livestock. Conversely, the Noisy Miner can be uncommon or even absent from plantings characterized by a dense understory [20] as well as from densely stocked regrowth vegetation. Indeed,

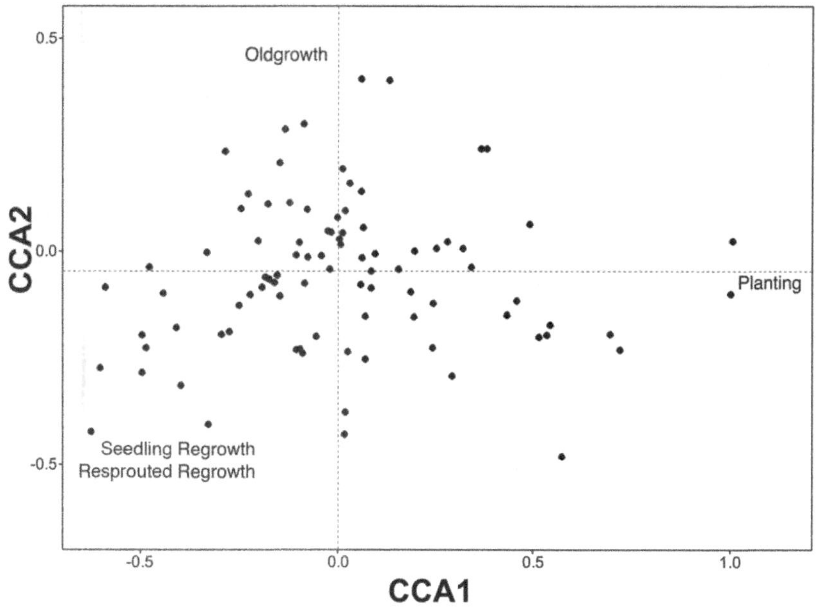

Figure 3. Biplot of the first two canonical axes showing species and growth types. Distances between species approximate the chi-squared distance between species distributions (see [59] for details of the approach used in data analyses).

our data (Appendix S2) indicated that the Noisy Miner occurred significantly less often in plantings than other growth types ($P<0.001$) and significantly more often in old growth than other growth types ($P<0.001$). Hence, the prevalence of the Noisy Miner in old growth woodland might be one of key the factors contributing to the paucity of birds of conservation concern in those areas.

3.2 The value of replanted areas

A key result of our study was quantification of the value of plantings for a number of bird species (Appendix S2). Perhaps most importantly, our extensive empirical data indicated that plantings supported several bird species of conservation concern. These included the Flame Robin and the Speckled Warbler. The Red-capped Robin also was found to be closely associated with plantings (Appendix S2), corroborating the findings of Major et al. [77,78], which showed that areas with a high stem density (e.g. dense shrubs and young saplings; see Appendix S3) were important nest sites for the species. Notably, other work on bird breeding success (that will be reported elsewhere; S Bond et al., unpublished data) has indicated that a number of species of conservation concern (e.g. Speckled Warbler, Southern Whiteface) successfully breed in planted areas. Our findings for birds of conservation concern therefore contrast markedly with those of other workers [22,79] who have suggested that the biodiversity value of plantings is marginal and such areas are suitable largely for taxa tolerant of disturbed (cleared) open-country landscapes. This, in turn, suggests that the biodiversity values of planted areas

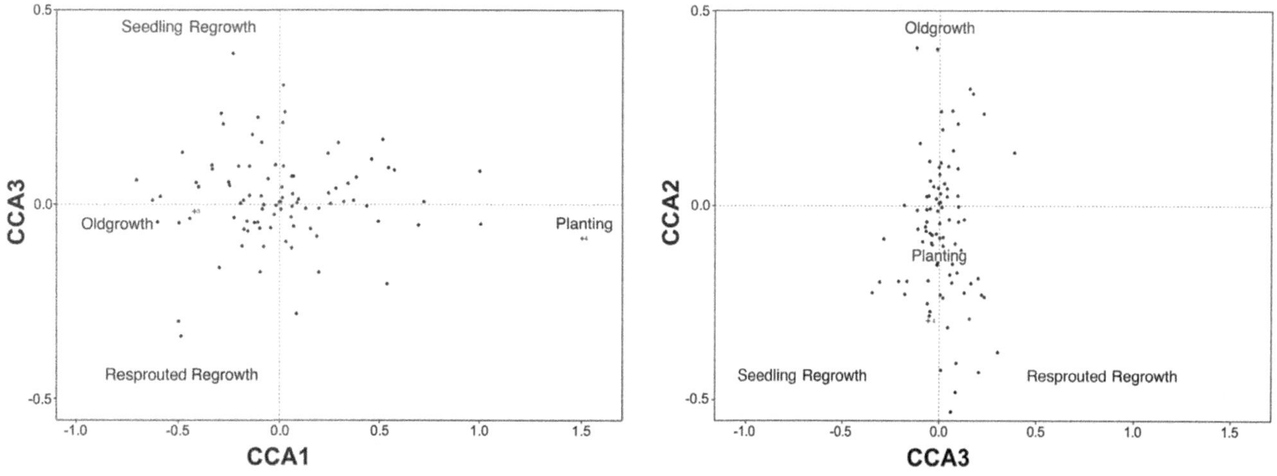

Figure 4. Correspondence analysis biplots of bird species and growth type. The diagrams are: (left) first versus third dimensions from correspondence analysis and (right) second and third dimensions from correspondence analysis. Distances between species approximate the chi-squared distance between species distributions (see [59] for details of the approach used in data analyses).

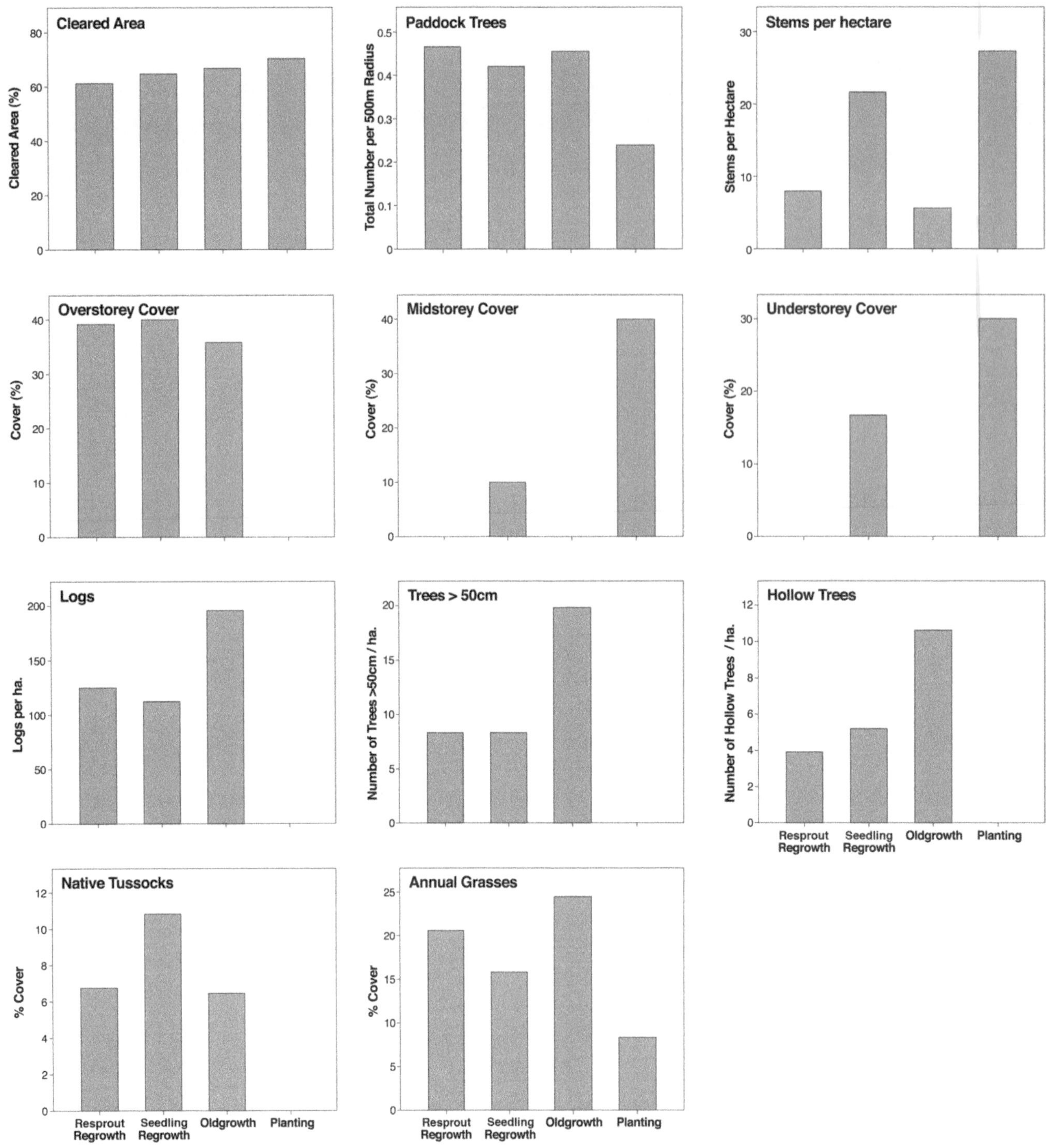

Figure 5. Median values for vegetation characteristics in different growth types.

may be regionally variable, possibly as a function of factors like the regional amount of potentially suitable habitat.

3.3 Management implications

Our study clearly indicates that different growth types of temperate woodland support different assemblages of native birds (Figures 3 and 4). Thus, our findings indicate that it would be inappropriate for bird conservation to clear old growth woodland and replace it with plantings – a conclusion similar to those we have recommended for the mammal and reptile conservation in semi-cleared agricultural landscapes in south-eastern Australia [9]. Rather, our findings suggest that a range of kinds of native vegetation encompassing old growth woodland, regrowth and plantings are likely to be required on a given area of farmland to support the diverse array of bird species that have the potential to occur in temperate woodland ecosystems.

Several recent studies have highlighted how a reduction in grazing pressure by domestic livestock can stimulate the development of seedling regrowth, e.g. [18,80,81]. However, it is not always possible for seedling regrowth to establish on some parts of farms such as where there has been long history of clearing and there is no seed bank [82]. In such places, revegetation requires deliberate planting [21]. Indeed, our work strongly indicated that plantings supported a bird assemblage that was significantly different from that of the other growth types (Figures 3 and 4). Different revegetation strategies that are likely to be required on some farms will, in turn, generate different kinds of woodland growth types and support different assemblages of birds.

Of the growth types of temperate woodland that we have examined on farms, areas of resprout and seedling regrowth are those of particular management and conservation concern. This is because while such areas support a range of bird species (including a number of conservation concern; see Appendix S2) and they are an obvious successional stage toward the development of old growth [18,83], temperate regrowth woodland can nevertheless be partially cleared under existing government legislation such as the *Native Vegetation Act 2003* in New South Wales [39] as well as in Queensland. However, the work we have reported here highlights the importance of regrowth vegetation for a wide range of bird species, including a number of species of conservation concern, as defined by Reid [38]. There is also considerable evidence of the important role that regrowth and other dense woody vegetation plays in key landscape functions such as reducing erosion and the maintenance of the integrity of soil biota [84].

Many landholders are deeply concerned about resprout and seedling regrowth vegetation on their farms, especially in terms of the impacts on grass growth and suitability for grazing as well as potential risks of unplanned fires. Consequently, there has been widespread clearing under the New South Wales *Native Vegetation Act 2003* as well as extensively developed proposals to actively thin resprout and seedling regrowth temperate woodland such as in the prescriptions for the Australian Government's Environmental Stewardship Program [44]. Moreover, several workers have discussed how high-stem-density vegetation may have negative impacts on thermal environments for groups such as reptiles at a local spatial scale [15,85]. Based on the results of this study, together with other work we (and others) have completed [11,15,85,86], we suggest that thinning must be guided by the habitat requirements and foraging patterns of particular species as well as the management objectives of a given area [87]. The appropriateness of particular management actions in a particular area will depend on which groups of species (e.g. reptiles versus birds), as well as which individual species within a given taxonomic group, have management priority in a given area. Differences in responses to different woodland growth types and woodland structural attributes therefore strongly suggest a need for spatial variation in management practices so that the different requirements of different species might be met in different parts of a given landscape. This is a common response in landscape approaches to biodiversity conservation: that is, "not to do the same thing everywhere" [88].

As outlined above, we have completed a major comparative study of bird responses to different revegetation growth types. Further key work that we have planned will include documenting longitudinal changes, such as those associated with woodland succession on bird assemblages. Presently the bird assemblages of plantings and resprout and seedling regrowth are markedly different (Figures 3 and 4) and therefore, a key question is whether the bird assemblages of these areas will eventually come to resemble one another or whether they will continue to be different and perhaps even diverge. This will be an important part of the continuation of sampling as highlighted by findings from other studies. For example, Wilkins et al. [89] evaluated the success of revegetation treatments on mined Australian coastal sand plains. They showed that revegetated areas were on a trajectory toward development of a new ecological community that differed significantly in species composition from pre-mining vegetation and adjacent un-mined vegetation. Thus, it will be important to maintain the work we have summarised in this paper as a true longitudinal study [90].

Acknowledgments

We thank Peter Vesk, Damon Oliver, Michaela Bowen, Clive McAlpine, Tara Martin, Sue McIntyre, Martine Maron and Andrew Bennett for suggestions about bird responses to regrowth vegetation. We thank Emmo Willinck, David Costello and Jack Chubb from the Murray CMA for their support and encouragement of this work. We also thank Clive Hilliker for his help with the figures in this paper. Comments by the Editor Matt Hayward, Hugh Possingham and an anonymous referee greatly improved an earlier version of this manuscript.

Author Contributions

Conceived and designed the experiments: DBL RMD MC DM SO PG. Performed the experiments: DBL RMD MC DM SO PG. Analyzed the data: DBL ARNM RMD MC DM SO PG. Wrote the paper: DBL ARNM RMD MC DM SO PG.

References

1. Bowen ME, McAlpine CA, House AP, Smith GC (2007) Regrowth forests on abandoned agricultural land: A review of their habitat values for recovering forest fauna. Biol Conserv 140: 273–296.
2. Broome A, Clarke S, Peace A, Parsons M (2011) The effect of coppice management of moth assemblages in an English woodland. Biodivers Conserv 20: 729–749.
3. Chazdon R, Peres CA, Dent DH, Sheil D, Lugo AE, et al. (2009) The potential for species conservation in tropical secondary forests. Conserv Biol 23: 1406–1417.
4. Dent DH, Wright SJ (2009) The future of tropical species in secondary forests: A quantitative review. Biol Conserv 142: 2833–2843.
5. Foster DR, Orwig DA (2006) Preemptive and salvage harvesting of New England forests: when doing nothing is a viable alternative. Conserv Biol 20: 959–970.
6. Odum EP (1959) Fundamentals of Ecology. Philadelphia, Pennslyvannia: W. B. Saunders Company. 546 p.
7. Barlow J, Mestre LAM, Gardner TA, Peres CA (2007) The value of primary, secondary and plantation forests for Amazonian birds. Biol Conserv 136: 212–231.
8. Bunnell FL, Dunsworth GB, eds. (2009) Forestry and Biodiversity: Learning How to Sustain Biodiversity in Managed Forests. Vancouver: UBC Press.
9. Cunningham RB, Lindenmayer DB, Crane M, Michael D, MacGregor C (2007) Reptile and arboreal marsupial response to replanted vegetation in agricultural landscapes. Ecol Appl 17: 609–619.
10. Felton A, Wood J, Felton AM, Lindenmayer DB (2008) Bird community responses to reduced-impact logging in a certified forestry concession in lowland Bolivia. Biol Conserv 141: 545–555.

11. Montague-Drake RM, Lindenmayer DB, Cunningham RB (2009) Factors affecting site occupancy by woodland bird species of conservation concern. Biol Conserv 142: 2896–2903.
12. Swanson ME, Franklin JF, Beschta RL, Crisafulli CM, DellaSala DA, et al. (2011) The forgotten stage of forest succession: early-successional ecosystems on forest sites. Front Ecol Environ 9: 117–125.
13. Woinarski J, Rankmore B, Hill B, Griffiths AD, Stewart A, et al. (2009) Fauna assemblages in regrowth vegetation in tropical open forests of the Northern Territory, Australia. Wildlife Res 36: 675–690.
14. Gibson L, Lee ML, Koh LP, Brook BW, Gardner TA, et al. (2011) Primary forests are irreplaceable for sustaining tropical biodiversity. Naturedoi:10.1038/nature10425.
15. Michael D, Cunningham RB, Lindenmayer DB (2011) Regrowth and revegetation in temperate Australia presents a conservation challenge for reptile fauna in agricultural landscapes. Biol Conserv 144: 407–415.
16. Posa MR, Sodhi NS (2006) Effects of anthropogenic land use on forest birds and butterflies in Subic Bay, Philippines. Biol Conserv 129: 256–270.
17. Bowen ME, McAlpine CA, Seabrook LM, House AP, Smith GC (2009) The age and amount of regrowth forest in fragmented brigalow landscapes are both important for woodland dependent birds. Biol Conserv 142: 3051–3059.
18. Fischer J, Stott J, Zerger A, Warren G, Sherren K, et al. (2009) Reversing a tree regeneration crisis in an endangered ecoregion. Proc Natl Acad Sci U S A 106: 10386–10391.
19. Kavanagh RP, Stanton MA, Herring MW (2007) Eucalypt plantings on farms benefit woodland birds in south-eastern Australia. Austral Ecol 32: 635–650.
20. Lindenmayer DB, Knight EJ, Crane MJ, Montague-Drake R, Michael DR, et al. (2010) What makes an effective restoration planting for woodland birds? Biol Conserv 143: 289–301.
21. Munro N, Lindenmayer DB (2011) Planting for Wildlife: A Practical Guide to Restoring Native Woodlands. Melbourne: CSIRO Publishing.
22. Selwood K, Mac Nally R, Thomson JR (2008) Native bird breeding in a chronosequence of revegetated sites. Oecologia 159: 435–446.
23. Paton DC, Prescott AM, Davies R, Heard LM (2000) The distribution, status and threats to temperate woodlands in South Australia. In: Hobbs RJ, Yates CJ, eds. Temperate Eucalypt Woodlands in Australia: Biology, Conservation, Management and Restoration. Chipping Norton: Surrey Beatty and Sons. pp 57–85.
24. Vesk P, Nolan R, Thomson JW, Dorrough JW, Mac Nally R (2008) Time lags in the provision of habitat resources through revegetation. Biol Conserv 141: 174–186.
25. Lindenmayer DB, Bennett AF, Hobbs RJ, eds. (2010) Temperate Woodland Conservation and Management. Melbourne: CSIRO Publishing.
26. Ford HA, Barrett GW, Saunders DA, Recher HF (2001) Why have birds in the woodlands of southern Australia declined? Biol Conserv 97: 71–88.
27. Ford HA, Walters JR, Cooper CB, Debus SJ, Doerr VA (2009) Extinction debt or habitat change? Ongoing losses of woodland birds in north-eastern New South Wales, Australia. Biol Conserv 142: 3182–3190.
28. Martin TG, Possingham HP (2005) Predicting the impact of livestock grazing on birds using foraging height data. J Appl Ecol 42: 400–408.
29. Sekercioglu CH (2006) Increasing awareness of avian ecological function. Trends Ecol Evol 21: 464–471.
30. MacArthur RH, MacArthur JW (1961) On bird species diversity. Ecology 42: 594–598.
31. Connell JH (1978) Diversity in tropical forests and coral reefs. Science 199: 1302–1310.
32. Franklin JF, Lindenmayer DB, MacMahon JA, McKee A, Magnuson J, et al. (2000) Threads of continuity. Conserv Biol in Practice 1: 8–16.
33. Egler FE (1954) Vegetation science concepts. I. Initial floristic composition – a factor in old field vegetation development. Vegetation 4: 412–417.
34. Fischer J, Lindenmayer DB, Montague-Drake R (2008) The role of landscape texture in conservation biogeography: a case study on birds in south–eastern Australia. Divers Distrib 14: 38–46.
35. Holling CS (1992) Cross-scale morphology, geometry, and dynamics of ecosystems. Ecol Monogr 62: 447–502.
36. Watson J, Freudenberger D, Paull D (2001) An assessment of the focal species approach for conserving birds in variegated landscapes in southeastern Australia. Conserv Biol 15: 1364–1373.
37. Johnson EA, Miyanishi K (2008) Testing the assumptions of chronosequences in succession. Ecol Lett 11: 419–431.
38. Reid J (1999) Threatened and declining birds in the New South Wales sheep-wheat belt. I. Diagnosis, characteristics and management. Canberra: Unpublished report to NSW National Parks and Wildlife Service, CSIRO Wildlife and Ecology.
39. NSW Office of Environment and Heritage (2011) Native vegetation management: Clearing. Available: http://www.environment.nsw.gov.au/vegetation/clearing.htm. Accessed 26 Sept 2011.
40. Cribb J (2010) The Coming Famine. The Global Food Crisis and What We Can Do to Avoid It. Melbourne and Oakland: CSIRO Publishing and University of California Press.
41. Millennium Ecosystem Assessment (2005) Ecosystems and Human Well-Being: Biodiversity Synthesis. Washington, D.C.: World Resources Institute.
42. Chappell MJ, Lavalle LA (2011) Food security and biodiversity: can we have both? Agr Hum Values 28: 3–26.
43. Department of Environment Climate Change and Water (2009) NSW Annual Report on Native Vegetation 2009. Sydney, Australia: Department of Environment Climate Change and Water.
44. Commonwealth of Australia (2009) Environmental Stewardship Strategic Framework. Canberra: Commonwealth of Australia.
45. Hobbs RJ, Yates CJ, eds. (2000) Temperate Eucalypt Woodlands in Australia: Biology, Conservation, Management and Restoration. Chipping Norton: Surrey Beatty and Sons.
46. Keith DA (2004) Ocean shores to desert dunes. The native vegetation of New South Wales and the ACT. Sydney: NSW Department of Environment and Conservation.
47. Pyke GH, Recher HF (1983) Censusing Australian birds: a summary of procedures and a scheme for standardisation of data presentation and storage. In: Davies SJ, ed. Methods of Censusing Birds in Australia. Perth: Proceedings of a symposium organised by the Zoology section of the ANZAAS and the Western Australian Group of the Royal Australasian Ornithologists Union. Department of Conservation and Environment. pp 55–63.
48. Lindenmayer DB, Wood JT, MacGregor C (2009) Do observer differences in bird detection affect inferences from large-scale ecological studies? Emu 109: 100–106.
49. Field SA, Tyre AJ, Possingham HP (2002) Estimating bird species richness: how should repeat surveys be organized in time? Austral Ecol 27: 624–629.
50. Cunningham RB, Lindenmayer DB, McGregor C, Crane M, Michael D (2008) The combined effects of remnant vegetation and replanted vegetation on farmland birds. Conserv Biol 22: 742–752.
51. Lindenmayer DB, Cunningham RB, MacGregor C, Crane M, Michael D, et al. (2008) Temporal changes in vertebrates during landscape transformation: a large-scale "natural experiment". Ecol Monogr 78: 567–590.
52. Lindenmayer DB, Wood JT, Cunningham RB, MacGregor C, Crane M, et al. (2008) Testing hypotheses associated with bird responses to wildfire. Ecol Appl 18: 1967–1983.
53. Tyre AJ, Tenhumberg B, Field SA, Niejalke D, Parris K, et al. (2003) Improving precision and reducing bias in biological surveys: estimating falsenegative error rates. Ecol Appl 13: 1790–1801.
54. ter Braak CJF (1986) Canonical Correspondence Analysis: A new eigenvector technique for multivariate direct gradient analysis. Ecology 67: 1167–1179.
55. Legendre P, Legendre L (1998) Numerical Ecology. Amsterdam, The Netherlands: Elsevier Science.
56. Palmer MW (1993) Putting things in even better order: The advantages of Canonical Correspondence Analysis. Ecology 74: 2215–2230.
57. Titeux N, Dufrêne M, Jacob J, Paquay M, Defourny P (2004) Multivariate analysis of a fine-scale breeding bird atlas using a geographical information system and partial canonical correspondence analysis: environmental and spatial effects. J Biogeogr 31: 1841–1856.
58. Zuur F, Ieno EN, Smith GM (2007) Analysing Ecological Data. New York: Springer Science.
59. Lepš J, Šmilauer P (2003) Multivariate Analysis of Ecological Data using CANOCO. Cambridge, UK: Cambridge University Press.
60. MacFaden SW, Capen DE (2002) Avian habitat relationships at multiple scales in a New England forest. Forest Sci 48: 243–253.
61. Howes AL, Maron M (2009) Interspecific competition and conservation management of continuous subtropical woodlands. Wildlife Res 36: 617–626.
62. Agresti A (2002) Categorical Data Analysis. New York: Wiley.
63. Nelson PR, Wludyka PS, Copeland KAF (2005) Analysis of Means: A Graphical Method for Comparing Means, Rates, and Proportions. Philadelphia, Pennslyvannia: SIAM Press.
64. Dwass M (1960) Some k-sample rank-order tests. In: Olkin I, Ghurye SG, Hoeffding H, Madow WG, Mann HB, eds. Contributions to Probability and Statistics Stanford: Stanford University Press. pp 198–202.
65. Steel RGD (1960) A rank sum test for comparing all pairs of treatments. Technometrics 2: 197–207.
66. Steel RGD (1961) Some rank sum multiple comparison tests. Biometrics 17: 539–552.
67. Fuller RJ, Green GH (1998) Effects of woodland structure on breeding bird populations in stands of coppiced lime (Tilia cordata) in western England over a 10-year period. Forestry 71: 199–218.
68. Dhondt AA, Wrege PH, Cerretani J, Sydenstricker KV (2007) Avian species richness and reproduction in short-rotation coppice habitat in central and western New York. Bird Study 54: 12–22.
69. Priday SD (2010) Beyond the "woody remnant" paradigm in conservation of woodland birds: habitat requirements of the Hooded Robin (Melanodryas cucullata cucullata). Emu 110: 118–124.
70. Lindenmayer DB, Franklin JF (2002) Conserving Forest Biodiversity: A Comprehensive Multiscaled Approach. Washington DC: Island Press.
71. Doerr VA, Doerr ED, Jenkins SH (2006) Habitats selection in two Australasian treecreepers: what cues should they use? Emu 106: 93–103.
72. Laven NH, Mac Nally R (1997) Association of birds with fallen timber in Box-Ironbark forest of central Victoria. Corella 22: 55–60.
73. Gibb H, Cunningham SA (2010) Revegetation of farmland restores function and composition of epigaeic beetle assemblages. Biol Conserv 143: 677–687.
74. Gibbons P, Briggs SV, Murphy DY, Lindenmayer DB, McElhinny C, et al. (2010) Benchmark stem densities for forests and woodlands in south-eastern Australia under conditions of relatively little modification by humans since European settlement. Forest Ecol Manage 260: 2125–2133.

75. Prober S, Thiele KR (1995) Conservation of the Grassy White Box Woodlands: Relative contributions of size and disturbance to floristic composition and diversity of remnants. Aust J Bot 43: 349–366.
76. Montague-Drake R, Lindenmayer DB, Cunningham RB, Stein J (2011) A reverse keystone species affects the landscape distribution of woodland avifauna: a case study using the Noisy Miner (Manorina melanocephala) and other Australian birds. Landscape Ecol 26: 1383–1394.
77. Major RE, Christie FJ, Gowing G (1999) Age structure and density of Red-capped Robin populations vary with habitat size and shape. J Appl Ecol 36: 901–908.
78. Major RE, Christie FJ, Gowing G (2001) Influence of remnant and landscape attributes on Australian woodland bird communities. Biol Conserv 102: 47–66.
79. Mac Nally R, Dr Vries L, Thomson JR (2010) Are replanted floodplain forests in southeastern Australia providing bird biodiversity benefits. Restor Ecol 18: 85–94.
80. Briggs SV, Taws NM, Seddon JA, Vanzella B (2008) Condition of fenced and unfenced remnant vegetation in inland catchments in south-eastern Australia. Aust J Bot 56: 590–599.
81. Weinberg A, Gibbons P, Briggs SV, Bonser S (2010) The extent and pattern of *Eucalyptus* regeneration in an agricultural landscape. Biol Conserv 144: 227–233.
82. Dorrough J, Moxham C (2005) Eucalypt establishment in agricultural landscapes and implications for landscape-scale restoration. Biol Conserv 123: 55–66.
83. Manning AD, Fischer J, Lindenmayer DB (2006) Scattered trees are keystone structures – implications for conservation. Biol Conserv 132: 311–321.
84. Eldridge DJ, Bowker MA, Maestre FT, Roger E, Reynolds JF, et al. (2011) Impacts of shrub encroachment on ecosystem structure and functioning: towards a global synthesis. Ecol Lett 14: 709–722.
85. Pike DA, Webb JK, Shine R (2011) Removing forest canopy cover restores a reptile assemblage. Ecol Appl 21: 274–280.
86. Antos MJ, Bennett AF, White JG (2008) Where exactly do ground-foraging woodland birds forage? Foraging sites and microhabitat selection in tenperate woodlands of southern Australia. Emu 108: 201–211.
87. Oliver DL (2011) To thin or not to thin? The value of dense native woody vegetation for birds. Woodland Wanderings 8: 8–16.
88. Lindenmayer DB, Hunter M (2010) Some guiding concepts for conservation biology. Conserv Biol 24: 1459–1468.
89. Wilkins S, Keith DA, Adam P (2003) Measuring success: Evaluating the restoration of a grassy eucalypt woodland on the Cumberland Plain, Sydney, Australia. Restor Ecol 11: 489–503.
90. Lindenmayer DB, Likens GE (2010) Effective Ecological Monitoring. Melbourne and London: CSIRO Publishing and Earthscan.

17

Local Crop Planting Systems Enhance Insecticide-Mediated Displacement of Two Invasive Leafminer Fly

Yulin Gao[1]*, Stuart R. Reitz[2], Qingbo Wei[1], Wenyan Yu[1], Zhi Zhang[1], Zhongren Lei[1]*

1 State Key Laboratory for Biology of Plant Diseases and Insect Pests, Institute of Plant Protection, Chinese Academy of Agricultural Sciences, Beijing, P. R. China,
2 Department of Crop and Soil Sciences, Malheur County Extension, Oregon State University, Ontario, Oregon, United States of America

Abstract

Liriomyza sativae and *L. trifolii* are highly invasive leafminer pests of vegetable crops that have invaded southern China in recent years. *Liriomyza sativae* was the first of these species to invade China, but it is now being displaced by *L. trifolii*. The rate and extent of this displacement vary across southern China. In Hainan, monocultures of highly valuable cowpea are planted and treated extensively with insecticides in attempts to control leafminer damage. In Guangdong, cowpea fields are interspersed with other less valuable crops, such as towel gourd (*Luffa cylindrica*), which receive significantly fewer insecticide applications than cowpea. To determine how differences in cropping systems influence the *Liriomyza* species composition, we conducted field trials in 2011 and 2012 in Guangdong where both species were present. We replicated conditions in Hainan by planting cowpea monocultures that were isolated from other agricultural fields, and we replicated conditions in Guangdong by planting cowpea in a mixed crop environment with towel gourd planted in neighboring plots. We then compared leafminer populations in cowpea treated with the insecticide avermectin and untreated cowpea. We also monitored leafminer populations in the untreated towel gourd. Untreated cowpea and towel gourd had comparatively low proportions of *L. trifolii*, which remained relatively stable over the course of each season. Avermectin applications led to increases in the proportions of *L. trifolii*, and after three weekly applications populations were >95% *L. trifolii* in both crop systems. However, the rate of change and persistence of *L. trifolii* in the mixed crop system were less than in the monocrop. These results indicate that *L. trifolii* is much less susceptible to avermectin than is *L. sativae*. Further, *L. sativae* was able to persist in the untreated towel gourd, which probably enabled it to recolonize treated cowpea.

Editor: Raul Narciso Carvalho Guedes, Federal University of Viçosa, Brazil

Funding: This work was supported in part by the National Key Basic Research Program (973 Projects Grant 2009CB119004) and the China Agriculture Research System (CARS-25-B-07). The funders had no role in study design, data collection and analysis, decision to publish, or preparation of the manuscript.

Competing Interests: The authors have declared that no competing interests exist.

* E-mail: ylgao@ippcaas.cn (YG); leizr@sina.com (ZL)

Introduction

Leafmining flies in the genus *Liriomyza* Mik (Diptera: Agromyzidae) are among the most important invasive pests of vegetable crops throughout the world. Damage occurs primarily as larvae feed within host plant foliage [1]. The most widespread and successful invasive *Liriomyza* species are *L. sativae* Blanchard, *L. trifolii* (Burgess) and *L. huidobrensis* (Blanchard) [2], and these have recently emerged as particularly damaging pests in China, where they have had a complex history of invasion and interspecific interactions [3,4]. *Liriomyza sativae* was found first in Hainan Province in October 1993, and then spread throughout China within a few years [5]. By 2005, the distribution of *L. sativae* encompassed more than 30 provinces, and it has since been regarded as an established exotic species in China. *Liriomyza trifolii* is a more recent invasive that was first recorded in Guangdong and Hainan Provinces in 2005 and 2006, respectively [6].

After their respective invasions, both of these highly polyphagous species became the most important pests of cowpea (*Vigna uniguilata*), towel gourd (*Luffa cylindrica*), cucumber (*Cucumis sativus*), and many other vegetable crops in southern China. However since its establishment, *L. trifolii* has begun to displace *L. sativae* in certain areas across southern China [3,7]. Gao et al. [7] proposed that *L. trifolii* in southern China are less susceptible to insecticides, specifically avermectin and cyromazine, than is *L. sativae*, and this difference in susceptibility is a major factor in the displacement of *L. sativae* by *L. trifolii*.

Although the displacement of *L. sativae* by *L. trifolii* has been rapid in certain areas of southern China, such as on Hainan Island, in other regions the displacement has not been as rapid or drastic. Displacements of one species by another are likely the outcome of interactions among multiple mechanisms [8]. Therefore, geographic differences in the demographic composition of *Liriomyza* populations could be a result of differences in cropping systems and interactions between cropping systems and insecticide use patterns.

In southern China, mixed plantings of cowpea, towel gourd, and cucumber are common, but the relative abundance of cowpea to non-cowpea crops is highly variable across regions. For example, the period from December to April corresponds to the time of when *Liriomyza* populations peak in Hainan Province, and during this time, cowpea is intensively planted so that it comprises 70–90% of the total acreage of *Liriomyza* susceptible crops (i.e. the proportion of cowpea, calculated by dividing the acreage of cowpea by the total acreage of the main host crops of *Liriomyza* in a region). Because cowpea is the most economically valuable vegetable crop in Hainan, farmers attempt to control leafminer populations through intensive insecticide use. Grower surveys have shown that avermectin and cyromazine account for >95% of the total insecticides used against leafminers [3]. Of these two, avermectins are the most commonly used insecticides in southern China.

In contrast to the predominance of cowpea in Hainan, in Guangdong province, mixed plantings of cowpea, towel gourd, cucumber and other *Liriomyza*-susceptible crops are common during the peak time for *Liriomyza* populations there, which is from April to June. In particular, mixed plantings of cowpea and towel gourd are a typical cropping system in Guangdong. On average, cowpea comprises about 15% of the acreage of *Liriomyza* hosts in Guangdong, and towel gourd comprises about 40% of the acreage. Also in contrast to Hainan Province, overall insecticides use is not as intensive in Guangdong Province [9]. With mixed cropping systems, farmers do not treat all crops with the same amount and frequency of insecticides. Less valuable crops, such as towel gourd, tend to be treated less than the high-value cowpeas [10]. Therefore, some non-cowpea host crops, such as towel gourd, may serve as refuge hosts for *Liriomyza sativae* when cowpea is intensively treated with insecticides.

To characterize how insecticide use and cropping systems interact to affect the demographic composition of the *Liriomyza* species complex in southern China, we conducted a series of surveys to monitor species composition across southern China and field experiments in different cropping systems. We were specifically interested in testing the hypotheses that use of the insecticide avermectin would increase the proportion of *L. trifolii* in populations, but that the effect of the insecticide would be affected by cropping system.

Results

Field surveys in Guangdong and Hainan Provinces

In the 2011 survey, we collected 28 samples of leafminers (>30 individuals per sample) from cowpea crops in the eight locations of Guangdong Province. Almost all samples contained both species of leafminers. Overall, *L. sativae* comprised 58% of the total *Liriomyza* population (n = 1055), with the remaining 42% all being *L. trifolii*. Although these results show clearly that both *L. trifolii* and *L. sativae* were prevalent in Guangdong Province, *L. sativae* made up a significantly greater proportion of the *Liriomyza* complex than did *L. trifolii* ($x^2 = 27.1$, df = 1, P<0.0001).

From January through March of 2011, we collected 24 samples of leafminers (>30 individuals per sample) from cowpea crops in eight locations of Hainan Province. The survey results indicated that almost all of the leafminers (n = 804) were *L. trifolii* (99.8%). No *L. sativae* individuals were found, and only two *L. huidobrensis* individuals were found. These results show clearly that *L. trifolii* has replaced *L. sativae* as the dominant *Liriomyza* species on this cowpea-intensively planting system ($x^2 = 798.0$, df = 1, P<0.0001).

Field population experiments in 2011

Field trials were conducted in a region of Guangdong Province where both *L. sativae* and *L. trifolii* were present. In 2011 trials, the use of avermectin led to significant increases in the proportion of *L. trifolii* in cowpea (Figure 1A and 1B). In the mixed plantings of cowpea and towel gourd, there was a significant treatment x time interaction demonstrating that the composition of populations changed with continued insecticide applications (F = 7.80 df = 3, 16, P = 0.002). On the sample date before insecticide treatments began (day 1 of the experiment), the mean percentage of *L. trifolii* in the cowpea plots to be treated with avermectin was 46.5±5.0%, with all other leafminers being *L. sativae*. In the sample collections following the use of avermectin (days 7, 13, 19 of the experiment) the proportion of *L. trifolii* increased to 95.1±2.4% on day 19 (test for linear increase: t = 6.06, df = 16, P<0.0001). In the cowpea plots not treated with avermectin, the initial percentage of *L. trifolii* was 40.2±4.9%, with all other leafminers being *L. sativae*. This percentage was not significantly different from that in the avermectin treated plots on day 1 (P = 0.38). However, there was no significant change in the proportion of *L. trifolii* over time in the untreated cowpea plots (t = 1.93, P = 0.07). The proportion of *L. trifolii* in the towel gourd plots actually declined significantly over time, going from 34.7±5.5% of the population at day 1 to 14.8±3.0% at day 49 of the experiment (F = 5.20, df = 8, 18, P = 0.0018).

In the cowpea monoculture plantings, there also was a significant treatment x time interaction (F = 9.95, df = 3, 16, P<0.0001), indicating that the proportions of *L. trifolii* in populations were affected by insecticide treatments. On day 1 of this experiment (before insecticide applications began), there was no significant difference in the composition of the leafminer populations between the insecticide and untreated plots (P = 0.55). In the avermectin treatment, 42.8±4.2% of the leafminers were *L.*

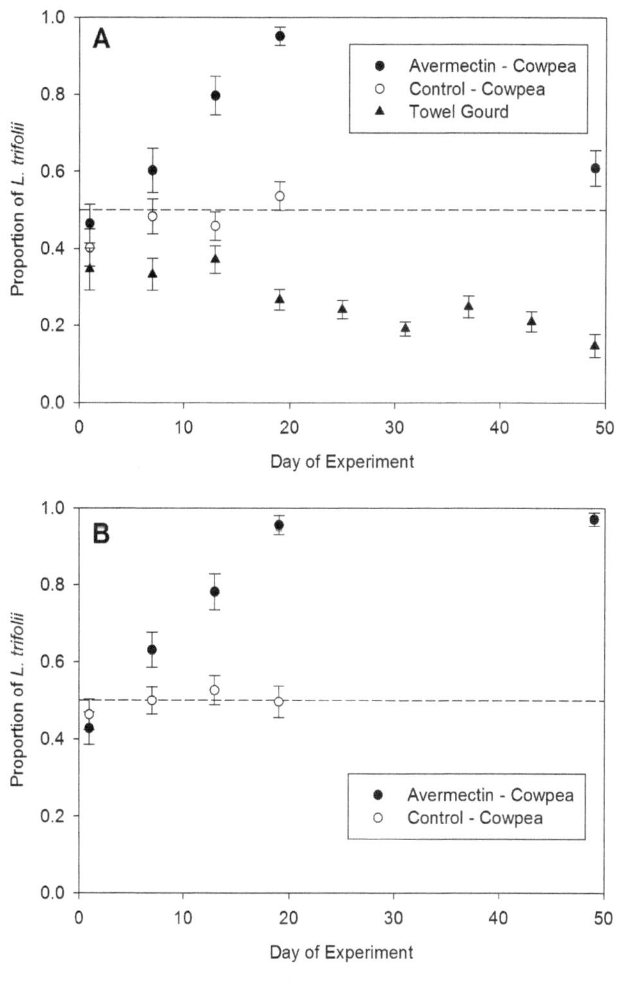

Figure 1. *Liriomyza* species composition during 2011. (A) mixed cropping systems of avermectin-treated and untreated cowpea (*Vigna uniguilata*), and adjoining plantings of towel gourd (*Luffa cylindrica*), and (B) plantings of cowpea that were treated with avermectin or were untreated controls. Avermectin treatments were made on days 1, 7, and 13 of the experiment following *Liriomyza* sample collections. Data points represent means ± SEM.

trifolii, and 46.3±3.9% of the leafminers were *L. trifolii* in the untreated plots. There was no significant change in the proportion of *L. trifolii* over time in the untreated plots (test for linear trend: t = 0.71, df = 16, P = 0.49). In contrast, there was a significant linear increase in the proportion of *L. trifolii* over days 1–19 in the plots treated with avermectin (t = 5.78, df = 16, P<0.0001).

The last avermectin application was made after samples were collected on day 13 of the experiments. In the mixed culture plantings, the proportion of *L. trifolii* in the avermectin treatment declined significantly from day 19 to day 49, when the population was only 60.9±4.7% *L. trifolii* (Figure 1A; t = 4.61, df = 4, P = 0.01). However, in the cowpea monoculture, the proportion of *L. trifolii* remained at very high levels (97. 1±1.7%) after insecticide applications were discontinued (Figure 1B; comparison of proportions between day 19 and day 49: t = −0.51, df = 4, P = 0.64).

Field population experiments in 2012

The 2012 trials gave similar results to the 2011 trials; however, the initial proportions of *L. trifolii* were higher in 2012 than in 2011. The overall percentage of *L. trifolii* in cowpeas at the beginning of the 2012 experiments was 54.9±4.0% in the mixed plantings and 62.6±4.1% in the cowpea monoculture compared with 43.3±3.5% and 44.5±2.9%, respectively at the beginning of the 2011 experiments. The overall percentage of *L. trifolii* in towel gourd plots was 40.0±6.9% in 2012 compared with 34.7±5.5% in 2011.

In 2012, the use of avermectin led to significant increases over time in the proportions of *L. trifolii* in cowpea (Figures 2A and 2B). In mixed plantings of cowpea and towel gourd, there was a significant treatment x time interaction demonstrating that the composition of populations changed with continued insecticide applications (F = 9.23, df = 3, 16, P = 0.0009). There was a significant linear increase in the mean percentage of *L. trifolii* in the cowpea plots that were treated with avermectin, as the percentage in these plot went from 54.2±5.9% on day 1 to 96.6±2.4% on day 19 (test for linear increase: t = 4.74, df = 16, P = 0.0002). In the untreated cowpea plots, the initial percentage of *L. trifolii* was 55.6±5.9%, with all other leafminers being *L. sativae*. This percentage was not significantly different from that in the avermectin treated plots on day 1 (P = 0.87). However, there was no significant change in the proportion of *L. trifolii* over time in the untreated cowpea plots (Figure 3; test for linear change in proportion of *L. trifolii*: t = −1.14, P = 0.27). The percentage of *L. trifolii* in the towel gourd plots ranged from 30–40% over the entire trial (F = 0.92, df = 8, 18, P = 0.52).

In the cowpea monoculture plantings, there also was a significant treatment x time interaction (F = 9.24, df = 3, 16, P = 0.0009), indicating that the proportions of *L. trifolii* in populations were affected by insecticide treatments, as occurred in the mixed crop experiment. On day 1 of this experiment (before insecticide applications began), there was no significant difference in the composition of the leafminer populations between the insecticide-treated and untreated plots (P = 0.59). The proportion of *L. trifolii* in the untreated plots remained relatively constant from day 1 to day 19, as there was no significant linear trend in this treatment (t = 0.67, df = 16, P = 0.51). In contrast, there was a significant linear increase in the proportion of *L. trifolii* over days 1–19 in the plots treated with avermectin (t = 4.15, df = 16, P = 0.0007).

The last avermectin application was made after samples were collected on day 13 of the experiments. As occurred in 2011, in the mixed culture plantings, the proportion of *L. trifolii* in 2012 in the avermectin treatment declined significantly from day 19 to day 49, when the population was only 57.6±3.2% *L. trifolii* (Figure 2A; t = 4.16, df = 4, P = 0.016). However, in the cowpea monoculture, the proportion of *L. trifolii* remained at very high levels (98. 1±1.3%) after insecticide applications were discontinued (Figure 3B; comparison of proportions between day 19 and day 49: t = 0.01, df = 4, P = 0.99).

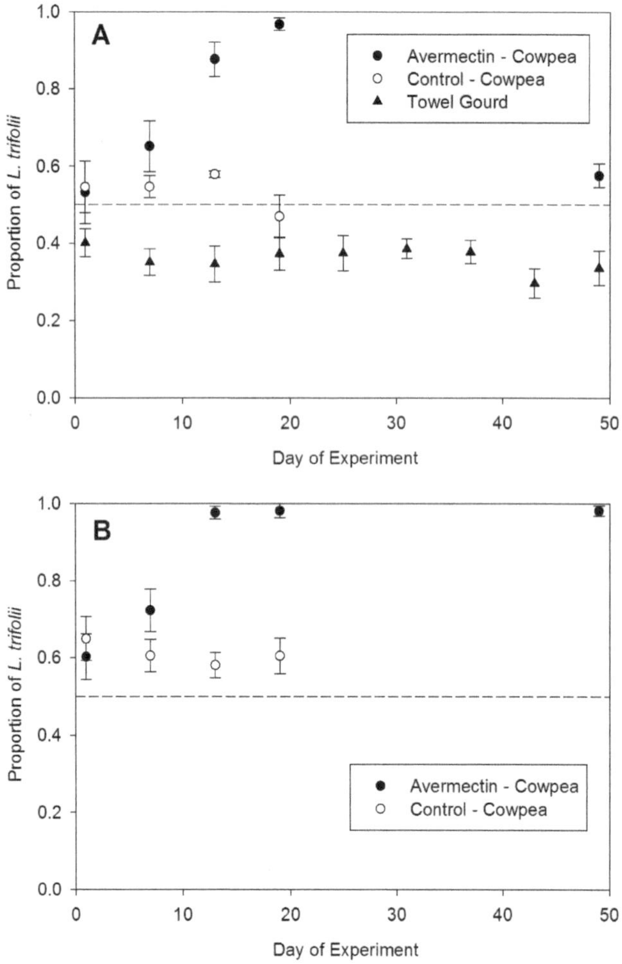

Figure 2. *Liriomyza* species composition during 2012. (A) mixed cropping systems of avermectin-treated and untreated cowpea, and adjoining plantings of towel gourd (*Luffa cylindrica*), and (B) plantings of cowpea that were treated with avermectin or were untreated controls. Avermectin treatments were made on days 1, 7, and 13 of the experiment following *Liriomyza* sample collections. Data points represent means ± SEM.

Discussion

Our surveys showed that *L. trifolii* has become the predominant leafminer species within cowpea crops in Hainan Province and that *L. sativae* is rare, if present at all. Although *L. trifolii* has been established for a longer time in Guangdong Province than in Hainan, it has not been as successful in invading Guangdong and displacing *L. sativae*. *Liriomyza sativae* was still the predominant species in Guangdong where cropping systems are more diversified and complex, and insecticide use is less intensive than in Hainan. The results of our field experiments demonstrate that insecticide use can rapidly and dramatically alter the species composition of *L. sativae* and *L. trifolii*, and therefore, insecticide use patterns may be responsible for shifting the species balance in favor of *L. trifolii*.

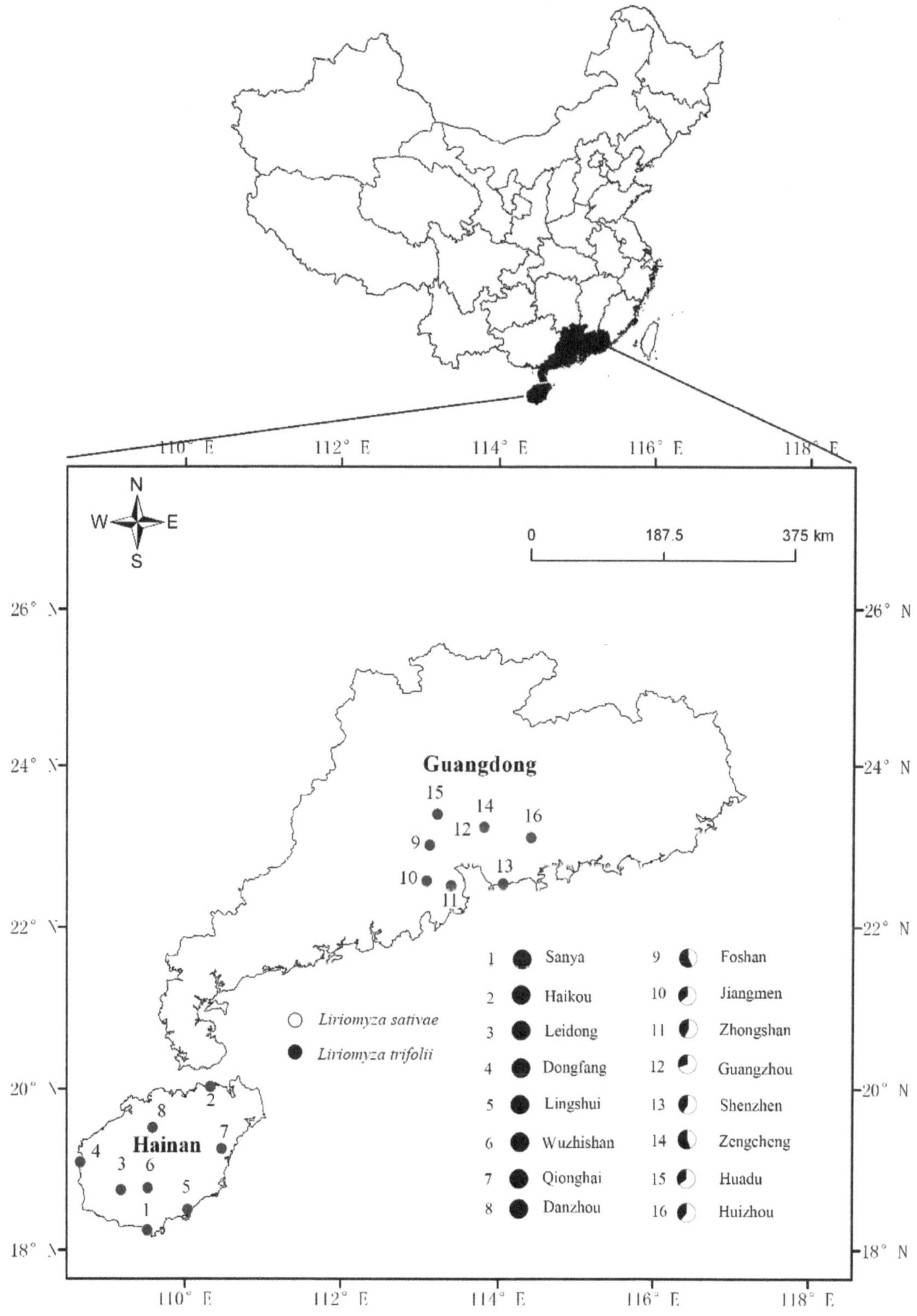

Figure 3. Map showing field sites in Hainan Province and Guangdong Province where cowpeas (*Vigna uniguilata*) were surveyed for *Liriomyza* species during 2011.

However, the continued predominance of *L. sativae* in Guangdong suggests that environmental differences allow it persist there.

In our field trials, we sought to replicate cropping environments similar to those in Hainan and Guangdong to assess how insecticide treatments and cropping systems may affect the demographics of these two invasive pest species. To this end, we established field sites with a cowpea monoculture, as is typically found in Hainan, and a more diversified cropping system that included cowpea and towel gourd, as is typically found in Guangdong. Conducting the studies in a location with significant abundance of both species allowed us to better characterize interspecific dynamics and more easily track demographic changes than at an extremely early or late stage in the process when one or the other species may be relatively rare.

In the absence of insecticide pressures, the two *Liriomyza* species coexisted over the two years of our study. In both years, there were greater proportions of *L. sativae* than *L. trifolii* in the towel gourd plantings. This result is not surprising as the species show distinct host plant affinities [1,11,12], and *L. sativae* often predominates over *L. trifolii* in cucurbit crops [13]. However, *L. trifolii* comprised a greater proportion of the population in the towel gourd plantings in 2012 (x±95% confidence interval: 36.1±2.5% for the season) than it did in 2011 (26.4±4.6%), suggesting that the overall population composition at the study sites had changed over time.

The most dramatic changes in species composition that we observed were results of avermectin treatments. The leafminer complex rapidly shifted from approximately equal proportions of the two species to overwhelming proportions of *L. trifolii* within three weeks of avermectin applications. However, in the mixed crop system, *L. sativae* began to recolonize the insecticide-treated cowpea plots within 30 days after insecticide treatments had ceased, but recolonization of insecticide-treated plots by *L. sativae* took longer in the monocrop cowpea system.

Recolonization by *L. sativae* was also less successful following the 2011 experiment. The proportions of *L. trifolii* in cowpea plantings were higher at the beginning of the 2012 season than at the beginning of the 2011 season in both the mixed crop and monocrop systems. The change from the beginning of the first season to the beginning of the second season was greater for the monocrop cowpea system than for the mixed cropping system. These results provide evidence that *L. trifolii* is less susceptible than *L. sativae* to avermectin, and that differential effects of insecticides can drive significant long term changes in populations. Palumbo et al. [14] also argued for differential susceptibility driving demographic shifts in leafminers. They observed that *L. sativae* was the most common leafminer in experimental plantings of lettuce, but that *L. trifolii* predominated in commercial lettuce fields that were subject to intensive insecticide applications.

Gao et al. [7] specifically compared the differential effects of avermectin on *L. sativae* and *L. trifolii*, using populations collected from Hainan Province. They reported that *L. trifolii* was significantly more tolerant to avermectin than was *L. sativae*. Other comparative studies have generally shown that *L. trifolii* is less susceptible to insecticides than is *L. sativae* [14–16]. An exception to these general trends is a study conducted with Japanese populations of these species that found either no significant difference in susceptibilities, or lower susceptibilities of *L. sativae*, to 25 different insecticides [17].

Japan is one region where *L. sativae* has successfully displaced *L. trifolii*. This different displacement pattern has been attributed to the greater fecundity of *L. sativae* compared with *L. trifolii* and a lower susceptibility to a common parasitoid, in addition to the differences in insecticide susceptibility [3,18].

Multiple mechanisms are likely to contribute to the displacement of one species by another, and it is difficult to separate effects of different mechanisms in field experiments. Therefore, we do not preclude differential parasitism as a contributing factor. It has been well documented that overuse of insecticides induces outbreaks of *Liriomyza* spp. by eliminating parasitoid populations that can suppress *Liriomyza* [19–21]. In general, avermectin has proven to be toxic to parasitoids of *Liriomyza* spp., but impacts of its use on leafminer pest management have been equivocal [22,23]. However, differential parasitism would explain the increased predominance only if *L. trifolii* were more susceptible to parasitization than *L. sativae*. Surveys show that *L. sativae* has a rich and diverse parasitoid community in China [24] and other studies have shown that *L. sativae* may be less susceptible to some common parasitoids than is *L. trifolii* [18,25].

Regardless of the specific underlying mechanism(s), insecticide use can lead to significant changes in the *Liriomyza* species complex in China. Our results show that these changes can be influenced by the general cropping system. *Liriomyza sativae* is more susceptible to displacement in intensively managed cowpea, especially where it is grown as a monoculture. It appears to be more persistent in more complex systems with hosts where it has reproductive advantages over *L. trifolii*. For the situation in southern China, maintaining untreated hosts that are better suited for *L. sativae* may increase the likelihood of its persistence in the presence of *L. trifolii*. Given the generally greater pest problems experienced with *L. trifolii* compared with *L. sativae*, this may aid in overall leafminer pest management.

Materials and Methods

Subheading Ethics Statement

No specific permissions were required for these locations/activities.

The location is not privately-owned or protected in any way.

The field studies did not involve endangered or protected species.

Surveys of *Liriomyza* spp. in commercial cowpea fields

In southern China, cowpea is mainly planted in Hainan Province, and the further away from the Hainan Province, the less cowpea is planted. Therefore, we surveyed *Liriomyza* spp. in two regions that differ in the abundance of cowpea. Hainan Province was the region with high cowpea abundance and Guangdong Province was the region with relatively low cowpea crop abundance. We conducted field surveys during 2011. Because of inherent differences in the cropping seasons, we surveyed Hainan Province from January to March, and we surveyed Guangdong Province from April to June. We surveyed eight locations (Sanya, Dongfang, Haikou, Leidong, Lingshui, Wuzhisan, Qionghai, and Danzhou Cities) across Hainan Province (Fig.1), and we surveyed eight locations (Foshan, Jiangmen, Zhongshan, Guangzhou, Shenzhen, Zengcheng, Huadu, and Huizhou Cities) across Guangdong Province (Fig.3).

Cowpea foliage containing mature larvae and/or puparia of *Liriomyza* spp. was collected at each location. Leaves were excised from plants and placed on plastic containers (40×40×20 cm) to collect puparia. Pupae were held in glass scintillation vials for adult emergence. Adults were counted and identified to species.

On each sample date at each location, we collected foliage with >30 individuals per sample from each of at least three cowpea plots that were 100–500 meters away from each other. Foliage was held in environmental chamber at 26°C,16:8 (L:D) photoperiod where the larvae and pupae were reared to adulthood. Adults were counted and identified to species. To confirm species identifications, representative adult males were identified based on the morphology of genitalia.

Insecticide and cropping system effects on *Liriomyza* populations

Field experiments were conducted at two sites in Siling County, located in Guangzhou City, Guangdong Province in 2011 and 2012, where *L. sativae* and *L. trifolii* were known to co-occur at the time of the experiments. One site was selected for planting cowpea only, with no other crops planted within 1 km of the site. This site was also surrounded by many tall apartment buildings. This setting is similar to the intensively-planted cowpea agroecosystems on Hainan Island. A second site in Siling County was selected for planting cowpea with towel gourd planted near the cowpea field. This arrangement is common in most regions of Guangdong Province. In early of February of both years, these same two sites in Siling County were selected for planting cowpea, where *L. trifolii* and *L. sativae* were observed occurring simultaneously.

The cowpea experiments at both field sites were arranged as randomized complete blocks with three replication of each of the two treatments, the avermectin treatment and the control. Each plot had four rows with approximately 0.002 ha and seeded at a rate expected to produce 140 plants per plot (70,000 per hectare), as is common in this farming system. A 2 m space was set up between plots to minimize insecticide contamination among treatment plots. Cowpeas were maintained using the standard agronomic practices of the local areas, and no other insecticides were used throughout the season. The insecticide treatment was avermectin applied at 4.5 ppm [A.I.], and the control consisted of distilled water. A hand-pump type sprayer used to apply the treatments. It was calibrated to deliver 1300 L/ha at 20–30 kPa, with 90 µm openings in the nozzles.

For towel gourd, three plots were planted near the cowpea crops at the second site in Siling County. Each towel gourd plot contained five rows of plants, with a total area of approximately 0.01 ha. Plots were seeded at a rate to produce 150 plants per plot (15,000 plants per hectare), as is common in this type of farming system. Towel gourd was maintained using the standard agronomic practices of the local areas, and no insecticides were used throughout the season.

For the cowpea crop at both sites, sampling for leafminers and insecticide applications began in early April, approximately 8 wk after sowing. Sampling for leafminers was conducted on days 1, 7, 13, and 19 of the experiments. Insecticide applications were made after leafminer samples were collected on days 1, 7 and 13 of the experiment. Therefore, samples were collected before insecticide treatments began, and then every six days after each of the three applications. Five sites were randomly chosen within each plot on each sampling date, with a total of 50–200 leaves containing late-instar larvae collected per plot per date. Leaves were excised and placed in plastic containers (40×40×20 cm) to collect puparia. Pupae were held in glass scintillation vials for adult emergence. Adults were counted and identified to species.

After samples were collected on day 19, very few *L. sativae* were found in the insecticide treated cowpeas compared with the control treatments. To determine if leafminer populations changed after avermectin was no longer applied, we removed all of the cowpea plants in the control plots at both sites, and kept the cowpea plants in the insecticide treated plots. Thirty days later, we sampled for leafminers at both sites in the manner described above. This sampling enabled us to determine how leafminer populations changed following the removal of insecticide pressure and if the residual populations of *L. sativae* would reestablish.

For towel gourd, sampling for leafminers began in early April, approximately 8 wk after sowing. Sampling was conducted on days 1, 7, 13, 19, 25, 31, 37, 43, and 49 of the experiments. Five sites were randomly chosen within each plot on each sampling date, with a total of 25 leaves (5 leaves/site) containing late-instar larvae randomly collected per plot per date. Leaves were excised and placed in plastic containers (40×40×20 cm) to collect puparia. Puparia were held in glass scintillation vials for adult emergence. Adults were counted and identified to species.

Statistical analyses

We used χ^2tests to determine if there were biases in proportions of *Liriomyza* species in the field surveys of Guangdong and Hainan provinces. We used logistic regression analyses to determine if the proportion of *L. trifolii* in cowpea changed over time in association with insecticide application treatments in the different cropping systems. Because each experiment was conducted in a unique environment, data for each experiment were analyzed separately. To determine if there was an insecticide effect, we used a model with the two levels of insecticide (avermectin-treated and untreated), sample day (day 1, day 7, day 13, day 19) and the insecticide x day interaction. We used a set of linear trend contrasts to determine if the proportion of *L. trifolii* in each insecticide treatment changed in a linear manner with time. We also used contrasts to determine if the mean proportion of *L. trifolii* differed between insecticide treatments on specific sample dates. To determine if the proportions of *L. trifolii* changed after insecticide applications were halted, we compared proportions of *L. trifolii* from the avermectin treated plots on day 19 and day 49 (36 days after the final insecticide application). We also used a linear trend analysis to determine if the proportion of *L. trifolii* changed over time in the towel gourd plots. All logistic analyses were conducted using the Proc Glimmix procedure (SAS version 9.4, SAS Institute, 2012).

Author Contributions

Conceived and designed the experiments: YG SR ZL. Performed the experiments: YG QW WY. Analyzed the data: YG SR. Contributed reagents/materials/analysis tools: YG SR ZL. Wrote the paper: YG SR ZZ.

References

1. Parrella MP (1987) Biology of *Liriomyza*. Annual Review of Entomology 32: 201–224.
2. Reitz SR, Gao Y, Lei Z (2013) Insecticide use and the ecology of invasive *Liriomyza* leafminer management. In: Trdan S, editor. Insecticides - Development of Safer and More Effective Technologies. Rijeka, Croatia: InTech.
3. Gao Y, Lei Z, Abe Y, Reitz SR (2011) Species displacements are common to two invasive species of leafminer fly in China, Japan, and the United States. Journal of Economic Entomology 104: 1771–1773.
4. Kang L (1996) Ecology and sustainable control of *Liriomyza* species. Beijing: Science Press.
5. Lei ZR, Wang Y (2005) *Liriomyza sativae*. In: Wan FH, Zheng XB, Guo JY, editors. Biology and management of invasive alien species in agriculture and forestry. Beijing: Science Press. pp. 177–205.
6. Lei ZR, Zhu CJ, Zhang CQ (2007) Risk analysis of alien invasive *Liriomyza trifolii* (Burgess) in China. Plant Protection 33: 37–41.
7. Gao Y, Reitz SR, Wei Q, Yu W, Lei Z (2012) Insecticide-mediated apparent displacement between two invasive species of leafminer fly. PLoS ONE 7: e36622, doi:36610.31371/journal.pone.0036622.
8. Reitz SR, Trumble JT (2002) Competitive displacement among insects and arachnids. Annual Review of Entomology 47: 435–465.
9. Zhang W, Jiang F, Ou J (2011) Global pesticide consumption and pollution: with China as a focus. Proceedings of the International Academy of Ecology and Environmental Sciences 1: 125–144.
10. PingSheng Z, Qi L (2010) Investigation on organophosphorus pesticide residue in fresh melon and legume vegetables in Huizhou. Guangxi Agricultural Sciences 41: 338–340.
11. Zehnder GW, Trumble JT (1984) Intercrop movement of leafminers. California Agriculture 38: 7–8.
12. Zehnder GW, Trumble JT (1984) Host selection of *Liriomyza* species (Diptera: Agromyzidae) and associated parasites in adjacent plantings of tomato and celery (*Liriomyza sativae, Liriomyza trifolii*). Environmental Entomology 13: 492–496.
13. Johnson MW (1987) Parasitization of *Liriomyza* spp. (Diptera: Agromyzidae) infesting commercial watermelon plantings in Hawaii. Journal Of Economic Entomology 80: 56–61.

14. Palumbo JC, Mullis CH Jr, Reyes FJ (1994) Composition, seasonal abundance, and parasitism of *Liriomyza* (Diptera: Agromyzidae) species on lettuce in Arizona. Journal of Economic Entomology 87: 1070–1077.
15. Mason GA, Johnson MW, Tabashnik BE (1987) Susceptibility of *Liriomyza sativae* and *Liriomyza trifolii* (Diptera: Agromyzidae) to permethrin and fenvalerate. Journal of Economic Entomology 80: 1262–1266.
16. Parrella MP, Keil CB (1985) Toxicity of methamidophos to four species of Agromyzidae. Journal of Agricultural Entomology 2: 234–237.
17. Tokumaru S, Kurita H, Fukui M, Abe Y (2005) Insecticide susceptibility of *Liriomyza sativae, L. trifolii,* and *L. bryoniae* (Diptera: Agromyzidae). Japanese Journal of Applied Entomology and Zoology 49: 1–10.
18. Abe Y, Tokumaru S (2008) Displacement in two invasive species of leafminer fly in different localities. Biological Invasions 10: 951–995.
19. Purcell MF, Johnson MW, Tabashnik BE (1995) Effects of insecticide use on abundance and diversity of tomato pests and associated natural enemies in Hawaii. Proceedings of the Hawaiian Entomological Society 32: 45–59.
20. Mujica N, Kroschel J (2011) Leafminer fly (Diptera: Agromyzidae) occurrence, distribution, and parasitoid associations in field and vegetable crops along the Peruvian coast. Environmental Entomology 40: 217–230.
21. Oatman ER, Kennedy GG (1976) Methomyl induced outbreak of *Liriomyza sativae* on tomato. Journal of Economic Entomology 69: 667–668.
22. Kaspi R, Parrella MP (2005) Abamectin compatibility with the leafminer parasitoid *Diglyphus isaea.* Biological Control 35: 172–179.
23. Weintraub PG (2001) Effects of cyromazine and abamectin on the pea leafminer *Liriomyza huidobrensis* (Diptera: Agromyzidae) and its parasitoid *Diglyphus isaea* (Hymenoptera: Eulophidae) in potatoes. Crop Protection 20: 207–213.
24. Chen X-X, Lang F-Y, Xu Z-H, He J-H, Ma Y (2003) The occurrence of leafminers and their parasitoids on vegetables and weeds in Hangzhou area, Southeast China. BioControl 48: 515–527.
25. Musundire R, Chabi-Olaye A, Salifu D, Krüger K (2012) Host plant-related parasitism and host feeding activities of *Diglyphus isaea* (Hymenoptera: Eulophidae) on *Liriomyza huidobrensis, Liriomyza sativae,* and *Liriomyza trifolii* (Diptera: Agromyzidae). Journal of Economic Entomology 105: 161–168.

18

Optimizing Hill Seeding Density for High-Yielding Hybrid Rice in a Single Rice Cropping System in South China

Danying Wang[1], Song Chen[1], Zaiman Wang[2], Chenglin Ji[1], Chunmei Xu[1], Xiufu Zhang[1]*, Bhagirath Singh Chauhan[3]

1 State Key Laboratory of Rice Biology, China National Rice Research Institute, Hangzhou, Zhejiang, China, **2** Key Laboratory of Key Technology on Agricultural Machine and Equipment, South China Agricultural University, Guangzhou, Guangdong, China, **3** Queensland Alliance for Agriculture and Food Innovation, University of Queensland, Queensland, Australia

Abstract

Mechanical hill direct seeding of hybrid rice could be the way to solve the problems of high seeding rates and uneven plant establishment now faced in direct seeded rice; however, it is not clear what the optimum hill seeding density should be for high-yielding hybrid rice in the single-season rice production system. Experiments were conducted in 2010 and 2011 to determine the effects of hill seeding density (25 cm×15 cm, 25 cm×17 cm, 25 cm×19 cm, 25 cm×21 cm, and 25 cm×23 cm; three to five seeds per hill) on plant growth and grain yield of a hybrid variety, Nei2you6, in two fields with different fertility (soil fertility 1 and 2). In addition, in 2012 and 2013, comparisons among mechanical hill seeding, broadcasting, and transplanting were conducted with three hybrid varieties to evaluate the optimum seeding density. With increases in seeding spacing from 25 cm×15 cm to 25 cm×23 cm, productive tillers per hill increased by 34.2% and 50.0% in soil fertility 1 and 2. Panicles per m^2 declined with increases in seeding spacing in soil fertility 1. In soil fertility 2, no difference in panicles per m^2 was found at spacing ranging from 25 cm×17 cm to 25 cm×23 cm, while decreases in the area of the top three leaves and aboveground dry weight per shoot at flowering were observed. Grain yield was the maximum at 25 cm×17 cm spacing in both soil fertility fields. Our results suggest that a seeding density of 25 cm×17 cm was suitable for high-yielding hybrid rice. These results were verified through on-farm demonstration experiments, in which mechanical hill-seeded rice at this density had equal or higher grain yield than transplanted rice.

Editor: Jose Luis Gonzalez-Andujar, Instituto de Agricultura Sostenible (CSIC), Spain

Funding: The authors are grateful for grants from the National Rice Industry Technology System (CARS-01-04A), the 3 Nong 6 Fang Program, and the Five-twelfth National Science and Technology Support Program (2011BAD16B14). The authors state that the funders had no role in study design, data collection and analysis, decision to publish, or preparation of the manuscript.

Competing Interests: The authors have declared that no competing interests exist.

* Email: chais.zju@gmail.com

Introduction

Although transplanting has been a major traditional method of rice establishment in most Asian countries, the rising labor cost and developments in rice production technology have improved the desirability of direct-seeded rice [1]. Direct seeding of rice is a low-cost establishment technology. It avoids nursery raising, seedling uprooting, and transplanting, and has the benefits of saving labor, facilitating timely establishment of rice, and earlier crop maturity (by 7–10 days) [2–4]. Direct seeding helps in solving the labor scarcity problem, which is now very critical in China's agricultural development. Simultaneously, the chemical weed control method has made such a switch technically viable. In 2008, for example, 8.3% of the rice-growing area in China was direct-seeded; in some provinces of south China, it was more than 30% [5].

In the south China irrigated rice ecosystem, broadcasting in wet conditions (wet seeding) is the principal method of rice establishment. In this method, pregerminated seeds are sown onto wet (saturated) puddled soils. But, some difficulties are faced in wet seeding. First, no specific varieties have been developed in China for wet seeding. The existing varieties used for transplanting do not appear to be well-adapted for seedling growth in an initially oxygen-depleted micro environment. Second, the extent of laser leveling in China is currently extremely smaller compared with that in other countries (e.g., 50–80% of the rice land in Australia is laser-leveled) [4]. Due to lack of uniform water distribution associated with unevenness of the land, the problem of excess- or no-water-caused yield variability within a field is common, which leads to poor establishment of direct-seeded rice [3,6]. Third, surface-sown seeds are damaged by birds and rodents. As a result, farmers often resort to the costly practice of increasing the seeding rate for direct-seeded rice, even by two to three times [7]. High seeding rates can result in large yield losses due to excessive number of tillers, increased proportion of ineffective tillers, higher spikelet sterility, and fewer grains per panicle [8]. Moreover, a dense canopy and less ventilation around the plants at high seeding rates can create favorable conditions for diseases (e.g., sheath blight) [9,10] and insects (e.g., brown planthoppers) and can make plants more prone to lodging [11]. High seeding rates using broadcasting restrict individual plant growth, which is not suitable for the production of high-yielding hybrid rice.

High-yielding hybrid rice, which is crucial to meet expanding food demand, had large panicles and high aboveground biomass [12,13]. Hybrid rice varieties grow well with proper planting density, but their lodging resistance decreases dramatically under dense seeding [14]. The use of high seeding rate in a broadcast culture may increase the gap between farmers' potential and actual yield.

Mechanical hill seeding of rice is one of the options used to solve the problem of high seeding rate and uneven plant establishment. Here, rice seeds are sown uniformly in the field at designated spacing and seed number per hill. Combined with a harrow or furrower opener, many hill seeders have land-preparation functions to improve soil evenness. Some hill seeders enable seeding and puddling to be performed simultaneously [15]; others can make furrows and plant the crop at the same time [16]; still others can effectively incorporate seeds into puddled soils at depths of 5–20 mm, which helps decrease losses caused by birds [15]. Even soil conditions and uniform seeding result in an even plant establishment. A well-ventilated micro-environment not only reduces pest and disease incidence but also enhances photosynthesis [17]. In previous studies, lodging resistance of hill-seeded rice was found to be higher than mechanically transplanted rice, row-seeded rice, and broadcast-seeded rice [15,18].

However, the practical use of mechanical hill seeders had not been seen in China until the 2000s. The adoption of this implement was hampered by the belief that broadcasting is not a suitable method for hybrid rice varieties and both government and farmers still have misgivings about the use of mechanical hill seeding for this purpose. Research is thus imperative to shed light on this technology. There is limited information on hill seeding density for high-yielding hybrid rice varieties, and results for optimum hill seeding density in a single-season rice production system varied from 20 cm×14 cm to 30 cm×20 cm spacing [17,19–21]. It is difficult to draw a concrete conclusion about what hill seeding density is optimum for high-yielding hybrid rice varieties.

The aim of this study was to evaluate the optimum hill seeding density to support the development of mechanical hill seeders for hybrid rice. In this study, 2-year field experiments and 2-year on-farm demonstration experiments were conducted in Zhejiang, China. In the field experiments, a high-yielding hybrid rice variety (Nei2you6) was hill-seeded manually at five seeding densities in two fields with different fertility profiles. The objectives were to evaluate the effects of hill seeding density on plant growth and grain yield, the interrelation between seeding density and soil fertility, and the optimum hill seeding density for high-yielding hybrid rice. In the on-farm demonstrations, three high-yielding hybrid rice varieties were planted using three methods: mechanical hill-seeded with a hill seeder at optimum seeding density, broadcasting, and transplanting following standard farmers' practice. The objective was to verify the optimum seeding density in the field by comparing the yields of hill-seeded, broadcast, and transplanted rice.

Materials and Methods

Field experiments and on-farm demonstrations were conducted; field experiments were done in 2010 and 2011 and the on-farm demonstration experiments were carried out in 2012 and 2013. Field experiments were conducted at the experimental farm of the China Rice Research Institute and on-farm demonstration experiments were conducted at the agricultural demonstration base of the China Rice Research Institute in Fuyang and Yuhang. No permission was needed to use fields at these facilities. Authors also confirmed that the study did not involve endangered or protected species.

Field experiments

The 2010 and 2011 experiments were conducted in two paddy fields with different fertility levels (soil fertility 1 and 2) at the China National Rice Research Institute (CNRRI) (120.2′ E, 30.3′ N, 11 m above sea level), Zhejiang Province, China. The area has a subtropical monsoon climate with an annual mean temperature of 13–20°C (range from 2°C in January to 39°C in July). Mean annual precipitation is 1200–1600 mm, with about 80% rainfall occurring between April and September. Soil at the experimental site is classified as ferric-accumulic stagnic anthrosol [22]. The cropping system of soil fertility 1 was monoculture rice-fallow without rice straw incorporation, and the cropping system of soil fertility 2 was rice-potato (*Solanum tuberosum*) with rice and potato straw incorporation. After 10 years of continuous farming, the soil properties in the two fields differed and this was noted at the start of the experiment in 2010. The soil in field 1 had 2.46 g total N kg^{-1}, 0.53 g total P kg^{-1}, 0.25 g available K kg^{-1}, 10.73 C mol kg^{-1} cation exchange capacity (CEC), and pH 5.78 (soil fertility 1). The corresponding soil properties in field 2 were 2.96 g total N kg^{-1}, 0.72 g total P kg^{-1}, 0.31 g available K kg^{-1}, 10.80 C mol kg^{-1} CEC, and pH 5.15 (soil fertility 2). The soil test was based on samples taken from the upper 10 cm of the soil. Soil chemical properties at the 10–20 cm depth were also different [23].

Pregerminated seeds of a high-yielding hybrid rice variety, Nei2you6, were hill-seeded manually in the two fields with different fertility levels using five spacing schemes (between and within rows: 25 cm×15 cm, 25 cm×17 cm, 25 cm×19 cm, 25 cm×21 cm, and 25 cm×23 cm). Each hill was planted with three to five seeds. The experiment was replicated four times and subplot size was 30 m^2. In both years, the fields were puddled in ponded conditions and drained 2 d before seeding. Pre-germinated seeds were sown on the surface of puddled soil on June 2, 2010 and May 30, 2011. Before sowing, seeds were soaked for 24 h and incubated for 24 h for proper germination, and seeds dressed with chemical pesticides to avoid the attack of soil insects, birds, and rodents.

To the crop, 225 kg N ha^{-1} (as urea) was applied in three splits in a ratio of 5:3:2 at basal, tillering, and earing stage, respectively. Potassium at 180 kg K_2O ha^{-1} was applied in two splits: 50% as basal dressing and 50% as topdressing at earing in the form of KCl. Phosphorus at 75 kg P_2O_5 ha^{-1} in the form of calcium superphosphate was applied as basal. Water management followed standard farmers' practice, in which the field was drained 2 d before seeding. There was no standing water on the soil surface from sowing to the three-leaf stage of rice, and then the field was reflooded and a water depth of 1 to 3 cm was maintained until the end of the tillering period. After this, water was drained for 7–10 days to control unproductive tillers and later, a water depth of 3–5 cm was maintained until the grain-filling stage. Weed and insects were intensively controlled by chemicals to avoid biomass and yield loss.

In 2010, the number of tillers from 30 hills of rice was counted at 1-week intervals, starting from 20 d after sowing (DAS) to panicle initiation stage. Tillers with at least one visible leaf were included. In 2011, 30 hills of rice plants were sampled at heading and maturity to calculate average panicle number per hill. In both years, six representative hills of rice plants from each plot were separately sampled at heading and divided into leaf blades, stem plus sheath, and panicles. Green leaves were grouped into the top

Table 1. Grain yield of a high-yielding hybrid rice variety Nei2you6 grown at five seeding spacings in soil fertility 1 and soil fertility 2 at the China National Rice Research Institute, Zhejiang Province, China, in 2010 and 2011.

Year	Field	Seeding spacing	Grain yield (kg ha^{-1})	Grain yield (g $hill^{-1}$)	Grain yield (g $shoot^{-1}$)
2010	Soil fertility 1	25 cm×15 cm	7804 c	29.37 d	3.09 d
		25 cm×17 cm	8348 a	36.01 c	3.49 b
		25 cm×19 cm	8152 ab	38.05 bc	3.41 bc
		25 cm×21 cm	7892 bc	41.65 b	3.36 c
		25 cm×23 cm	7938 b	46.05 a	3.68 a
	Soil fertility 2	25 cm×15 cm	9133 b	34.69 d	3.64 b
		25 cm×17 cm	9433 a	40.63 c	3.84 a
		25 cm×19 cm	8801 c	42.08 bc	3.68 b
		25 cm×21 cm	8437 d	45.09 b	3.50 c
		25 cm×23 cm	8311 d	48.35 a	3.38 d
2011	Soil fertility 1	25 cm×15 cm	7842 c	29.79 e	3.14 d
		25 cm×17 cm	8573 a	36.80 d	3.56 b
		25 cm×19 cm	8218 b	39.61 c	3.55 b
		25 cm×21 cm	8121 b	43.24 b	3.48 c
		25 cm×23 cm	8140 b	47.19 a	3.70 a
	Soil fertility 2	25 cm×15 cm	9380 b	35.39 d	3.71 bc
		25 cm×17 cm	9611 a	41.13 c	3.88 a
		25 cm×19 cm	8942 c	42.91 bc	3.76 b
		25 cm×21 cm	8831 c	47.14 ab	3.66 c
		25 cm×23 cm	8582 d	49.94 a	3.49 d
Analysis of variance					
Seeding density (A)			*	*	*
Soil fertility (B)			*	*	*
Year (C)			NS	NS	NS
A×B			*	*	*
A×C			NS	NS	NS
B×C			NS	NS	NS
A×B×C			*	*	*

Within a column for each year, means followed by the same letters are not significantly different according to the LSD test (0.05).
*Significance at the 0.05 level based on analysis of variance.
NS denotes nonsignificance based on analysis of variance.

three green leaves and others, and their leaf area was measured with a leaf area meter (LI-3000A, LICOR, Lincoln, NE, USA). LAI was calculated as the total green leaf area of the plant, divided by the corresponding ground sampling area. All samples were oven-dried at 80°C until constant weight for determination of aboveground dry weight at flowering. At maturity, 30 hills were sampled diagonally from each subplot, panicle numbers were counted for each hill to determine the number of panicles per m^2. Six representative hills of rice plants were selected to determine aboveground total dry weight and harvest index. Grain yield per hill was measured from the sampled 30 hills and grain yield per shoot was calculated as the ratio of yield per hill to panicle numbers per hill. In each subplot, grain yield was determined from a harvesting area of 30 m^2 and converted to kg ha^{-1} at 14% moisture content. Panicles were hand-threshed and filled spikelets were separated from unfilled spikelets by submerging them in tap water. Dry weights of straw, rachis, and filled and unfilled spikelets were determined after oven drying at 80°C to constant weight. Three subsamples of 20 g of filled spikelets were taken and counted to determine grain weight (20 divided by the counted filled spikelets number); and the number of filled spikelets was calculated as total dry weight of filled spikelets divided by grain weight. All unfilled spikelets were counted and the number of total spikelets (filled and unfilled) was calculated. Aboveground total dry weight at maturity included the total dry matter of straw, rachis, and filled and unfilled spikelets. Dry weight accumulation during the grain-filling phase was calculated as the difference in total aboveground dry weight between flowering and maturity stages. Spikelets per panicle (ratio of total spikelets number to panicles number), grain-filling percentage (ratio of filled spikelets number to total spikelets number), and harvest index (ratio of dry grain yield to total aboveground biomass dry weight) were calculated.

On-farm demonstration experiments

On-farm demonstrations were conducted in 2012 and 2013 in Fuyang (120.0′ E, 30.1′ N; 14 m above sea level) and Yuhang (120.3′ E, 30.4′ N; 40 m above sea level), Zhejiang Province. Soil at both sites is classified as ferric-accumulic stagnic anthrosol [22]. The soil properties in Fuyang were 2.48 g total N kg-1, 0.56 g total P kg-1, 0.26 g available K kg-1, and pH 5.63. The soil

Table 2. Yield components of a high-yielding hybrid rice variety Nei2you6 grown at five seeding spacings in soil fertility 1 and soil fertility 2, China National Rice Research Institute, Zhejiang Province, China, 2010 and 2011.

Year	Field	Seeding spacing	Spikelets panicle^{-1} (no.)	Grain filling (%)	Grain weight (mg)	Panicles m^{-2} (no.)	Panicles hill^{-1} (no.)
2010	Soil fertility 1	25 cm×15 cm	115.52 b	76.82 e	30.30 a	240.97 a	9.03 c
		25 cm×17 cm	125.92 a	77.74 d	30.24 a	228.08 b	9.71 bc
		25 cm×19 cm	126.70 a	78.85 c	30.46 a	223.75 b	10.60 b
		25 cm×21 cm	125.42 a	79.99 b	30.72 a	226.41 b	11.91 a
		25 cm×23 cm	127.55 a	81.51 a	30.52 a	210.76 c	12.11 a
	Soil fertility 2	25 cm×15 cm	135.01 a	81.63 b	30.73 a	247.01 a	9.25 cd
		25 cm×17 cm	135.78 a	83.15 a	30.44 a	238.86 b	10.17 c
		25 cm×19 cm	130.83 a	81.44 b	30.22 a	229.02 b	10.86 c
		25 cm×21 cm	135.80 a	78.06 c	30.56 a	234.84 b	12.36 b
		25 cm×23 cm	133.09 a	74.11 d	30.33 a	236.59 b	13.59 a
2011	Soil fertility 1	25 cm×15 cm	127.68 b	84.90 c	33.48 a	266.33 a	9.98 c
		25 cm×17 cm	142.00 a	86.66 b	34.10 a	257.20 ab	10.95 bc
		25 cm×19 cm	140.04 a	87.15 b	33.66 a	247.31 b	11.72 b
		25 cm×21 cm	135.88 a	86.65 b	33.28 a	245.27 b	12.91 a
		25 cm×23 cm	140.97 a	90.09 a	33.74 a	232.94 c	13.39 a
	Soil fertility 2	25 cm×15 cm	143.37 a	86.67 b	32.63 a	262.29 a	9.83 e
		25 cm×17 cm	147.10 a	90.07 a	32.98 a	258.76 a	11.01 d
		25 cm×19 cm	144.61 a	90.02 a	33.40 a	253.12 a	12.00 c
		25 cm×21 cm	147.12 a	84.56 c	33.10 a	254.42 a	13.40 b
		25 cm×23 cm	147.09 a	81.91 d	33.53 a	261.49 a	15.03 a

Within a column for each site, means followed by the same letters are not significantly different according to the LSD test (0.05).

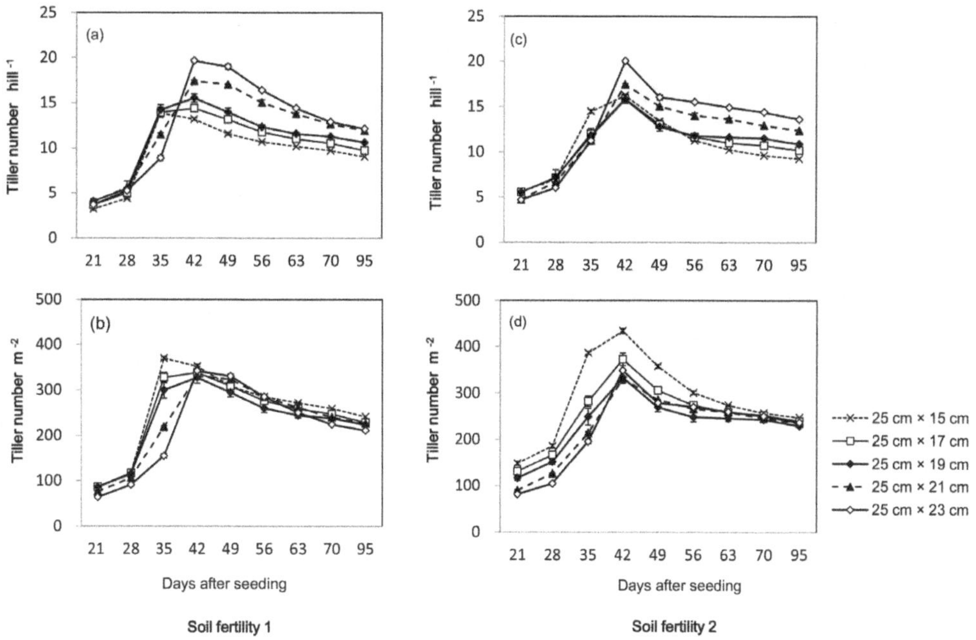

Figure 1. Tillers per hill (a, c) and tillers per m² (b, d) of a high-yielding hybrid rice variety Nei2you6 grown at five plant spacings in soil fertility 1 (a, b) and soil fertility 2 (c, d). China National Rice Research Institute, Zhejiang Province, China, 2010.

properties in Yuhang were 2.66 g total N kg-1, 0.62 g total P kg-1, 0.32 g available K kg-1, and pH 5.75. The hybrid rice variety used in Fuyang was Nei2you6. In Yuhang, high-yielding hybrid rice varieties Yongyou9 and Yongyou12 were grown. These high-yielding hybrid rice varieties are used widely in south China under a single rice cropping system. Three treatments were included at each site: mechanical hill seeding, broadcasting, and transplanting. Hill seeding was done by using a newly developed hill-seeder ZB-10 (Wang et al 2010) at a spacing of 25 cm×17 cm, three to five seeds per hill, and the seed rate was 21.0 to 25.5 kg ha^{-1}. This spacing was chosen because the highest grain yield was noted in this treatment in the 2010 and 2011 experiments. Broadcasting and transplanting were standard farmers' practices. In the broadcasting treatment, sprouted seeds were broadcast at a seeding rate of 33.8 kg ha^{-1}. In the transplanting treatment, seedlings with four to five leaves were transplanted at a spacing of 30 cm×20 cm; two to three seedlings per hill. In the mechanical hill seeding and broadcasting treatments, seed treatment was the same as used in field experiments in 2010 and 2011; all varieties were seeded on the same day (June 2) in both years. Because direct seeding could facilitate timely establishment of rice and earlier crop maturity by 7–10 d, the nursery for the transplanting treatment was raised on May 25 and transplanted on June 18 in both years. The planting area of each treatment was 3000 m^2, a total of 9000 m^2 for each variety. Fertilizer and water management in the three treatments were the same as that described for the field experiments in 2010 and 2011. Management operations such as weeding, irrigation, and plant protection measures were done as needed.

At maturity stage, a 5-m^2 area was harvested diagonally from six places in each treatment for yield determination, and grain yield was calculated at 14% moisture content. Ten hills were sampled diagonally from each 5-m^2 harvest area to determine yield components in the mechanical hill seeding and transplanting treatments, and a 0.5-m^2 area from each 5-m^2 harvest area in the broadcasting treatment. Panicles were hand-threshed and filled and unfilled grains were separated. Dry weights of filled and unfilled spikelets were determined after oven drying samples at 80°C to constant weight. Three subsamples of 20 g of filled spikelets were taken and counted to determine weight and total number of filled spikelets. Unfilled spikelets were counted and grain-filling percentage was calculated.

Statistical analyses

In the field experiments, normal distribution of data was tested for grain yield, yield components, productive tiller percentage, aboveground dry weight, and harvest index by Q-Q plots using SPSS 11.5 software, and no data transformation was needed to meet the assumptions of normality and constant variance. Data were analyzed following analysis of variance (ANOVA), and differences among treatments were compared based on the least significant difference (LSD) test at the 5% level of probability. The statistical model used in the analysis of grain yield per hectare, per hill, and per shoot included the sources of variation due to replication, seeding density, soil fertility, year, and the interactions of seeding density×soil fertility, seeding density×year, soil fertility×year, and seeding density×soil fertility×year. The bivariate correlations between grain yield and yield components, including grain weight, panicles per hectare and per hill, spikelets per panicle, and grain-filling percentage, were conducted using the Pearson correlation coefficient (two-tailed).

In the on-farm demonstration experiments, ANOVA and mean comparisons were based on the LSD test at the 0.05 probability level for each variety and year (SPSS 11.5).

Results

Field experiments

In both years, a significant grain yield response to seeding density was observed (Table 1). Grain yield per hill increased consistently with decreases in seeding density. The plants grown at 25 cm× 23 cm spacing had the highest grain yield per hill and

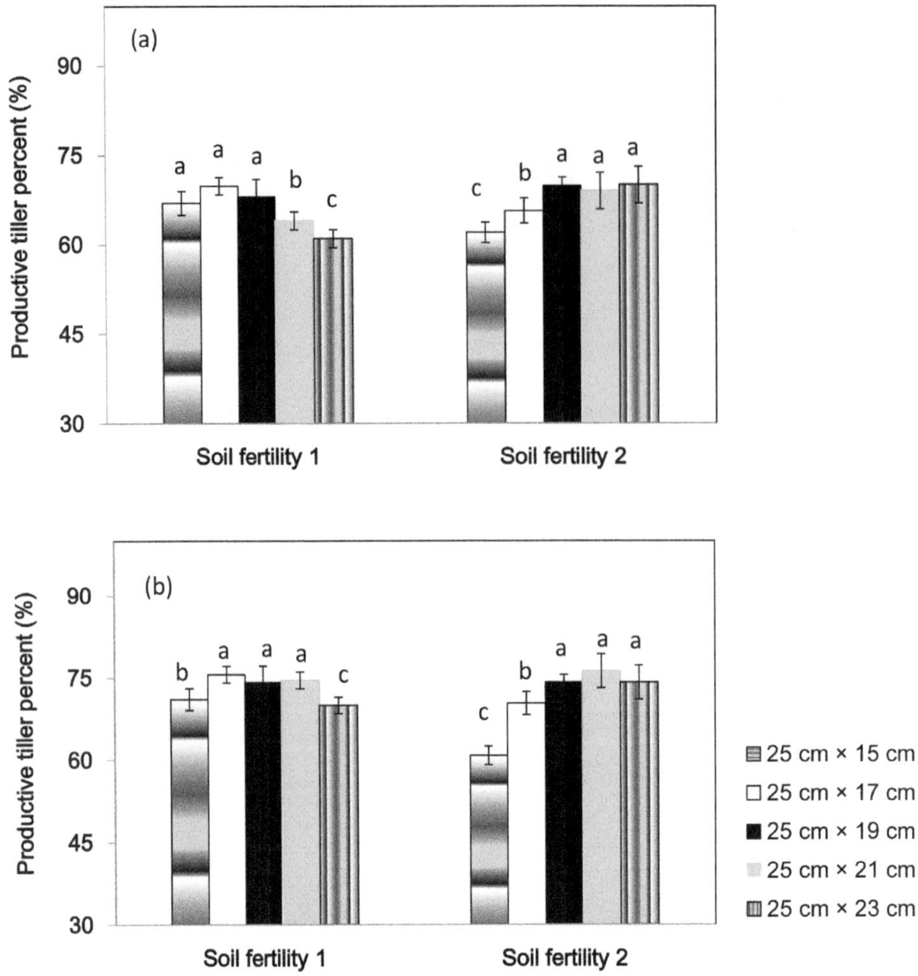

Figure 2. Productive tiller percentage of a high-yielding hybrid rice variety Nei2you6 grown at five seeding spacings in soil fertility 1 and soil fertility 2. China National Rice Research Institute, Zhejiang Province, China, 2010 (a) and 2011 (b).

plants grown at 25 cm×15 cm spacing had the lowest. The rate of change in grain yield per hectare with seeding spacing, however, was different between soil fertility 1 and 2. In soil fertility 1, 25 cm×17 cm spacing had the highest grain yield per ha, while 25 cm×15 cm spacing had the lowest value. In soil fertility 2, highest grain yield per ha was obtained at 25 cm×17 cm spacing, but, beyond this, grain yield decreased with an increase in seeding spacing and rice plants grown at 25 cm× 23 cm produced the lowest grain yield per ha. There were large differences in grain yield per shoot among the five seeding spacings in both soil fertility fields. In soil fertility 1, the spacing of 25 cm× 23 cm produced the maximum grain yield per shoot and the lowest value was observed at 25 cm×15 cm spacing. In soil fertility 2, the response of grain yield per shoot to seeding density was the same as that for grain yield per ha; 25 cm×17 cm had the maximum and 25 cm× 23 cm had the minimum grain yield per shoot. The difference in grain yield between the two years was nonsignificant (Table 1).

In both years, there was no significant difference in grain weight among different spacing treatments in both fields (Table 2). The number of panicles per hill consistently increased with increases in seeding spacing in both years and both fields. Across years, with increases in seeding spacing from 25 cm×15 cm to 25 cm×23 cm, panicles per hill increased by 34.2% in soil fetility 1 and by 50.0% in soil fertility 2. In both years, the number of panicles per m^2 was not significantly different at spacing ranging from 25 cm×17 cm to 25 cm× 21 cm in soil fertility 1; the maximum and minimum number of panicles per m^2 was observed at 25 cm×15 cm and 25 cm× 23 cm spacing, respectively. In soil fertility 2, 25 cm×15 cm spacing had higher number of panicles per m^2 in 2010, while no difference was observed among other treatments in this year and among all spacings in 2011. In both years, 25 cm×15 cm spacing had the minimum number of spikelets per panicle in soil fertility 1, while no difference was found among treatments in soil fertility 2. Grain-filling percentage in 2011 was higher than that in 2010 in both soil fertility fields. The plants grown at 25 cm× 23 cm spacing had the highest and 25 cm×15 cm spacing had the lowest grain-filling percentage in soil fertility 1. Meanwhile, in soil fertility 2, 25 cm× 17 cm spacing had the highest grain filling percentage and 25 cm× 23 cm spacing had the lowest value in both years.

In 2010, maximum tiller number per hill was observed at 25 cm×15 cm spacing at 35 DAS in soil fertility 1, while at 42 DAS at other seeding spacings in soil fertility 1 and all treatments in soil fertility 2 (Figure 1). A decreasing trend of tiller number per hill was observed after the reproductive stage, irrespective of treatment. Maximum tiller numbers per hill and per m^2 were affected by seeding density. In both soil fertility fields, maximum tiller number per hill increased with increases in seeding spacing;

Table 3. Leaf area index (LAI), leaf area per hill, and leaf area per shoot at flowering of a high-yielding hybrid rice variety Nei2you6 grown at five seeding spacings in soil fertility 1 and soil fertility 2, China National Rice Research Institute, Zhejiang Province, China, 2010 and 2011.

Year	Field	Seeding spacing	LAI	Leaf area (m^2 per hill)			Leaf area ($\times 10^{-3}$ m^2 per shoot)		
				Total	Top 3 leaves	Others	Total	Top 3 leaves	Others
2010	Soil fertility 1	25 cm×15 cm	5.72 a	0.21 e	0.12 d	0.11 c	26.87 c	14.16 c	12.71 b
		25 cm×17 cm	5.39 c	0.23 d	0.13 d	0.11 c	27.76 c	15.18 c	12.58 b
		25 cm×19 cm	5.42 bc	0.26 c	0.14 c	0.12 b	28.78 b	16.15 b	12.63 b
		25 cm×21 cm	5.49 b	0.29 b	0.16 b	0.13 a	29.08 b	16.46 b	12.62 b
		25 cm×23 cm	5.53 b	0.32 a	0.18 a	0.13 a	31.18 a	17.80 a	13.35 a
	Soil fertility 2	25 cm×15 cm	7.03 a	0.27 e	0.16 c	0.10 d	35.66 a	21.02 a	13.64 c
		25 cm×17 cm	7.07 a	0.30 d	0.17 bc	0.13 c	35.55 a	20.97 a	14.57 b
		25 cm×19 cm	6.68 b	0.32 c	0.18 b	0.13 c	34.67 b	19.59 b	15.07 b
		25 cm×21 cm	6.66 b	0.35 b	0.18 b	0.16 b	34.06 c	17.89 c	16.17 a
		25 cm×23 cm	6.46 c	0.37 a	0.19 a	0.17 a	32.43 d	17.05 c	15.37 b
2011	Soil fertility 1	25 cm×15 cm	6.20 a	0.23 d	0.12 d	0.11 d	29.11 c	15.34 c	13.77 ab
		25 cm×17 cm	5.61 d	0.23 d	0.13 d	0.11 d	28.90 c	15.80 c	13.10 b
		25 cm×19 cm	5.88 c	0.28 c	0.16 c	0.12 c	31.18 b	17.49 b	13.69 ab
		25 cm×21 cm	5.83 c	0.31 b	0.18 b	0.13 b	31.88 b	17.48 b	13.40 b
		25 cm×23 cm	5.99 b	0.34 a	0.20 a	0.15 a	33.78 a	19.28 a	14.47 a
	Soil fertility 2	25 cm×15 cm	7.77 a	0.29 e	0.18 c	0.12 e	38.30 a	23.24 a	15.08 c
		25 cm×17 cm	7.51 b	0.32 d	0.19 b	0.13 d	37.75 a	22.27 ab	15.47 c
		25 cm×19 cm	7.24 c	0.34 c	0.20 ab	0.15 c	37.55 a	21.23 b	16.33 b
		25 cm×21 cm	7.01 d	0.37 b	0.19 b	0.17 b	35.81 b	18.81 c	17.00 a
		25 cm×23 cm	6.86 d	0.39 a	0.21 a	0.19 a	34.43 b	18.11 c	16.33 b

Green leaves per hill were subdivided into top three green leaves and others. Leaf area per shoot was calculated as the ratio of leaf area per hill to productive tiller number per hill.
Within a column for each site, means followed by the same letters are not significantly different according to the LSD test (0.05).

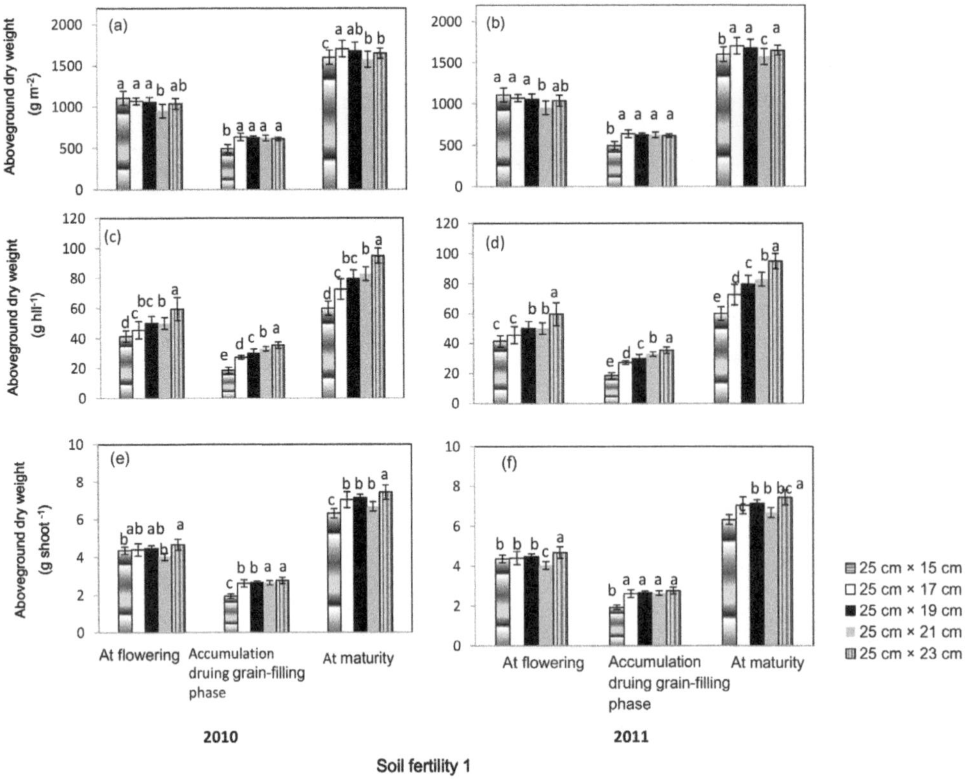

Figure 3. Aboveground dry weight at flowering and maturity, and dry weight accumulation during grain-filling phase of a high-yielding hybrid rice variety Nei2you6 grown at five seeding spacings in soil fertility 1 at the China National Rice Research Institute, Zhejiang Province, China, in 2010 and 2011. Aboveground dry weight was calculated as dry weight per m^2 (a, b), dry weight per hill (c, d), and dry weight per shoot (e, f). Dry weight per shoot was the ratio of dry weight per hill to productive tiller number per hill.

25 cm× 23 cm spacing had the highest tiller number per hill and 25 cm×15 cm spacing had the lowest. However, maximum tiller number per m^2 in both fields decreased with increases in seeding spacing; it was highest at 25 cm×15 cm spacing. From the maximum tillering to flowering stage, the difference in tiller number per m^2 among treatments gradually decreased with the reduction in unproductive tillers.

The effects of seeding density on the percentage of productive tillers were also different between the two soil fertility fields (Figure 2). In soil fertility 1, 25 cm× 23 cm spacing had the lowest productive tiller percentage in both years, and no significant difference was observed among spacings ranging from 25 cm×15 cm to 25 cm×19 cm in 2010 and from 25 cm×17 cm to 25 cm×21 cm in 2011. In both years, in soil fertility 2, 25 cm×15 cm spacing had the lowest productive tiller percent, followed by 25 cm×17 cm spacing, and both these spacings had significantly lower productive tillers than the other ones.

Although leaf area per hill at flowering consistently increased in both years with increases in seeding spacing in both soil fertility fields, the change in leaf area per shoot was different between the two fields (Table 3). In both years, total leaf area per shoot and top three leaf blades per shoot increased with increases in seeding spacing in soil fertility 1, while they decreased in soil fertility 2. The highest leaf area index (LAI) in soil fertility 1 was obtained at 25 cm×15 cm spacing, followed by that at 25 cm×23 cm spacing. The highest LAI in soil fertility 2 was also observed at 25 cm×15 cm spacing, while it gradually decreased with increases in seeding spacing; 25 cm×23 cm spacing had the lowest LAI in both years.

Aboveground dry weight per hill at flowering and maturity, and dry weight accumulation per hill during grain-filling phase increased with increases in seeding spacing; 25 cm× 23 cm spacing had the highest values and 25 cm×15 cm spacing had the lowest in both soil fertility fields in both years (Figures 3 and 4). In terms of aboveground dry weight per shoot, a difference existed between the two soil fertility fields. In both years, the highest aboveground dry weight per shoot at flowering and maturity was observed at 25 cm× 23 cm spacing in soil fertility 1; rice plants grown at 25 cm×15 cm spacing had the lowest dry weight accumulation per shoot during the grain-filling phase and aboveground dry weight per shoot at maturity. In soil fertility 2, no significant difference in dry weight accumulation during the grain-filling phase was observed in spacing ranging from 25 cm×17 cm to 25 cm× 23 cm in both years; and aboveground dry weight per shoot at flowering and maturity decreased with increases in spacing from 25 cm×17 cm to 25 cm× 23 cm. Significant differences also existed in aboveground dry weight per m^2 between the two soil fertility fields. In soil fertility 1, no significant difference was found for aboveground dry weight per m^2 at flowering at spacings ranging from 25 cm×15 cm to 25 cm×19 cm. Rice plants grown at 25 cm×15 cm spacing had the lowest dry weight accumulation during grain filling, and no significant difference was observed at spacings ranging from 25 cm×17 cm to 25 cm×23 cm. Aboveground dry weight at maturity was highest at 25 cm×17 cm spacing in both years and lowest at 25 cm×15 cm spacing in 2010 and at 25 cm×15 cm spacing in 2011. In soil fertility 2, no significant difference in dry weight accumulation during grain-filling phase was observed among the treatments; aboveground dry weights per m^2 at

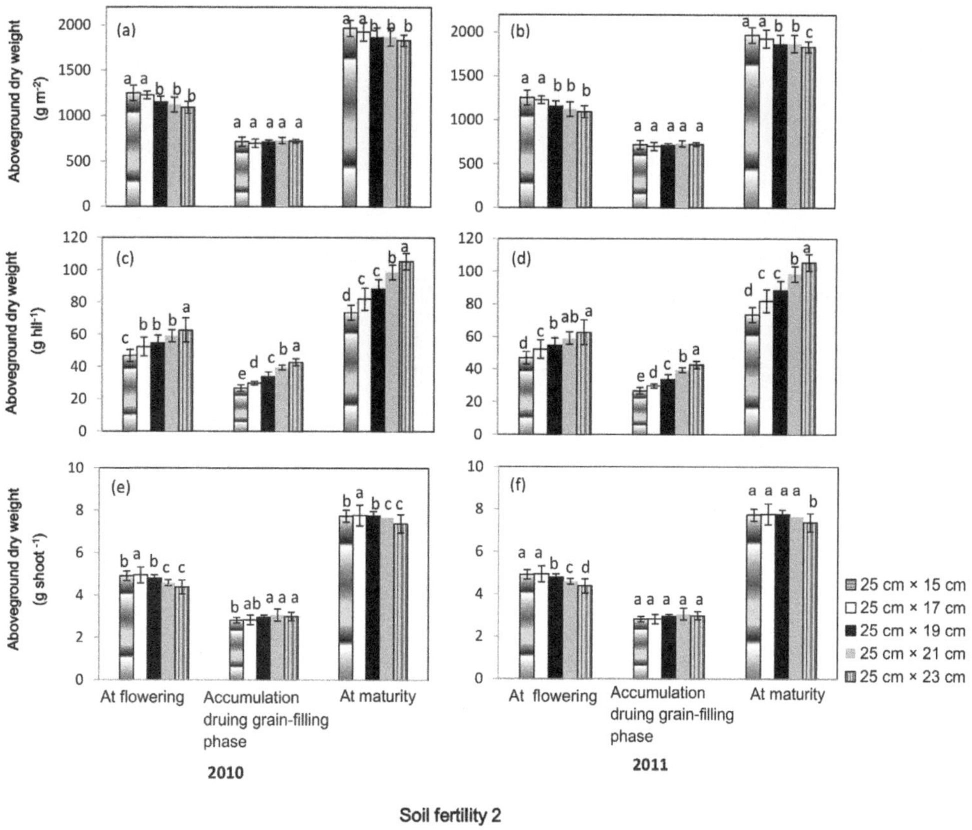

Figure 4. Aboveground dry weight at flowering and maturity, and dry weight accumulation during grain-filling phase of a high-yielding hybrid rice variety Nei2you6 grown at five seeding spacings in soil fertility 2 at the China National Rice Research Institute, Zhejiang Province, China, in 2010 and 2011. Aboveground dry weight was calculated as dry weight per m^2 (a, b), dry weight per hill (c, d), and dry weight per shoot (e, f). Dry weight per shoot was the ratio of dry weight per hill to productive tiller number per hill.

flowering and maturity were significantly higher at 25 cm×15 cm and 25 cm×17 cm spacing than at spacing wider than 25 cm×19 cm.

The effect of seeding density on harvest index (HI) was also different between the two soil fertility fields (Figure 5). In soil fertility 1, 25 cm×23 cm spacing and 25 cm×21 cm spacing had higher HI than that at the other three narrower spacings in both years. In soil fertility 2, plants grown at spacings ranging from 25 cm×17 cm to 25 cm×21 cm had significantly higher HI than plants grown at 25 cm×15 cm and 25 cm×15 cm spacings.

On-farm demonstration experiments

In the on-farm demonstration experiments in 2012 and 2013, mechanical hill seeding, using seeder ZB-10 at a spacing of 25 cm×17 cm, outperformed broadcasting in terms of grain yield for all three high-yielding hybrid rice varieties, increasing grain yield by 10.4% in Yongyou 9 and 12.3% in Yonyou 12 across the 2 years. In 2012, variety Nei2you6 produced 26.6% higher grain yield in mechanical hill seeding compared with broadcasting because more than 30% of the plants lodged in the broadcast culture due to heavy rain during the later grain-filling phase (Figure 6). In both years, no significant difference in grain yield between mechanical hill-seeded rice and transplanted rice was observed for Nei2you 6 and Yonyou 9. Yongyou 12 produced more than 1000 kg ha^{-1} grain yield in the mechanical hill seeding treatment across 2 years, increasing grain yield by 9.1% compared with that of transplanted rice.

Mechanical hill-seeded rice had more spikelets per panicle and fewer panicles per m^2 compared with broadcasting, but fewer spikelets per panicle and more panicles per unit area than transplanted rice (Table 4). Across the years, mechanical hill-seeded hybrid rice varieties Nei2you 6, Yongyou 9, and Yonyou 12 had 36.0%, 28.3%, and 33.1% higher number of spikelets per panicle, and 17.4%, 16.7%, and 14.3% lower number of panicles per m^2, respectively, compared with the broadcast culture. Compared with the transplanted treatment, 7.3%, 15.5%, and 13.4% decline in spikelets per panicle, and 11.2%, 20.7%, and 28.5% increase in panicle number per m^2 were observed in mechanical hill-seeded hybrid rice varieties Nei2you 6, Yongyou 9, and Yonyou 12, respectively.

Discussion

Our study showed that high-yielding hybrid rice variety Nei2you6 produced maximum grain yield at a seeding spacing of 25 cm×17 cm in two soil fertility fields in both 2011 and 2012. Optimal hill seeding density for hybrid rice has also been reported in previous studies: Peilianyou 986 had the highest grain yield at a spacing of 25 cm×18 cm [20]; Jinyou 207 had the highest grain yield when hill-seeded at 30 cm×20 cm spacing [17]; and hill-seeded Peizataifen and Yuyouxiangzhang [19, 20] gave the highest grain yield at 25 cm×14 cm. However, these studies were conducted at one location in a single season and used only one hybrid rice variety with one to three seeding rates. It is thus difficult to conclude which seeding density can fullfill the yield

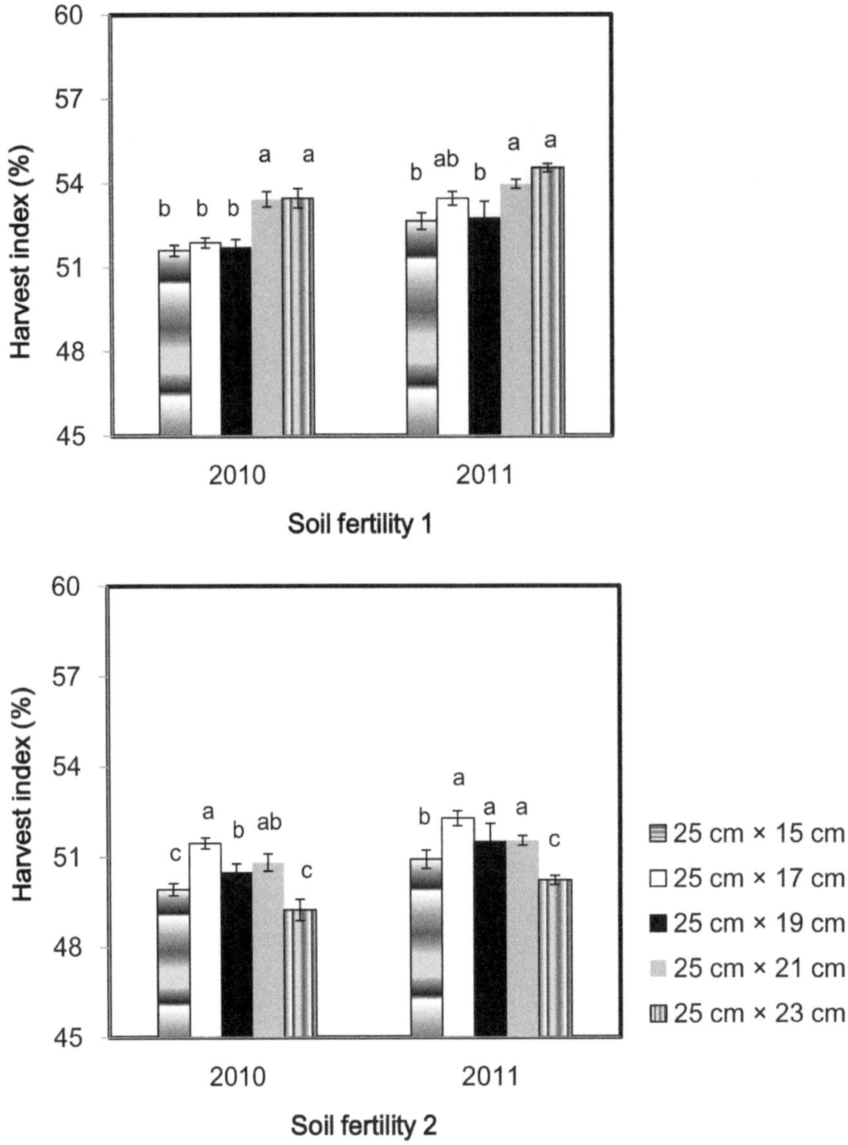

Figure 5. Harvest index of a high-yielding hybrid rice variety Nei2you6 grown at five seeding spacings in soil fertility 1 and soil fertility 2 at the China National Rice Research Institute, Zhejiang Province, China, in 2010 and 2011. Dry weight per shoot was calculated as the ratio of dry weight per hill to productive tiller number per hill.

potential of most high-yielding hybrid rice varieties. In our study, we compared plant yield and dry weight at five hill seeding spacings in two fields of different fertilities during a 2-year period. Besides that, three high-yielding hybrid rice varieties were mechanically seeded with a hill seeder at 25 cm × 17 cm spacing in the on-farm demonstration experiments and compared with broadcasting and transplanting. Mechanical hill seeding outperformed broadcasting in terms of grain yield for all three high-yielding hybrid rice varieties; they also had grain yield equal to or higher than that of transplanted rice. We conclude that, for the south China region, a hill seeding spacing of 25 cm × 17 cm, with three to five seeds per hill, is the appropriate density to be used for single cropping of high-yielding hybrid rice.

There have been extensive studies on the relationships between yield and plant density under nonstressed conditions [24–29]. For decades, rice reseachers in China have emphasized the importance of regulating the compensatory and competitive relationships among plants to increase grain yield by modifying planting density. In the 1960s, dense planting to increase panicle number per unit area was recommended in inbred rice production [30,31]. In the 1980s, with the development of inbred and hybrid rice varieties with large panicles, the advice on rice planting density changed. It was reported that planting at wide spacing decreased plant number per unit area, which improved the morphological traits of the rice plants such as plant height, tiller angle, and leaf angle [32]. It has been widely accepted at present in high-yielding hybrid rice production systems that increasing panicle size, rather than panicle number per unit area, increases yield [32–34]. In considering best management practices for rice, the basic concept applied is the use of young seedlings at optimal spacing to stimulate individual plant growth [35]. Recently, on-farm surveys conducted in Zhejiang, China, showed that some farmers transplanted hybrid rice using extremely wide spacing because of rising labor cost. They opined that the reduced plant number

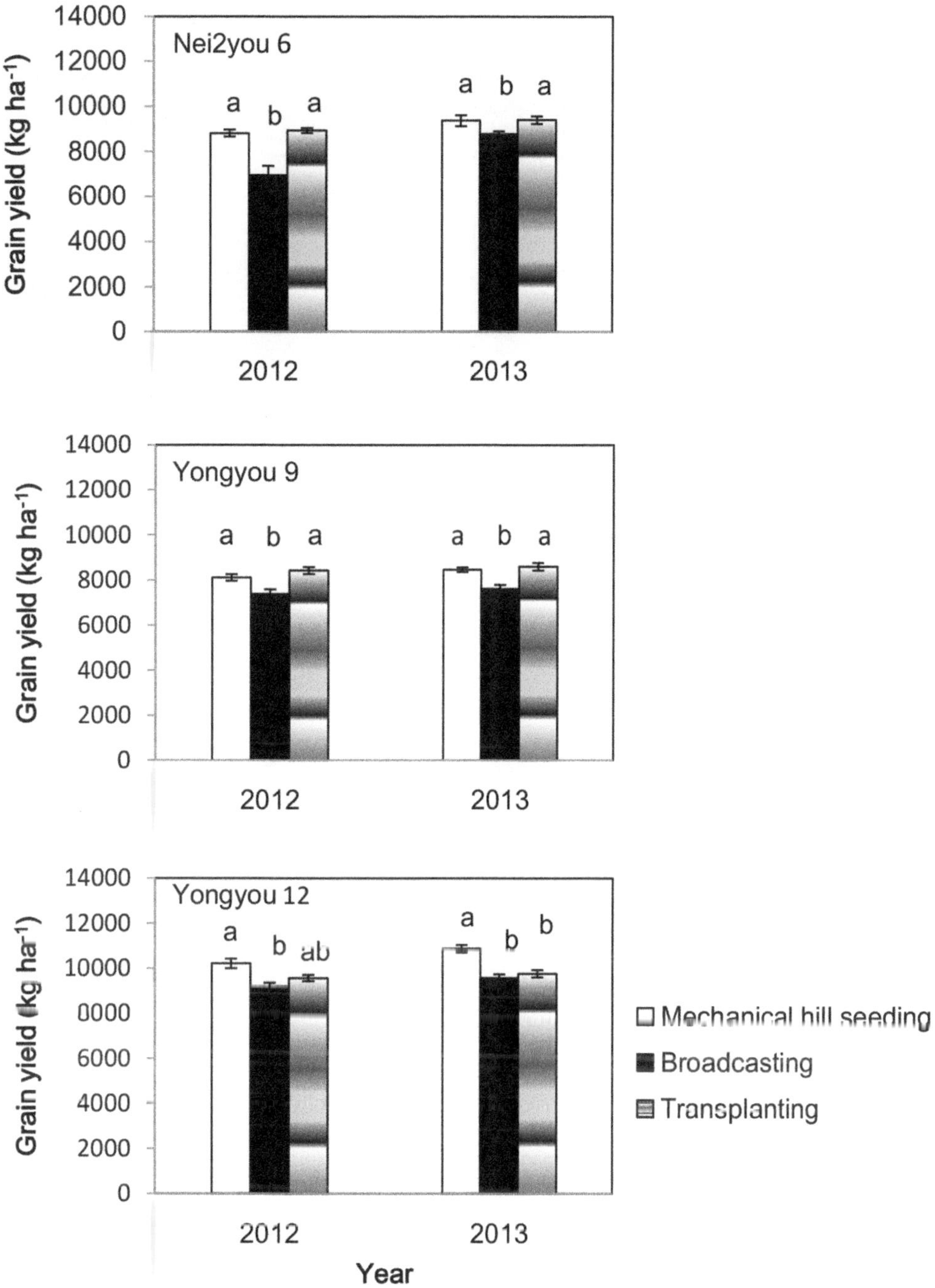

Figure 6. Grain yield of three high-yielding hybrid rice varieties under three plant establishment methods in Fuyang and Yuhang, Zhejiang Province, in 2012 and 2013. Plant spacing for mechanical hill seeding was 25 cm×17 cm; broadcasting and transplanting done according to standard farmer's practice.

caused by low planting density could be compensated for by the improvement in growth of the individual plants [36]. In our study, although grain yield per hill across two years consistently increased with increases in seeding spacing, grain yield per hill of hybrid rice variety Nei2you6 at 25 cm×23 cm spacing was 57.6% and 40.3% higher than at 25 cm×15 cm spacing in soil fertility 1 and 2, respectively. Grain yield per area across two years was highest at 25 cm×17 cm spacing, which was 5.3% and 12.7% higher than at 25 cm×23 cm spacing in soil fertility 1 and 2, respectively. These results suggest that yield per unit area cannot be fully compensated for by the yield increase of individual plants. Therefore, enough seedlings are necessary to achieve high yield. The on-farm surveys in Zhejiang Province also showed that the lowest grain yield was obtained with the widest planting spacing because of the lack of adequate plant population [36].

Rice grain yield per area or per hill is ultimately determined by the number of panicles per area or per hill and the grain yield per panicle. It was reported that yield was positively related to panicle number per unit area [36–37]. In our study, there was a significantly positive correlation between yield per hill and panicles per hill ($r = 0.98^{**}$), and a significantly negative correlation between grain yield per ha and panicles per hill ($r = -0.89^{*}$)

Table 4. Grain yield components of three high-yielding hybrid varieties under three plant establishment methods in Fuyang and Yuhang, Zhejiang Province, China, in 2012 and 2013.

Variety	Year	Treatment	Spikelets panicle^{-1} (no.)	Grain filling (%)	Grain weight (mg)	Panicles m^{-2} (no.)
Nei2you 6	2012	Mechanical hill seeding	176.03 b	83.25 a	26.76 b	235.56 b
		Broadcasting	130.82 c	80.32 b	26.72 b	283.65 a
		Transplanting	193.86 a	84.41 a	27.18 a	213.09 c
	2013	Mechanical hill seeding	174.03 b	85.75 a	27.42 a	250.60 b
		Broadcasting	126.57 c	82.40 b	26.84 b	305.00 a
		Transplanting	183.58 a	86.52 a	27.41 a	224.30 b
Yongyou 9	2012	Mechanical hill seeding	169.87 b	85.16 b	21.88 a	251.52 b
		Broadcasting	133.13 c	83.37 c	21.77 a	303.61 a
		Transplanting	203.08 a	87.63 a	22.17 a	210.60 c
	2013	Mechanical hill seeding	166.79 b	87.55 ab	22.79 a	262.00 b
		Broadcasting	129.37 c	85.49 b	22.45 a	313.00 a
		Transplanting	195.33 a	89.61 a	22.62 a	214.90 c
Yongyou 12	2012	Mechanical hill seeding	388.02 b	82.16 a	19.46 b	171.93 b
		Broadcasting	294.69 c	80.24 b	19.79 ab	202.85 a
		Transplanting	457.15 a	84.45 a	20.21 a	136.71 c
	2013	Mechanical hill seeding	389.09 b	84.46 ab	20.71 a	182.90 b
		Broadcasting	289.35 c	82.40 b	20.62 a	211.30 a
		Transplanting	439.71 a	86.52 a	20.62 a	139.50 c

Plant spacing for mechanical hill seeding was 25 cm×17 cm; broadcasting and transplanting done according to standard farmer's practice.

(Table 5). Panicle number is largely determined by the number of tillers that develop during the vegetative stage [38] and there is a large variation in tillering capacity among rice varieties [39,40]. Environmental factors and agronomic practices also alter tiller production and survival [41–43]. Plant density and fertilization are important factors influencing tiller production in rice [39,44–45]. LAI probably affected tillering by attenuation of light intensity and/or by influencing light quality at the base of the canopy where tiller buds and young tillers are located [46]. Furthermore, there are evidences showing that plant N status and LAI depend on each other in determining tillering [42]. A higher plant N concentration was needed to prevent tillers from dying when LAI was high [42]. In our study, maximum tiller number per hill increased with increases in seeding spacing, suggesting that plant tillering capacity increased with a decrease in seeding density. Compared with soil fertility 1, 17.2% and 10.1% increase in maximun tiller number per hill was obtained at 25 cm×15 cm and 25 cm×17 cm spacing, respectively, in soil fertility 2. However, no difference in maximum tiller number per hill was found between the two soil fertility fields at spacings ranging from 25 cm×19 cm to 25 cm×23 cm, suggesting that the negative effect of dense seeding on plant tillering capacity could be partly compensated for by an increase in nutrition supply in soil fertility 2. However, the rate of change in productive tiller percentage against seeding spacing was different between the two soil fertility fields. The lower productive tiller percentage at 25 cm×15 cm in soil fertility 2 offset its advantage of maximun tiller number per hill, and no significant difference in productive tillers per hill was observed at this spacing between the two fields. A 7% higher productive tillers at 25 cm×23 cm spacing in soil fertility 2 resulted in significantly higher panicle number in this field compared with soil fertility 1. The number of surviving tillers depends mainly on the extent of competition among tillers for carbohydrates produced and N absorbed from the maximum tillering stage to heading [40]. The differences in productive tiller percentage among seeding densities and between two soil fertility fields indicate that nutrition supply could increase panicle number at wide seeding spacing, but it cannot compensate for the negative effect of high seeding density on panicle number per hill because of the reduced productive tiller percentage.

There are compensatory and competitive relationships among tillers and their yield components over a wide range of plant densities [39]. In soil fertility 2, when seeding spacing was increased from 25 cm×17 cm to 25 cm×23 cm, the number of panicles per hill increased by 33.6% in 2010, and by 22.3% in 2011, while grain yield per shoot declined by 12.0% and by 10.0%, respectively. Grain filling percentage declined with increases in seeding spacing in both years in soil fertility 2, and no significant differences in spikelets per panicle and grain weight were found among treatments, suggesting that grain filling percentage decreased with the increase in productive tillers per hill, which resulted in the decline in yield per shoot. However, there was an interaction between seeding density and soil fertility. Less change in grain yield per shoot was in soil fertility 1 than soil fertility 2 at seeding spacings ranging from 25 cm×17 cm to 25 cm×21 cm. Rice plants grown at 25 cm×23 cm spacing had the highest grain yield per shoot in soil fertility 1, but had the the lowest grain yield per shoot in soil fertility 2. It was reported that high-yielding varieties usually have high biomass accumulation during the reproductive phase [47–51]. In our study, although aboveground dry weight per hill at flowering and maturity increased with increases in seeding spacing, aboveground dry weight per shoot was different between the two fertility fields. Rice plants seeded at 25 cm×23 cm spacing had the lowest aboveground dry weight per shoot both at flowering and maturity in soil fertility 2, while they had the highest aboveground dry weight per

Table 5. Correlation coefficients between grain yield and yield components of a high-yielding hybrid rice variety Nei2you6 grown at five seeding spacings at the China National Rice Research Institute, Zhejiang Province, China, in 2010 and 2011.

Yield and yield components	Spikelets per panicle	Grain filling percentage	Grain weight	Panicles per unit area	Panicles per hill	Grain yield per ha	Grain yield per hill
Grain filling percentage	−0.218						
Grain weight	−0.001	−0.802					
Panicles per unit area	0.207	−0.105	−0.449				
Panicles per hill	0.229	−0.842	0.973**	−0.378			
Grain yield per ha	0.159	0.828	−0.947*	0.409	−0.891*		
Grain yield per hill	0.291	−0.729	0.942*	−0.502	0.980**	−0.824	
Grain yield per shoot	0.041	0.965**	−0.791	−0.098	−0.774	0.868	−0.635

*Significant at $p<0.05$.
**Significant at $p<0.01$.

shoot in soil fertility 1. A different response of leaf area per shoot to seeding spacing was one of the factors that influence dry matter accumulation. The top three-leaf area per shoot increased with increases in seeding spacing in soil fertility 1, while it decreased in soil fertility 2. Considering that there are more panicles per hill in soil fertility 2 than in soil fertility 1 at wider seeding spacing and no extra fertilizer application, the lower area of the top three leaves and aboveground dry weight per shoot at 25 cm×23 cm spacing in soil fertility 2 may be attributed to limited N supply and/or N uptake. A recent study reported low N concentrations in shoots due to dilution as a result of more biomass production [43], which suggests that rice plants seeded at wider spacing might need more fertilizer to obtain high yield than those subjected to dense spacing. More fertilizer is needed not only to improve productive tiller percentage at wide spacing but also to increase N concentrations in plants and to improve dry weight accumulation and grain yield per shoot. Nevertheless, we think that the application of extra N to compensate for the reduced grain yield due to low planting density is not a good option because of potential environmental problems and increased cost for rice farmers.

In the on-farm demonstration experiments, mechanical hill seeding of rice at 25 cm×17 cm spacing had equal or higher grain yield than transplanted rice, indicating that this spacing was the optimum seeding density for high-yielding hybrid rice varieties. Hill-seeded rice had more panicles per unit area and less spikelets per panicle compared with transplanted rice.

Acknowledgments

The authors are grateful to Ms. Priscilla Grace Cañas of the International Rice Research Institute, for providing valuable comments on the manuscript.

Author Contributions

Conceived and designed the experiments: DYW. Performed the experiments: DYW SC ZMW CLJ XFZ. Analyzed the data: DYW CMX. Contributed reagents/materials/analysis tools: XFZ. Contributed to the writing of the manuscript: DYW BSC.

References

1. Pandey S, Velasco L (2002) Economics of direct seeding in Asia: patterns of adoption and research priorities. In: Pandey S, Mortimer M, Wade L, Tuong TP, Lopez K, editors. Seeding: research strategies and opportunities. Los Baños: International Rice Research Institute. pp. 3–14.
2. Farooq M, Siddique KHM, Rehman H, Aziz T, Lee DJ, et al. (2011) Rice direct seeding: experiences, challenges and opportunities. Soil Till Res 111: 87–98.
3. Chauhan BS, Mahajan G, Sardana V, Timsina J, Jat ML (2012) Productivity and sustainability of the rice-wheat cropping system in the Indo-Gangetic Plains of the Indian subcontinent: problems, opportunities, and strategies. Adv Agron 117: 315–369.
4. Joshi E, Kumar D, Lal B, Nepalia V, Gautam P, et al. (2013) Management of direct seeded rice for enhanced resource use efficiency. Plant Knowledge J 2: 119–134.
5. Chen F, Chen P (2011) Status and economic evaluation of direct-seeded rice production in south China-based on farmer's investigation. China Rice 17: 1–5 (in Chinese).
6. Chauhan BS (2013) Strategies to manage weedy rice in Asia. Crop Prot. 48: 51–56.
7. Wang D, Zhang X, Lu Y, Xu C, Jin B, et al. (2010) Grain yield difference investigation and development trends discussion of direct seeding rice production in Zhejiang province. China Rice 16: 22–26 (in Chinese).
8. Kabir MH, Saha A, Mollah IU, Kabir MS, Rahman F (2008) Effect of crop establishment methods and weed management practices on the productivity of boro rice in lowland ecosystem. Int J Biol Res 5: 42–51.
9. Mithrasena YJPK, Adikari WP (1986) Effect of density on sheath blight (ShB) incidence. Int Rice Res Notes 11: 20–21.
10. Sittisuang P (1995) Extent of problems and policies in wet-seeded rice production in Thailand. In: Moody K, editor. Constraints, opportunities, and innovations for wet-seeded rice. Los Baños: International Rice Research Institute, pp. 1–5.

11. Islam MF, Sarkar MAR, Islam MS, Parveen S, Hossain MS (2008) Effects of crop establishment methods on root and shoot growth, lodging behavior of Aus rice. Int J Biol Res 5: 60–64.
12. Ying J, Peng S, He Q, Yang H, Yang C, et al. (1998) Comparison of high-yield rice in tropical and subtropical environments. I. Determinants of grain and dry matter yields. Field Crops Res 57: 71–84.
13. Wu W, Zhang H, Qian Y, Cheng Y, Wu G, et al. (2008) Analysis on dry matter production characteristics of super hybrid rice. Rice Sci 15: 110–118.
14. Yang S, Xie L, Zheng S, Li J, Yuan J (2009) Effects of nitrogen rate and transplanting density on physical and chemical characteristics and lodging resistance of culms in hybrid rice. Acta Agron Sin 35: 93–103 (in Chinese, with English abstract).
15. Yoshinaga S (2005) Improved lodging resistance in rice (*Oryza sativa* L.) cultivated by submerged direct seeding using a newly developed hill seeder. Jpn Agric Res Q 39: 147–152.
16. Wang Z, Luo X, Tang X, Ma G, Zhang G, et al. (2010) Precision rice hill-direct-seeding technology and machine based on the combination of agricultural machinery and agronomic technology. J South China Agric Univ 31: 91–95 (in Chinese, with English abstract).
17. Zhu D, Wang J, Xu W, Wang Y, Xu W, et al. (2008) Rice mechanical precision hill seeded experiments. Anhui Agri Sci Bull 14: 133–135 (in Chinese).
18. Lei X, Liu L, Gou W, Ma R, Ren W (2013) Effects of planting methods on culm lodging resistance of indica hybrid rice (*Oryza sativa* L.). Acta Agron Sin 39: 1814–1825 (in Chinese with English abstract).
19. Shu S, Zheng T, Jia X, Luo X, Li G, et al. (2009) Effects of nitrogen fertilizer and density on precision hill-direct-seeding rice I – Yield formation characteristics. Chin Agric Sci Bull 25: 142–146 (in Chinese with English abstract).
20. Cheng J, Luo X, Fan Q, Zhang J, Wu J, et al. (2010) Influence of different planting methods on growth and development characteristics and yield of rice. J Huazhong Agric Univ 29: 1–5 (in Chinese with English abstract).
21. Zeng S, Huang Z, Wang Z, Luo X, Tang X (2014) Effects of different planting density on grain yield of precision hill-drop drilling rice. J Huazhong Agric Univ 33: 12–18 (in Chinese with English abstract).
22. Gong Z, Zhang G, Luo G (1999) Diversify of Anthrosols in China. Pedosphere 9: 193–204.
23. Cheng S, Zheng X, Wang D, Chen L, Xu C, et al. (2012) Effect of long-term paddy-upland yearly rotations on rice (*Oryza sativa*) yield, soil properties, and bacteria community diversity. Sci World J 279641. doi:10.1100/2012/279641.
24. Counce PA, Moldenhauer KAK, Marx DB (1989) Rice yield and plant yield variability responses to equidistant spacing. Crop Sci 29: 175–179.
25. Miller BC, Hill JE, Roberts SR (1991) Plant population effects on growth and yield in water-seeded rice. Agron J 83: 291–297.
26. Gravios KA, Helms RS (1982) Path analysis of rice yield and yield components as affected by seeding rate. Agron J 84: 1–4.
27. Baloch AW, Soomro AM, Javad MA, Ahmed M, Bughio HR, et al. (2002) Optimum plant density for high yield in rice (*Oryza sativa* L.). Asian J Plant Sci 1: 25–27.
28. Oziegbe M, Faluyi JQ (2007) Effects of seeding rate on the yield components of an enhanced rice cultivar (DTPMFe+) *Oryza sativa* Linn. Int J Bot 3: 317–320.
29. Lin X, Zhu D, Chen H, Zhang Y (2009) Effects of plant density and nitrogen application rate on grain yield and nitrogen uptake of super hybrid rice. Rice Sci 16: 138–142.
30. Ding Y (1961) China rice cultivation. Beijing: China Agriculture Press. pp. 53–56 (in Chinese).
31. Chen Y (1959) Chen Yongkang rice cultivation experience. Shanghai: Shanghai Scientific and Technical Publishers. pp. 1–54 (in Chinese).
32. Jiang P, Feng L (1987) The effect of cultivation techniques on plant type in view of the high yield result induced by TFS cultivation method in paddy rice. Chin J Rice Sci 1: 111–117.
33. Lin Q, Zhang H, Cheng G, Zhu J, Lu W (1982) Technical approaches to achieve large-area high-yield cultivation of IR 24 and discuss on the small groups, strong individual cultivation patterns. Jiangsu Agric Sci 9: 1–10 (in Chinese).
34. Ling QH (2000) Crop population quality. Shanghai. Shanghai Scientific and Technical Publishers. pp. 42–107 (in Chinese).
35. Uphoff N (2003) Higher yields with fewer external inputs? The system of rice intensification and potential contributions to agricultural sustainability. Int J Agric Sustain 1: 38–50.
36. Wang D, Zhang X, Zhou C, Zheng G, Zhang G, et al. (2010) Grain yield difference investigation and reasonable planting density analysis of rice production in Zhejiang Province. Acta Agric Zhejiang 22: 330–336 (in Chinese with English abstract).
37. Huang M, Zou Y, Jiang P, Xia B, Ibrahim M, et al. (2011) Relationship between grain yield and yield components in super hybrid rice. Agric Sci China 10: 1537–1544.
38. De Datta SK (1981) Principles and practices of rice production. New York: John Wiley & Sons. pp. 220–358.
39. Wu G, Wilson LT, McClung AM (1998) Contribution of rice tillers to dry matter accumulation and yield. Agron J 90: 317–323.
40. Nuruzzaman M, Yamamoto Y, Nitta Y, Yoshida T, Miyazaki A (2000) Varietal differences in tillering ability of fourteen japonica and indica rice varieties. Soil Sci Plant Nutr 46: 381–391.
41. Hanada K (1993) Tillers. In: Matsuo T, Hoshikawa K, editors. Science of the rice plant. Tokyo: Morphology, Food and Agricultural Policy Center. pp. 222–258.
42. Zhong X, Peng S, Sanico AL, Liu H (2003) Quantifying the interactive effect of leaf nitrogen and leaf area on tillering of rice. J Plant Nutr 26: 1203–1222.
43. Huang M, Zou Y, Jiang P, Xia B, Feng Y, et al. (2011) Yield component differences between direct-seeded and transplanted super hybrid rice. Plant Prod Sci 14: 331–338.
44. Yoshida S, Hayakawa Y (1970) Effects of mineral nutrition on tillering of rice. Soil Sci Plant Nutr 16: 186–191.
45. Fagada SO, De Datta SK (1971) Leaf area index, tillering capacity, and grain yield of tropical rice as affected by plant density and nitrogen level. Agron J 63: 503–506.
46. Zhong X, Peng S, Sheehy JE, Visperas RM, Liu H (2002) Relationship between tillering and leaf area index: quantifying critical leaf area index for tillering in rice. J Agric Sci 138: 269–279.
47. Katsura K, Maeda S, Horie T, Shiraiwa T (2007) Analysis of yield attributes and crop physiological traits of Liangyoupeijiu, a hybrid rice recently bred in China. Field Crops Res 103: 170–177.
48. Horie T, Lubis I, Takai T, Ohsumi A, Kuwasaki K, et al. (2003) Physiological traits associated with high yield potential in rice. In: Mew TW, Brar DS, Peng S, Dawe D, Hardy B, editors. Rice science: innovations and impact for livelihood. Proceedings of the International Rice Research Conference. Beijing: International Rice Research Institute. pp. 117–145.
49. Takai T, Matsuura S, Nishio T, Ohsumi A, Shiraiwa T, et al. (2006) Rice yield potential is closely related to crop growth rate during later reproductive period. Field Crops Res 96: 328–335.
50. Wu W, Zhang H, Qian Y, Cheng Y, Wu G, et al. (2008) Analysis on dry matter production characteristics of super hybrid rice. Rice Sci 15: 110–118.
51. Zhang Y, Tang Q, Zou Y, Li D, Qin J, et al. (2009) Yield potential and radiation use efficiency of "super" hybrid rice grown under subtropical conditions. Field Crops Res 114: 91–98.

19

Early Response of Soil Properties and Function to Riparian Rainforest Restoration

Rose Gageler[1], Mark Bonner[1], Gunnar Kirchhof[1], Mark Amos[2], Nicole Robinson[1], Susanne Schmidt[1]*, Luke P. Shoo[3]

1 School of Agriculture and Food Sciences, The University of Queensland, St Lucia, Australia, **2** Lake Baroon Catchment Care Group, Maleny, Australia, **3** School of Biological Sciences, The University of Queensland, St Lucia, Australia

Abstract

Reforestation of riparian zones is increasingly practiced in many regions for purposes of biodiversity conservation, bank stabilisation, and improvement in water quality. This is in spite of the actual benefits of reforestation for recovering underlying soil properties and function remaining poorly understood. Here we compare remnant riparian rainforest, pasture and reforestation plantings aged 2–20 years in an Australian subtropical catchment on ferrosols to determine the extent to which reforestation restores key soil properties. Of the nine soil attributes measured (total nitrogen, nitrate and ammonium concentrations, net nitrification and ammonification rates, organic carbon, bulk density, fine root biomass and water infiltration rates), only infiltration rates were significantly lower in pasture than remnant riparian rainforest. Within reforestation plantings, bulk density decreased up to 1.4-fold and infiltration rates increased up to 60-fold with time post-reforestation. Our results suggest that the main outcome of belowground processes of early reforestation is the recovery of the soils' physical structure, with potential beneficial ecosystem services including reduced runoff, erosion and associated sediment and nutrient loads in waterways. We also demonstrate differential impacts of two commonly planted tree species on a subset of soil properties suggesting that preferential planting of select species could accelerate progress on specific restoration objectives.

Editor: Kurt O. Reinhart, USDA-ARS, United States of America

Funding: LPS was supported by funding from the Australian Government's National Environmental Research Program (Tropical Ecosystems and Environmental Decisions Hubs). The funders had no role in study design, data collection and analysis, decision to publish, or preparation of the manuscript.

Competing Interests: The authors have declared that no competing interests exist.

* Email: Susanne.Schmidt@uq.edu.au

Introduction

Globally, riparian zones are widely recognized for their critical role in water regulation [1–5], and conservation of biodiversity [6–9]. Clearing of vegetation from stream banks can cause increased soil erosion and decreased water quality through the loss of filtration services, in addition to the loss of critical species habitat [10]. In direct response, degraded stream-banks are increasingly being restored to reinstate ecological functions [11]. Riparian zones are also regarded as foci of restoration opportunity as they are often preferentially renounced from agricultural production over agronomically more valuable areas [9].

Soil physical, chemical and functional properties change with succession from pasture to secondary forest including reforestation plantings in the tropics, subtropics and temperate regions [12–16]. Soil properties such as bulk density, soil water content, fine root biomass, porosity, stocks of nitrogen, carbon and phosphorus, nitrogen cycling, availability of macro- and micronutrients, as well as toxins (metals, salts), have been studied in context of land use change [17–19]. However, most studies focus on mid- to long-term effects of reforestation with less consideration of the short-term effects (<10–20 years) of reforestation plantings, and associated reinstatement of soil functions that might motivate landholders to become actively engaged in restoration. Another knowledge gap is the relative contribution of tree species to outcomes of riparian restoration plantings, and such information could inform decisions on species selection and reforestation goals.

In Australia, riparian reforestation is a major focus of habitat restoration in the tropics and subtropics; over 70% of rainforest restoration projects have targeted banks of rivers and streams [20]. However, much evaluation of the success of reforestation plantings has focused on aboveground attributes [21–23], with less regard to belowground processes including soil function (but see Paul [14] for the recovery of soils via different rainforest restoration pathways). Understanding belowground processes is important to quantify the capacity of reforestation to recover ecosystem function.

Soil properties and functions would intuitively be expected to change over time following reforestation towards a state convergent on intact rainforest ecosystems. Studying riparian soil under remnant rainforest, pasture, and within 20 years post reforestation, we addressed three questions (i) how do soil attributes differ between pasture and remnant riparian rainforest, (ii) how do soils change with time following reforestation, and (iii) do tree species affect soils differentially following reforestation?

The study was motivated by the need for restoration of forest cover in water catchments due to the improvements in water quality that can be achieved [24]. The region targeted for our study is located in the Australian subtropics and has experienced large-scale clearing of rainforest since the mid 19th century. The region is an important water catchment that supplies drinking water to more than 100000 people in the region [25], and numerous land owners and stakeholders are committed to ongoing restoration of riparian zones. Our research aimed to generate knowledge of soil properties to direct future reforestation activities.

Materials and Methods

2.1. Study site

This study focused on riparian areas in the Maleny region in southeastern Queensland, Australia (26°45′S, 152°48′E). The climate is classified as subtropical with an annual rainfall of 1900 mm and annual daily mean temperature maximum and minimum of 23.2°C and 14.2°C, respectively. The soils in the region are classified as permeable red clay-loam ferrosols [26,27] or Nitisol in the US soil taxonomy. Prior to European settlement in the mid-1800s and subsequent land clearing for timber production and agriculture (especially dairy production), the region was dominated by cool-subtropical rainforest classified as complex notophyll vine forest [28,29].

2.2. Sampling design

Field measurements and soil sampling were carried out between May and June 2013. The sampling design was based around riparian reforestation projects established in the region by the Lake Baroon Catchment Care Group (LBCCG). Reforestation sites were grouped into five spatial clusters across the Maleny Plateau (range of elevation 196 to 441 m a.s.l.) to capture broad variation in landform and land use history. A subset of one to three reforestation sites ranging in age between 2 and 20 years were then selected from each cluster for a total of 10 sites (Table 1). Sites were also established in remnant riparian rainforest where available (2 of 5 clusters) and samples were collected from pasture (adjacent to plantings or remnant forest) at all sites in each cluster.

At each reforestation or remnant site, five subplots were established. Each subplot was deliberately located under a different tree species. The five tree species targeted here were derived from a pool of species commonly used in local rainforest plantings, which in turn were selected to include early to late successional species, nitrogen fixing and/or waterlogging tolerant species. The tree species chosen to investigate species-specific effects on soil properties included *Acacia melanoxylon* R.Br. (Fabaceae, N_2 fixer, early succession species), *Ficus coronata* Spin (Moraceae, nitrophile, waterlogging tolerant, early succession species), *Flindersia schottiana* F. Muell. (Rutaceae, mid-late succession), *Homalanthus nutans* (G.Forst.) Guill. (Euphorbiaceae, early succession species) and *Podocarpus elatus* R.Br. ex Endl (Podocarpaceae, mid-late succession, gymnosperm). In the event that one of the chosen tree species was absent from a site, a closely related species was substituted or, if not available, subplots were located under a random tree within the site. Subplots were positioned under the canopy of each selected tree (mean distance from stem 89.5 cm, SD 45.2 cm).

At each reforestation or remnant forest site, we performed the following procedures: (1) collected a single 8 cm diameter core to a depth of 30 cm (hereafter "textural core") and divided it into 0–15 cm and 15–30 cm fractions for textural analysis; (2) performed an infiltration test; and, (3) collected two samples of the surface soil using a 12.5 cm diameter, 8.5 cm tall, 1043 cm^3 volume bulk density ring (hereafter "bulk density ring") from each of five subplots for analysis of physical and chemical properties of the soil. In addition, a single texture core and two to four bulk density ring subplots were collected from adjacent pasture and infiltration measurements performed. For the purposes of analysis, samples and infiltration measures from pasture were calculated as means for each cluster.

Overall, we sampled 10 reforestation sites, 2 remnant riparian rainforest sites (Table 1) and 5 pasture clusters comprising 24 cores for soil texture analysis, 150 cores for soil chemical and physical analyses and 24 infiltration tests. All sites were private land holdings and owners approval was obtained by Mark Amos (Lake Baroon Catchment Care Group, Maleny, info@lbccg.org.au). Upon returning from the field, all soil samples were stored at 4 °C in sealed plastic bags prior to respective analyses.

Table 1. GPS coordinates and ages of reforestation and remnant sites.

Site type	Age (years)	Site location
Reforestation	3	26°46′02 S, 152°49′38 E
Reforestation	3	26°46′17 S, 152°49′51 E
Reforestation	4	26°45′06 S, 152°50′46 E
Reforestation	6	26°42′01 S, 152°53′58 E
Reforestation	7	26°46′10 S, 152°49′57 E
Reforestation	8	26°44′05 S, 152°50′54 E
Reforestation	11	26°44′46 S, 152°51′13 E
Reforestation	11	26°46′12 S, 152°49′39 E
Reforestation	12	26°46′08 S, 152°49′43 E
Reforestation	20	26°42′01 S, 152°53′58 E
Remnant	-	26°45′07 S, 152°51′02 E
Remnant	-	26°44′50 S, 152°51′11 E

All sites were located on private land in the Maleny region. Pasture sites were located immediately adjacent to each forested site. Ages are based on the year that they were planted relative to 2013.

2.3. Texture analysis

Changes in soil texture mostly occur due to weathering over very long time periods and as demonstrated by Maloney [13]. Unless there is erosion or deposition of soil material, texture is unlikely to vary over time since reforestation. For this reason, the purpose of conducting a soil texture analysis at each site was not to determine changes due to reforestation, but to characterize and evaluate background variation in soil types among sites that might warrant differential treatments of sites in the analysis. Soil samples were not dispersed prior to analysis and as such, the calculated percentages represent the relative size of secondary particles (aggregates) present rather than primary particle size *sensu stricto*. This was considered a more relevant measure of soil function than particle size determined using standard methods, as it gives a relative measure of soil aggregation and structure. Furthermore, the ferrosols in this study are recognized as being subplastic, with clay-sized particles being primarily composed of sesquioxides that do not readily break down into primary particles [30]. These analyses of soil aggregation were conducted on air-dried soil from 0–15 and 15–30 cm depth cores according to the hydrometer method for soil texture analysis [31,32]. For each sample, 40 g of air-dried soil was ground and mixed with 100 ml 50 g l^{-1} Sodium hexametaphosphate solution and filled to 500 ml with deionized water. Samples were shaken overnight and then emptied into a 1 l cylinder for analysis. Cylinders were filled to 1 l with deionized water, mixed with a plunger and left to settle. Temperature and specific gravity readings were taken 40 seconds and 7 hours after plunging [32]. Using temperature and specific gravity readings, the fraction of sand, silt and clay sized particles in each soil sample was then calculated as a percentage.

2.4. Soil physical properties: bulk density and infiltration rates

Soil bulk density was determined from the second bulk density ring in each subplot. Fresh soil was oven dried to a constant weight at 105 °C. Soil bulk density is expressed as grams of dry soil per volume. To quantify fine root biomass, soil from one bulk density ring per subplot was passed through a sieve (8 mm mesh) and roots were manually removed. Roots were then oven dried at 65 °C to a constant weight, and results expressed as grams of fine roots per kilogram of dry soil.

Infiltration rates were measured using a ponded disk permeameter [33]. Soils were assumed to be at field capacity as the region had received in excess of 30 mm in the week preceding the first set of measurements and over 100 mm in the week prior to the second sampling visit. Infiltration tests were performed at a random, relatively flat area representative of the site. Tests in the adjacent pasture were performed at a random point beyond the farthest extent of the neighbouring forest canopy. Upon setting up of the permeameter, grass and other ground covers were clipped to just above the soil surface and the soil pre-moistened with rainwater to reduce the time needed to reach steady state infiltration. The permeameter was filled with rainwater collected from water tanks on local landholder properties to ensure that water quality was similar to that occurring under natural condition. Once the disk permeameter was in place, measurements of water level changes were recorded until the time interval between measurements was stable. As soils were close to field capacity at the start of measurements, steady state was reached quickly. Infiltration rate was calculated as the average volume change per second and then converted into millimeters per hour.

2.5. Soil chemical properties: nitrogen and organic carbon

Total concentrations of soil nitrogen (TN) and organic carbon (SOC) were determined using a TruSpec CHN analyzer (LECO Australia Pty. Ltd., Castle Hill, NSW 2154, Australia), using dried, ground and sieved soil (2 mm sieve). Nitrate (NO_3^-) and ammonium (NH_4^+) were extracted with potassium chloride (KCl) and quantified via colourimetric assays. To remove NO_3^- and NH_4^+ from soil exchange sites, 15 g of fresh sieved soil was mixed with 1 M KCl in a 15 ml falcon tube and shaken on an orbital shaker for 1 hour. Samples were centrifuged at 4000 rpm for 4 minutes and 1 ml of the supernatant was frozen until analysis via colourimetric assays according to standard procedures [34,35]. Net nitrification and ammonification rates were calculated from the difference in concentration between fresh soil samples and those incubated for 20 days at 28°C. Briefly, 15 g of soil were weighed into 50 ml falcon tubes and incubated for 20 days. Soils were kept at field moisture by watering daily with deionized water. Seeds that germinated during the incubation period were removed. Concentrations of NO_3^- and NH_4^+ in the incubated soils following KCl extraction were quantified using colourimetric assays as described above. Net nitrification and ammonification rates are expressed as mg nitrate or ammonium generated per kg dry soil per day.

2.6. Statistical analysis

Statistical analyses were performed in R version 3.0.1 [36]. To avoid non-independence among subsamples, we derived the aggregated mean of soil attributes for each site (reforestation or remnant rainforest) or cluster (pasture) and used these values as inputs in all analyses, excepting those determining species-specific effects which used values from a single subplot within each site. Pairwise correlation tests (Spearman's) were performed to examine associations between tested soil attributes. Soil attributes under both pasture and remnant riparian rainforest were directly compared using a one-way analysis of variance (ANOVA). Developmental trajectories of soil attributes within reforested sites (site level and separately for specific tree species within sites) were examined using linear regressions between either age or log-transformed age and individual soil attributes. The age term was log-transformed to account for potential non-linear developmental trajectory of reforestation plantings; often showing a rapid change immediately post planting followed by diminishing returns as the system reaches natural equilibrium. For the purposes of quantifying response trajectories with age, we assume that pasture represents 0 years post reforestation.

Results

Of the 56 subplots within sites, 47 (84%) were associated with one of the five target trees. *Homalanthus nutans* was the species most commonly absent, being present in 50% of reforested sites. Soil aggregation showed only minor variation among sites and land uses so was not considered further in subsequent analyses (sand sized aggregates: mean = 60%, range = 46–76%, clay sized aggregates: mean = 30%, range = 14–46%).

3.1. Correlations between variables

Infiltration rates were inversely correlated with bulk density ($P = 0.005$) but were not correlated with soil particle size (percentage sand, silt and clay) at either 0–15 cm or 15–30 cm depth ($P > 0.5$, not shown). Bulk density was correlated with both SOC and TN ($P = 0.02$ and 0.01 respectively) (Table 2). Pairwise

Table 2. Pairwise correlations (Spearman's) between soil attributes across pasture, remnant rainforest and reforestation sites (n = 5, 2, 10 respectively).

Variable	Variable	Spearman's correlation coefficient	P value
Bulk density	SOC	−0.554	0.023
Bulk density	TN	−0.615	0.010
Infiltration	Bulk density	−0.654	0.005
TN	SOC	0.784	<0.001
Ammonification rate	Nitrification rate	−0.664	0.005
Ammonification rate	Fine root biomass	0.662	0.005
Nitrate	Ammonification rate	−0.529	0.031
Nitrate	TN	0.495	0.045
Nitrate	SOC	0.630	0.008

Only significant pairwise correlations (P<0.05) are shown.

correlations revealed strong associations between soil nitrogen-related variables.

3.2. Differences between pasture and remnant rainforest

There was no apparent difference between pasture and remnant riparian rainforest in the tested soil attributes except infiltration rates (Table 3). Infiltration rates were significantly greater in the remnant rainforest (1421±995 mm h^{-1}) than in pasture sites (220±151 mm h^{-1}) ($F = 9.526$, $DF = 5$, $P = 0.027$, Table 3).

3.3. Recovery of soil properties under reforestation plantings

Only bulk density and infiltration rates showed a relationship with age within 20 years post reforestation (Table 4). In each case, models with a log-transformed age term explained more of the variation than models with an un-transformed age term. Bulk density of the soils ranged from 0.312 to 0.991 g cm^{-3}, and decreased significantly with time post reforestation ($F = 6.312$, $DF = 13$, $P = 0.026$, Figure 1). Conversely, infiltration rates significantly increased ($F = 10.56$, $DF = 13$, $P = 0.006$) within 20 years post reforestation (Figure 1). Infiltration rates in reforestation sites ranged from 149 mm h^{-1} at a 7-year-old reforestation site to over 1800 mm h^{-1} at an 11-year-old reforestation site. Infiltration rates of pasture sites varied from 3 mm h^{-1} to over 650 mm h^{-1}, although this variation was slightly less when data was aggregated by cluster.

3.4. Species specific effects on the recovery of soils under reforestation plantings

Species-specific effects were apparent for two of five tree species when tested soil attributes were modeled against log-transformed age (Figure 2). Ammonium concentrations in soil associated with *Flindersia schottiana* decreased with time post reforestation ($F = 4.97$, $DF = 8$, $P = 0.056$). Fine root biomass in soil associated with *Ficus coronata* decreased through time within 20 years post reforestation ($F = 4.921$, $DF = 8$, $P = 0.057$).

Discussion

Replanting riparian areas can be driven by a variety of objectives including increasing landscape connectivity for wildlife movements and biogeochemical functions such as trapping nutrients, sediment, pesticides, bank stabilisation, improved water quality and flood mitigation [10]. Here, we studied the trajectory of riparian reforestation to identify key changes in soil attributes through the transition from pasture to riparian tree plantings of increasing age and remnant rainforest. We showed that the major

Table 3. ANOVA results from a comparison of soil attributes between remnant riparian rainforest (n = 2) and pasture (n = 5).

Variable	ANOVA P value	Mean pasture ±1 SD	Mean remnant ±1 SD	Contrast remnant (R) vs. pasture (P)
Fine root biomass (g kg^{-1})	0.09	6.64±4.05	13.00±0.38	NS
Bulk density (g cm^{-3})	0.36	0.69±0.08	0.61±0.12	NS
SOC (%)	0.60	6.15±1.83	7.31±4.15	NS
TN (%)	0.85	0.71±0.14	0.74±0.26	NS
Nitrate (mg kg^{-1})	0.26	11.8±7.24	22.7±17.9	NS
Nitrification rate (mg kg^{-1} soil d^{-1})	0.65	0.67±0.75	0.39±0.03	NS
Ammonium (mg kg^{-1})	0.46	4.36±4.27	7.85±8.10	NS
Ammonification rate (mg kg^{-1} soil d^{-1})	0.73	0.65±0.98	0.65±0.98	NS
Infiltration rate (mm h^{-1})	**0.03**	**220±151**	**1421±995**	**R>P**

Bold text indicates significance at 0.05 level, NS not significant if P>0.05, SD standard deviation.

Table 4. Relationship between tested soil attributes and the age of the reforestation sites (n = 15).

Variable	Age				Log Age			
	Slope	R^2 value	F statistic	P value	Slope	R^2 value	F statistic	P value
Fine root biomass (g kg^{-1} soil)	0.02	0.001	0.02	0.90	1.14	0.03	0.34	0.57
Bulk density (g cm^{-3})	−0.009	0.28	5.16	**0.04**	−0.12	0.32	6.31	**0.03**
SOC (%)	−0.03	0.01	0.15	0.70	−0.03	<0.001	0.002	0.97
TN (%)	−0.001	0.004	0.05	0.83	0.005	<0.001	0.004	0.95
Nitrate (mg kg^{-1})	0.09	0.002	0.03	0.86	3.58	0.03	0.38	0.55
Nitrification rate (mg kg^{-1} d^{-1})	0.03	0.13	1.93	0.19	0.30	0.07	0.95	0.35
Ammonium (mg kg^{-1})	−0.09	0.04	0.51	0.49	−1.72	0.11	1.55	0.24
Ammonification rate (mg kg^{-1} d^{-1})	−0.03	0.11	1.54	0.24	−0.47	0.15	2.25	0.16
Infiltration rate (mm h^{-1})	54.9	0.39	9.77	**0.008**	678	0.45	10.6	**0.006**

Pasture is assumed to be 0 years post reforestation. Results from linear models with both age and log-transformed age are shown. P values in bold are significant at the 0.05 level.

belowground outcomes within 3 to 20 years since planting are physical and associated functional attributes of the soil.

4.1. Differences in soil properties between pasture and remnant rainforest

Infiltration rate was the only soil variable studied here that differed significantly between pasture and remnant riparian rainforest. Pasture infiltration rates were ≈15% of those of riparian rainforest, consistent with the finding of an infiltration rate in pasture 20% of that in remnant rainforest by Peñuela and Drew [15]. We detected no statistically significant differences in other soil properties between pasture and remnant rainforest soils. Limited effects of land use on selected chemical properties are consistent with previous findings in Australian on ferrosol soils [14]. However, it was expected that attributes such as bulk density and fine roots would differ between land use types [14]. It is possible that these attributes differ at a scale inscrutable beyond background variability between sites, possibly necessitating longitudinal study of permanent plots following reforestation. Limited availability of remnant forest within sampling clusters, a consequence of extensive clearing of riparian rainforest in the region, was also a contributing factor in the comparatively low statistical power for detecting changes in measured attributes. Nevertheless, the data also indicate that there is little degradation of soil with respect to N and C stock in the studied systems.

4.2. Recovery of soils under reforestation plantings

Bulk density and infiltration rates both changed significantly with time under the reforestation plantings. Bulk density showed a negative trend with age whereas infiltration increased with time in the reforestation plantings. The observed trend for decreasing bulk density is consistent with previous studies globally [9,14,37,38]. The higher bulk density and lower infiltration rates of the pastures may be partially attributed to compaction by livestock and/or machinery [39]. Lower bulk densities are often associated with soil of forests and reforestation plantings because greater presence of roots and soil biota and higher concentrations of organic matter enhance soil porosity. In remnant forest and reforestation plantings, the high turnover of roots results in more open soil pores as roots grow and subsequently die [40]. Additionally, pastures generally have more shallow and non-woody roots while forests have a broader range of fine to large woody roots including with larger, woody roots that leave conduits in the soil post death [41–43].

Higher densities and diversities of soil organisms in forests and reforestation plantings [44,45] contribute to the aeration of soil and reduce bulk density. High soil organic matter is also associated with low bulk density [18], a trend seen in this study as bulk density was negatively correlated with SOC. This was expected as SOC is indicative of litter dynamics, turnover and decomposition rates and is responsible for providing low-density mass to the soil and improving soil structure [15,18,37,46]. The bulk density values in our study (0.65 g cm^{-3}) are low in comparison to average soils (>1.2 g cm^{-3}). Paul [14] also reported low bulk densities (0.6–0.8 g cm^{-3}) for similar ferrosols in reforestation plantings and remnant rainforest, confirming the low bulk density of this soil type. We attribute these low bulk densities to the friable nature of the ferrosols in the study region, high porosity and the high organic matter content in the topsoil. We also propose that the low bulk density contributes to the trend of increasing infiltration rates with time following reforestation as the two variables were inversely correlated.

Similar to prior studies, we found a significant increase in infiltration rates with reforestation age, with infiltration rates of the

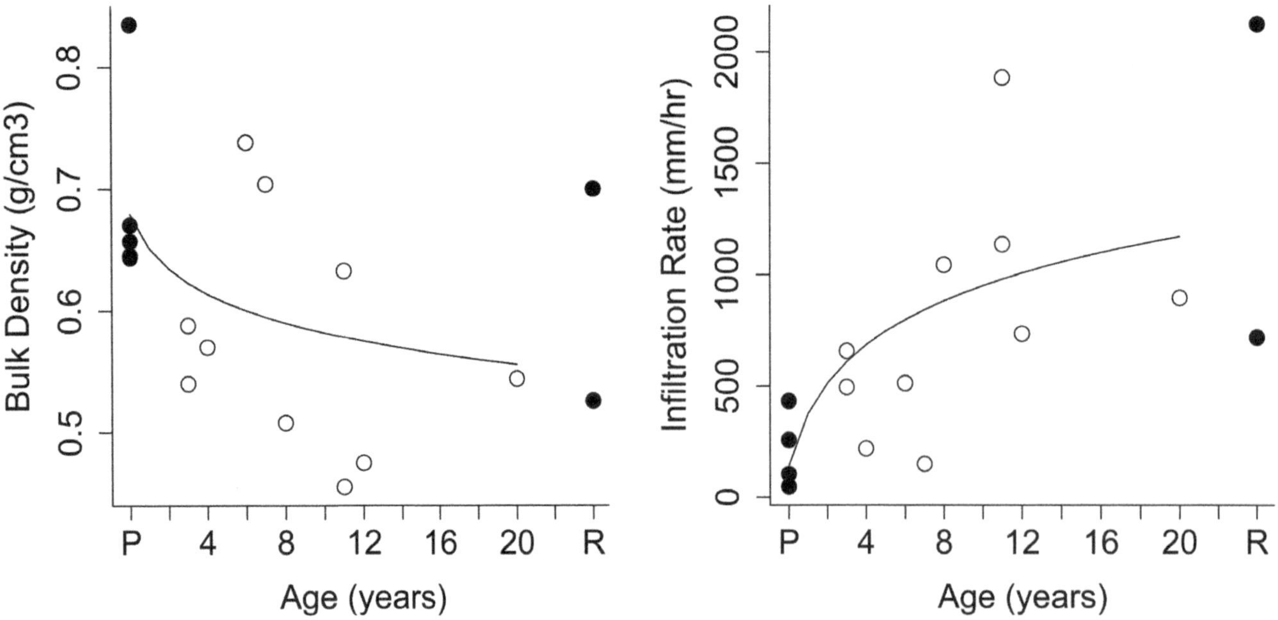

Figure 1. Differences in bulk density (left panel) and infiltration rates (right panel) among pasture (P), reforestation sites aged 2 to 20 years, and remnant rainforest (R) (n = 5, 10 and 2 respectively). Lines show fitted values of the modeled relationship between soil properties and log-transformed age of reforestation. Open circles show reforestation plantings, filled circles indicate pasture clusters and remnant riparian rainforest.

pastures being always lower than those of adjacent reforestation sites. We note a strong positive pattern with age within the first 5–10 years post reforestation, indicating that even young tree plantings are able to promote increased infiltration rates [37,47]. The trajectory in infiltration rates from the plantings (up to 20 years old) indicate that increases of 182 mm h^{-1} can be expected for each year in the first decade since planting, with most gains seen within the first 10 years post-reforestation. This trend is consistent with findings of Bharati [48] of short term changes in infiltration rates under riparian buffers when compared to adjacent cultivated fields and pastures, noting a 5-fold increase in infiltration rates under riparian buffers after 6 growing seasons. Our finding that infiltration rates are strongly enhanced soon after reforestation has meaningful consequences for decision making on land use, water quality and conservation with reduced run-off, slowed surface erosion and reduced non-point source pollution as beneficial outcomes [49].

Here, two of the tested nine soil attributes varied between pasture and reforestation sites. One possible explanation is the recentness of the plantings. Reforestation sites were relatively recent compared to studies that focused on plantings of much greater age (up to 60 years) [13,38,43] and this may have contributed to the uniformity between sites in our study. However, the common lack of differences between the studied pasture and remnant riparian rainforest sites suggests that inherent high soil fertility may be a larger contributor to soil properties than land cover, as has been suggested previously [43,50].

4.3. Species specific effects on the recovery of soils under reforestation plantings

In previous studies, species-specific effects were not used to help elucidate changes in soil properties post-reforestation [14,51], although it is known that species-specific effects can modify soil properties [52]. We detected two species-specific effects in the five tree species and eight variables examined. *Ficus coronata* and *Flindersia schottiana* showed negative relationships with time since reforestation for fine root biomass and soil ammonium levels respectively. We hypothesized that as reforested sites become more established, fine root biomass increases to access nutrients in the surface soil layer of rainforests [41,43,53]. The studied *F. coronata* specimens displayed the opposite trend to that expected, showing a negative fine root relationship with time post-reforestation. This decreasing trend exhibited by *F. coronata* may be explained by a need for greater structural support through large woody roots in which they invest carbon at the expense of fine surface roots, or the ability to access nutrients from the deeper soil, however greater sampling at depth would be needed to confirm these hypotheses. Nevertheless, if verified, an ability to shift root production into in deeper layers could be a useful trait providing physical impediment in the subsoil, particularly where greater soil stability is a desired goal of reforestation.

Another species-specific effect was observed with *F. schottiana* showing a decreasing relationship with ammonium concentrations and time post-reforestation. This result was not entirely unusual as ammonium is readily converted to nitrate in some soils and nitrate may be preferentially used by plants [54]. However, we did not detect higher nitrate concentration in soil under *F. schottiana*, which would be expected if nitrification is simulated. This may be due to high N uptake and demand of *F. schottiana,* which was not investigated here. However, this finding conforms broadly to the results of prior studies [14,38,55], which saw that ammonium levels were highest in pasture and lowest in forest.

It is noteworthy that the species-specific effects detected were not the same as those that were seen at a site level. When means at the site level were considered, only infiltration and bulk density were significant, however at a species level, both *F. schottiana* and *F. coronata* showed significant variations through time for ammonium levels and fine root biomass respectively. This suggests that while individual species may be able to preferentially alter their immediate surroundings, their influence is diminished at a

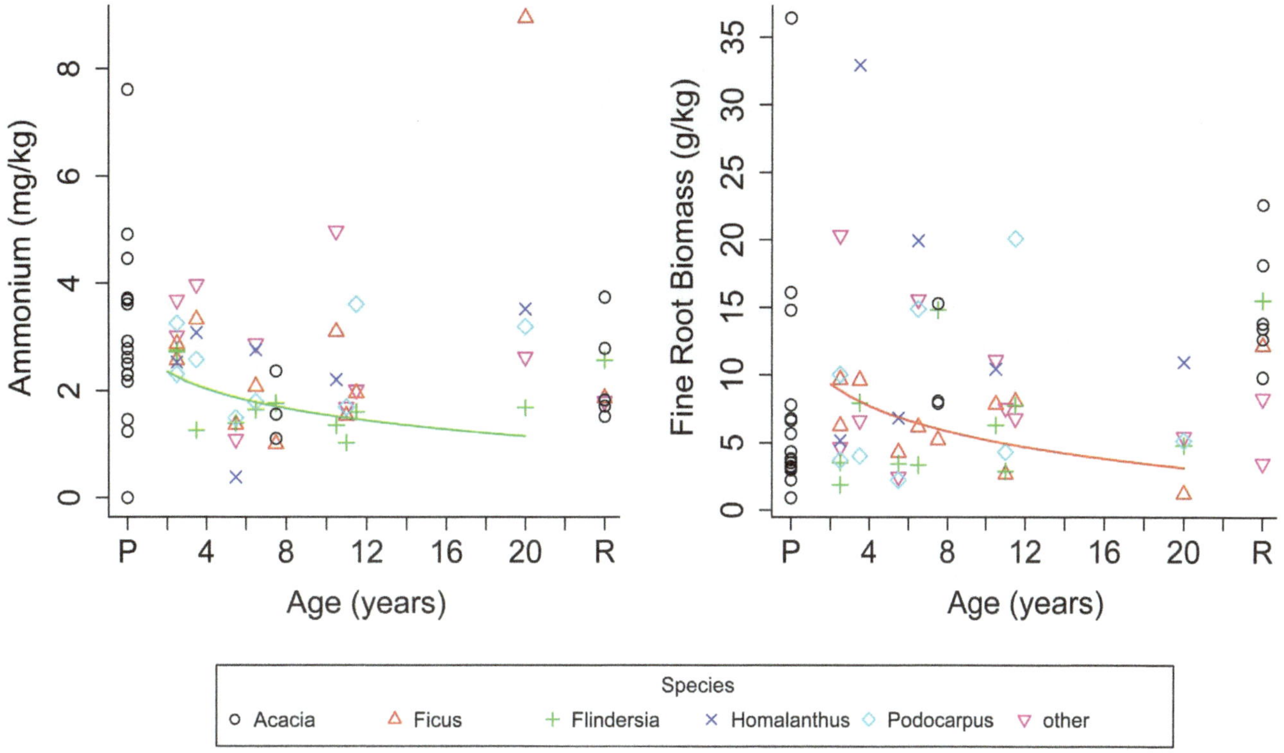

Figure 2. Differences in ammonium (left panel) and fine root biomass (right panel) over time under five tree species in pasture (P), reforestation sites aged 2 to 20 years, and remnant rainforest (R). Genus names refer to the selected tree species under which the soil samples were taken. "Other" refers to soil samples taken from non-specified tree species or pastoral sites. Lines show fitted values of the modeled relationship between individual soil properties and log-transformed age for species with significant models - *Flindersia schottiana* (a) and *Ficus coronata* (b).

site level. A possible explanation is that the diverse mix of native tree species, representing a variety of physiotypes, and the individual species specific effects on soil are less influential than the effects of the overall plantings on soil properties. Had the reforestation planting been of a simpler species mix or the relative proportion of select species been greater, soil traits may have reflected more strongly particular species-specific effects. This indicates that particular reforestation outcomes may be facilitated by explicit selection of certain tree species. Further studies have to determine the efficacy and benefits of a less species-rich reforestation species mixture.

Conclusion

Reduced bulk density and enhanced infiltration rates were the key immediate outcomes resulting from riparian reforestation plantings. It is noteworthy that the most change occurred within just 5 to 10 years post-planting suggesting that this form of active restoration can reinstate important soil functions within a comparatively short time. Given the linkages between increased infiltration and ecosystem services such as nutrient buffering, filtration of runoff and sediments, we conclude that the reforestation is, by extension, achieving the objective of improving water quality. This finding has consequences for land managers and conservationists interested in lowering runoff, non-point source erosion and limiting loss of topsoil, all of which are major concerns and motivators for reforestation. Further, soil infiltration rates are intricately linked with several soil properties influenced by slight changes in soil structure well before changes are perceivable in other soil properties [56]. This signifies that the reforestation plantings are on a positive trajectory to restoring other soil properties and ecosystem function with infiltration rates as front-runner in these changes.

Acknowledgments

We thank the landholders that participated in the study for generously allowing access to their land. We also thank the Lake Baroon Catchment Care Group (LBCCG) for liaising with landholders, being so knowledgeable and motivated, and accommodating to our needs. We acknowledge the advice and thank Jessica Vogt and Richard Brackin on sample processing.

Author Contributions

Conceived and designed the experiments: RG GK MA SS LPS. Performed the experiments: RG MB MA SS LPS. Analyzed the data: RG GK LPS. Contributed reagents/materials/analysis tools: RG MA GK SS LPS NR. Contributed to the writing of the manuscript: RG MB GK MA SS LPS.

References

1. Connor S, Nelson PN, Armour JD, Hénault C (2012) Hydrology of a forested riparian zone in an agricultural landscape of the humid tropics. Agr Ecosyst Environ. 180: 111–122.
2. Meynendonckx J, Heuvelmans G, Muys B, Feyen J (2006) Effects of watershed and riparian zone characteristics on nutrient concentrations in the River Scheldt Basin. Hydrol Earth Syst Sci Discuss 3: 653–679.
3. Orzetti LL, Jones RC, Murphy RF (2010) Stream condition in Piedmont streams with restored riparian buffers in the Chesapeake Bay watershed. J Am Water Resour As 46: 473–485.
4. Russo TA, Fisher AT, Roche JW (2012) Improving riparian wetland conditions based on infiltration and drainage behavior during and after controlled flooding. J Hydrol 432: 98–111.
5. Vidon P (2012) Towards a better understanding of riparian zone water table response to precipitation: surface water infiltration, hillslope contribution or pressure wave processes? Hydrol Processes 26: 3207–3215.
6. Charron I, Lalonde O, Roy AG, Boyer C, Turgeon S (2008) Changes in riparian habitats along five major tributaries of the saint Lawrence river, Québec, Canada: 1964–1997. River Res Appl 24: 617–631.
7. Griscom HP, Griscom BW, Ashton MS (2009) Forest Regeneration from Pasture in the Dry Tropics of Panama: Effects of Cattle, Exotic Grass, and Forested Riparia. Restor Ecol 17: 117–126.
8. Scott SJ, McLaren G, Jones G, Harris S (2010) The impact of riparian habitat quality on the foraging and activity of pipistrelle bats (Pipistrellus spp.). J Zool 280: 371–378.
9. Rodrigues Nogueira Jr L, de Moraes Goncalves JL, Lex Engel V, Parrotta J (2011) Soil dynamics and carbon stocks 10 years after reforestation of degraded land using Atlantic Forest tree species. Forest Syst 20: 536–545.
10. Naiman RJ, Décamps H, McClain ME (2005) Riparia: ecology, conservation and management of streamside communities. New York: Academic Press.
11. Sweeney BW, Bott TL, Jackson JK, Kaplan LA, Newbold JD, et al. (2004) Riparian deforestation, stream narrowing, and loss of stream ecosystem services. Proc Nat Acad Sci USA 101: 14132–14137.
12. Guo LB, Gifford M (2002) Soil carbon stocks and land use change: a meta analysis. Glob Change Biol 8: 345–360.
13. Maloney KO, Garten CT, Ashwood TL (2008) Changes in soil properties following 55 years of secondary forest succession at Fort Benning, Georgia, USA. Restor Ecol 16: 503–510.
14. Paul M, Catterall CP, Pollard PC, Kanowski J (2010) Recovery of soil properties and functions in different rainforest restoration pathways. Forest Ecol Manage 259: 2083–2092.
15. Peñuela MC, Drew AP (2004) A model to assess restoration of abandoned pasture in Costa Rica based on soil hydrologic features and forest structure. Restor Ecol 12: 516–524.
16. Zhu B, Li Z, Li P, Liu G, Xue S (2010) Soil erodibility, microbial biomass, and physical-chemical property changes during long-term natural vegetation restoration: a case study in the Loess Plateau, China. Ecol Res 25: 531–541.
17. Bronick CJ, Lal R (2005) Soil structure and management: a review. Geoderma 124: 3–22.
18. Murty D, Kirschbaum MUF, McMurtrie RE, McGilvray H (2002) Does conversion of forest to agricultural land change soil carbon and nitrogen? a review of the literature. Glob Change Biol 8: 105–123.
19. Schoenholtz SH, Van Miegroet H, Burger JA (2000) A review of chemical and physical properties as indicators of forest soil quality: challenges and opportunities. Forest Ecol Manag 138: 335–356.
20. Catterall C, Harrison DA (2006) Rainforest Restoration Activities in Australia's Tropics and Subtropics. Cairns, Australia: Cooperative Research Centre for Tropical Rainforest Ecology and Managemen, Rainforest CRC. 0864437692 0864437692.
21. Catterall CP, Kanowski J, Wardell-Johnson GW, Proctor H, Reis T, et al. (2004) Quantifying the biodiversity values of reforestation: perspectives, design issues and outcomes in Australian rainforest landscapes. In: Conservation of Australia's Forest Fauna (ed Lunney, D), Vol. 2. Royal Zoological Society of New South Wales, Sydney 2: 359–393.
22. Kanowski J, Catterall C, Wardell-Johnson G, Proctor H, Reis T (2003) Development of forest structure on cleared rainforest land in eastern Australia under different styles of reforestation. Forest Ecol Manage 183: 265–280.
23. Kanowski J, Catterall CP (2010) Carbon stocks in above-ground biomass of monoculture plantations, mixed species plantations and environmental restoration plantings in north-east Australia. Ecol Manage Restor 11: 119–126.
24. Wickham JD, Wade TG, Riitters KH (2011) An environmental assessment of United States drinking water watersheds. Landscape Ecol 26: 605–616.
25. Australian Bureau of Statistics (2011) Census.
26. Ellison L, Coaldrake JE (1954) Soil mantle movement in relation to forest clearing in Southeastern Queensland. Ecology 35: 380–388.
27. Willmott WF (2007) Rocks and landscapes of the Sunshine Coast. Brisbane: Geological Society of Australia, Queensland Division.
28. Nakamura A, Catterall CP, House APN, Kitching RL, Burwell CJ (2007) The use of ants and other soil and litter arthropods as bio-indicators of the inpacts of rainforest clearing and subsequent land use. J Insect Conserv 11: 177–186.
29. Smith I (2001) Juggling-balancing Tourism and Biodiversity in a Small, Valuable Reserve - Mary Cairncross Scenic Reserve. In: Boyes B, editor. ***Biodiversity Conservation "From Vision to Reality"***. Gatton, Queensland: Lockyer Watershed Management Association (LWMA) Inc. - Lockyer Landcare Group. 81–84.
30. McIntyre D (1976) Subplasticity in Australian soils. I. Description, occurrence, and some properties. Aust J Soil Res 14: 227–236.
31. DeForest JL (2011) Soil texture - Hydrometer method. Lab protocols for testing soils Ohio University, Athens, Ohio: The Department of Environmental and Plant Biology.
32. Robertson GP, Coleman DC, Bledsoe CS, Sollins P (1999) Standard soil methods for long-term ecological research. New York: Oxford University Press.
33. Perroux KM, White I (1988) Designs for Disc Permeameters. Soil Sci Soc Am J 52: 1205–1215.
34. Kandeler E, Gerber H (1988) Short-term assay of soil urease activity using colorimetric determination of ammonium. Biol Fert Soils 6: 68–72.
35. Miranda KM, Espey MG, Wink DA (2001) A rapid, simple spectrophotometric method for simultaneous detection of nitrate and nitrite. Nitric Oxide 5: 62–71.
36. R Core Team (2013) R: A language and environment for statistical computing. Vienna, Austria: R Foundation for Statistical Computing.
37. Neris J, Jiménez C, Fuentes J, Morillas G, Tejedor M (2012) Vegetation and land-use effects on soil properties and water infiltration of Andisols in Tenerife (Canary Islands, Spain). Catena 98: 55–62.
38. Reiners WA, Bouwman AF, Parsons WFJ, Keller M (1994) Tropical rain forest conversion to pasture: changes in vegetation and soil properties. Ecol Appl 4: 363–377.
39. Bell LW, Kirkegaard JA, Swan A, Hunt JR, Huth NI, et al. (2011) Impacts of soil damage by grazing livestock on crop productivity. Soil Tillage Res 113: 19–29.
40. Beven K, Germann P (1982) Macropores and water flow in soils. Water Resour Res 18: 1311–1325.
41. Finér L, Ohashi M, Moguchi K, Hirano Y (2011) Factors causing variation in fine root biomass in forest ecosytems. Forest Ecol Manage 261: 265–277.
42. Jiménez EM, Moreno FH, Peñuela MC, Patiño S, Lloyd J (2009) Fine root dynamics for forests on contrasting soils in the Colombian Amazon. Biogeoscience 6: 2809–2827.
43. Powers JS, Perez-Aviles D (2012) Edaphic factors are a more important control on surface fine roots than stand age in secondary tropical dry forests. Biotropica 45: 1–9.
44. Colloff MJ, Pullen KR, Cunningham SA (2010) Restoration of an ecosystem function to revegetation communities: The role of invertebrate mscropores in enhancing soil water infiltration. Restor Ecol 18: 65–72.
45. Nakamura A, Proctor H, Catterall CP (2003) Using soil and litter arthropods to assess the state of rainforest restoration. Ecol Manage Restor 4: S20–S28.
46. Laub BG, McDonough OT, Needelman BA, Palmer MA (2013) Comparison of designed channel restoration and riparian buffer restoration effects on riparian soils. Restor Ecol: 1–9.
47. Wood HB (1977) Hydrologic differences between selected forested and agricultural soils in Hawaii. Soil Sci Soc Am J 41: 132–136.
48. Bharati L, Lee K-H, Isenhart TM, Schultz RC (2002) Soil-water infiltration under crops, pasture, and established riparian buffer in Midwestern USA. Agrofor Syst 56: 249–257.
49. Schultz R, Isenhart T, Simpkins W, Colletti J (2004) Riparian forest buffers in agroecosystems–lessons learned from the Bear Creek Watershed, central Iowa, USA. Agrofor Syst 61: 35–50.
50. Lamb D (1980) Soil nitrogen mineralisation in a secondary rainforest succession. Oecologia 47: 257–263.
51. Paul M, Catterall CP, Pollard PC, Kanowski J (2010) Does soil variation between rainforest, pasture and different reforestation pathways affect the early growth of rainforest pioneer species? Forest Ecol Manage 260: 370–377.
52. Bezemer TM, Lawson CS, Hedlund K, Edwards AR, Brook AJ, et al. (2006) Plant species and functional group effects on abiotic and microbial soil properties and plant–soil feedback responses in two grasslands. J Ecol 94: 893–904.
53. Claus A, George E (2005) Effect of stand age on fine-root biomass and biomass distribution in three European forest chronosequences. Can J Forest Res 35: 1617–1625.
54. Piccolo MC, Neill C, Cerri CC (1994) Net nitrogen mineralization and net nitrification along a topical forest-to-pasture chronosequence. Plant Soil 162: 61–70.
55. Rasiah V, Florentine SK, Williams BL, Westbrooke ME (2004) The impact of deforestation and pasture abandonment on soil properties in the wet tropics of Australia. Geoderma 120: 35–45.
56. Radke JK, Berry EC (1993) Infiltration as a tool for detecting soil changes due to cropping, tillage, and grazing livestock. Am J Alternative Agr 8: 164–174.

Effect of Late Planting and Shading on Cellulose Synthesis during Cotton Fiber Secondary Wall Development

Ji Chen[1], Fengjuan Lv[1], Jingran Liu[1], Yina Ma[1], Youhua Wang[1], Binglin Chen[1], Yali Meng[1], Zhiguo Zhou[1]*, Derrick M. Oosterhuis[2]

1 Key Laboratory of Crop Physiology & Ecology, Ministry of Agriculture, Nanjing Agricultural University, Nanjing, Jiangsu Province, PR China, **2** Department of Crop, Soil, and Environmental Sciences, University of Arkansas, Fayetteville, Arkansas, United States of America

Abstract

Cotton-rapeseed or cotton-wheat double cropping systems are popular in the Yangtze River Valley and Yellow River Valley of China. Due to the competition of temperature and light resources during the growing season of double cropping system, cotton is generally late-germinating and late-maturing and has to suffer from the coupling of declining temperature and low light especially in the late growth stage. In this study, late planting (LP) and shading were used to fit the coupling stress, and the coupling effect on fiber cellulose synthesis was investigated. Two cotton (*Gossypium hirsutum* L.) cultivars were grown in the field in 2010 and 2011 at three planting dates (25 April, 25 May and 10 June) each with three shading levels (normal light, declined 20% and 40% PAR). Mean daily minimum temperature was the primary environmental factor affected by LP. The coupling of LP and shading (decreased cellulose content by 7.8%–25.5%) produced more severe impacts on cellulose synthesis than either stress alone, and the effect of LP (decreased cellulose content by 6.7%–20.9%) was greater than shading (decreased cellulose content by 0.7%–5.6%). The coupling of LP and shading hindered the flux from sucrose to cellulose by affecting the activities of related cellulose synthesis enzymes. Fiber cellulose synthase genes expression were delayed under not only LP but shading, and the coupling of LP and shading markedly postponed and even restrained its expression. The decline of sucrose-phosphate synthase activity and its peak delay may cause cellulose synthesis being more sensitive to the coupling stress during the later stage of fiber secondary wall development (38–45 days post-anthesis). The sensitive difference of cellulose synthesis between two cultivars in response to the coupling of LP and shading may be mainly determined by the sensitiveness of invertase, sucrose-phosphate synthase and cellulose synthase.

Editor: Xianlong Zhang, National Key Laboratory of Crop Genetic Improvement, China

Funding: This work was funded by the National Natural Science Foundation of China (30971735), China Agriculture Research System (CARS-18-20), and the Special Fund for Agro-scientific Research in the Public Interest (Impact of climate change on agriculture production of China, 200903003). The funders had no role in study design, data collection and analysis, decision to publish, or preparation of the manuscript.

Competing Interests: The authors have declared that no competing interests exist.

* Email: giscott@njau.edu.cn

Introduction

Cotton fiber development is delineated into four stages: fiber initiation, elongation, secondary wall thickening and maturation [1]. Cotton fiber, which deposit almost pure cellulose into secondary cell walls, are referred to as a primary model system for cell wall biogenesis [2,3], and many of the textile properties of cotton fiber are directly dependent on the amount and property of cellulose, which is mainly formed during secondary wall development [4–7].

The deposition of fiber secondary wall cellulose begin at about 16 days post anthesis (DPA) (at least 5 days prior to elongation cessation) and last around 15–35 d [6,8], and the period would be prolonged by cool temperature [9]. Fiber cellulose synthesis is believed to be carried out by the plasma membrane-associated rosette structure [10]. In the rosette structure, sucrose synthase (SuSy) associated with the plasma membrane (M-SuSy) may form a complex with cellulose synthase (CesA) to channel carbon from sucrose into cellulose [2,11]. In the process, sucrose is degraded by SuSy to provide uridine diphosphate glucose (UDP-glucose) for cellulose synthesis [2,5,12], and a portion of fructose maybe recycled to sucrose through sucrose-phosphate synthase (SPS) [5,13]. The energy and hexoses required for the maintenance of cell growth is provided by the soluble SuSy (S-SuSy) in the cytosol [11]. Sucrose can be converted at high rates to both cellulose and callose (β-1,3-glucan) [14]. In 20 DPA cotton boll, fiber callose is codistributed with abundantly present SuSy in the fiber cell wall region (CW-SuSy) [11,15]. The distribution of SuSy is consistent with its having a dual role in cellulose and callose synthesis in secondary-wall-stage cotton fiber [15]. In addition to SuSy, acidic invertase (either tightly bound to the cell wall or inside the vacuole) and alkaline invertase (a nonglycosylated cytosolic invertase) (INV) can also catalyze hydrolysis of sucrose [16,17].

Cotton fiber development is restricted by declining temperature or low light in many cotton-growing areas [18–21]. However, these two climatic factors often appear as a combined one. Multiple cropping cotton areas (such as cotton-rapeseed or cotton-wheat double cropping systems) are popular in the Yangtze River Valley and Yellow River Valley of China [22]. Due to the competition of temperature and light resources during the growing season of double cropping system, cotton is generally late-germinating and late-maturing and has to suffer from the coupling of declining temperature and low light especially in the late growth stage, e.g., in the Yangtze River Valley, cotton often suffers from rainy and overcast weather during the early stage of flowering and boll formation, as well as from declining temperature and overcast weather during the late stage of flowering and boll formation. These sub-optimal environmental condition during fiber development may hinder cellulose synthesis in fiber [5,23,24], and have a negative impact on fiber quality [25–27].

Declining temperature hinders cellulose synthesis within cotton fiber [5]. Sucrose synthesis is a particularly cool temperature-sensitive step in the partitioning of carbon to cellulose [23], and the activities of related enzymes in cellulose synthesis (SuSy, SPS, INV) are also affected by declining temperature, and leaded to restrain cellulose synthesis and sucrose metabolism [5,23,28]. The activities of SuSy, SPS and INV in various plants or organs are also affected by shading, and result in the decline of biomass and yield [29–31], cotton grown in reduced light environments produced inferior fiber with a lower quality [32,33].

Studies on the enzymological mechanism of carbon partitioning to cellulose synthesis have been carried out in plants under declining temperature or low light [5,33,34], but little is reported about the response of carbon partitioning to cellulose synthesis to the coupling of declining temperature and low light. Therefore, in the study, late planting (LP) and shading were used to fit the combined situation which cotton generally suffers from in the cotton-rapeseed or cotton-wheat double cropping systems, the impact of declining temperature and low light (formed by LP and shading) on sucrose metabolism, cellulose synthesis and related enzymes activity change during fiber secondary wall development (FSWD) were studied, and the physiological and biochemical mechanism of carbon partitioning to cellulose synthesis in response to adverse environmental conditions of declining temperature with low light would be elucidated.

Materials and Methods

Experimental Design

Field experiments were conducted at Pailou experimental station of Nanjing Agricultural University at Nanjing, China (32°02′N, 118°50′E), in the Yangtze River Valley in 2010 and 2011. The experimental soil was clay, mixed, thermic, Typic alfisols (udalfs; FAO luvisol) with 18.3 and 18.1 g kg^{-1} organic matter, 1.1 and 1.0 g kg^{-1} total N, 64.5 and 70.2 mg kg^{-1} available N, 17.9 and 20.3 mg kg^{-1} available P, and 102.3 and 111.1 mg kg^{-1} available K contained in 20 cm depth of the soil profile before sowing cotton in 2010 and 2011, respectively.

Two cotton (*Gossypium hirsutum* L.) cultivars, Kemian 1 which was cool temperature-tolerant and Sumian 15 which was cool temperature-sensitive [6,35] were selected based on the categorization of cultivars widely grown in the Yangtze River Valley in its low temperature sensitivity. In the field, different environmental condition during fiber development were provided by planting cotton in different dates [6], 25 April, 25 May and 10 June in 2010 and 2011. Planting date of 25 April is comparatively appropriate to grow cotton in the Yangtze River Valley, and 25 May and 10 June are belong to late planting dates (LPD). Cotton seeds were sown in a nursery bed, and seedlings with three true leaves were transplanted to field at a spacing of 80 cm×25 cm.

When approximately 50% of flowers in the first fruiting node of the 6–7th sympodial branches of plants in each planting date bloomed, three shading treatments were imposed for the plots of each planting date, including an unshaded control (*CRLR* (crop relative light rate) 100%), mild shading (*CRLR* 80%), severe shading (*CRLR* 60%) achieved with white nylon cloth (12 m length, 7 m width, 2 m height, and two different kinds of cloth which reduced the incident light by 20% and 40%, respectively). Shading cloths were removed after cotton bolls in the first fruiting node of the twelfth synpodial branches opened. Experiments were arranged as a randomized complete block design in the field with three replications and each plot was 6 m wide and 11 m long. Furrow-irrigation was applied as needed during both seasons. Conventional insect and weed control methods were utilized as needed.

Sampling and processing

Cotton flowers in the first or the second fruiting node of the 6–7th sympodial branches with the same anthesis date were tagged with small plastic tags listing the flowering date. About 6–8 cotton bolls in the similar size with the same anthesis date for each treatment were collected from once every 7 days starting from 10 DPA until boll opening. Cotton bolls were collected at 9:00–11:00 am, and cotton fiber were excised from bolls with a scalpel and were immediately put into liquid nitrogen for subsequent measurement.

Weather data

Weather data were collected from the Nanjing weather station located about 6 km from the plot area. Table 1 shows the mean daily maximum temperature (MDTmax), mean daily temperature (MDT), mean daily minimum temperature (MDTmin) and mean daily radiation (MDR) during FSWD for different planting dates. FSWD was calculated from the initiation date of fiber biomass rapid-accumulation (data not shown) to the boll opening date [7,36].

Field microclimate measurement

Field microclimate were measured at 15, 30 and 45 DPA (15, 30 and 45 days after initiation of shading). Air temperature, relative humidity and photosynthetically active radiation (PAR) were measured every two hours from 6:00am to 6:00pm, using a Hygro-Thermometer Psychrometer (DT-8892, CEM, Shenzhen, China) to measure air temperature and relative humidity at the position of 6–7th fruiting branches. PAR was measured at the position about 0.2 m above the canopy (PAR_0, below the shading cloth) by a Decagon AccuPAR LP-80 Ceptometer (Decagon Devices, Logan, Utah, USA). Measurements were taken only when the direct sunlight was not blocked by clouds.

Cellulose content, sucrose content and callose content analyses

Cotton fiber was digested in an acetic-nitric reagent, and then the cellulose content was measured with anthrone according to Updegraff [37].

Sucrose was extracted and quantified by a modified method of Pettigrew [33]. About 0.3 g dry weight (DW) fiber samples were extracted with three successive 5 ml washes of 80% ethanol [5]. The ethanol samples were incubated in an 80°C water bath for 30 min.Then the samples were centrifuged at 10,000 g for 10 min, and three aliquots of supernatant were collected together for sucrose measurement [5]. The sucrose assay was conducted according to Hendrix [38].

Callose (β-1,3-glucan) was extracted and quantified by a modified method of Köhle [39]. Fiber was soaked for 2–3 h in 5 ml of ethanol to remove autofluorescent soluble material and

Table 1. Weather factors during cotton fiber secondary wall development (FSWD) under different planting dates.

Years	Planting dates	Flowering dates	Starting date of FSWD	Boll opening date	Duration of FSWD	Weather factors			
	(dd−mm)	(dd−mm)	(dd−mm)	(dd−mm)	(d)	MDT (°C)	MDTmax (°C)	MDTmin (°C)	MDR (MJ m^{-2})
2010	25-Apr	28-Jul	11-Aug	8-Sep	29	28.3	32.4	25.3	15.8
	25-May	19-Aug	8-Sep	13-Oct	41	21.4	25.5	18.5	11.5
	10-Jun	4-Sep	28-Sep	5-Nov	39	16.7	21.3	13.2	11.1
					CV(%)	26.51	21.32	31.82	20.34
2011	25-Apr	27-Jul	12-Aug	14-Sep	34	25.9	29.5	23.0	13.0
	25-May	25-Aug	14-Sep	24-Oct	41	19.9	24.5	16.6	12.5
	10-Jun	10-Sep	30-Sep	11-Nov	44	17.4	21.4	14.4	10.2
					CV(%)	20.63	16.15	24.83	12.56

MDT, MDTmax, MDTmin and MDR stand for mean daily temperature, mean daily maximum temperature, mean daily minimum temperature and mean daily radiation.

then oven dried. The above sample (200 mg) was ground into a fine powder in liquid nitrogen followed by 5 ml of 1 N NaOH. The resulting suspension was incubated at 80°C for 30 min to solubilize the callose and centrifuged (15 min, 380 g). The supernatant was used for the callose assay. 0.6 ml of supernatant was mixed with 1.2 ml of 0.1% (w/v) aniline blue WS in water, resulting in a violet-red color. After addition of 0.63 ml of 1N HCl the color changes to deep blue, indicating neutral to acidic pH values [39]. The final pH value was adjusted by addition of 1.77 ml 1M glycine/NaOH buffer (pH 9.5) and the tubes were mixed vigorously. During the following incubation for 20 min at 50°C and further 30 min at room temperature, the aniline blue becomes almost completely decolorize [39]. Fluorescence of the assay was read in a Tecan Infinite M200 microplate reader (Tecan, Männedorf, Switzerland, excitation 400 nm, emission 510 nm, slit 5 nm). Calibration curves were established using a freshly prepared solution of the β-1,3-glucan in 1N NaOH [39].

Enzymatic analyses

Enzyme extraction and assay were according to King [40] with minor modifications. Fiber cell samples, about 0.5 g fresh weight (FW), were ground into a fine powder in liquid nitrogen followed by grinding in cold extraction buffer (5:1, v/w), which contained 50 mM N-(2-hydroxyethyl) piperazine -N'-(2-ethanesulfonic acid)-NaOH (Hepes-NaOH) (pH 7.5), 10 mM $MgCl_2$, 1 mM ethylenediamine tetraaceticacid (EDTA), 1 mM ethyleneglycol bis-(2-aminoethylether)-tetraacetic acid (EGTA), 0.5% (w/v) bovine serum albumin (BSA), 2% (w/v) polyvinylpyrrolidone (PVP), 0.1% (v/v) Triton X-100, 2 mM dithiothreitol (DTT), and 1 mM phenylmethylsulfonyl fluoride (PMSF) as described in the previous research [5]. The resulting homogenate was centrifuged at 15,000 g for 20 min, and the supernatant was stored at 4°C for analysis [5]. All extraction procedures were carried out at 0–4°C.

SuSy activity was assayed by measuring the cleavage of sucrose [40]. Each reaction contained 20 mM piperazine-N,N'-bis (2-ethanesulfonic acid)-KOH (Pipes-KOH) (pH 6.5), 100 mM sucrose, 2 mM UDP, and 200 μl of extract in a total volume of 650 μl as described previously [5]. Reactions were started by incubating at 30°C for 30 min. The reactions were stopped with 250 μl of 0.5 M N-tris-(hydroxy-methyl) methylglycine-KOH (Tricine-KOH) (pH 8.3), which were heated for 10 min in boiling water, and the amount of fructose in SuSy reactions was determined as described before [5].

Soluble acid and alkaline invertases' activities were measured by incubation of 100 ml of extract with 1M sucrose in 200 mM acetic acid-NaOH (pH 5.0) (acid invertase), or 100 mM sodium acetate-acetic acid (pH 7.5) (alkaline invertase), in a total volume of 2.5 ml [5]. Reactions were started by incubating at 30°C for 30 min and were stopped with 1 ml of 3,5-dinitro salicylic acid (DNS), and boiling for 5 min [5,40]. Glucose content was measured spectrophotometrically at 540 nm.

SPS activity was assayed by measuring the synthesis of sucrose-6-P from UDP-glucose and fructose-6-P [41]. Each reaction contained 14 mM UDP-glucose, 50 mM fructose-6-P, 50 mM extraction buffers, 50 mM $MgCl_2$ and 200 ml extract in a total volume of 650 ml as described in previous research [5]. Reactions were started by incubating the enzyme extracts at 30°C for 30 min and were stopped with 100 ml of 2N NaOH and 10 min of heating at 100°C to destroy unreacted hexoses and hexose phosphates [5]. After adding 1 ml of 0.1% (w/v) resorcin in 95% (v/v) ethanol, reactions were incubated for 30 min at 80°C. Sucrose-6-P content was calculated from a standard curve measured at 480 nm [5].

Semi-quantitative RT-PCR analyses

Fiber sampling in 2011 was used in the semi-quantitative RT-PCR analyses. Total RNA was isolated from cotton fiber according to Jiang and Zhang [42]. For each reaction, 2 μg of RNA was reverse transcribed to cDNA with oligo(dT)$_{15}$. PCR was performed in a final volume of 25 μL containing 2U of *Taq* DNA polymerase. The gene-specific primers of two cellulose synthase catalytic subunit, CesA1 and CesA2 (*GhCesA1*, U58283 and *GhCesA2*, U58284),were designed to unique regions of both cDNA (Table 2), the cotton 18srRNA gene (*Gh18srR*,U42827) [43] was used as an internal control. PCR reaction was initially denatured at 94°C for 5 min and 30 cycles at 94°C for 30 s, proper annealing temperature for 30 s and72°C for extending 30–60 s (Table 2), a final extension of RT-PCR products at 72°C for 10 min. PCR products were size-separated by electrophoresis in a 1.8% agarose gel. All photographs were statistically analyzed with the software Quantity One.

Data analysis

OriginPro 8.0 was adopted for data processing and drawing of figures. An analysis of variance was performed using SPSS statistic package Version 17.0. The means were separated using the least significant difference (LSD) test at 5% of probability level. The coefficient of variation (CV) was calculated as the ratio of the standard deviation to the mean. Changing amplitude (Δ%) = (Treatment-Control)/Control × 100%, control is the plot of 25 April plus *CRLR* 100%. The fiber growing period delineated by the two dotted lines in figures stood for the period of fiber secondary wall rapid-thickening (FSWR), which was calculated from the initiation date to the termination date of fiber biomass rapid-accumulation. The formation of cotton fiber biomass could be described by the logistic regression model and then the initiation and termination date of the fiber biomass rapid-formation were obtained (data not shown), which represented the initiation and termination date of FSWR, respectively, and the duration was the period of FSWR [7,36].

Results

Field environmental condition and field microclimate

Environmental condition of normal planting date (NPD) of 25 April was advantageous to develop cotton, and was the optimal planting date in the Yangtze River Valley [44], delaying the planting date would prolong FSWD from 29 to 41 days in 2010 and from 34 to 44 days in 2011 (Table 1). In two experimental years, MDT, MDTmax, MDTmin and MDR during FSWD decreased as planting date delayed. Fiber microclimate data in Table S1 were expressed as the mean of data measured from 6:00am to 6:00pm at 15, 30 and 45 DPA, air temperature and PAR$_0$ in cotton field also decreased as planting dates delayed, but the CVs of MDT, MDTmax and MDTmin were higher than that of MDR (Table 1), it was indicated that effect of LP on temperature factors was greater than sunshine factors during FSWD. MDTmin was the primary environmental factor affected by LP, and reduced from 25.9°C to 13.2°C in 2010 and from 25.3°C to 14.4°C in 2011 (Table 1).

During 6:00am–6:00pm, PAR$_0$ peaked at midday, shading significantly reduced PAR$_0$ by 18%–25% for *CRLR* 80% and 35%–44% for *CRLR* 60% treatments as experimental design (Figure 1); air temperature peaked at 12:00am–2:00pm and was not significantly different between shading and normal light treatments, except one or two determination points and their numbers of each planting date with only small deviations of no more than 1.5°C (Figure 2); mean relative humidity decreased until 12:00am–2:00pm and thereafter increased (Figure 2), and shading treatments was statistically different around 10:00am–2:00pm, but the deviation of these treatments were less than 7% (Figure 2). Field microclimate data measured at 15, 30 and 45 DPA in Table S1 also showed that difference of air temperature or mean relative humidity among different shading treatments were no more than 1°C or 5%, respectively. These small differences in temperature and relative humidity would probably only be a minor effect on carbohydrate concentrations compared with the effect of 20%–40% PAR reduction, which reduced the carbohydrate levels in fiber under shading by lower photosynthetic rates because of the lower light levels as shown by Pettigrew [33]. Therefore, the decline of PAR$_0$ was the key reason for the adverse effect on cotton fiber development caused by shading.

Sucrose, cellulose and callose contents in cotton fiber

During fiber development, sucrose contents in cotton fiber declined from 10 DPA under NPD$_{25\text{-Apr}}$, but there were a peak value which occurred at 17 and 24 DPA under LPD$_{25\text{-May}}$ and LPD$_{10\text{-Jun}}$, respectively (Figure 3). LP enhanced sucrose content in cotton fiber compared to the normal planting. Compared to *CRLR* 100%, under NPD$_{25\text{-Apr}}$, sucrose contents of *CRLR* 80% and *CRLR* 60% decreased. Sucrose in cotton fiber after 38 DPA under NPD$_{25\text{-Apr}}$ had already been depleted, but there was surplus sucrose in the developing fiber of 59 DPA in LPD$_{10\text{-Jun}}$ (Figure 3). Under LPD, fiber sucrose contents of *CRLR* 80% and *CRLR* 60% decreased before 31 DPA, but after 31 DPA the sucrose contents of *CRLR* 100% were lower than that of *CRLR* 80% and *CRLR* 60% (Figure 3). The maximum sucrose content in cotton fiber can reflect the amount of available sucrose in developing fiber, and the minimum sucrose content show the residual sucrose content in mature fiber [5]. Shading could decrease the maximum sucrose content by 3.5%–15.7%, while the maximum sucrose and minimum sucrose content under LPD increased 10.4%–48.5%

Table 2. Primers, T_m, extension time and cycles in RT-PCR program.

Gene	Accession No.	Primer	T_m (°C)	No. of cycles	Extension time (s)	Length of amplified DNA (bp)
GhCesA1	U58283	Forward: 5'- TGGGTTGAATGTTAATGGT-3'	58	30	60	632
		Reverse: 5'- CAGGATACCACTTAGGGAACT-3'				
GhCesA2	U58284	Forward: 5'- CTGGCTTTGGTTCACTTGC-3'	58	30	60	529
		Reverse: 5'- CCGCCATTATCGTTGCTTA-3'				
Gh18srR	U42827	Forward: 5'-CTGAGAAACGGCTACCACAT-3'	53	25	30	500
		Reverse: 5'-CTATGAAATACGAATGCCCC-3'				

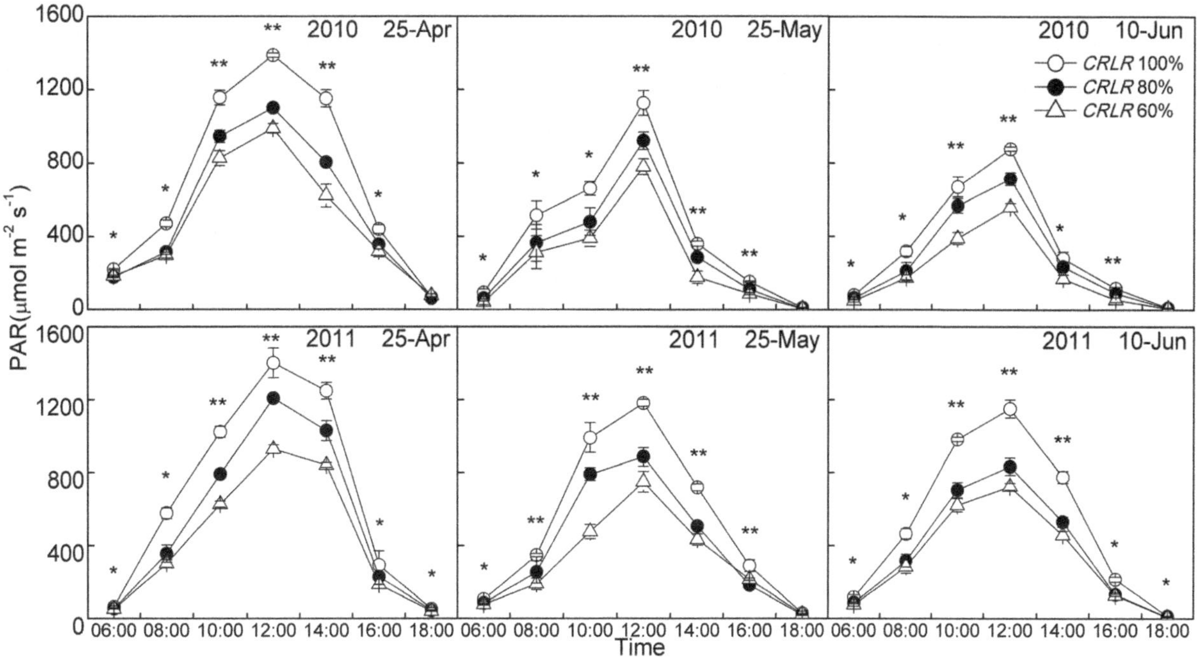

Figure 1. Changes of photosynthetically active radiation measured at the position about 0.2 m above the canopy (PAR_0) at 30 DPA under the coupling of planting date and shading in 2010 and 2011. * and ** mean significant difference among three shading treatments at 0.05 and 0.01 probability levels, respectively.

and 125.8%–1349.0%, respectively, compared to normal planted cotton (Table 3, Table 4). The maximum sucrose content under the coupling of LP and shading did not increase as much as that under the same planting date without shading. In contrast, the minimum sucrose content under $LPD_{10\text{-Jun}}$ plus *CRLR* 60% reached the highest (Table 3). The CVs of the maximum and minimum sucrose contents in two cultivars caused by LP were higher than that caused by shading, it was indicated that the effect of LP on fiber sucrose content was greater than that of shading, and the CVs caused by the coupling of LP and shading were similar to LP. Compared to the maximum sucrose content, the minimum sucrose content was more susceptible to LP or shading, and the trend was consistent between two cultivars (Table 3). The CVs and its changing amplitude (Δ%) of the maximum sucrose content response to the coupling of LP and shading in Sumian 15 was higher than that of Kemian 1, but the CVs and its Δs of the minimum sucrose content between two cultivars was different in two years (Table 3, Table 4).

Carbon from sucrose can be converted at high rate to both cellulose and callose (β-1,3-glucan) [14]. In this study, β-1,3-glucan content in cotton fiber was low at 10 DPA and rose abruptly at approximately the time as the onset of secondary wall cellulose synthesis as also reported by Maltby et al. [45]. Callose content increased from 10 DPA to 17–24 DPA when the peak appeared

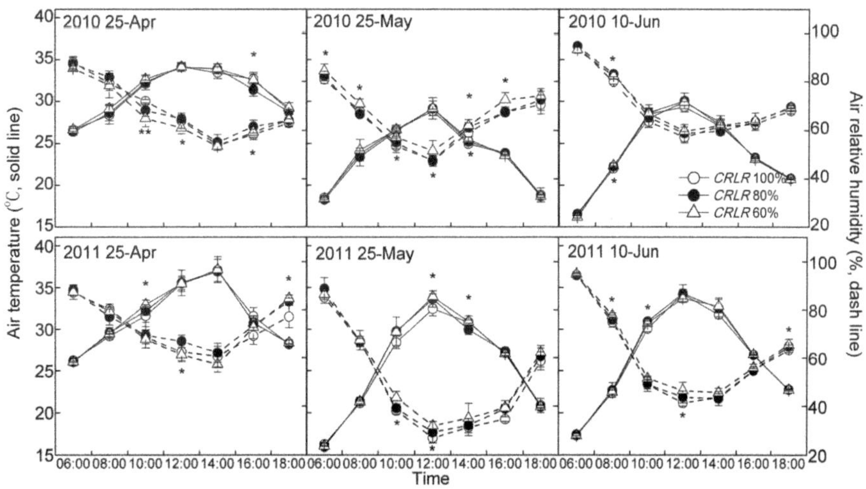

Figure 2. Changes of mean air temperature and mean relative humidity at 30 DPA under the coupling of planting date and shading in 2010 and 2011. * and ** mean significant difference among three shading treatments at 0.05 and 0.01 probability levels, respectively.

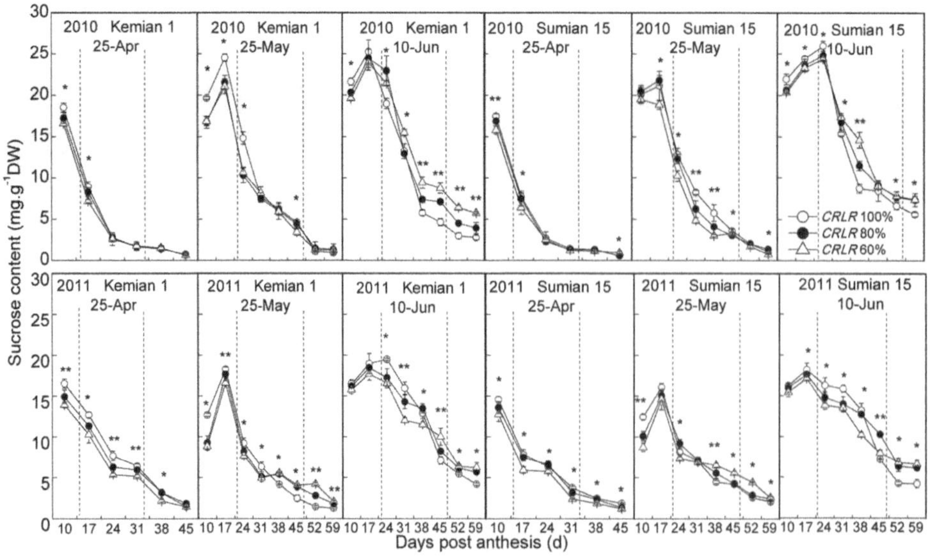

Figure 3. Changes of sucrose contents in cotton fiber of two cultivars under the coupling of planting date and shading in 2010 and 2011. The fiber growing period delineated by the two dotted lines stand for the period of fiber secondary wall rapid-thickening (FSWR). * and ** mean significant difference among three shading treatments at 0.05 and 0.01 probability levels, respectively.

(Figure 4), and then declined significantly. LP and shading could increased fiber callose content, which reached the maximum under the coupling of LP and shading (Figure 4, Table 3). The period of FSWR was 17–31 DPA under NPD$_{25\text{-Apr}}$, and were 24–45 DPA and 24–52 DPA under LPD$_{25\text{-May}}$ and LPD$_{10\text{-Jun}}$, respectively. Meanwhile, the fiber cellulose rapid-accumulation was almost the same (Figure 4). Cellulose content in cotton fiber increased from 10 DPA, and compared to NPD$_{25\text{-Apr}}$ which the initiation of fiber cellulose fast-accumulation began at 17 DPA, LPD$_{10\text{-Jun}}$ delayed to 24 DPA and restrained cellulose synthesis. Shading under all planting dates decreased fiber cellulose contents (Figure 4), but the decreasing extents were different in different fiber development stages and planting dates. Shading under NPD$_{25\text{-Apr}}$ had greater impacts on cellulose synthesis during 24–31 DPA, while shading under LPD were during 38–45 DPA. At the mature stage (59 DPA), *CRLR* 80% and *CRLR* 60% decreased cellulose content by 0.7%–2.8% and 1.1%–5.6% under NPD$_{25\text{-Apr}}$, respectively, while LPD$_{25\text{-May}}$ and LPD$_{10\text{-Jun}}$ decreased by 6.7%–10.2% and 15.0%–20.9%, respectively (Table 4). The final cellulose content under the coupling of LPD$_{10\text{-Jun}}$ and *CRLR* 60% was the lowest which decreased 18.2%–25.5% (Table 3). The CVs of the maximum callose contents of two cultivars caused by shading were higher than that caused by LP and the coupling of LP and shading, it was indicated that the effects of shading on maximum callose contents were greater than LP and the coupling of LP and shading. In contrast, the effects of LP on final cellulose contents were greater than shading, but close to the coupling of LP and shading (Table 3). The CVs and Δs of the final cellulose content response to the coupling of LP and shading in Sumian 15 were higher than Kemian 1 (Table 3, Table 4), it was indicated that cellulose content of Sumian 15 was more sensitive to the coupling of LP and shading than Kemian 1.

Activity changes of related cellulose synthesis enzymes during fiber secondary wall development

Fiber SuSy activity decreased during fiber development (Figure S1). During FSWR, SuSy activity was the lowest under LPD$_{25\text{-May}}$ and rose again under LPD$_{10\text{-Jun}}$ (Table 5). Shading decreased the SuSy activities under all planting dates. In the period of FSWR, the decline caused by shading under NPD$_{25\text{-Apr}}$ and LPD$_{10\text{-Jun}}$ were larger than LPD$_{25\text{-May}}$ (Figure S1) and the effect of shading under LPD$_{10\text{-Jun}}$ on the Δs of SuSy activities were opposite in two years (Table 6).

Acidic INV in cotton fiber was higher than alkaline invertase, from the point of view of CVs, acidic INV was more susceptible to LP or shading, and both of their activities decreased during fiber development (Figure S2–S3, Table 5). Under LPD, the fiber acidic and alkaline INV activities increased and remained at a high rate at the end of fiber development. During FSWR, shading decreased acidic and alkaline INV activities of Kemian 1, shading also decreased acidic and alkaline INV activities of Sumian 15 under planting date of 25 April and 25 May, but increased the activities under LPD$_{10\text{-Jun}}$ (Figure S2–S3).

Fiber SPS activity increased and peaked at 24 or 31 DPA, and then decreased during fiber development. LP delayed the peaks to 38, 45 or 52 DPA and decreased the peak values (Figure S4). During FSWR, SPS activity decreased under shading and the peak was delayed under LPD$_{25\text{-May}}$ in 2010. SPS activity under NPD$_{25\text{-Apr}}$ was similar to LPD$_{25\text{-May}}$ during FSWR, however decreased notably under LPD$_{10\text{-Jun}}$ (decreasing 19.6%–37.8%). Under the coupling of shading and LPD$_{10\text{-Jun}}$, SPS activities decreased by 24.3%–43.0%, which were more than shading or LP alone, and the decreasing amplitude of Sumian 15 (decreasing 37.8%–43.0%) was larger than that of Kemian 1 (decreasing 24.3%–37.4%) (Figure S4, Table 5, Table 6).

After analysing the CVs of mean sucrose metabolism enzyme activities (such as SuSy, SPS and acidic/alkaline INV), it was found that the effects of LP on sucrose metabolism enzymes were greater than shading during FSWR, and were similar to the effects of the coupling of LP and shading. Among the four kinds of sucrose metabolism enzymes, SPS was the most significant affected by shading, and acidic INV was the most significant affected by LP and by the coupling of LP and shading (Table 5).

Table 3. Maximum/minimum sucrose, maximum callose and final cellulose contents in cotton fiber and their analysis of variance under the coupling of planting date and shading in 2010 and 2011.

Planting dates	CRLR	Maximum sucrose content		Minimum sucrose content		Maximum callose content		Final cellulose content	
(dd-mm)	(%)	(mg g^{-1}DW)		(mg g^{-1}DW)		(mg g^{-1}DW)		(%)	
		2010	2011	2010	2011	2010	2011	2010	2011
Kemian 1									
25-Apr	100	18.5 a	16.5 a	0.7 a	1.5 a	7.7 b	5.6 c	92.2 a	91.6 a
	80	17.3 b	14.9 b	0.7 a	1.9 a	7.6 b	9.6 b	91.2 a	89.0 a
	60	16.6 c	13.9 c	0.6 a	1.4 a	11.4 a	11.5 a	87.0 b	90.6 a
25-May	100	24.5 a	18.3 a	4.6 a	3.9 a	8.3 b	7.0 c	85.3 a	82.3 a
	80	21.5 b	17.6 b	4.4 a	4.1 a	8.1 b	9.7 b	85.0 a	77.0 b
	60	21.1 b	16.6 c	3.4 a	2.5 b	12.2 a	12.6 a	83.1 b	76.8 b
10-Jun	100	25.2 a	19.5 a	4.6 c	7.1 c	9.4 b	7.0 c	78.4 a	76.6 a
	80	24.4 b	19.0 a	7.0 b	8.2 b	9.3 b	11.6 b	76.8 b	73.3 b
	60	23.9 c	17.7 b	8.[illegible] a	10.0 a	12.8 a	13.3 a	75.4 c	73.1 b
	CV_P(%)	16.23	8.25	68.92	67.66	10.39	13.09	8.09	9.07
	CV_S(%)	5.60	8.64	6.[illegible]7	14.50	24.11	34.11	3.90	1.45
	$CV_{P\times S}$(%)	15.53	10.66	73.90	70.52	20.86	27.90	7.09	9.19
>**Sumian 15**									
25-Apr	100	17.5 a	14.6 a	0.[illegible] a	1.9 a	8.4 b	8.1 b	86.2 a	88.0 a
	80	16.9 b	13.6 b	0.[illegible] a	1.3 a	12.7 a	8.2 b	85.6 ab	87.4 a
	60	15.8 c	12.8 c	0.[illegible] a	1.2 a	13.1 a	13.0 a	84.7 b	86.0 a
25-May	100	21.2 a	16.1 a	3.[illegible] a	4.3 b	9.3 c	9.4 b	80.4 a	81.1 a
	80	21.8 a	15.0 b	3.[illegible] a	4.2 b	14.2 b	9.2 b	77.2 ab	79.1 ab
	60	19.5 b	14.6 b	3.[illegible] a	5.6 a	15.1 a	13.8 a	76.5 b	77.1 b
10-Jun	100	26.0 a	18.2 a	8.[illegible] a	7.2 b	10.7 c	10.3 c	72.0 a	69.6 a
	80	24.8 b	17.6 ab	9.[illegible] a	10.3 a	14.7 b	11.1 b	70.0 ab	66.5 ab
	60	24.4 b	17.1 b	9.[illegible] a	8.0 b	16.2 a	14.4 a	68.0 b	65.6 b
	CV_P(%)	19.74	11.31	9[illegible].45	59.86	12.20	12.13	8.98	11.68
	CV_S(%)	5.05	6.57	3[illegible].95	24.40	22.86	28.57	0.88	1.18
	$CV_{P\times S}$(%)	17.73	12.07	8[illegible].28	65.09	21.34	21.96	8.83	11.29

DW, dry weight; Values followed by the different letters within a column are significantly different at 0.05 probability level; CV_S was calculated from the data of three shading treatments in normal planting date of 25 April; CV_P was calculated from the data of three planting dates and $CV_{P\times S}$ was calculated from the data of all treatments; * and ** mean significant difference at 0.05 and 0.01 probability levels, respectively; NS means non-significant differences.

Table 4. The changing amplitude (Δ%) of maximum/minimum sucrose, maximum callose and final cellulose contents in cotton fiber under the coupling of planting date and shading in 2010 and 2011.

Treatments	Maximum sucrose content		Minimum sucrose content		Maximum callose content		Final cellulose content	
	2010	2011	2010	2011	2010	2011	2010	2011
Kemian 1								
CRLR 80%	−6.8	−9.6	9.2	25.6	−1.3	73.0	−1.1	−2.8
CRLR 60%	−10.4	−15.7	−4.7	−2.6	47.6	106.7	−5.6	−1.1
25-May	32.4	10.7	584.6	161.0	7.2	26.6	−7.5	−10.2
25-May+*CRLR* 80%	16.3	6.8	554.0	174.4	4.5	75.0	−7.8	−15.9
25-May+*CRLR* 60%	13.8	0.5	405.7	68.2	58.4	127.8	−9.9	−16.2
10-Jun	36.4	18.0	584.9	376.5	22.4	26.8	−15.0	−16.4
10-Jun+*CRLR* 80%	32.0	14.8	944.7	455.2	20.4	108.4	−16.7	−20.0
10-Jun+*CRLR* 60%	29.2	7.3	1191.3	573.4	66.8	138.9	−18.2	−20.2
Sumian 15								
CRLR 80%	−3.5	−6.6	−0.9	−28.9	51.7	1.8	−0.7	−0.7
CRLR 60%	−9.5	−12.3	64.5	−36.1	55.6	60.6	−1.7	−2.3
25-May	21.1	10.4	513.6	125.8	10.8	16.6	−6.7	−7.8
25-May+*CRLR* 80%	24.8	3.3	422.1	121.8	69.8	13.5	−10.4	−10.1
25-May+*CRLR* 60%	11.6	0.2	466.5	195.3	80.5	70.2	−11.3	−12.4
10-Jun	48.5	25.2	1349.0	281.9	27.3	27.7	−16.5	−20.9
10-Jun+*CRLR* 80%	41.9	21.2	1468.0	446.5	75.5	37.0	−18.8	−24.4
10-Jun+*CRLR* 60%	39.5	17.7	1445.5	322.1	92.9	77.9	−21.1	−25.5

Δ% = (Treatment-Control)/Control × 100%, control is the plot of 25 April + *CRLR* 100%.

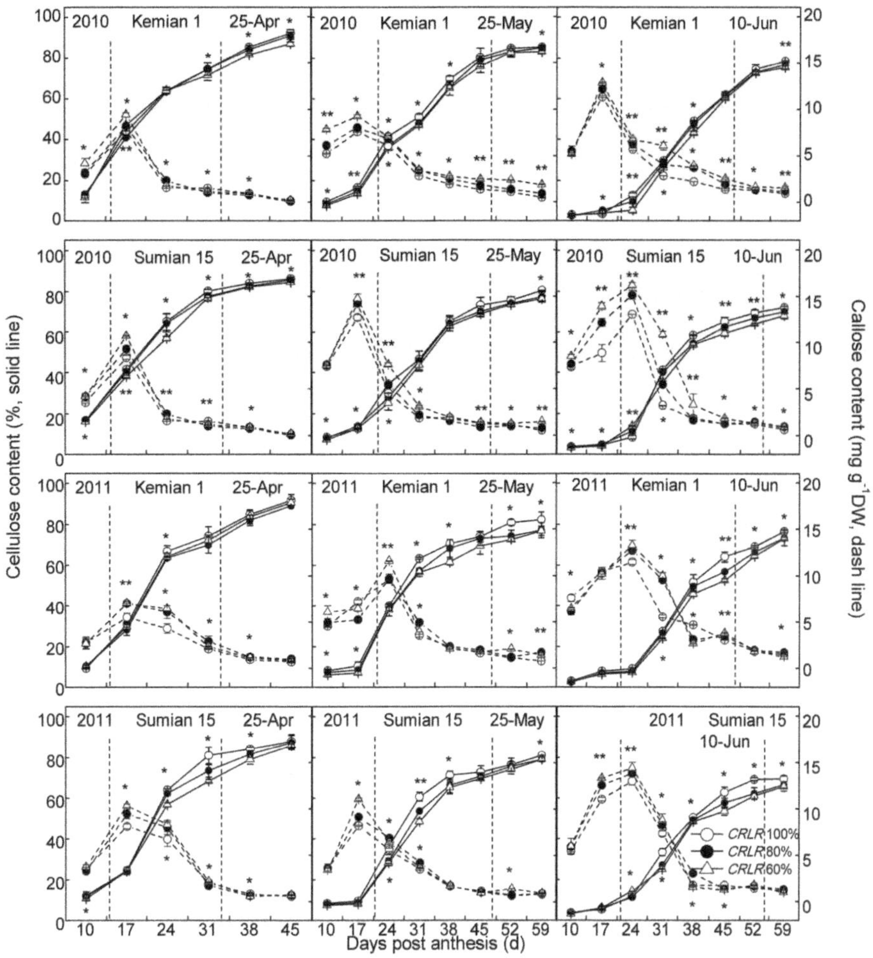

Figure 4. Changes of cellulose and callose contents in cotton fiber of two cultivars under the coupling of planting date and shading in 2010 and 2011. The fiber growing period delineated by the two dotted lines stand for the period of fiber secondary wall rapid-thickening (FSWR). * and ** mean significant difference among three shading treatments at 0.05 and 0.01 probability levels, respectively.

Gene expression of cellulose synthase in cotton fiber

The peak of *GhCesA1* and *GhCesA2* expression under $NPD_{25\text{-Apr}}$ appeared around 17 DPA, and delayed to 24 DPA under LPD (Figure 5). Under $NPD_{25\text{-Apr}}$ and $LPD_{25\text{-May}}$, shading delayed the *GhCesA1* expression in Kemian 1, and then shading restrained the expression in $LPD_{10\text{-Jun}}$ (Figure 5–6). Compared to Kemian 1, shading delayed the *GhCesA1* expression of Sumian 15 under $NPD_{25\text{-Apr}}$ and shading restrained the expression under LPD.

Gene expression intensity and duration of *GhCesA1* were lower than that of *GhCesA2* in two cultivars, it was indicated that the coupling of LP and shading might have a more adverse effect on *GhCesA1* than *GhCesA2*. Under $NPD_{25\text{-Apr}}$, shading delayed the *GhCesA2* expression, and appeared to restrain the *GhCesA2* expression under LPD. The postponing or restraining extent under *CRLR* 60% was more serious than under *CRLR* 100% and *CRLR* 80% (Figure 5–6).

Discussion

Cotton fiber development is affected by cool temperature [5,36,46] and low light [19,20,33,47], which often occur together [7,48]. In order to fit the coupling situation of declining temperature and low light which cotton generally suffers from in the cotton-rapeseed or cotton-wheat double cropping systems, the couping of LP and shading experiments were designed and carried out in 2010 and 2011. The results showed that MDTmin was the primary environmental factor affected by LP during FSWD (Table 1), and shading under three planting dates significantly reduced PAR_0 by 18%–25% and 35%–44% under *CRLR* 80% and *CRLR* 60%, respectively (Figure 1). $LPD_{25\text{-May}}$ and $LPD_{10\text{-Jun}}$ ($MDTmin_{FSWD}$ of 18.5 and 13.2°C in 2010, 16.6 and 14.4°C in 2011) plus *CRLR* 80% and *CRLR* 60% resulted in an extending of cotton fiber development period and affected cellulose synthesis (Tables 1 and 3).

Toward the end of fiber elongation, secondary wall deposition begins via enhanced cellulose synthesis, while fiber callose synthesis reaches the peak in this period [2,14,45]. Sucrose is the substrate that is required for high-rate callose and secondary-wall cellulose synthesis, and sucrose metabolism is sensitive to cool temperature or shading [2,14,41,33]. In this study, the coupling of LP and shading had more adverse impacts on cellulose synthesis during 38–45 DPA than before 31 DPA (Figure 4). Under LPD, shading decreased sucrose content before 31 DPA, and then increased it after 31 DPA in fiber (Figure 3, Table 3). The initial decline of sucrose content under shading was probably caused by lower photosynthetic rates [33], and the next increasing of sucrose content under LPD was a self-regulating phenomenon and the reduced sucrose transformation rate [5]. In developing fiber

Table 5. Average activities of sucrose phosphate synthase (SPS), sucrose synthase (SuSy), acidic invertase (Acidic INV) and alkaline invertase (Alkaline INV) in cotton fiber during fiber secondary wall rapid-thickening (FSWR) and their analysis of variance under the coupling of planting date and shading in 2010 and 2011.

Planting dates	*CRLR*	SPS		SuSy		Acidic INV		Alkaline INV	
(dd-mm)	(%)	(mg sucrose $g^{-1}FW\ h^{-1}$)		(mg fructose $g^{-1}FW\ h^{-1}$)		(mg glucose $g^{-1}FW\ h^{-1}$)		(mg glucose $g^{-1}FW\ h^{-1}$)	
		2010	2011	2010	2011	2010	2011	2010	2011
Kemian 1									
25-Apr	100	13.9 a	10.0 a	71.2 a	72.8 a	63.7a	78.1 a	54.8 a	70.8 a
	80	14.2 a	8.6 b	68.4 b	63.8 b	62.2 b	74.9 b	52.6 ab	65.1 b
	60	12.7 b	7.8 c	64.3 c	63.1 b	57.8 c	71.2 c	50.8 b	64.8 b
25-May	100	13.7 a	9.8 a	51.7 a	50.4 a	70.9 a	73.7 a	55.9 a	69.5 a
	80	13.6 a	9.2 b	51.8 a	45.1 b	66.1 b	68.4 b	52.7 b	66.2 b
	60	12.9 b	8.4 c	48.0 b	43.6 c	61.6 c	64.1 c	52.0 c	63.3 c
10-Jun	100	11.2 a	7.8 a	94.6 a	66.7 a	118.3 a	104.4 a	73.8 a	88.6 a
	80	10.5 a	6.8 b	88.7 b	59.8 b	110.0 b	101.3 b	69.7 b	85.5 b
	60	9.8 b	6.3 c	88.8 b	59.7 b	103.2 c	95.6 c	68.7 b	85.6 b
	CV_P(%)	11.68	13.30	29.61	18.35	35.20	19.46	17.37	14.02
	CV_S(%)	6.04	12.50	5.11	8.14	4.95	4.65	3.77	5.06
	$CV_{P\times S}$(%)	12.86	15.38	25.33	17.06	30.21	18.51	15.30	14.04
Sumian 15									
25-Apr	100	14.4 a	9.6 a	66.3 a	73.1 a	73.2 a	77.0 a	50.3 a	70.7 a
	80	12.7 b	9.3 a	62.6 b	69.5 b	71.0 b	71.3 b	48.4 b	68.3 b
	60	11.0 c	8.3 b	59.5 c	66.0 c	67.6 c	66.7 c	47.3 b	68.1 b
25-May	100	12.8 a	9.6 a	50.1 a	49.6 a	73.4 a	81.4 a	61.5 a	65.3 a
	80	12.5 ab	9.1 b	47.2 b	47.9 b	73.9 a	77.0 b	59.8 b	62.4 b
	60	11.9 b	9.1 b	47.0 b	43.9 c	67.6 b	70.4 c	57.0 c	62.4 b
10-Jun	100	9.0 a	6.6 a	88.0 a	68.6 a	105.6 c	130.2 a	71.0 c	77.5 c
	80	8.8 ab	5.9 b	84.7 a	63.8 b	107.3 b	124.2 b	72.8 b	79.8 b
	60	8.2 b	5.8 b	74.8 b	59.4 c	108.5 a	123.9 b	75.3 a	84.4 a
	CV_P(%)	23.24	19.66	27.95	19.54	22.16	30.68	17.01	8.59
	CV_S(%)	13.53	7.15	5.38	5.08	4.00	7.23	3.12	2.09
	$CV_{P\times S}$(%)	19.12	19.36	23.98	17.61	21.85	28.98	17.76	11.08

FW, fresh weight; Values in the same planting date followed by the different letters within a column are significantly different at 0.05 probability level; CV_S was calculated from the data of three shading treatments in normal planting date of 25 April, CV_P was calculated from the data of three planting dates and $CV_{P\times S}$ was calculated from the data of all treatments; * and ** mean significant difference at 0.05 and 0.01 probability levels, respectively; NS means non-significant differences.

Table 6. The changing amplitude (Δ%) of average activities of sucrose phosphate synthase (SPS), sucrose synthase (SuSy), acidic invertase (Acidic INV) and alkaline invertase (Alkaline INV) in cotton fiber during fiber secondary wall rapid-thickening (FSWR) under the coupling of planting date and shading in 2010 and 2011.

Treatments	SPS		SuSy		Acidic INV		Alkaline INV	
	2010	2011	2010	2011	2010	2011	2010	2011
Kemian 1								
CRLR 80%	1.8	−13.5	−3.9	−12.4	−2.4	−4.1	−3.9	−8.1
CRLR 60%	−9.2	−21.8	−9.7	−13.4	−9.2	−8.9	−7.2	−8.5
25-May	−1.7	−1.5	−27.3	−30.8	11.4	−5.7	2.0	−1.9
25-May+*CRLR* 80%	−2.3	−8.4	−27.3	−38.1	3.8	−12.4	−3.8	−6.6
25-May+*CRLR* 60%	−7.5	−15.7	−32.6	−40.1	−3.2	−18.0	−5.1	−10.7
10-Jun	−19.6	−21.9	32.9	−8.4	85.8	33.7	34.7	25.1
10-Jun+*CRLR* 80%	−24.3	−32.1	24.6	−17.9	72.7	29.6	27.2	20.7
10-Jun+*CRLR* 60%	−29.9	−37.4	24.7	−18.1	62.0	22.4	25.5	20.9
Sumian 15								
CRLR 80%	−11.9	−2.4	−5.6	−4.9	−3.0	−7.4	−3.8	−3.4
CRLR 60%	−23.8	−12.8	−10.2	−9.7	−7.7	−13.4	−6.0	−3.7
25-May	−10.8	0.3	−24.5	−32.1	0.3	5.7	22.2	−7.7
25-May+*CRLR* 80%	−13.4	−4.9	−28.7	−34.5	1.0	0.0	18.9	−11.8
25-May+*CRLR* 60%	−17.5	−5.1	−29.2	−39.9	−7.7	−8.6	13.3	−11.7
10-Jun	−37.8	−30.5	32.8	−6.1	44.3	69.0	41.2	9.5
10-Jun+*CRLR* 80%	−38.8	−37.8	27.8	−12.7	46.6	61.3	44.6	12.9
10-Jun+*CRLR* 60%	−43.0	−39.5	12.9	−18.7	48.2	60.9	49.8	19.4

Δ% = (Treatment-Control)/Control × 100%, control is the plot of 25 April + *CRLR* 100%.

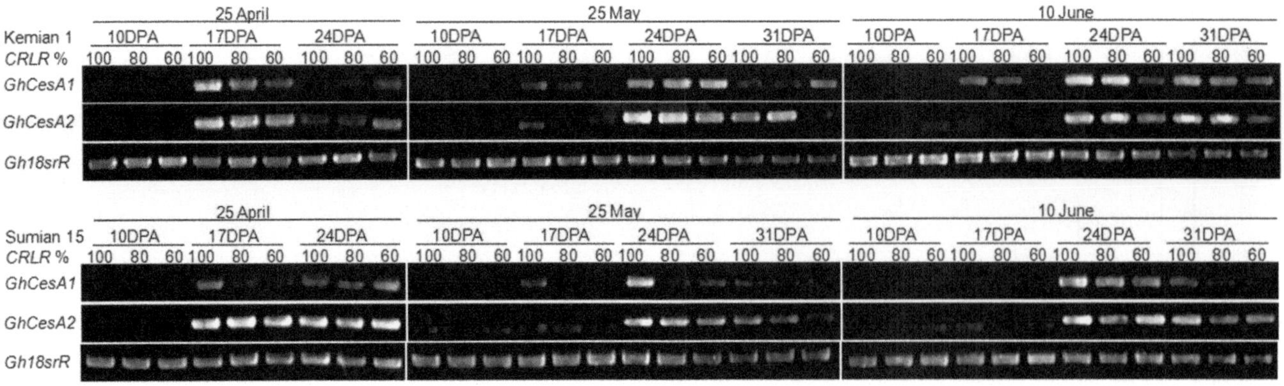

Figure 5. Gene expressions of two cellulose synthase catalytic subunits (CesA1 and CesA2) in cotton fiber of two cultivars under the coupling of planting date and shading in 2011.

during 38–45 DPA under the coupling of LP and shading, the greater increasing range of residual sucrose content and decreasing range of cellulose content indicated that compared to single stress, the sucrose transforming ability declined more under combined stress, and more obviously during 38–45 DPA. On the other hand, sucrose could not be effectively used for cellulose synthesis under cool temperature [5] and callose synthesis can replace cellulose synthesis after wounding [2]. Under $NPD_{25\text{-}Apr}$, the transformation from sucrose to cellulose and callose advanced simultaneously, and fiber cellulose started to accumulate rapidly before 17 DPA (cellulose contents were about 40%, Figure 4), but sucrose converted to callose more than to cellulose under shading. In contrast, under LPD, although fiber sucrose content was high during 10–17 DPA, the conversion rate of sucrose to callose was much greater than to cellulose, with lower cellulose contents about 10% before 24 DPA, shading under LPD exacerbated the situation (Figure 4). The result indicated that in the early stage of FSWD (before 24 DPA), compared to LP or shading, there was more abundant sucrose in cotton fiber under the coupling of LP and shading, however, carbon from sucrose was converted mainly to callose instead of cellulose. Whereas in the later stage of FSWD (after 38 DPA), carbon in fiber stagnated in the form of sucrose and the partitioning to cellulose synthesis decreased, and the coupling of LP and shading made it more serious.

The postponing or restraining trend of *GhCesA* expression under the coupling of LP and shading were corresponding with the downward trend of cellulose (Figure 4 and Table 4). The cellulose synthase complex was sensitive to adverse environmental effects and which directly affected cellulose synthesis. As to cool temperature-sensitive cultivar Sumian 15, shading appeared to restrain *GhCesA* expression under $LPD_{25\text{-}May}$, earlier than shading under $LPD_{10\text{-}Jun}$ restraining the expression in relative cool temperature-tolerant cultivar Kemian 1, it was indicated that the *GhCesA* in Sumian 15 was more sensitive to environmental change, and the postponing or restraining extent was more serious as increasing shading degree (Figure 5).

In cotton fiber, there were many sucrose metabolism enzymes contributing to cellulose synthesis besides CesA [2,12,49]. Sucrose is degraded to provide UDP-glucose for cellulose synthesis by SuSy, which is the critical partner in high-rate secondary-wall cellulose synthesis [2,12,49]. Only M-SuSy or CW-SuSy protein associated with CesA in the plasma membrane-associated rosette structure possess β-1,4-glucan (cellulose) synthesis activity [10,11]. Compared to $NPD_{25\text{-}Apr}$, the restrained activity of SuSy during FSWR under $LPD_{25\text{-}May}$ was caused by shading and declining temperature, and the increasing activity under the coupling of shading and $LPD_{10\text{-}Jun}$ (MDTmin was the lowest in this experiment) might be due to a large part of M-SuSy becoming

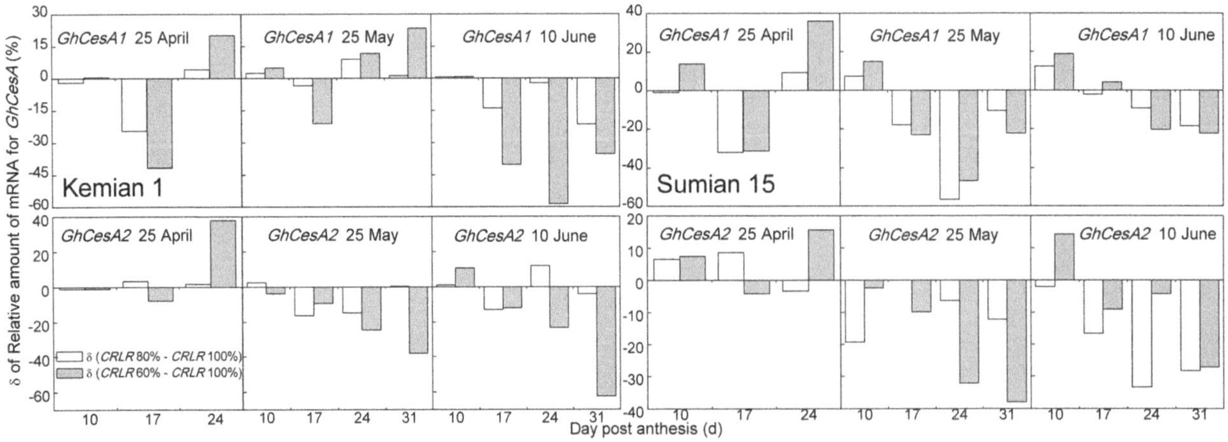

Figure 6. δ of relative amount of mRNA for two cellulose synthase catalytic subunits (CesA1 and CesA2) in cotton fiber of two cultivars under the coupling of planting date and shading in 2011. δ = (*CRLR* 80% - *CRLR* 100%) or (*CRLR* 60% - *CRLR* 100%).

S-SuSy. The enhanced S-SuSy degraded sucrose for maintenance and survival metabolism through glycolysis instead of contributing to cellulose synthesis [2,5,11].

Consistent with the results of Shu et al. [5], activities of another sucrose degrading enzymes acidic/alkaline INV in cotton fiber of two cultivars increased under LPD, and was reduced by shading under the planting date of 25 April and 25 May, but shading under $LPD_{10\text{-}Jun}$ ($MDTmin_{FSWD}$ reached the lowest of 16.6 and 14.4°C in 2010 and 2011) increased INV activities of Sumian 15. In contrast, shading under $LPD_{10\text{-}Jun}$ reduced INV activities of Kemian 1 (Table 5), the probable reason was that Sumian 15 was a cool temperature-sensitive cultivar, while acidic INV was the most affected enzyme by the coupling of LP and shading (Table 5), INV in fiber was more sensitive to the coupling of LP and shading than Kemian 1, the increasing extent of INV activity in Sumian 15 was greater as the coupling stress became heavier (Table 6). However, abundant fructose produced by the increased INV activity under the coupling of LP and shading would inhibit the ability of M-SuSy [2,5] and have an adverse effect on cellulose synthesis.

As the enzyme synthesizing sucrose in cotton fiber, SPS is very sensitive to cool temperature and the activity is hindered under adverse environmental conditions [2,5]. SPS was also the most sensitive to shading among four sucrose metabolism enzymes (SuSy, SPS, acidic/alkaline INV, Table 5). Flux from fructose to sucrose might be hindered due to the decline of SPS activity, leading to fructose increasing in fiber, which further suppressed M-SuSy activity, and resulting in an adverse effect on cellulose synthesis. Under the coupling of LP and shading, SPS activity in fiber decreased, and the activity peak was delayed to 38, 45 or 52 DPA, similar to the time when the coupling of LP and shading had greater effects on cellulose synthesis (Figure 4 and Figure S4), it was indicated that the decline of SPS activity and its peak delay may be the reason why cellulose synthesis was sensitive to the combined stress during the later stage of FSWD (38–45 DPA).

The basic mechanisms regulating cellulose synthesis in different cotton cultivars are believed to be similar [6], but cotton cultivars have different levels of sensitivity in response to adverse environmental stress [4,33]. It has been shown that Sumian 15 which was cool temperature-sensitive, and Kemian 1 was partially tolerant to cool temperature [6,35]. In our research, Sumian 15 had a higher negative response of fiber cellulose to the coupling of LP and shading compared to Kemian 1 (Table 4). The decreasing range of SPS activity in Sumian 15 was greater, and CesA was more sensitive to the coupling of LP and shading than Kemian 1 (Figure 5, Table 6). In contrast, Kemian 1 had a greater ability to form higher cellulose content under the coupling of LP and shading (Table 3) [5], it was indicated that the relative cool temperature-tolerant cultivar, such as Kemian 1, still had a greater adaptability to the coupling of LP and shading than the cool temperature-sensitive cultivar, such as Sumian 15.

Conclusions

(1) The coupling of LP (mainly MDTmin decreased) and shading (*CRLR* 80% and *CRLR* 60%) affected the key enzymes activities (SuSy, SPS, acidic/alkaline INV and CesA) involved in fiber sucrose metabolism and cellulose synthesis and hindered the flux from sucrose to cellulose during FSWD. As for the four sucrose metabolism enzymes (SuSy, SPS, acidic/alkaline INV), effects of LP were greater than shading. The decline of SPS activity and its peak delay probably caused cellulose synthesis being more sensitive to the coupling stress during the later stage of FSWD (38–45 DPA).

(2) LP and shading combined to produce a more severe impact on cellulose synthesis than either stress alone. In the earlier stage of cotton FSWD (before 24 DPA), sucrose contents in cotton fiber under the coupling of LP and shading were mainly used for synthesizing callose instead of synthesizing cellulose. In the later stage of cotton FSWD (after 38 DPA), carbon in fiber stagnated in the form of sucrose and the partitioning to cellulose synthesis decreased, and the coupling of LP and shading made it more serious.

(3) Due to a less sensitive INV, SPS and CesA, the relative cool temperature-tolerant cultivar Kemian 1 had a relatively higher tolerance to the coupling of LP and shading compared to the cool temperature-sensitive cultivar Sumian 15.

Supporting Information

Figure S1 Changes of sucrose synthase activities in cotton fiber of two cultivars under the coupling of planting date and shading in 2010 and 2011. The fiber growing period delineated by the two dotted lines stand for the period of fiber secondary wall rapid-thickening (FSWR).

Figure S2 Changes of acidic invertase activities in cotton fiber of two cultivars under the coupling of planting date and shading in 2010 and 2011. The fiber growing period delineated by the two dotted lines stand for the period of fiber secondary wall rapid-thickening (FSWR).

Figure S3 Changes of alkaline invertase activities in cotton fiber of two cultivars under the coupling of planting date and shading in 2010 and 2011. The fiber growing period delineated by the two dotted lines stand for the period of fiber secondary wall rapid-thickening (FSWR).

Figure S4 Changes of sucrose phosphate synthase activities in cotton fiber of two cultivars under the coupling of planting date and shading in 2010 and 2011. The fiber growing period delineated by the two dotted lines stand for the period of fiber secondary wall rapid-thickening (FSWR).

Table S1 Variance analysis of mean air temperature, mean relative humidity and photosynthetically active radiation (PAR) in the cotton field under the coupling of planting date and shading in 2010 and 2011. Data in Table S1 are averaged by measurement data from 6:00am to 6:00pm and PAR was measured at the position about 0.2 m above the canopy. *CRLR* and DPA stand for crop relative light rates and days post anthesis, respectively. Values followed by a different small letter within the same column in the same planting date are significantly different at 0.05 probability level.

Author Contributions

Conceived and designed the experiments: YW BC ZZ. Performed the experiments: JC FL JL Y. Ma. Analyzed the data: JC FL. Contributed reagents/materials/analysis tools: YW BC Y. Meng. Wrote the paper: JC DMO.

References

1. Lee JJ, Woodward AW, Chen ZJ (2007) Gene expression changes and early events in cotton fibre development. Ann Bot 100: 1391–1401.
2. Haigler CH, Ivanova-Datcheva M, Hogan PS, Salnikov VV, Hwang S, et al. (2001) Carbon partitioning to cellulose synthesis. Plant Mol Biol 47: 29–51.
3. Kim HJ, Triplett BA (2001) Cotton fiber growth in planta and in vitro: models for plant cell elongation and cell wall biogenesis. Plant Physiol 127: 1361–1366.
4. Haigler CH (2007) Substrate supply for cellulose synthesis and its stress sensitivity in the cotton fiber. In: Brown RM Jr, Saxena IM, editors. Cellulose: Molecular and Structural Biology. New York: Springer. pp. 147–168.
5. Shu HM, Zhou ZG, Xu NY, Wang YH, Zheng M (2009) Sucrose metabolism in cotton (*Gossypium hirsutum* L.) fibre under low temperature during fibre development. Europ J Agron 31: 61–68.
6. Wang YH, Shu HM, Chen BL, McGiffen ME, Zhang WJ, et al. (2009) The rate of cellulose increase is highly related to cotton fibre strength and is significantly determined by its genetic background and boll period temperature. Plant Growth Regul 57: 203–209.
7. Zhao WQ, Wang YH, Zhou ZG, Meng YL, Chen BL, et al. (2012) Effect of nitrogen rates and flowering dates on fiber quality of cotton (*Gossypium hirsutum* L.). Am J Exp Agric 2: 133–159.
8. Meinert MC, Delemer DP (1977) Change in biochemical composition of the cell wall of the cotton fiber during development. Plant Physiol 59: 1088–1097.
9. Xie W, Trolinder NL, Haigler CH (1993) Cool temperature effects on cotton fibre initiation and elongation clarified using in vitro cultures. Crop Sci 33: 1258–1264.
10. Fujii S, Hayashi T, Mizuno K (2010) Sucrose synthase is an integral component of the cellulose synthesis machinery. Plant Cell Physiol 51: 294–301.
11. Ruan YL (2007) Rapid cell expansion and cellulose synthesis regulated by plasmodesmata and sugar: insights from the single-celled cotton fibre. Funct Plant Biol 34: 1–10.
12. Delmer DP, Amor Y (1995) Cellulose biosynthesis. Plant Cell 7: 987–1000.
13. Babb VM, Haigler CH (2001) Sucrose phosphate synthase activity rises in correlation with high-rate cellulose synthesis in three heterotrophic systems. Plant Physiol 127: 1234–1242.
14. Amor Y, Haigler CH, Johnson S, Wainscott M, Delmer DP (1995) A membrane-associated form of sucrose synthase and its potential role in synthesis of cellulose and callose in plants. Proc Natl Acad Sci 92: 9353–9357.
15. Salnikov VV, Grimson MJ, Seagull RW, Haigler CH (2003) Localization of sucrose synthase and callose in freeze-substituted secondary-wall-stage cotton fibers. Protoplasma 221: 175–184.
16. Tang GQ, Luscher M, Sturm A (1999) Antisense repression of vacuolar and cell wall invertase in transgenic carrot alters early plant development and sucrose partitioning. Plant Cell 11: 177–189.
17. Wäfler U, Meier H (1994) Enzyme activities in developing cotton fibers. Plant Physiol Biochem 32: 697–702.
18. Dong HZ, Li WJ, Tang W, Li Z, Zhang DM, et al. (2006) Yield, quality and leaf senescence of cotton grown at varying planting dates and plant densities in the Yellow River Valley of China. Field Crops Res 98: 106–115.
19. Wang QC, Wang ZL, Song XL, Li YJ, Guo Y, et al. (2005) Effects of shading at blossoming and boll-forming stages on cotton fiber quality. Chin J Appl Ecol 16: 1465-1468. (in Chinese with English abstract)
20. Wang QC, Sun XZ, Song XL, Guo Y, Li YJ, et al. (2006) Effect of shading at different developmental stages of cotton bolls on cotton fibre quality. Acta Agron Sin 32: 671–675. (in Chinese with English abstract)
21. Yeates SJ, Constable GA, McCumstie T (2010) Irrigated cotton in the tropical dry season. III: Impact of temperature, cultivar and planting date on fibre quality. Field Crops Res 116: 300–307.
22. Dai JL, Dong HZ (2014) Intensive cotton farming technologies in China: Achievements, challenges and countermeasures. Field Crops Res 155: 99–110.
23. Martin LK, Haigler CH (2004) Cool temperature hinders flux from glucose to sucrose during cellulose synthesis in secondary wall stage cotton fibers. Cellulose 11: 339–349.
24. Bradow JM, Davidonis GH (2010) Effects of environment on fiber quality. In: Stewart JMcD, Oosterhuis D, Heitholt JJ, Mauney JR, editors. Physiology of Cotton. New York: Springer. pp. 229–245.
25. Liaktas A, Roussopulos D, Whittington WJ (1998) Controlled-temperature effects on cotton yield and fiber properties. J Agric Sci 130: 463–471.
26. Eaton FM, Ergle DR (1953) Effects of shade and partial defoliation on carbohydrate levels and the growth, fruiting, and fiber properties of cotton plants. Plant Physiol 29: 39–49.
27. Pettigrew WT (1995) Cotton responses to shade at different growth stages: growth, lint yield and fiber quality. Agron J 87: 947–952.
28. Khayat E, Zieslin N (1987) Effect of night temperature on the activity of sucrose phosphate synthase, acid invertase, and sucrose synthase in source and sinktissues of Rosa hybrida cv Golden Times. Plant Physiol 84: 447–449.
29. Ren WJ, Yang WY, Xu JW, Fan GQ, Ma ZH (2003) Effect of low light on grains growth and quality in rice. Acta Agron Sin 29: 785–790. (in Chinese with English abstract)
30. Li T, Ohsugi R, Yamagishi T, Sasaki H (2006) Effects of low light on rice sucrose content and sucrose degradation enzyme activities at grain-filling stage. Acta Agron Sin 32: 943–945. (in Chinese with English abstract)
31. Zhang JW, Dong ST, Wang KJ, Hu CH, Liu P (2008) Effects of shading in field on key enzymes involved in starch synthesis of summer maize. Acta Agron Sin 34: 1470–1474. (in Chinese with English abstract)
32. Zhao D, Oosterhuis DM (2000) Cotton responses to shade at different growth stages: growth, lint yield and fibre quality. Expl Agric 36: 27–39.
33. Pettigrew WT (2001) Environmental effects on cotton fiber carbohydrate concentration and quality. Crop Sci 41: 1108–1113.
34. Bian HY, Wang YH, Chen BL, Shu HM, Zhou ZG (2008) Effects of the key enzymes activity on the fiber strength formation under low temperature condition. Sci Agric Sin 41: 1235–1241. (in Chinese with English abstract)
35. Wang YH, Shu HM, Chen BL, Xu NY, Zhao YC, et al. (2008) Temporal-spatial variation of cotton fiber strength and its relationship with temperature in different cultivars. Sci Agric Sin 41: 3865–3871. (in Chinese with English abstract)
36. Haigler CH, Rao NR, Roberts EM, Huang JY, Upchurch DR, et al. (1991) Cultured ovules as models for cotton fiber development under low temperatures. Plant Physiol 95: 88–96.
37. Updegraff DM (1969) Semimicro determination of cellulose inbiological materials. Anal Biochem 32: 420–424.
38. Hendrix DL (1993) Rapid extraction and analysis of nonstructural carbohydrates in plant tissues. Crop Sci 33: 1306–1311.
39. Köhle H, Jeblick W, Poten F, Blaschek W, Kauss H (1985) Chitosan-elicited callose synthesis in soybean cells as a Ca^{2+}-dependent process. Plant Physiol 77: 544–551.
40. King SP, Lunn JE, Furbank RT (1997) Carbohydrate content and enzyme metabolize in developing canola siliques. Plant Physiol 114: 153–160.
41. Winter H, Huber SC (2000) Regulation of sucrose metabolism in higher plants: localization and regulation of activity of key enzymes. Crit Rev Plant Sci 19: 31–67.
42. Jiang JX, Zhang TZ (2003) Extraction of total RNA in cotton tissues with CTAB-acidic phenolic method. Cott Sci 15: 166–167. (in Chinese with English abstract)
43. Wang L, Li XR, Lian H, Ni DA, He YK, et al. (2010) Evidence that high activity of vacuolar invertase is required for cotton fiber and Arabidopsis root elongation through osmotic dependent and independent pathway, respectively. Plant Physiol 154: 744–756.
44. Liu JR, Ma YN, Lv FJ, Chen J, Zhou ZG, et al. (2013) Changes of sucrose metabolism in leaf subtending to cotton boll under cool temperature due to late planting. Field Crops Res 144: 200–211.
45. Maltby D, Carpita NC, Montezinos D, Kulow C, Delmer DP (1979) β1,3-Glucan in Developing Cotton Fibers. Plant Physiol 63: 1158–1164.
46. Roberts EM, Rao NR, Huang JY, Trolinder NL, Haigler CH (1992) Effects of cycling temperatures on fiber metabolism in cultured cotton ovules. Plant Physiol 100: 979–986.
47. Zhao D, Oosterhuis DM (1998) Physiologic and yield responses of shaded cotton to the plant growth regulator PGR-IV. J Plant Growth Regul 17: 47–52.
48. Roussopoulos D, Liakatas A, Whittington WJ (1998) Cotton responses to different light-temperature regimes. J Agric Sci 131: 277–283.
49. Ruan YL, Llewellyn DJ, Furbank RT (2003) Suppression of sucrose synthase gene expression represses cotton fiber cell initiation, elongation and seed development. Plant Cell 15: 952–964.

21

To Burn or Not to Burn? Effect of Management Strategy on North American Prairie Vegetation for Public Urban Areas in Germany

Anja Schmithals*, Norbert Kühn

Department of Landscape Architecture and Environmental Planning, Chair of Vegetation Technology and Planting Design, Berlin Institute of Technology, Berlin, Germany

Abstract

North American prairie vegetation has been a role model for designing highly attractive plantings for German urban green spaces for the past decade. In combination with gravel mulch top layers on planting sites and non-selective maintenance techniques like mowing or burning, prairie plantings are considered to be cost-effective and low-maintenance. This study was undertaken to assess the impact of different maintenance strategies and especially the necessity of fire management on the development success of ornamental prairie plantings in central Europe. A four factorial split-plot-block design was set up for investigation of different mixtures of prairie species under varying management conditions (mow-only, mowing plus selective weeding, mowing plus weeding and burning) on two differing soil types (in-situ topsoil and in-situ topsoil with a graywacke gravel mulch top layer) over three years. Significant effects of maintenance strategy on mortality rates and vitality were documented for a number of target species, which responded species specifically, either being slightly affected by the burning or thriving on it. Those effects were mostly restricted to topsoil sites. A strong impact on weed species presence and abundance and resulting maintenance times was found on both soil types. On topsoil sites, mow-only treatment resulted in a short-term loss of the original planting due to extensive weed growth. Corresponding gravel mulch sites were generally less colonised and visually dominated by weeds. Differences between weeded and weeded plus burned sites were minor. Unexpectedly, weed species populations were mostly unaffected by the additional burning treatment, while maintenance times and costs increased. No overall benefit of fire management for the establishment of prairie plantings was documented. The most effective management combination proved to be mowing plus regular selective weeding measures on gravel mulched planting sites.

Editor: Anil Shrestha, California State University, Fresno, CA, United States of America

Funding. This study was funded by the Deutsche Forschungsgemeinschaft (DFG, German Research Foundation) (www.dfg.de). The funders had no role in study design, data collection and analysis, decision to publish, or preparation of the manuscript.

Competing Interests: The authors have declared that no competing interests exist.

* Email: anja.schmithals@tu berlin.de

Introduction

For the past few decades in Germany, structurally diverse and sophisticated plantings- particularly herbaceous plantings- have been declining in popularity, due to high maintenance costs and the lack of personnel with the relevant expertise [1,2]. Inspired by the knowledge gained from ecological analysis of natural plant communities [3,4], a new, more naturalistic approach to planting design was developed and is supported by an increased awareness of the beauty of native plant communities like meadows, calcareous grasslands or woodland edges [5]. To correspond with this "New German Style" [6], new maintenance concepts have been developed [7,8] and economized by a high degree of mechanization [9].

The North American prairie first became a new model for German planting design concepts around the year 2000. The prairie idea was widely promoted in German landscape architecture and garden design magazines [10–17]. In addition to its great abundance of attractive species, the prairie ecosystem itself inspired a new approach to planting designs. A prairie planting was believed to be a self-sustainable, highly attractive plant community that could be maintained with very little effort [18–21]. In subsequent years, numerous ornamental prairie projects were initiated in public green spaces by parks departments (e. g. [22–25]), but none were monitored systematically to demonstrate whether the expected community stability and associated low maintenance requirement actually prevailed.

In natural prairie communities, disturbance on a regular basis is known to be a necessity for maintenance [26–28], with fire being a particularly high impact disturbance [29]. When burning is omitted from management, a loss of species with low growth form and small seeds, in particular legumes, results [30]. Prairie species are fire-adapted [31] and compensate for aboveground loss of biomass with their extensive root systems which allow them to regenerate successfully [32].

In the management of German public green spaces, prescribed burning could allow for effective control of native wintergreen weedy vegetation, while leaving the target species mostly unharmed. To date it remains unknown whether German prairie plantings can in fact be managed as extensively as believed and if fire is a mandatory disturbance. Reliable data on the actual care requirements are needed in terms of both time required for maintenance and appropriate techniques. Additionally, for prairie plantings to be a suitable alternative for German low-maintenance public planting designs, the prairie species in question must not only establish well, but also must be long-lived, vigorous and successful at reproducing in order to create visually attractive and truly sustainable plantings. In the face of climate and growing conditions that differ from their native habitats, these characteristics should not be taken for granted. None of these aspects have been empirically investigated before.

The effect of fire on natural prairie communities depends highly on when burning occurs during the year and the intervals between fire events. Burning in spring decimates wintergreen species alien to the prairie ecosystem [33] and reduces species richness, favouring the warm-season grasses [34,35]. Accumulated ground level biomass is removed, resulting in increased soil temperatures, optimum sunlight exposure, a temporary nutrient surplus [36,37] and improved conditions for germination [38]. Vegetation development is commonly stimulated, and late-developing grass species benefit the most [39]. Burning during the summer promotes subordinate species, increases species diversity and weakens the competitiveness of the *Gramineae* species [40–42]. Maximum forb species diversity is achieved with a burning regime every two to three years [43]. Shorter intervals between fires promote grass development and abundance. Longer intervals induce a general species impoverishment with higher forb ratios [26,44]. Selective weed control is important in newly established North American prairie plantings, which are usually not burned during the first growing season [45].

With the picture of a diverse planting rich in forb species in mind, it is expected that the effect on community development caused by a biennial summer burning would suit the long-term design goals best [42,43]. Unfortunately, this would destroy the visual effect of the planting at the peak of seasonal development for the rest of the growing season, which would not be acceptable in the context of public green space. Alternatively, an early spring burning should affect native wintergreen species disproportionately, without conflicting with public interests, although it risks directing development away from flowering plants towards a grass-dominated prairie plant community.

Mowing of ornamental prairie plantings in the public green could be another suitable maintenance technique. As with fire management, the effect of mowing on plant communities depends greatly on the time and frequency of cutting. Mowing of natural prairie communities facilitates forb species [46–48]. Diversity is generally reduced because the remaining litter layer acts as an establishment barrier against germination [49], with the small-seeded grasses being more heavily affected than many of the forb species [50].

Luz [51] tested various mowing techniques for extensive ornamental herbaceous plantings and found mowing with a brush cutter, a rotary mower or a flail mower (provided that the planting lacks a gravel mulch layer) to be time-saving alternatives to traditional cutting-back. He determined the optimum time for mowing to be late winter before the shoots of geophytes emerge. The time saved over cutting flowers back manually significantly reduces maintenance costs [9].

Again, in the context of managing German public plantings, dormant-season mowing together with subsequent raking away of the voluminous clippings might be an adequate maintenance strategy for urban prairie plantings. In this case, Central European species would be expected to prevail without selective suppression via burning. It is of interest whether selective weeding might prove necessary for long-term successful maintenance, despite the promised self-sustainability of prairie plant communities.

Our objective in this study was the investigation of the effects of different management techniques (fire, mowing and selective weeding) on the establishment of a mixed-planting of prairie forbs and grasses in an urban context. We specifically focused on the effects of additional burning measures vs. weeding-only management on establishment success, on individual target species' vitality, on control of weedy species and resulting maintenance effort.

Work Schedule and Methods

Ethics statement

The field studies did not involve endangered or protected species.

Experimental setup and preparation

The experiment site was located on property of the Berlin Institute of Technology in the south-western part of Berlin (52.455215°N 13.298584°E). The Chair of Vegetation Technology and Planting Design at the Department of Landscape Architecture and Environmental Planning is the assigned administrative and decision-making authority for the study site.

The site had been used for experimental vegetable gardening and was abandoned in the summer of 2008. After clearing, the site was rototilled 40 cm deep in late 2008. In 2009, *Sinapis alba* was sown as an intermediate crop. The site was cleared again at the end of summer 2009.

A four factorial split-plot-block design (A/(B+C)/D−Bl) was established (total size 43.20 m×22.60 m) in September 2009) as described by Thomas [52] (pp. 156–157).

The factors were as follows:

A. Substrate (in-situ topsoil, 7 cm greywacke chippings 5/11 on top of in-situ topsoil)
B. Species diversity (32, 16, 8 species)
C. Grass-to-forb ratio (80:20, 50:50, 20:80)
D. Management type [mowing (m), m + selective weeding (s), m + s + burning in early spring (b)].

Nine species mixes (three different species diversities combined with three different grass-to-forb ratios) were established in a split block design (factor combination B+C; 3.60 m×4.80 m each). Each mixture plot was subdivided into three subplots (split plot; 3.60 m×1.60 m each) which were assigned to the different management types. This basic unit was planted on i) topsoil and ii) topsoil with a layer of greywacke chipped gravel mulch on top (factor A), giving the study design 2×3×3×3 individual factor combination units. The study factors "soil type" and "management" forbid a randomization for practical reasons; therefore it was established as a fully balanced, non-randomized design. Three replicates were established. Plots were marked by wooden pegs at the corners. A 100 cm geomembrane was installed around the experiment site to inhibit clonal immigration of surrounding turf grasses and to reduce seed input. The same was done between soil types to prevent the gravel mulch of the elevated area from moving into the topsoil area.

Seedling plants with miniature root-balls (ø 4 cm) were planted at a density of 25 plants/m^2 in an 8×18 grid mixed planting. The planting was followed over three years.

The study design was developed to allow for many factors to be considered economically. Substrate effects on management issues are the focus of this article. Research into the effects of the other factors, grass-to-forb ratio and species diversity, is ongoing and will be presented in the future.

Species choice

Species were selected on the basis of suitability for dry-mesic to mesic soils, medium size and growth, attractiveness of flowers or seed heads, wild-growing origin and known or assumed horticultural potential. Transient forb species were included for the initial effect of the plantings, slow developers for long-term aspects. Altogether, three C4 grasses and 29 forb species were chosen.

Two nurseries carried out the cultivation according to their species expertise and logistical capacities. Twenty-five of the target species were planted in 2009. The remainder of the delivered species were either of insufficient quality or quantity and needed to be substituted. Additional cultivation was started in early 2010. Those seedlings were planted in September 2010. See table 1 for final species choice and species mixture compositions.

Study site maintenance

Each year in late winter when the soil was still frozen and the weather was dry (usually mid-February), the site was mowed to a height of 10 cm with a sickle bar mower attached to a walking-tractor (cutting width 120 cm). The clippings were removed by raking. Burning was usually undertaken in mid-March, when native vegetation came into leaf. A three-wheeled mobile weed burner (Ecoflame Select 500) fuelled by a propane gas cylinder with an effective burning width of 50 cm and a gas consumption rate of 3 kg/h was used to burn the plots until standing vegetation and surface litter were carbonised.

Selective weeding was undertaken monthly between April and September. Weeds growing above the general plant canopy and weeds that were visually disturbing were pulled by hand and removed from the plots. All woody vegetation (mainly *Betula pendula, Acer platanoides, Salix* spp.) and certain herbaceous species (*Cirsium arvense, Humulus lupulus, Solidago canadensis*) were consistently removed regardless of their size using weeding tools if necessary.

Weeds that grew outside of the selective weeding plots (s- and b-treatments) were allowed to develop. No further maintenance was undertaken.

Data collection

The time required for maintenance treatments was recorded plot-wise, except for mowing and raking, which were measured at the level of substrate units.

Survival of planted individuals was recorded in early June and September for three years starting in 2011, as were cover values of grasses, forbs, and bare ground. Identity and cover of weedy species were determined for all plots in June and September prior to selective weeding measures. Vitality criteria (e. g. basal diameter, number of generative shoots) were assessed in July and September for selected species based on the seasonal development of the traits to be recorded. For each species to be assessed, two different vitality criteria were determined. Assessment of vitality aspects was only undertaken within the 32-species plots, where all of the species to be assessed were present. Sample size per species was capped at five individuals per study plot. For all of the grass species and *Echinacea angustifolia*, this required a permanent marking of the selected individuals from a generally greater stock. For the remaining species, this included all planted individuals.

Limitations of the study

Planting material of some species was partly of insufficient quality (*Echinacea angustifolia, Oligoneuron album*), which may have accounted for high mortality rates during the establishment phase.

Because study site establishment occurred over two years, it was difficult to compare first year mortality rates, since establishment conditions differed greatly between years (two weeks of unfavourable weather after planting in 2009; safe-site effects for seedling establishment due to standing vegetation in 2010 accompanied by moderate temperatures). The species used in this study were a non-representative selection of the natural diversity found in native prairie plant communities. Results found within this study should not be applied to German urban ornamental prairie plantings in general but are most likely species specific.

Statistical analysis

Statistical tests were undertaken using R 3.0.2 [53].

Where mortality data, total coverage values and vitality data of target species were not sufficiently normal for parametric analysis, they were either successfully logit-transformed or non-parametric methods were used (Kruskal-Wallis test in lieu of ANOVA). Count data of numbers of weedy species were analysed via generalised linear mixed modelling with a Poisson error distribution. Coverage values of weedy species were analysed via generalised linear modelling with either a Gamma or a Tweedie error distribution. Similarity analysis of weed species communities was done by permutational multivariate ANOVA using distance matrices (Bray-Curtis indexes for abundance data and Jaccard indexes for presence/absence data). Management duration data were analysed via Wilcoxon-Mann-Whitney tests.

Results

Effect of maintenance type on target species establishment

At the level of species mixture, no significant effect of maintenance type on mortality rates was found between June 2010 and June 2013, either for the whole of the study layout or at the level of substrate units.

At the individual species level, seven species showed a significant response to management type: *Agastache foeniculum, Asclepias tuberosa, As. verticillata, Bouteloua curtipendula, Penstemon hirsutus, Schizachyrium scoparium* and *Sporobolus heterolepis* (see Fig. 1; for *Penstemon hirsutus* data assessment started in Jun 2011). Some of these species were heavily damaged by being burned, others seem to have profited from the treatment.

Agastache foeniculum had a tendency to higher average mortality rates under fire management (treatment b) during several assessments. When analysis was undertaken at the substrate unit level, those differences were found to be significant on topsoil sites only.

Bouteloua curtipendula had a general tendency to higher mortality rates on burned sites. At the substrate unit level, this was verified for topsoil plots only. Mortality rates of *Schizachyrium scoparium* were generally significantly higher under burning management compared to weeding management (treatment s). Analysis at the substrate level confirms this trend for both soil types, but mostly without verifiable significance. *Sporobolus heterolepis* mortality rates were lowest with weeding management and increased by additional burning on topsoil sites, whereas no differences were observed on gravel mulch sites.

Both *Asclepias* species had significantly lower average mortality rates after the second growing season on burned sites compared to sites managed by mowing and weeding or mow-only sites (treatment m). Significant differences between s-management and b-management were found in topsoil sites only. *Penstemon hirsutus* performed best on mow-only sites and regularly had

Table 1. Complete list of study species and mixture compositions.

Mixture	1	2	3	4	5	6	7	8	9
# of species in the mixture	**32**	**32**	**32**	**16**	**16**	**16**	**8**	**8**	**8**
Grass-to-forb ratio	**80:20**	**50:50**	**20:80**	**80:20**	**50:50**	**20:80**	**80:20**	**50:50**	**20:80**
Species	**Quantity in the mix**								
Agastache foeniculum	1	2	3						
Amorpha canescens	1	2	2	1	1	2			
Asclepias tuberosa	1	2	6	2	6	10	7	15	29
Asclepias verticillata	1	3	4						
Aster oblongifolius	1	3	5	2	6	9			
Baptisia australis	1	1	2	1	1	2	1	2	3
Bouteloua curtipendula	38	24	9	38	24	9	38	24	9
Dalea purpurea	1	2	1						
Echinacea angustifolia	1	4	7	4	9	14	9	24	36
Echinacea pallida	1	3	4						
Echinacea paradoxa	1	2	4						
Eryngium yuccifolium	1	3	5						
Liatris aspera	1	3	4						
Liatris ligulistylis	1	4	6	4	8	12			
Monarda bradburiana	1	2	3						
Oenothera macrocarpa	1	2	4	3	7	11			
Oligoneuron album	1	2	3	2	6	9	7	18	30
Parthenium integrifolium	1	2	4						
Penstemon grandiflorus	1	2	3						
Penstemon hirsutus	1	3	4	2	8	12			
Penstemon ovatus	1	3	4						
Pycnanthemum pilosum	1	3	4						
Pycnanthemum tenuifolium	1	3	5						
Ratibida columnifera var. pulcherrima	1	2	4						
Rudbeckia hirta	1	2	4						
Rudbeckia missouriensis	1	2	5	2	7	11			
Rudbeckia triloba	1	3	4	2	4	6	5	13	17
Schizachyrium scoparium	38	24	10	38	24	10	38	24	10
Solidago missouriensis	1	2	3	2	4	7			
Sporobolus heterolepis	39	24	10	39	24	10	39	24	10
Verbena hastata	1	2	4	2	5	10			
Verbena stricta	1	3	4						

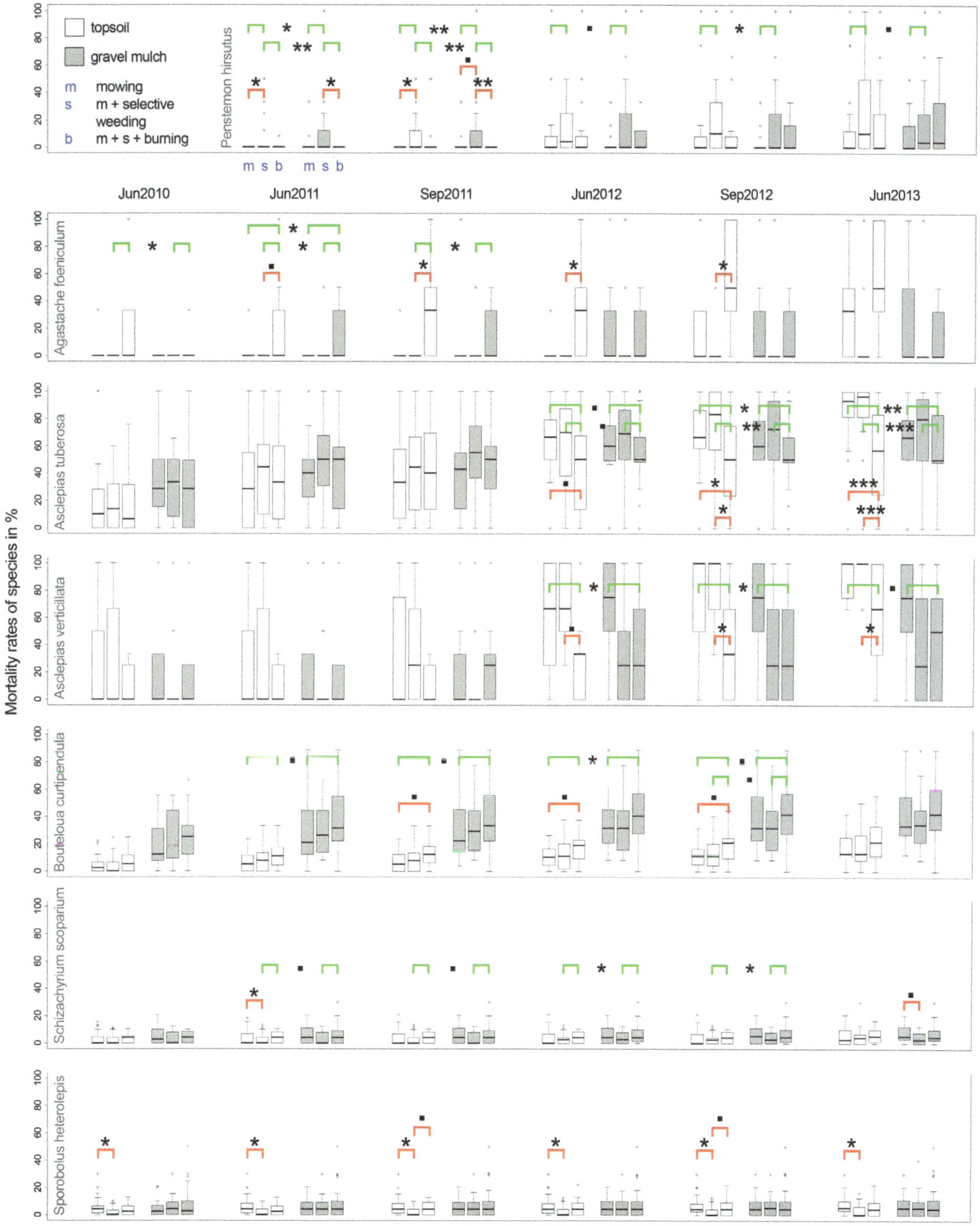

Figure 1. Effects of three different maintenance strategies [mowing (m), mowing + selective weeding (s), mowing + selective weeding + burning (b)] on mortality rates of planted target species between June 2010 and June 2013 differentiated by soil type. Significant differences between management types are indicated by red brackets and asterisks. P<0.1 * P<0.05 ** P<0.01 *** P<0.001. Asterisks on green brackets indicate significant differences between management types for the whole of the planting regardless of soil type. These results could not always be verified at the level of substrate units, but are mostly in accordance with the trend suggested by data distribution.

significantly higher average mortality rates under weeding management on both types of substrate.

The remaining 25 species did not respond significantly differently to the different treatments.

Management type also did not have any measurable influence on target species total coverage values.

Effect of maintenance type on target species vitality

Original sample sizes for detailed surveying of individual species ranged from 72 to 270, depending on the mixture-related frequency of each species. For a complete list of species, assessed criteria and sample sizes see table 2. Sample populations of *Asclepias verticillata*, *Dalea purpurea*, *Echinacea angustifolia*, *Liatris aspera*, *Penstemon ovatus* and *Ratibida columnifera var. pulcherrima* either died out or the numbers dropped below levels needed for statistical analysis. Significant differences in vitality criteria due to management treatment were found in three species: *Amorpha canescens*, *Monarda bradburiana* and *Schizachyrium scoparium*.

Amorpha canescens produced significantly fewer vegetative shoots on burned plots regardless of soil type in 2011 and 2012 (significance levels varying between $p<0.1$. and $p<0.01$**). There was no measurable effect on the number of generative shoots. The original sample size was 90 individuals, of which 5 had died by the first assessment in September 2011. By the end of the study, 66 individuals remained, the majority of which grew on topsoil plots (56%).

Monarda bradburiana produced significantly more generative shoots on burned topsoil plots during all assessments ($p<0.05$* in 2011 and 2012, $p<0.01$** in 2013). No differences were observed on gravel mulched plots. Diameters of clumps were significantly larger on burned topsoil plots after 2011 ($p<0.1$. in 2012, $p<0.01$** in 2013). The original sample size was 108, which fell to 77 individuals towards the end of the study (44% growing on topsoil).

While basal diameters of *Schizachyrium scoparium* clumps were independent of management type, the number of flowering shoots was higher on burned topsoil plots during all years ($p<0.05$*). No effect was observed on gravel mulch sites.

Effect of maintenance type on weed species presence and abundance and weed species community composition

Analysis of data on weed species presence and abundance clearly showed a significant influence of maintenance type. The greatest quantities of assessed weed species and the highest cover values were found on topsoil sites with m-maintenance, the lowest values were the result of s-management on gravel-mulched sites.

Study plots that received m-treatment mostly had significantly higher weed species numbers and cover values than plots with s- or b-treatment (see Fig. 2), although the level of significance varied. Minor differences were found for the comparison of weed numbers of s- and b-treatments in 2012. Only the last assessment revealed significant differences between cover values of s- and b-treatments.

Differences in weed species community composition between m-plots vs. s-plots and b-plots were highly significant for all assessments ($p<0.001$***). The species that predominantly accounted for those differences generally included woody species (*Betula pendula*, *Clematis vitalba*, *Populus tremula*, *Salix aurita*, *S. caprea*) and herbaceous species of high growth and visual dominance (*Cirsium arvense*, *Conyza canadensis*, *Epilobium lamyi*, *Erigeron annuus*, *Solidago canadensis*, *Sonchus olearceus*).

Between s- and b-treatments, no significant differences in weed species community composition were found at any assessment.

Comparison of labour input between different maintenance treatments

On topsoil sites, total weeding times were highest for s-sites and lower on b-sites. On gravel mulch sites, weeding times on b-sites were mostly equal to or even higher than on s-sites. However, none of these differences were significant. Fig. 3 shows the total maintenance times for mowing, weeding and burning activities for the three maintenance strategies, separated by year and substrate. We included machine set-up and clean-up times (approximately 30 minutes per burning occasion) and increased personnel needed for burn control during the burning process (doubling of assessed burning times). We calculated this effort on the basis of one-third of the size of the total study site (equals 311 m^2) and added these values in min/m^2. We excluded fuel costs for the burner (depending on the runtimes), labour time needed for the acquisition and return of the burning equipment, fuel costs and operating costs of the transport vehicle and daily rental costs for the burning equipment from the analysis.

When all labour times related to burning activities were summed, the total duration of average maintenance on b-sites considerably exceeded s-site maintenance times, except in the case of topsoil sites in 2013, where total labour times were about equal.

Discussion

Effect of maintenance type on target species establishment

Only a few of the target species showed significant reactions to burning treatment in terms of mortality. Half of those species had higher average mortality rates; the other half had lower rates.

No obvious benefit of a fire management for the establishment was documented that would outweigh the increased costs of burning.

Asclepias tuberosa and *A. verticillata* had lower mortality rates on burned sites, especially on topsoil. Their rhizomes, which lie deep in the soil, protected them effectively from fire damage. No apparent reason accounts for the better performance on topsoil sites, though. *Agastache foeniculum* generally performed best on mowed but unburned sites. It is an early developing species with close-to-the-surface buds, which may have been significantly harmed by burning.

The higher mortality rates of both grass species on burned sites regardless of soil type were consistent for the duration of the study. It was clearly not a failure of initial establishment, but rather a continuous tendency to higher death rates under fire management that accounted for this.

At first glance, this appears to be inconsistent with current knowledge about the general facilitation of prairie grasses in natural prairies by dormant-season fire management, e.g. [33,39,43,54]. None of these studies, however, evaluated the longevity of individual plants with regard to management, but instead they assessed the performance of an overall grass community by means of total productivity. It is not unlikely that substantially higher grass seedling rates on burned sites account for most of this facilitating effect [55], more than compensating for the shortened lifespan of the individual grasses under fire management. Since the horticultural use of prairie vegetation is not essentially different from other ornamental plantings in that they are focused on maintaining a distinct and fairly static design vision over the long term, species-specific reactions to alternative management regimes are critical and the documented fire effect was mostly counterproductive during this early development stage of the planting. It may become interesting again when the planting

Table 2. Criteria for vitality assessment and sample size development of 14 prairie species.

Species	Data collection		orig. SS	Avg survival rate (%)			Assessment criteria							
	June	Sept.		in 2011	in 2012	in 2013	basal Ø	clump Ø	# of gen. shoots	# of veg. shoots	hgt of longest shoot	# of flowers	# of open flowers	# of flowers on longest shoot
Amorpha canescens		x	90	93.3	80.0	73.3			x	x				
Asclepias verticillata	x		144	80.5	46.5	38.2			x			x		
Bouteloua curtipendula		x	270	98.9	90.0	35.5	x		x					
Dalea purpurea	x		72	61.1	5.5	1.3			x				x	
Echinacea angustifolia	x		216	50.5	25.0	17.6			x				x	
Eryngium yuccifolium	x		164	94.2	85.9	30.5			x		x			
Liatris aspera	x		144	76.4	46.5	35.4			x					x
Monarda bradburiana	x		108	77.7	76.8	71.3		x	x					
Parthenium integrifolium	x		126	92.1	91.3	75.4	x		x					
Penstemon ovatus	x		144	62.5	37.5	25.0		x	x					
Ratibida columnifera var. pulcherrima	x		126	30.2	4.3	0.8			x				x	
Schizachyrium scoparium		x	270	100.0	99.6	97.8	x		x					
Solidago missouriensis		x	108	98.1	90.7	-	x		x					
Sporobolus heterolepis		x	270	99.3	99.3	98.5	x		x					

SS = sample size, Avg = average, gen. = generative, veg. = vegetative, hgt = height, length information values in cm.

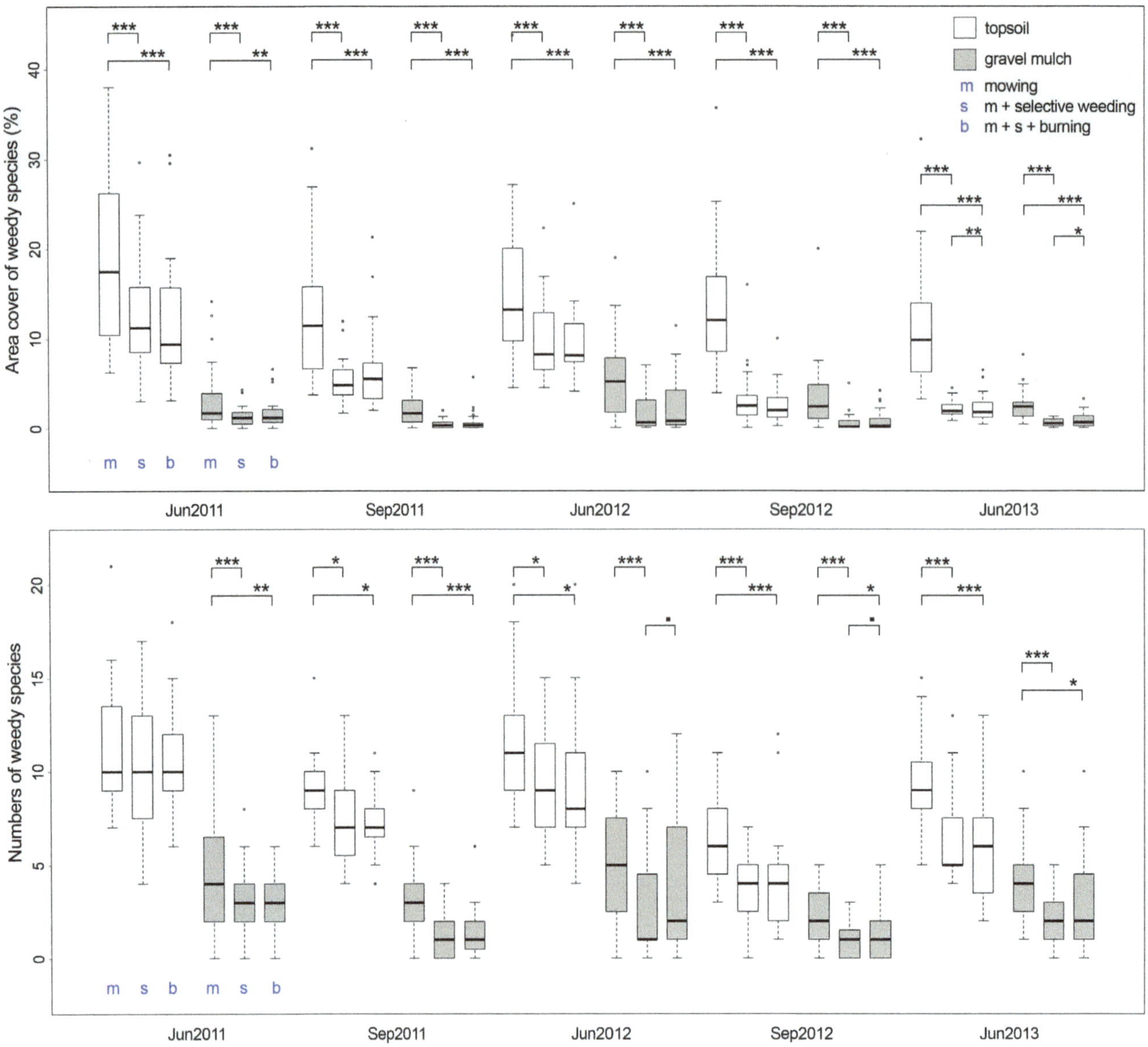

Figure 2. Effects of three different maintenance strategies on cover values and numbers of weedy species between June 2011 and June 2013 differentiated by soil type. Significant differences between management types are indicated by asterisks. P<0.1 * P<0.05 ** P<0.01 *** P<0.001.

is fully developed and in need of corrective measures to preserve the original design idea.

One reason for the predominant similarity in mortality rates between b-sites and s-sites may have been the study design. We compared unburned and burned plantings, both of which were mowed and raked before the burning treatment. No standing biomass or litter remained, which would have created an organic mulch layer and consequently significant differences in site conditions, as would be found natural prairie communities (light-, temperature- and establishment barrier created by the remaining litter in unburned areas vs. clean-swept dark burned surface). The effect of burning therefore may have been strongly reduced owing to the similarities in the site conditions [56].

A second reason for the minimal differences between b-sites and s-sites may have been the development stage of the planting material at the beginning of the study. Because we planted rather than seeded the study site, the individuals had a sufficiently well-developed root system from the start, with at least eight weeks of the growing season left after the planting process. The first burning took place five months after planting. By that time, most individuals were well grown-in. The burning effect on mortality rates most likely would have been more distinct if we had included naturally rejuvenated specimens in the analysis, which are more easily damaged by fire. We would expect the same at later stages of development, when competition between densely growing mature individuals becomes an important factor for survival and reproductive success. At that stage, fire will most likely have a strong influence on established competitive hierarchies and may be used to steer the planting towards the long-term development objectives by adjusting burning dates during the season. Our data indicate that fire is more harmful than helpful for survival of the target species during the post-seedling stage but still early in the

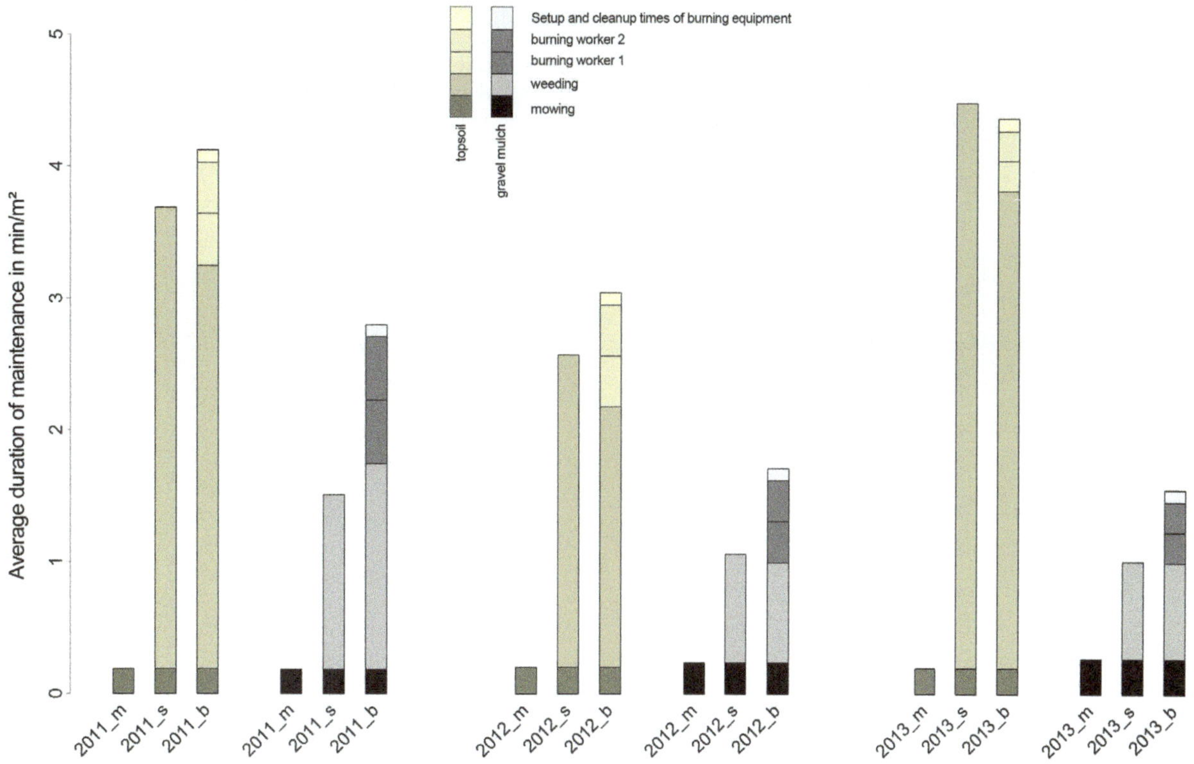

Figure 3. Maintenance times for mowing, selective weeding, and burning measures on the three different treatment sites (_m, _s, _b) and comparison of total maintenance duration for top soil and gravel mulch sites in 2011, 2012 and 2013.

establishment phase of a planting development. In the context of designed prairie plantings in an urban setting, this is a very important aspect with regard to the pursued longevity and potential (self-) sustainability of a public prairie planting and needs to be investigated further. We are interested in moderately dynamic communities, where persistent plants sustain the original design picture for a long time, and necessary rejuvenation is also able to take place. If gaps are needed for the grasses to rejuvenate, then sustainable management of ornamental prairie plantings needs to provide this. Obviously, this need not necessarily be done by burning. It may also be achieved by raking mowed sites, thus avoiding the negative effects of burning on the mortality rates of stock plants.

Effect of maintenance type on target species vitality

An influence of management type on species vitality was proven for only three species (*Amorpha canescens*, *Monarda bradburiana*, *Schizachyrium scoparium*). Burning increased basal diameters and numbers of flowers in two species. The third species developed fewer vegetative shoots, but the same number of generative shoots and therefore was not restricted in its reproductive potential. The overall effect of burning on vitality therefore was rated minimally positive, but negligible.

The general lack of a significant effect on the majority of the species (especially the grasses) was unexpected as common knowledge holds that burning has a favourable effect on the vigour and vitality of prairie species in general [57] and specifically on the vitality of prairie grass [33,58,59]. Again, the lack of significant effect was most likely caused by the similar preparations of s- and b-sites prior to the burning treatment. When standing biomass was removed by mowing and raking, no temporary nutrient surplus was available after burning to support more vigorous growth at the beginning of the growing season. Surface light conditions and soil surface temperatures after raking and subsequent burning were basically equal to those following a natural burning management, whereas nutrient availability was not. Temporary nutrient surplus is known to be an important factor in prairie species facilitation via burning, though [60,61]. Since similar growing conditions on s-sites and b-sites yielded mostly equal results with respect to individual plant vitality, s-management was rated adequate, while b-management was rated dispensable.

Effect of maintenance type on weed species presence and abundance and weed species community composition

The influence of management on the presence and abundance of weed species was highly significant. Omission of selective weeding measures resulted in increasingly weed-infested plots, lacking any visual quality of the original plantings, regardless of soil type. This rapid maldevelopment stood in contrast to the promoted low-maintenance of prairie plantings [18–21] during the establishment phase.

Differences between s- and b-plots were rare and of minor significance. We expected wintergreen species, especially wintergreen annuals, that were alien to the fire-adapted plant community to be significantly affected by the burning (as described by Hitchmough [62]), while perennial and geophytic species would be less affected. We either expected a substantial reduction in total weed coverage values or noticeable differences in the weed community composition; neither expectation was met.

A possible reason for this lack of difference may have been the early time of burning during the year. While native weedy vegetation was already in leaf, it was not fully developed and may not have been affected strongly enough by the burning treatment. We would expect a later date to be more effective (end of April, shortly before the prairie species shoots emerge). Where geophytes -an important component of public green spaces in Germany- are present, however, this might coincide with the emergence of their shoots or blooming. This needs to be considered when designing species mixtures.

A slight tendency to higher numbers and abundances of weeds on burned gravel mulch sites compared to unburned but weeded gravel mulch sites (see Fig. 2) was interpreted as a rather counterproductive effect of burning. This may have been due to a fertilising effect of the deposited ashes, which provided organic material and nutrients in gaps in the mulch and created a temporary nutrient surplus supporting germination. The mow-only treatment was abandoned in late summer of 2013 after the last data assessment due to an increasing *Calamagrostis epigejos* and *Cirsium arvense* infestation, which by June dominated most of the m-plots and started to influence neighbouring sites. Both species are known for their efficient vegetative spread as well as highly successful generative reproduction [63,64]. Therefore, a regular s-management was introduced in order to relieve remaining target vegetation within the former m-plots and to prevent infestation of the well maintained s-plots and b-plots.

Comparison of labour input between different maintenance treatments

While selective weeding proved to be essential for successful establishment and preservation of the original planting design, the additional expenditure of labour for burning had only a small influence on weeding times and planting development.

On topsoil sites, the average savings in weeding labour time on burned plots was generally smaller than the additional time investment for the burning treatment, resulting in higher total management times (Fig. 3).

On gravel mulch sites, the effect of burning on weeding times was only inconsistent and minor. Total maintenance times on burned gravel mulch plots were significantly higher than on s-sites.

On the whole, burning treatment significantly increased total maintenance times and costs, regardless of soil type, without any significant benefit for the plantings' development during the first three years of establishment.

Conclusions

To estimate the general effect of different management techniques and especially the potential benefit of a burning treatment on prairie-like herbaceous mixed-plantings in public open space, several points were considered in this study.

We analysed the effect of fire on community and individual species mortality. At the community level, no effect of burning was registered. At the species level, only a few species reacted significantly, and those reactions were of an ambivalent nature: some species showed higher mortality rates under a burning regime (*Agastache foeniculum*, *Bouteloua curtipendula*, *Schizachyrium scoparium*), others seem to have profited from regular burning of the site (*Asclepias tuberosa*, *A. verticillata*, *Penstemon hirsutus*).

The comparison of different vitality criteria revealed no measurable effect of regular fire management on the majority of target species. We found significant differences in only three species, of which one was slightly affected by the burning (*Amorpha canescens*) and two thrived on it (*Monarda bradburiana*, *Schizachyrium scoparium*).

Analyses of species numbers and cover values of weedy plants revealed the critical impact of selective weeding activities on the establishment and maintenance of the target species populations compared to a mow-only treatment. No consistent differences were found between sites with selective weeding and weeding plus burning management, though. Neither the composition of weed species communities nor the numbers and abundances of weed species were significantly affected by the additional burning measures.

Summing up, mow-only treatment of mixed prairie plantings resulted in the loss of the original design in just a short time due to uncontrolled aggressive weed expansion, especially on topsoil sites. To evaluate the benefit of an additional burning treatment, the slightly increased vitality and lower average mortality of a small number of target species need to be weighed against higher average mortality rates of other target species and significantly increased input in labour and financial resources. Therefore, we conclude that there is no overall positive effect of burning on the establishment and maintenance of mixed prairie plantings on topsoil sites and even less on gravel-mulched sites. Considering maintenance costs, we conclude that the most effective management combination is mowing and raking in mid-spring plus regular selective weeding measures on gravel-mulched plantings. To evaluate the long-term effect of different management techniques on urban ornamental prairie plantings, rejuvenation success of target species and community dynamics need to be assessed in further studies.

Acknowledgments

We would like to thank Christian Unger and Sebastian Unger for their constant and reliable assistance in the fieldwork, Kelaine Ravdin for her professional editing support and helpful comments on the manuscript.

Author Contributions

Conceived and designed the experiments: AS NK. Performed the experiments: AS. Analyzed the data: AS. Contributed reagents/materials/analysis tools: NK. Wrote the paper: AS NK.

References

1. Becker C (1995) Grünflächen - Folgekosten werden zum Problem. Stadt+Grün 7: 461–467.
2. Schmidt H (2005) Kommunale Grünflächenpflege. Stadt+Grün 3: 18–24.
3. Hansen R, Stahl F (1981) Die Stauden und ihre Lebensbereiche. Stuttgart: Eugen Ulmer GmbH & Co. 573 p.
4. Grime JP (2001) Plant Strategies, Vegetation Processes, and Ecosystem Properties. John Wiley & Sons.
5. Reif J, von Birgelen A, Edelmann T (2013) Die Rabatte ist tot, es lebe… Stadt+Grün 1: 59–61.
6. Hanzen T, Huxmann N (2013) New German Style Planting - Auf der Suche nach dem neuen(?) deutschen Exportschlager. Stadt+Grün 3: 19–25.
7. Schmidt C (2005) Neue Pflegekonzepte für nachhaltige Staudenpflanzungen. Stadt+Grün 3: 30–35.
8. Kühn N (2013) Von Hansens Lebensbereichen zu den Strategietypen. Garten und Landschaft 4: 18–21.
9. Kietsch U, Kircher W (2005) Stauden unterm Mähgerät. Gartenpraxis 8: 18–23.
10. Hertlein-Rieder V (2006) Neuland für die Südspitze Marzahn. Garten + Landschaft 1: 10–13.
11. Jakob F (2002) Hochgrasprärie: Attraktivität preiswert aus Saatgut. Gartenpraxis 6: 47–51.
12. Köhlein F (2002) Nordamerikanische Prärie im Garten. DEGA 11: 16–18.
13. Oehmichen F (2000) Ein Tunnel durch die Blumenwiese. Gartenpraxis 4: 39–44.
14. Schmidt C (2003) Präriestauden: Helfer für grüne Städte. GrünForumLA 12: 29–31.
15. Schmidt C (2004) Neue Pflanzen aus der Hochgrasprärie. DEGA 11: 12–14.
16. Schmidt C (2005) Ästhetik made in USA. GrünForumLA 2: 24–26.

17. Cascorbi U (2007) Transatlantische Mischungen - Präriestauden und -gräser in gärtnerischen Wiesen. Stadt+Grün 5: 49–53.
18. de la Fleur M (2001) Erfahrungen mit der Hochgrasprärie. Gartenpraxis 10: 29–33.
19. Rothmund S (2000) Kurzgrasprärien als Vorbild in der Gartenkultur. Gartenpraxis 12: 31–36.
20. Schmidt C (2003) Präriepflanzen im öffentlichen Grün. Billiger Jakob oder innovativer Impuls? GrünForumLA 3: 18–20.
21. Schmidt C (2007) Mittelhohe Präriegräser. Gartenpraxis 1: 32-39.
22. Guldner I (2005) Wagnis und Erfolg - Der Präriegarten im Berggarten Hannover-Herrenhausen. Stadt+Grün 7: 29–32.
23. Elsner AK (2007) Extensive Pflanzungen mit Präriestauden. Garten + Landschaft 5: 16–19.
24. Schmidt C (2004) Präriemischpflanzungen auf trockenen Böden. Gartenpraxis 5: 32–39.
25. Schmidt C (2004) Präriemischpflanzungen auf frischen Böden. Gartenpraxis 7: 17–25.
26. Collins SL (2000) Disturbance Frequency and Community Stability in Native Tallgrass Prairie. The American Naturalist 155: 311–325.
27. Collins SL, Barber SC (1985) Effects of disturbance on diversity in mixed-grass prairie. Plant Ecology 64: 87–94.
28. Steinauer EM, Collins SL (1996) Disturbance as a management component in tallgrass prairie. In: Samson FB, Knopf FL, editors. Prairie conservation: preserving North America's most endangered ecosystem. Covelo, CA: Island Press. pp. 39–52.
29. Collins SL, Steinauer EM (1998) Disturbance, diversity and species interactions in tallgrass prairie. In: Knapp AK, Briggs JM, Hartnett DC, editors. Grassland dynamics: long-term ecological research in tallgrass prairie. Oxford University Press. pp. 140–156.
30. Leach MK, Givnish TJ (1996) Ecological determinants of species loss in remnant prairies. Science 273: 1555–1558.
31. Vogl RJ (1974) Effects of Fire on Grasslands. In: Kozlowski TT, Ahlgren CE, editors. Fire and Ecosystems. New York: Academic Press, Inc. pp. 139–189.
32. Kucera CL (1970) Ecological Effects of Fire on Tallgrass Prairie. In: Schramm P, editor. Proceedings of a Symposium on Prairie and Prairie Restoration. Knox College, Galesburg, Illinois. pp. 12.
33. Hulbert LC (1988) Causes of fire effects in tallgrass prairie. Ecology 69: 46–58.
34. Collins SL (1987) Interaction of disturbances in tallgrass prairie: a field experiment. Ecology 68: 1243–1250.
35. Collins SL (1992) Fire frequency and community heterogeneity in tallgrass prairie vegetation. Ecology 73: 2001–2006.
36. Knapp AK (1984) Post-burn Differences in Solar Radiation, Leaf Temperature and Water Stress Influencing Production in a Lowland Tallgrass Prairie. American Journal of Botany 71: 220–227.
37. Knapp AK, Seastedt TR (1986) Detritus accumulation limits productivity of tallgrass prairie. Bioscience 36: 662–668.
38. Loydi A, Eckstein RL, Otte A, Donath TW (2013) Effects of litter on seedling establishment in natural and semi-natural grasslands: a meta-analysis. Journal of Ecology 101: 454–464.
39. Gibson DJ, Seastedt TR, Briggs JM (1993) Management Practices in Tallgrass Prairie: Large- and Small-Scale Experimental Effects on Species Composition. Journal of Applied Ecology 30: 247–255.
40. Biondini ME, Steuter AA, Grygiel CE (1989) Seasonal fire effects on the diversity patterns, spatial distribution and community structure of forbs in Northern Mixed Prairie, USA. Plant Ecology 85: 21–31.
41. Howe HF (1994) Response of early- and late-flowering plants to fire season in experimental prairies. Ecological Applications 4: 121–133.
42. Copeland TE, Sluis WJ, Howe HF (2002) Fire Season and Dominance in an Illinois Tallgrass Prairie Restoration. Restoration Ecology 10: 315–323.
43. Kucera CL, Koelling M (1964) The influence of fire on composition of central Missouri prairie. American Midland Naturalist 72: 142–147.
44. Bowles M, Jones M, McBride J (2002) Twenty-year changes in burned and unburned sand prairie remnants in northwestern Illinois and implications for management. American Midland Naturalist 149: 35–45.
45. Ohlenbusch PD (1997) Establishing Native Grasses. Kansas State University - Agricultural Experiment Station and Cooperative Extension Service. report number MF-2291. 4 p.
46. Rogers WE, Hartnett DC (2001) Vegetation responses to different spatial patterns of soil disturbance in burned and unburned tallgrass prairie. Plant Ecology 155: 99–109.
47. Briggs JM, Knapp AK (2001) Determinants of C3 forb growth and production in a C4 dominated grassland. Plant Ecology 152: 93–100.
48. Collins SL, Knapp AK, Briggs JM, Blair JM, Steinauer EM (1998) Modulation of Diversity by Grazing and Mowing in Native Tallgrass Prairie. Science 280: 745–747.
49. Foster BL, Gross KL (1998) Species richness in a successional grassland: effects of nitrogen enrichment and plant litter. Ecology 79: 2593–2602.
50. Foster BL (1999) Establishment, competition and the distribution of native grasses among Michigan old-fields. Journal of Ecology 87: 476–489.
51. Luz H (2005) Staudenwiesen am Badesee. Gartenpraxis 5: 51–54.
52. Thomas E (2006) Feldversuchswesen. Stuttgart: Eugen Ulmer Verlag. 387 p.
53. R Core Team (2014) R: A language and environment for statistical computing. R Foundation for Statistical Computing. Available: http://www.R-project.org/. Accessed 15.09.2014
54. Towne EG (1995) Influence of Fire Frequency and Burning Date on the Proportion of Reproductive Tillers in Big Bluestem and Indian Grass. In: Hartnett DC, editor. Proceedings of the Fourteenth North American Prairie Conference. Manhattan, Kansas: Kansas State University. pp. 75–78.
55. Glenn-Lewin DC, Johnson LA, Jurik TW, Akey A, Leoschke M, et al. (1990) Fire in Central North American Grasslands: Vegetative Reproduction, Seed Germination, and Seedling Establishment. In: Collins SL, Wallace LL, editors. Fire in North American Tallgrass Prairies. Norman: University of Oklahoma Press. pp. 28–45.
56. Tix BA, Hebberger JA, Vaughan E, Charvat I (2003) The Effects of Fire Versus Mowing on Prairie Plant Communities. University of Minnesota - Department of Plant Biology report number MN/RC – 2003-20. 49 p.
57. Daubenmire RF (1968) The ecology of fire in grasslands. Advances in Ecological Restoration 5: 209–266.
58. Towne G, Owensby CE (1984) Long-term Effects of Annual Burning at Different Dates in Ungrazed Kansas Tallgrass Prairie. Journal of Range Management 37: 392–397.
59. Schacht WH, Smart AJ, Anderson BE, Moser LE, Rasby R (1998) Growth responses of warm-season tallgrasses to dormant-season management. Journal of Range Management 51: 442–446.
60. Tix D, Hebberger JA, Charvat I (2006) Influence of aboveground biomass removal on nitrogen mineralization in a restored tallgrass prairie. Restoration Ecology 14: 561–568.
61. Ojima DS, Schimel DS, Parton WJ, Owensby CE (1994) Long- and Short-term Effects of Fire on Nitrogen Cycling in Tallgrass Prairie. Biogeochemistry 24: 67–84.
62. Hitchmough J, de la Fleur M (2006) Establishing North American prairie vegetation in urban parks in northern England: Effect of mangement and soil type on long-term community development. Landscape and Urban Planning 78: 386–397.
63. Mayer F (2000) Long distance dispersal of weed diaspores in agricultural landscapes - The Scheyern approach. Doctoral Thesis, Technische Universität München. 206 p.
64. Rebele F, Lehmann C (2001) Biological flora of Central Europe: Calamagrostis epigejos (L.) Roth. Flora 196: 325–344.

The Continuous Incorporation of Carbon into Existing *Sassafras albidum* Fine Roots and Its Implications for Estimating Root Turnover

Thomas S. Adams, David M. Eissenstat*

Department of Ecosystem Science and Management and the Ecology Graduate Program, the Pennsylvania State University, University Park, Pennsylvania, United States of America

Abstract

Although understanding the timing of the deposition of recent photosynthate into fine roots is critical for determining root lifespan and turnover using isotopic techniques, few studies have directly examined the deposition and subsequent age of root carbon. To gain a better understanding of the timing of the deposition of root carbon, we labeled four individual *Sassafras albidum* trees with 99% ^{13}C CO_2. We then tracked whether the label appeared in roots that were at least two weeks old and no longer elongating, at the time of labeling. We found that not only were the non-structural carbon pools (soluble sugars and starch) of existing first-order tree roots incorporating carbon from current photosynthate, but so were the structural components of the roots, even in roots that were more than one year old at the time of labeling.Our findings imply that carbon used in root structural and nonstructural pools is not derived solely from photosynthate at root initiation and have implications regarding the determination of root age and turnover using isotopic techniques.

Editor: Malcolm Bennett, University of Nottingham, United States of America

Funding: This work was supported by the National Science Foundation (IOS-0824704, IOS-1120482) to DM Eissenstat. The funders had no role in study design, data collection and analysis, decision to publish, or preparation of the manuscript.

Competing Interests: The authors have declared that no competing interests exist.

* E-mail: dme9@psu.edu

Introduction

The timing of the deposition of photosynthetically derived carbon into roots is of considerable scientific importance because it has implications both for factors controlling root physiology and lifespan and for the determination of root age and turnover using isotopic techniques. Despite this importance, and over two decades of work on the subject, there is still no clear consensus concerning the timing of the deposition of root carbon. Until recently, it has been assumed that both root nonstructural storage carbon and structural carbon are primarily deposited when a root is first initiated. Marshall and Waring [1] contended that nonstructural root storage carbohydrates, such as starch, are deposited solely when the root is first formed and it is the rate of depletion of these reserves that determines root longevity. Recent attempts to determine root age and root turnover using isotopic techniques, both bomb ^{14}C and ^{13}C labeling, have assumed that root structural tissue originates within the year that a root is formed and that no new carbon is subsequently added to the root [2–8]. However, to our knowledge, these assumptions concerning the timing of the origin of root structural and nonstructural carbon have not been adequately tested.

A number of studies have called into question the validity of the assertion by Marshal and Waring [1] that root nonstructural carbohydrate reserves (e.g., starch) are solely created when a root is first born. For example, Nguyen *et al.* [9] found that root carbohydrate levels increased late in the growing season, implying that root carbohydrate reserves were not solely laid down when a root is first formed. Additionally Kosola *et al.* [10] found in a study that tracked individual roots, that the root starch concentration of Eugenei hybrid poplars did not decline with root age. As a result, they concluded that the starch reserves in the fine roots of the poplar were not determined at root birth and that these reserves were "labile and dynamic".

Structural carbon has also traditionally been viewed as originating from recent photosynthate during root formation [2–8]. But recent studies using bomb ^{14}C have found that the origin, and subsequent age, of root structural carbon may involve more complicated processes than previously assumed. Evidence now suggests that stored carbon fixed prior to root formation may be incorporated in the structural tissue of newly formed roots [3,6,8,11]. Furthermore, the amount of stored carbon incorporated in the formation of larger diameter fine roots (1.5–2 mm), presumably of higher root order [12], is higher than in smaller diameter, lower order roots (<0.5 mm) [6]. Because root age is positively correlated with root diameter [13–15], it is difficult to disentangle the relationship between root age assessed by direct observation, root diameter, and estimated age based on carbon residence time.

In an attempt to gain a better understanding of the timing of the deposition of root carbon, we labeled *Sassafras albidum* trees with 99% $^{13}CO_2$. We then tracked whether the label appeared in first-order roots that were at least two weeks old and no longer elongating at the time of labeling. Based on previous studies, we hypothesized that the ^{13}C label would appear in the nonstructural carbon pools of the existing first-order roots, but not in their structural carbon pool.

Materials and Methods

This study was conducted at a common garden planting (mixed species planting) located at the Russell E. Larson Agricultural Research Center, Pennsylvania State University, USA (40.8°N, 77.9°W) [16]. For this study we labeled *Sassafras albidum* with 99% $^{13}CO_2$ (Cambridge Isotope Laboratories, Andover, MA USA) (Figure 1). We chose this species because of its comparatively small size (ca. 2.5 m tall) relative to other tree species in the garden, and the easily recognizable, relatively large diameter, fine roots which emit a distinctive odor when harvested thereby precluding confusion with the roots of neighboring trees. Each species plot represented six trees planted in a double row with spacing of 3 m between trees within the row, 3 m between the double rows and 5 m between the six-tree plots.

Soils at the common garden are relatively fertile Hagerstown silt loam, well-drained, with a pH ranging from 6.1–6.5 and with some areas high in calcium. Previous to planting the trees, the site was used as a grass hayfield. The entire area was fenced to keep out deer. *Sassafras albidum* trees were collected in the early spring of 1996 from seedlings around State College, PA and therefore were approximately 13 years old at the time of the experiment. Understory vegetation was controlled within a half-meter of the tree using weed barrier cloth and gravel mulch and sprayed with glyphosate to a distance of about two meters from the trunk. Further from the trees, grass was mowed weekly or longer, depending on climatic conditions.

Figure 1. Methods used in study. A. *Sassafras albidum* tree covered in a clear mylar balloon during the $^{13}CO_2$ labeling process. **B.** Example of a root box showing roots that were traced on a clear acetate window facing the study tree using different color paint pens.

Root boxes were installed in the spring of 2008 and again in the spring of 2009[17]. Briefly, a 0.6×0.5×0.4 m hole was dug in the ground approximately 0.5 m in front of each study tree. A 0.6×0.5×0.4 m box constructed of 1.9 cm treated plywood, with a clear 0.127 mm thick (5-mil) acetate window, was then placed in the ground with the acetate window facing the tree (Figure 1B). A removable piece of 1.3 cm thick foam insulation board was placed against the window when not in use to minimize temperature differences between the soil and the interior of the root box. In this fashion, roots growing against the acetate window could be followed from birth by tracing the roots on the acetate using different colored paint pens (Marvy DecoColor pens, Uchida of America Corp., Torrance, CA, USA) on different tracing dates. Three small first-order roots per tree were harvested prior to labeling and pooled into a single pre-label sample per tree to assess background ^{13}C levels. After labeling the trees with ^{13}C, preexisting roots of a known age were then excised by cutting through the acetate window and individual first-order roots were analyzed for ^{13}C in structural and nonstructural carbon pools.

Sassafras albidum trees, approximately 2.5 m in height, were labeled with 99% $^{13}CO_2$ by first surrounding the trees in a 2.4×1.2×1.2 m clear mylar balloon supported by a frame constructed of 1.9 cm internal diameter pvc pipe (Figure 1A). Mylar was also used to cover the soil surface to minimize non-target $^{13}CO_2$ uptake. During the labeling, air was constantly circulated in the balloon using three small battery powered fans hung within the canopy of the tree. Carbon dioxide concentrations within the balloon were measure during the labeling using a Li Cor 6200 (Li-Cor Biosciences, Lincoln, NE, USA). The labeling itself occurred by releasing a short pulse of 99% $^{13}CO_2$ into the balloon and waiting until the CO_2 concentration in the balloon dropped back below background levels (~385 ppm). This process took approximately 30 minutes per tree once the balloon was installed. Roots were collected immediately prior to labeling and at various times after labeling. Only first-order, distal, roots were sampled from the root boxes for ^{13}C analysis. All roots sampled from the trees were at least two weeks old at the time of labeling and were no longer actively elongating. Three trees were labeled on August 5th, 2008 and based on these results a fourth tree was labeled the following year on September 3rd, 2009 to enable the sampling of roots of greater age at the time of labeling as well as to examine in greater detail the fate of the label in both the structural and the non-structural root carbon pools. Only a single tree was labeled in 2009 due to a lack of additional suitably sized trees.

In 2008 all roots visible through the acetate windows were traced 19 days (d) prior to labeling for each of three trees labeled with 99% $^{13}CO_2$ and an unlabeled control tree. The unlabeled control tree was used to insure accurate background ^{13}C measurements. Root samples were excised from the root boxes immediately prior to labeling and three times after labeling (3, 6 and 19d). Additional larger diameter, higher-order roots (approximately 4th or 5th order) were dug from the soil directly beneath the three labeled trees and the unlabeled-control tree four days after labeling. The upper, non-distal, portion of two roots from each of these samples were then cut into segments and dissected under the microscope into three anatomical categories: late wood, early wood, and secondary cortex (Figure 2); where the late wood accounts for the inner most portion of the root cross section, surrounded by the early wood and finally the secondary cortex. Because these roots were not followed in the root boxes, their precise age was unknown, but based on the development of early and late wood; they were at least one year old at the time of labeling [18]. Individual sampled roots were freeze dried, ground, and then underwent acid-base-acid washing until the supernatant was clear to remove nonstructural carbon [4]. The remaining sample was oven dried at 60°C for 24 hrs, weighed, and placed into tin capsules.

In 2009, a single Sassafras tree was labeled with $^{13}CO_2$. Only one tree was labeled in 2009 and less $^{13}CO_2$ was used than in 2008 due to a decreased number suitable unlabeled trees. Because there was no statistical difference ($P=0.40$) in isotopic signature between the roots of the control tree and the roots sampled prior to labeling the labeled trees in 2008, we did not sample the roots of a second control tree in 2009. Roots were traced weekly prior to labeling, with three root flushes observed at 71d, 28d, and 16d prior to labeling. The weekly tracing interval did not allow for the determination of exact root ages in each flush, as roots could have been born on any of the previous seven days. For analysis, we assigned root birth as the day of initial observation, so a root might actually be as much as 7d older than the assigned birth date. Individual first-order roots were sampled immediately prior to labeling and twice after labeling, once at five days after labeling and once at 14 days after labeling. These individual sampled first-order roots were ground and then underwent a stepwise process to isolate soluble sugars, starch and structural carbon. This involved first boiling the root samples in Millipore water (Millipore, Billerica, MA USA) and sampling the resulting supernatant for soluble sugar ^{13}C content. The remaining solid sample was then re-suspended in Millipore water and digested with 0.5 M sodium acetate and 5 units of amyloglucosidase and 2.5 units alpha amylase [19]. The resulting supernatant was then analyzed for starch ^{13}C content. Finally, the remaining root sample underwent the acid-base-acid cleaning mentioned above, leaving structural carbon for ^{13}C analysis. Solid samples were oven dried at 60°C for 24 hrs, weighed, and placed in tin capsules. Liquid samples were pipetted into pre-weighed tin capsules, oven dried at 60°C for 24 hrs, and then re-weighed.

All samples from both years of the study were analyzed for ^{13}C content at the UC Davis Stable Isotope Facility. Statistical analyses were conducted using SAS JMP 9.02 (SAS Institute Inc., Cary, NC, USA). Results from each year were analyzed using an ANOVA with sampling date, tree, and labeling treatment as factors in 2008 and sampling date, root age, root carbon pool, and labeling treatment as factors in 2009, where only one tree was labeled. Non-significant factors were removed from the model. Results from the final model were considered statistically significant at $P\leq 0.05$ using a one tailed T-test.

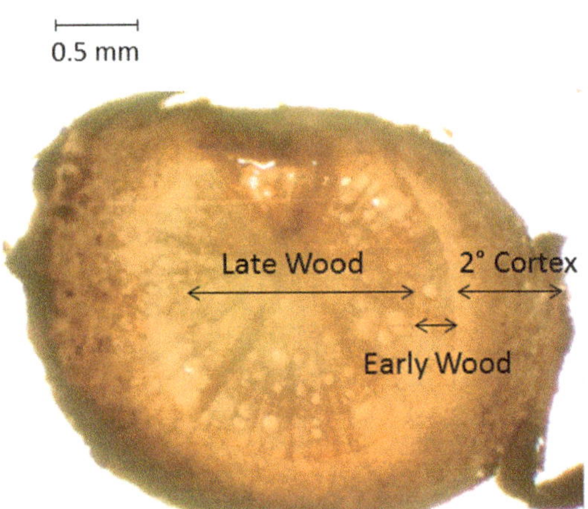

Figure 2. *Sassafras albidum* root cross-section. Micrograph depicting how higher-order roots (approximately 4th or 5th order) were dissected prior to $\delta^{13}C$ analysis in 2008.

Results

In both 2008 and 2009 structural carbon of the roots of the labeled *S. albidum* trees, which were at least two weeks old at the time of labeling, were significantly (2008 $n=35$, $P<0.0001$; 2009 $n=38$, $P=0.02$) enriched in ^{13}C compared with roots sampled from the same trees prior to 13CO2 labeling (Figure 3, Table 1). The lower enrichment observed in 2009 likely resulted from less $^{13}CO_2$ being used in the labeling process. In 2008, no significant differences were observed among the three labeled trees in the ^{13}C enrichment of the roots ($P=0.44$) nor in timing of the root sampling post labeling ($P=0.19$). Additionally, in 2008 all portions of the dissected roots (early wood (*ew*), late wood (*lw*), and secondary cortex (*sc*); were significantly enriched in ^{13}C by labeling with 99% $^{13}CO_2$ (Figure 4) (*ew*: $P=0.01$, *sc*: $P=0.01$, *lw*: $P=0.04$). In 2009, in addition to root structural carbon, both root soluble sugars and root starch were also significantly enriched with ^{13}C by labeling ($P=0.001$ and $P<0.0001$) (Figure 5), with soluble sugars being most enriched followed by starch and then structural carbon. The timing of the root sampling post labeling significantly affected the ^{13}C enrichment of the root sugars ($P=0.005$), with roots sampled later showing decreased ^{13}C enrichment. Root age at the time of labeling did not significantly affect ^{13}C enrichment of the root sugars ($P=0.09$). Neither of these factors significantly affected the ^{13}C enrichment of the structural carbon pool (root age $P=0.15$, sample date $P=0.14$), nor of the starch pool (root age $P=0.26$, sample date $P=0.26$) (Table 1).

Discussion

Our results clearly demonstrate that the timing of the deposition of photosynthetically derived carbon into roots involves more complicated processes than has previously been assumed. Not only were the non-structural carbon pools (soluble sugars and starch) of existing first-order tree roots incorporating carbon from current photosynthate (Figure 5), but so were the structural components of the roots (Figures 3 and 4). Our results are in direct contradiction with the assertion of Marshall and Waring [1] that root carbohydrates associated with starch formation are derived only at root initiation. Additionally, the incorporation of current photosynthate into the structural carbon of roots which were at least two weeks old, and in some roots more than a year old, at the time of labeling has implications regarding the determination of root age and turnover using isotopic techniques. For example, when using bomb ^{14}C analyses, if the incorporation of new carbon into the structure of existing roots continues to occur over sufficient time scales, an underestimation of the actual age of roots will result since the ^{14}C signature of the newly incorporated carbon would be depleted in ^{14}C compared to the carbon used to construct the root at initiation. Furthermore, in recent studies that have used the incorporation of ^{13}C from current photosynthesis into the structural carbon pools of labeled tree roots to assess root turnover [2,5], if new structural carbon is continually added to existing roots, root turnover will appear to occur more quickly since the depletion in the ^{13}C signal will not be solely derived from the death of labeled roots and the initiation of new root growth.

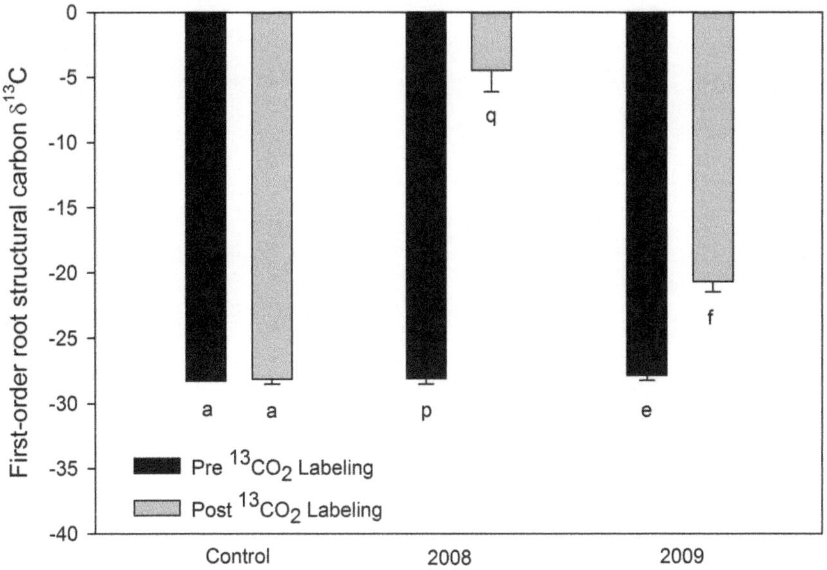

Figure 3. Mean *Sassafras albidum* first-order root structural carbon $\delta^{13}C$ values. Data from an unlabeled control tree, and trees labeled in 2008 and in 2009. All roots were at least 2 weeks old at the time of labeling and subsequent sampling and some were older than 71 days. Black bars represent roots sampled prior to whole-tree labeling and grey bars represent roots sampled after whole-tree labeling. Note that less negative values indicate greater isotopic enrichment. Error bars denote standard error between individual sampled roots. Different lower case letters denote significant differences at $P \leq 0.05$ using a one tailed T-test between pre- and post-labeled roots.

Our results demonstrate that current photosynthate is incorporated into the structural tissue of higher-order roots (Figure 4) as well as first-order roots (Figures 3, 5). We also found evidence, when examining the full range of root ages sampled (16d to > 1year), that this incorporation appears to decrease with root age (Figure 6). However, even the roots that were older than one year at the time of labeling never were depleted to the level of the non–labeled roots (Figure 6). Additionally, this trend of decreasing incorporation of C in structural tissues with age was not significant when just examining first-order roots sampled within a 2.5-month period.

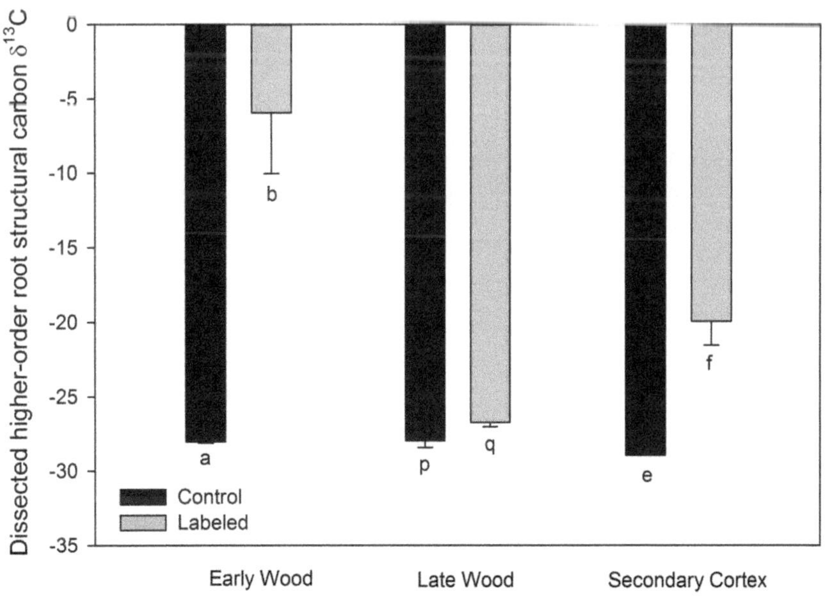

Figure 4. Mean *Sassafras albidum* root structural carbon $\delta^{13}C$ values from higher-order dissected roots. Trees were labeled and sampled in 2008. Control roots were sampled from unlabeled control trees and labeled roots were sampled from trees labeled with 99% $^{13}CO_2$. Late wood refers to the innermost portion of the dissected roots (see Figure 2). Early wood refers to the root tissue immediately surrounding the late wood. Secondary cortex refers to the outermost portion of the root surrounding the early wood. Note that less negative values indicate greater isotopic enrichment. Error bars denote standard error. Different lower case letters denote significant differences at $P \leq 0.05$ using a one tailed T-test between dissected root tissue from unlabeled control trees and labeled trees.

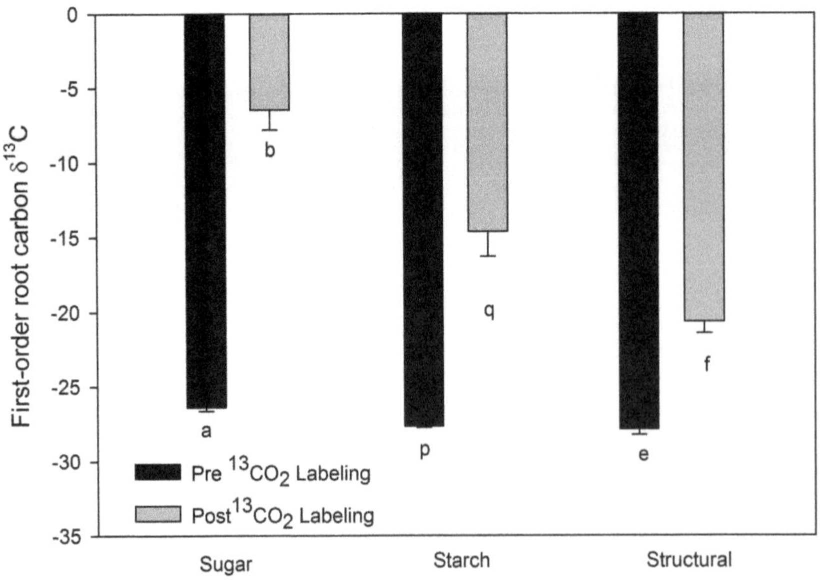

Figure 5. Mean *Sassafras albidum* first-order root carbon $\delta^{13}C$ values. Roots were at least two weeks old at the time of whole tree labeling in 2009. Black bars represent roots that were sampled immediately prior to whole-tree labeling. Light grey bars represent roots sampled after whole-tree labeling. Note that less negative values indicate greater isotopic enrichment. Error bars denote standard error. Different lower case letters denote significant differences at $P \leq 0.05$ using a one tailed T-test between pre- and post-labeled roots of each tissue type.

Because roots of higher branching order must in general be older than the lower order roots they support, the diminished incorporation of current photosynthate with root age over longer time scales suggests that higher-order, longer-lived roots have fewer errors associated with estimates of root lifespans and turnover based on isotopic techniques. However, because roots of higher branching order, unlike lower-order roots, often undergo secondary development, the potential for newly incorporated

Table 1. *Sassafras albidum* $\delta^{13}C$ root data from trees labeled with 99% $^{13}CO_2$.

Year	Root age (d)	Number of days post-labeling		
		3	6	19
Structural $\delta^{13}C$ (*Control* = −28.13±0.24 (*n* = 9))				
2008	**>14**	−2.18±3.38 (*n* = 14)	−5.75±1.28 (*n* = 13)	−6.23±3.89 (*n* = 8)
2009		**5**	**14**	
Structural $\delta^{13}C$ (*Control* = −27.85±0.38 (*n* = 2))				
	16	−16.84±3.73 (*n* = 3)	−18.93±3.76 (*n* = 2)	
	28	−19.99±1.46 (*n* = 13)	−21.87±1.22 (*n* = 13)	
	71	−21.99±1.91 (*n* = 4)	−22.04±1.87 (*n* = 3)	
Starch $\delta^{13}C$ (*Control* = −27.64±0.09 (*n* = 2))				
	16	−16.66±3.91 (*n* = 3)	−12.17±8.08 (*n* = 2)	
	28	−14.90±2.57 (*n* = 13)	−15.50±2.70 (*n* = 13)	
	71	−17.38±4.51 (*n* = 4)	−5.41±12.77 (*n* = 3)	
Sugar $\delta^{13}C$ (*Control* = −26.38±0.25 (*n* = 2))				
	16	−0.68±5.81(*n* = 3)	−9.11±1.98 (*n* = 2)	
	28	−0.18±1.37 (*n* = 13)	−10.86±1.77 (*n* = 13)	
	71	−2.93±9.11 (*n* = 4)	−14.91±1.04 (*n* = 3)	

Data showing the number of samples (*n*) analyzed as well as the average $\delta^{13}C$ and associated standard error for each root age class, sampling date, and carbon pool from trees labeled with 99% $^{13}CO_2$ in 2008 (n = 3) and 2009 (n = 1), where less negative values indicate greater isotopic enrichment. $^{13}CO_2$ labeling significantly enriched the root structural $\delta^{13}C$ in 2008 ($n = 35$, $P < 0.0001$) and 2009 ($n = 38$, $P = 0.02$). Additionally, both the root starch and root sugars were significantly enriched in 2009 ($n = 38$, $P_{starch} < 0.0001$, $P_{sugars} = 0.001$). The number of days sampling occurred post labeling did not significantly affect enrichment in root structural $\delta^{13}C$ in 2008 ($P = 0.19$) nor root structural $\delta^{13}C$ nor root starch $\delta^{13}C$ in 2009 ($P_{structural} = 0.14$, $P_{starch} = 0.26$). In 2009 root sugars were significantly affected by the days post labeling ($P = 0.005$), with roots sampled later showing decreased enrichment. The age of the roots at the time of labeling did not significantly affect enrichment in any of the carbon pools in 2009 ($P_{structural} = 0.15$, $P_{starch} = 0.26$, $P_{sugars} = 0.09$).

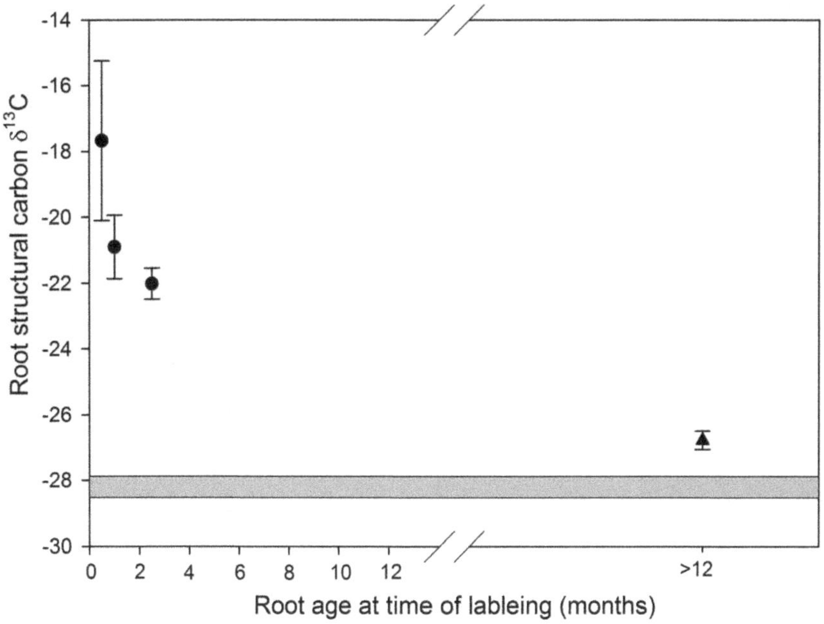

Figure 6. Mean *Sassafras albidum* root structural carbon $\delta^{13}C$ values verses root age at time of sampling. Circles represent roots sampled in 2009. The triangle represents the mean of the late wood (lw) structural root tissue from the 2008 dissection samples which were older than one year when sampled but the exact age is unknown. The grey horizontal bar shows the 95 percentiles for background $\delta^{13}C$ structural carbon values from 2008 and 2009. Note that less negative values indicate greater isotopic enrichment. Error bars denote standard error.

carbon to bias estimates of root lifespan and turnover may actually increase with increasing root order. One possible way to overcome this bias, based on our observation of decreased recent carbon assimilation in the older tissue of higher-order roots (Figure 4: late wood vs. early wood), would be to use the oldest, inner most root tissue of higher-order woody roots for isotopic analysis.

Although we did not specifically examine the physiological processes involved in the observed incorporation of recent photosynthate into the structural pools of existing roots, secondary cellular processes such as cell wall thickening could be responsible. In another tree species, first-order *Liriodendron tulipifera* roots of known age collected from the same common garden during the same time period showed significant secondary wall thickening between birth and 14 days in the hypodermal and endodermal cells of the primary tissue (Zadworny *et al.* unpublished data). Similar results have been observed among citrus rootstocks when comparing the primary tissue of first- with second-order roots [20]. Regardless of the exact physiological mechanism involved, we have clearly demonstrated that new carbon is incorporated into the carbon pools of existing first-order roots for months after birth and in higher order roots, for potentially years. Because this study involved only one species, further study is needed to determine if the observed assimilation of new carbon into existing non-elongating first-order roots is a broadly occurring process. Additionally, more work is needed to uncover the underlying mechanisms driving the observed assimilation as well as the exact extent to which new carbon may influence estimates of root age and turnover using isotopic techniques.

Acknowledgments

We would like to thank Marcin Zadworny for his help in dissecting root samples and providing the micrograph.

Author Contributions

Conceived and designed the experiments: TSA DME. Performed the experiments: TSA. Analyzed the data: TSA. Contributed reagents/materials/analysis tools: DME. Wrote the paper: TSA DME.

References

1. Marshall JD, Waring RH (1985) Predicting fine root production and turnover by monitoring root starch and soil temperature. Can J For Res 15:791–800
2. Fahey TJ, Jacobs KR, Sherman RE (2012) Fine root turnover in sugar maple estimated by 13C isotope labeling. Can JFor Res 42: 1792–1795.
3. Gaudinski JB, Torn MS, Riley WJ, Swanston C, Trumbore SE, et al. (2009) Use of stored carbon reserves in growth of temperate tree roots and leaf buds: analyses using radiocarbon measurements and modeling. Glob Chang Biol 15: 992–1014.
4. Gaudinski JB, Trumbore SE, Davidson EA, Cook AC, Markewitz D, et al. (2001) The Age of Fine-Root Carbon in Three Forests of the Eastern United States Measured by Radiocarbon. Oecologia 129: 420–429.
5. Keel SG,Campbell CD, Högberg MN, Richter A, Wild B, et al. (2012) Allocation of carbon to fine root compounds and their residence times in a boreal forest depend on root size class and season. New Phytol 194: 972–981.
6. Sah SP, Jungner H, Oinonen M, Kukkola M, Helmisaari HS (2011) Does the age of fine root carbon indicate the age of fine roots in boreal forests? Biogeochemistry 104: 91–102.
7. Trumbore S, Da Costa ES, Nepstad DC, Barbosa De Camargo P, Martinelli LA, et al. (2006) Dynamics of fine root carbon in Amazonian tropical ecosystems and the contribution of roots to soil respiration. Glob Chang Biol 12: 217–229.
8. Vargas R, Trumbore SE, Allen MF (2009) Evidence of old carbon used to grow new fine roots in a tropical forest. New Phytol 182: 710–718.
9. Nguyen PV, Dickmann DI, Pregitzer KS, Hendrick R (1990) Late-season changes in allocation of starch and sugar to shoots, coarse roots, and fine roots in two hybrid poplar clones. Tree Physiol 7: 95–105
10. Kosola KR, Dickmann DI, Parry D (2002) Carbohydrates in individual poplar fine roots: effects of root age and defoliation. Tree Physiol 22: 741–746.
11. Carbone MS, Czimczik CI, McDuffee KE, Trumbore SE (2007) Allocation and residence time of photosynthetic products in a boreal forest using a low-level (14)C pulse-chase labeling technique. Glob Chang Biol 13: 466–477.
12. Pregitzer KS, DeForest JL, Burton AJ, Allen MF, Ruess RW, et al. (2002) Fine root architecture of nine North American trees. Ecological Monographs 72: 293–309.

13. Gaul D, Hertel D, Leuschner C (2009) Estimating fine root longevity in a temperate Norway spruce forest using three independent methods. Funct Plant Biol 36: 11–19.
14. Riley WJ, Gaudinski JB, Torn MS, Joslin JD, Hanson PJ (2009) Fine-root mortality rates in a temperate forest: estimates using radiocarbon data and numerical modeling. New Phytol 184: 387–398.
15. Wells CE, Eissenstat DM (2001) Marked Differences in Survivorship among Apple Roots of Different Diameters. Ecology 82: 882–892.
16. McCormack ML, Adams TS, Smithwick EAH, Eissenstat DM (2012) Predicting fine root lifespan from plant functional traits in temperate trees. New Phytol 195: 823–831
17. Zadworny M, Eissenstat DM (2011) Contrasting the morphology, anatomy and fungal colonization of new pioneer and fibrous roots. New Phytol190: 213–221
18. Fayle DCF (1968) Radial growth in tree roots; distribution, timing, anatomy. Faculty of Forestry, University of Toronto, Tech. Rep. 9.
19. Comas LH, Anderson LJ, Dunst RM, Lakso AN, Eissenstat DM (2005) Canopy and environmental control of root dynamics in a long-term study of Concord grape. New Phytol 167: 829–840.
20. Eissenstat DM and Achor DS (1999) Anatomical characteristics of roots of citrus rootstocks that vary in specific root length. New Phytol 141: 309–321.

The Kill Date as a Management Tool for Cover Cropping Success

María Alonso-Ayuso, José Luis Gabriel, Miguel Quemada*

School of Agriculture Engineering, Technical University of Madrid, Madrid, Spain

Abstract

Integrating cover crops (CC) in rotations provides multiple ecological services, but it must be ensured that management does not increase pre-emptive competition with the subsequent crop. This experiment was conducted to study the effect of kill date on: (i) CC growth and N content; (ii) the chemical composition of residues; (iii) soil inorganic N and potentially mineralizable N; and (iv) soil water content. Treatments were fallow and a CC mixture of barley (*Hordeum vulgare* L.) and vetch (*Vicia sativa* L.) sown in October and killed on two different dates in spring. Above-ground biomass and chemical composition of CC were determined at harvest, and ground cover was monitored based on digital image analysis. Soil mineral N was determined before sowing and after killing the CC, and potentially mineralizable N was measured by aerobic incubation at the end of the experiment. Soil water content was monitored daily to a depth of 1.1 m using capacitance sensors. Under the present conditions of high N availability, delaying kill date increased barley above-ground biomass and N uptake from deep soil layers; little differences were observed in vetch. Postponing kill date increased the C/N ratio and the fiber content of plant residues. Ground cover reached >80% by the first kill date (~1250°C days). Kill date was a means to control soil inorganic N by balancing the N retained in the residue and soil, and showed promise for mitigating N losses. The early kill date decreased the risk of water and N pre-emptive competition by reducing soil depletion, preserving rain harvested between kill dates and allowing more time for N release in spring. The soil potentially mineralizable N was enhanced by the CC and kill date delay. Therefore kill date is a crucial management variable for maximizing the CC benefits in agricultural systems.

Editor: Anil Shrestha, California State University, Fresno, United States of America

Funding: This work was funded by the Spanish Comisión Interministerial de Ciencia y Tecnología (project AGL 2011-24732) and the Regional Government of Madrid (Project AGRISOST, S2009/AGR-1630). The funders had no role in study design, data collection and analysis, decision to publish, or preparation of the manuscript.

Competing Interests: The authors have declared that no competing interests exist.

* Email: miguel.quemada@upm.es

Introduction

The potential for cover crops (CC) to provide ecological services has been documented in diverse cover cropping systems and environments [1]. Replacing fallow periods with CC may enhance soil aggregate stability [2], water retention capacity [3], nutrient supply [4] and disease suppression [5]. Moreover, plant residues covering the soil following crop kill can improve soil protection over time and help control weeds, preserve soil moisture, ameliorate compacted soils, and reduce soil erosion [6]. However, improperly managed CC may have a detrimental effect on the cash crop either by competing for water and nutrients, building up diseases or retarding seed germination [6,7]. Overall, there are two crucial management factors that determine CC success: species selection and CC termination [8]. In this manuscript, we focus on the termination of CC incorporated into field crop rotations, particularly on kill date. Previous experiments have shown that changing kill date by a few weeks in spring may have a large effect on the water and N pre-emptive competition with the subsequent cash crop, nitrate leaching control, and soil moisture conservation [7,9–12], but there is no agreement on the advantages and limitations of postponing CC termination. Therefore, a better understanding of the effect of the living CC and its dead mulch on the dynamics of N and water might improve our ability to choose a kill date that maximizes the benefits of CC in the system.

The effect of kill date on soil water availability is a balance between the water extracted by the living CC and the evaporation prevented by the residue mulch [9]. Integrating CC into the cropping system increases above-ground biomass and resistance to decomposition, which enhances the development of a dead mulch covering the soil surface [13]. Some authors have found that, despite the soil water depletion, moisture conservation was improved when the CC was killed later [9,11]. However, other experiments have shown that an earlier CC kill reduced water pre-emptive competition, preserved top soil moisture, and increased water availability for the subsequent crop [7,12]. Similarly, postponing kill date enhanced N pre-emptive competition as soil available N is depleted by the CC uptake and only slowly replenished by the mineralization of a high C/N residue [6]. Legume species may compensate for this effect by fixing more N_2 when killed late [10]. The water and N balance is driven by various factors that are all affected by kill date, such as above-ground biomass accumulation, ground cover (GC), and N uptake

[6]. The chemical composition of crop residues, such as lignin content and the C/N ratio, are important characteristics governing the decomposition process and determining the persistence of the residue layer covering the soil [14]. However, there are few experiments that combine multiple approaches to study these variables, and to our knowledge, none that involves continuous monitoring of soil water dynamics.

We hypothesize that a combined study of CC growth, GC and water and N dynamics would contribute to understanding the effect of kill date and improve management practices. This experiment was conducted over two years to study the effect of kill date on the following factors: (i) the growth and N content of the CC; (ii) the chemical composition and residue quality of the CC; (iii) soil inorganic N dynamics and its potentially mineralizable N; and (iv) the soil water content. A CC mixture of barley and vetch was selected because mixing legumes and non-legumes is known to be an efficient technique to merge N and water management benefits of the individual species [9,15].

Materials and Methods

1. Field experiment

The experiment was conducted from October 2011 to November 2013 at La Chimenea field station (40°03′N, 03°31′W, 550 m a.s.l.), which is located in the central Taxus river basin in Aranjuez (Madrid, Spain). The soil is a silty clay loam (Typic Calcixerept) [16] with a high content of organic matter and carbonates, pH ~8 and low stone content throughout the profile. The most relevant soil characteristics are presented in Table 1. Total organic C was measured by the Walkley-Black method [17]. Calcium carbonate was determined by a volumetric calcimeter method [18]. The climate is semiarid Mediterranean [19] with high interannual variability (Figure 1). Annual rainfall averages ~350 mm with less rain in the summer and more in the autumn and a mean annual temperature of 14.2°C. Weather data were recorded throughout the experimental period by a data logger (CR10X. Campbell Scientific Ltd, Shepshed, UK) located ~100 m from the field site.

The experiment consisted of twelve plots (15 m×12 m) randomly assigned to three treatments (fallow and two different CC kill dates) with four replications. Cover crop treatments were sown with a barley (*Hordeum vulgare* L., cv. Vanessa) + vetch (*Vicia sativa* L., cv. Aitana) mixture in early October (October 6, 2011; October 8, 2012). The mixture, 30% (54 kg ha^{-1}) barley seeds and 70% (45 kg ha^{-1}) vetch, was broadcast by hand and buried (~5 cm) with a shallow cultivator. All treatments received ~18 mm irrigation on the sowing date with a sprinkle irrigation system (9.5 mm h^{-1}) to ensure uniform CC establishment. Plots from the first kill treatment (FK) were killed in mid-March (March 13, 2012; March 14, 2013) when barley was at the end of the booting growth stage and vetch at the stem elongation stage. Plots from the second kill treatment (SK) were killed in mid-April (April 9, 2012; April 13, 2013) when barley was at the emerging inflorescence growth stage and vetch was at the stem elongation stage. Cover crops were killed by application of glyphosate (N-phosphonomethyl glycine, 0.7 kg a.e. ha^{-1}) followed by a shredder. The last day of April was chosen as a hypothetical planting date (HPD) for both years, as it is a typical date for planting summer cash crops in the area. The HPD represented the beginning of the mulch period during which the CC residue remained on the ground. To compare the residue mulch effect on water conservation, SWC was homogenized by adding water to the drier plots with a drip irrigation system (30.5 mm to FK treatment and 35.5 mm to SK) in April 2012. As rainfall was abundant in March 2013, no supplemental irrigation was applied. Fallow plots were kept free of weeds by hand weeding during the trial period. The experiment was conducted in a field that was cultivated with triticale during the two previous years and had not received organic amendments or N fertilizer for four years prior to the beginning of the trial.

2. Cover crop analysis

To measure the above-ground biomass and GC in each plot, four 0.5 m×0.5 m squares were marked after sowing the CC. Ground cover was monitored in each square based on digital photos taken from a nadir perspective at 1.2 m height every other week [20]. The images were taken with a Nikon S210 Coolpix camera that had a lens resolution of 5 Mpixels. The images were saved at a resolution of 1200×800 pixels and the GC percentage was determined with the software SigmaScan Pro (Systat software, Chicago, IL, USA) by means of the "Turf analysis" macro developed by Karcher and Richardson [21]. The process consisted of creating a pixel layer by selecting the hue and saturation ranges that identified the surface covered by the crop. The GC was determined to be the number of pixels of the layer divided by the total number of pixels constituting the image of the marked squares [20]. The GC evolution was adjusted to the Gompertz function [22], and the thermal times (°C days) until the GC reached 30% (t_{30}) and 80% (t_{80}) were calculated.

Just before killing the CC, the above-ground biomass of the marked squares was hand harvested by cordless grass shear at the ground level, separated by species, oven dried for 48 h at 65°C,

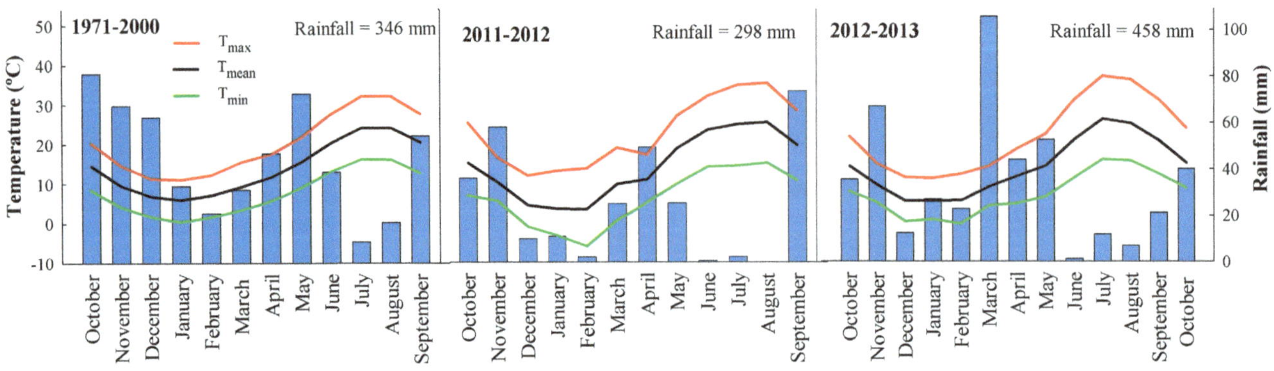

Figure 1. Monthly rainfall (bars) and average maximum, mean and minimum temperatures from 1971 to 2000 and during the two experimental seasons at Aranjuez (Madrid, Spain).

Table 1. Soil properties at the beginning of the experiment.

Depth (cm)	0–23	23–40	40–70	70–120
pH (1:2.5)	8.16	8.06	8.02	7.84
Organic Matter (g kg^{-1})	31.8	29.2	21.9	22.3
CO_3 (g CO_3^{2-} kg^{-1})	198.0	201.3	159.0	181.0
Sand (g kg^{-1})	260	250	250	250
Silt (g kg^{-1})	490	510	520	460
Clay (g kg^{-1})	250	240	230	290

and weighed to determine biomass of barley, vetch, and mixture in kg ha^{-1}. The C and N concentrations in the above-ground biomass were determined for a subsample of each species from each plot using the Dumas combustion method (LECO CHNS-932 Analyzer, St. Joseph, MI, USA). Total N content was calculated for each specie as the product of above-ground biomass times N concentration. The atmospheric N_2 fixation by the legume was estimated by the natural abundance method [23] based on the $\delta^{15}N$ (‰) determination (Europe Scientific 20–20 IRMS Analyzer, Crewe, UK) for subsamples from the vetch and a barley reference cultivated as a sole crop in an adjacent field. Cover crop residue quality was assessed by measuring neutral detergent fiber (NDF), acid detergent fiber (ADF), and lignin (L) with the Goering and Van Soest method [24] in subsamples of each crop species from each plot.

3. Soil inorganic nitrogen content (N_{min})

Four soil cores were taken from each plot to a depth of 1.2 m in 0.2 m intervals with an Eijkelkamp helicoidal auger (Eijkelkamp Agrisearch Equipment, Geisbeek, Netherlands) just before sowing the CC, after the second kill date each year, and at the end of the experiment. Soil cores were combined by depth to provide a composite profile of six samples. For each plot, soil N_{min} was calculated for each layer. Soil samples were placed in a plastic box and immediately firmly closed then transported, and refrigerated (4–6°C). Within five consecutive days, a soil subsample from each box was extracted with 1 M KCl (~30 g of soil: 150 ml of KCl), centrifuged, decanted, and a subsample of the supernatant volume was stored in a freezer until later analysis. Nitrate concentration in the extracts was determined by the Griess-Ilosvay method [25] after reduction of NO_3^- to NO_2^- with a Cd column. Ammonium in the soil extracts was determined by the salicylate-hypochlorite method [26].

4. Soil nitrogen mineralization potential (N_0)

Soil N mineralization potential was estimated by adapting the procedure proposed by Stanford and Smith [27]. A subsample of the top soil layer (0–0.2 m) free from residue contamination was collected from each plot after the second kill date of the second year, air-dried, and sieved (<6 mm). A homogeneous mixture of 30 g of that soil and an equal weight of sand was packed to a depth of 5 cm in plastic syringes (3.5 cm diameter, 10 cm long) to achieve a bulk density of 1.25 g cm^{-3}. Thin glass wool layers were placed between the soil and the bottom of the syringe and over the soil sample to avoid soil loss during leaching and to minimize moisture loss. Initial soil inorganic N was removed by leaching

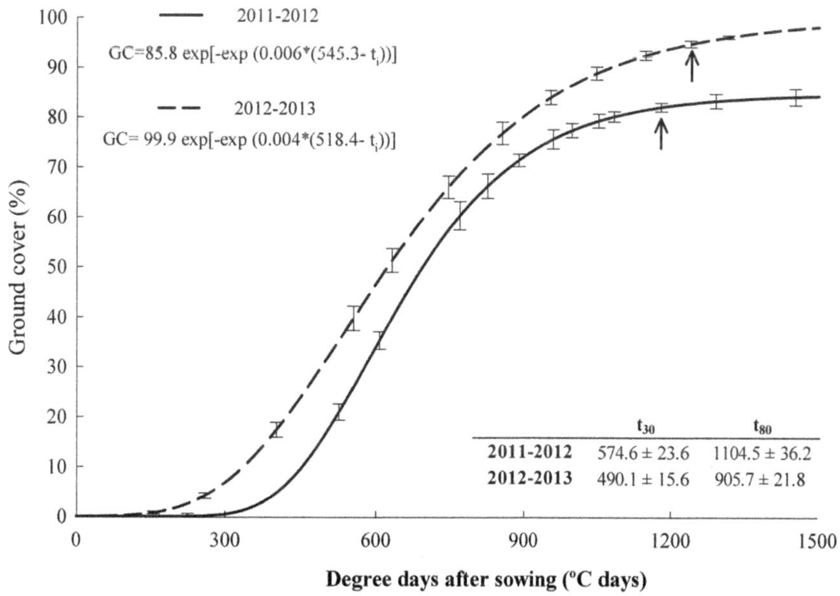

Figure 2. Ground cover (GC) development during the cover crop growth period in both experimental seasons. Arrows show the first kill date. Lines represent the Gompertz model adjusted to the observed values. Fitted models and the thermal time until the ground cover reaches 30 and 80% (t_{30} and t_{80}) are shown.

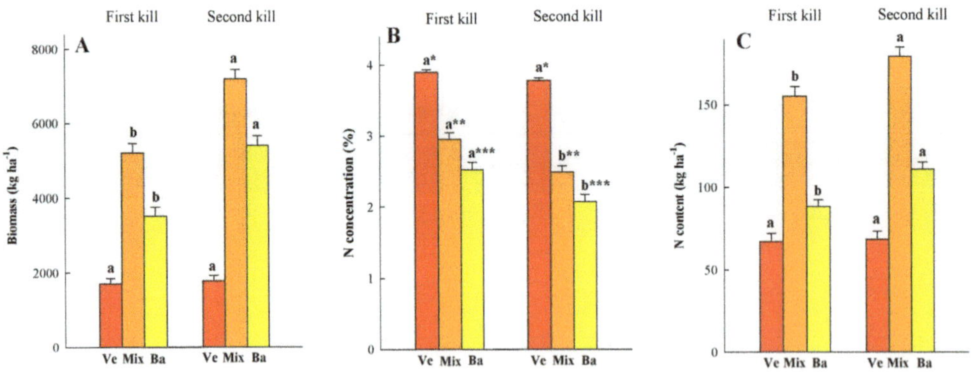

Figure 3. Biomass (A), N concentration (B) and N content (C) for vetch (Ve), barley (Ba) and the mixture (Mix) measured for the first and second kill dates. Values are the mean of two cropping seasons. Letters above bars show significant differences between kill dates for each species, and asterisks between species for each kill date. Small bars represent the standard error.

with 100 ml of 0.01 M $CaCl_2$ followed by 20 ml of a N-free nutrient solution (0.0095 M $CaSO_4$, 0.000047 M KH_2PO_4, 0.00138 M K_2SO_4, 0.0003 M $MgSO_4$). The excess water was removed using a vacuum pump until a weight within 2 g of that measured before leaching was reached. The syringes were covered with a porous parafilm and incubated aerobically at 35°C. Syringes were removed from the incubator at 14, 28, 42, 56, and 60 d after preparation and were leached with 100 ml of 0.01 M $CaCl_2$ solution followed by 20 ml of N-free solution. Leachates were made up to 100 ml with $CaCl_2$, and subsamples were stored in a freezer at −25°C until later analysis. After the leaching procedure, the cores were allowed to drain under vacuum until a weight within 2 g of that measured before leaching was reached. Nitrate concentrations in the leachates were determined by the Griess-Ilosvay method, ammonium by the salicylate-hypochlorite method and total N by the Dumas combustion method, as described above. The N_0 and the mineralization rate constant (k) were estimated after fitting a non-linear regression model ($Nt = N_0 \exp(-k\ t)$) for describing the cumulative N mineralized (Nt) with time (t) in each soil sample.

5. Soil volumetric water content (SWC)

The SWC was monitored daily during the field trial using the EnviroScan capacitance probes (Sentek Pty Ltd, Stepney, Australia) that has been described in detail elsewhere [28]. Seventy-two capacitance sensors were mounted on twelve plastic extrusions (four repetitions per treatment), introduced into access tubes located in the middle of each plot and connected to three data loggers. Sensor readings were automated and stored in the data loggers and downloaded weekly. To ensure the measurement reliability, a normalization procedure was conducted that obtained reference readings by exposing each sensor to air and water (~20°C). The sensors were centered at 10, 30, 50, 70, 90, and 110 cm below the soil surface in each plastic extrusion, and normalized readings were registered every 6 h. A daily average of the four readings from the 0–20 (10), 20–40 (30), 40–60 (50), 60–80 (70), 80–100 (90), 100–120 (110) cm-deep soil layers was transformed into SWC using a calibration equation that was obtained at the experimental site [29]. The SWC data set was comprised of two CC seasons, which started before sowing in October and lasted until August each year, and used to study the effect of the living CC and the dead mulch on soil moisture.

6. Statistical analysis

Analyses of variance (ANOVA) and t-test were performed in order to determine differences between kill dates for each variable. Year was considered a random effect and, interaction between kill date and year was also tested. Means were separated by Tukey's test at the 0.05 probability level ($P \leq 0.05$). Least significant differences (LSD) were calculated for SWC ($P \leq 0.05$). Prior to conducting the ANOVA, tests were conducted to verify if the assumptions of ANOVA were met. The Gompertz model was

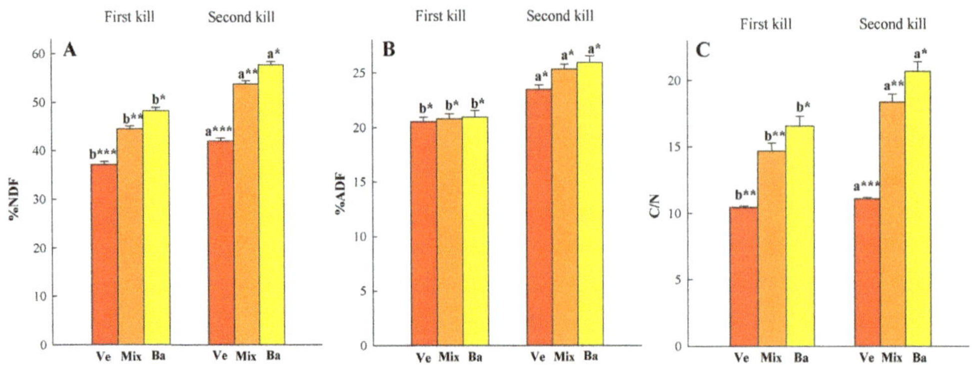

Figure 4. Neutral detergent fiber-NDF- (A) and acid detergent fiber -ADF- (B) fractions, and C/N ratio (C) in vetch (Ve), barley (Ba) and the mixture (Mix), measured for the first and second kill dates. Values are the mean of two cropping seasons. Letters above bars show significant differences between kill dates for each species, and asterisks between species for each kill date. Small bars represent the standard error.

Table 2. Soil inorganic N content (kg N ha^{-1}) in the different soil layers and the entire profile at different sampling dates for the fallow (Fa), first kill (FK) and second kill (SK) treatments.

	Oct 2011	Apr 2012	Oct 2012	Apr 2013	Oct 2013
0–40 cm					
Fa	94 (13.7)	126.8 (7.9) a	135.3 (15.4)	29.7 (3.8) b	71.54 (5.38) c
FK	94 (13.7)	46.9 (7.1) b	121.5 (16.6)	56.9 (3) a	119.37 (28.48) b
SK	94 (13.7)	37 (8.2) b	154 (8)	24.8 (4.5) b	171.72 (2.36) a
40–80 cm					
Fa	24.7 (9.5)	71.6 (10.6) a	86.1 (4.6) a	40.1 (1.8) a	72.22 (7.43)
FK	24.7 (9.5)	30.4 (5.3) b	48.3 (4.4) b	42.6 (3.5) a	65.38 (19.89)
SK	24.7 (9.5)	25.4 (5.1) b	48.1 (7.1) b	27.1 (4.1) b	39.39 (2.68)
80–120 cm					
Fa	26.1 (11.1)	60.6 (3.5)	66.7 (11)	68.8 (3.3) a	100.37 (11.74) a
FK	26.1 (11.1)	56.1 (24.5)	61.5 (29.3)	47 (7.4) b	61.59 (17.91) ab
SK	26.1 (11.1)	34.5 (3.7)	33.5 (6.3)	32.3 (8.3) b	39.24 (5.83) b
Soil profile (0–120 cm)					
Fa	144.8 (22.8)	259 (8.0) a	288.1 (25.4)	138.6 (5.1) a	244.13 (19.05)
FK	144.8 (22.8)	133.32 (27.5) b	231.3 (31.2)	146.6 (10.7) a	246.34 (52.87)
SK	144.8 (22.8)	96.9 (15.4) b	235.6 (14.2)	84.2 (13.3) b	247.34 (7.9)

Means with standard error in parentheses.
Within a date and depth, means with the same letter are not significantly different between kill dates at $P<0.05$.

fitted to the GC and the N mineralization potential model was fitted to the cumulative N mineralized using a non-linear regression procedure. The models were evaluated for their ability to simulate the observed data by comparing the mean of the lack of fit to the mean square due to pure error by using the variance ratio, or F test. When the lack of fit was significantly smaller than the pure error, the model fitted the data. Further discussion regarding this evaluation procedure can be found elsewhere [30,31]. Statistical analyses were accomplished using the PASW Statistics Software (SPSS, Chicago, IL, USA).

Results

1. Weather conditions

The first experimental year was drier than the second (Figure 1). In the first year, rainfall was substantially lower (298 mm) than the 30-year average although in the second, it was greater (408 mm). Differences mainly occurred in spring; from March to May, the rainfall was 202 mm in 2013 compared with 101 mm in 2012. The rest of the season was rather similar for both years except for intense precipitation in late September 2012. Air temperature followed a typical Mediterranean seasonal distribution over the two years. Winter was cooler in the first year with an average minimum temperature of −3 and −5°C in January and February, respectively.

2. Cover crop: GC, above-ground biomass and N content

The GC followed a classical Gompertz model (Figure 2). The F-test comparing the mean squares due to pure error and lack of fit was not significant at 0.01 level, therefore, the model was adjusted to the GC observations. The first kill occurred between 157 and 159 d after sowing (~1200°C days) whereas the second year occurred between 184 and 186 d after sowing (~1500°C days). The GC>30%, a crucial threshold for erosion control [32], was attained between 37 and 41 d after sowing (>490°C days). The GC>80%, considered to be full cover for erosion control and direct soil evaporation [32], was reached between 95 and 130 d after sowing (>900°C days). Differences were observed in the maximum GC attained, 85% in 2012 and 100% in 2013. In both

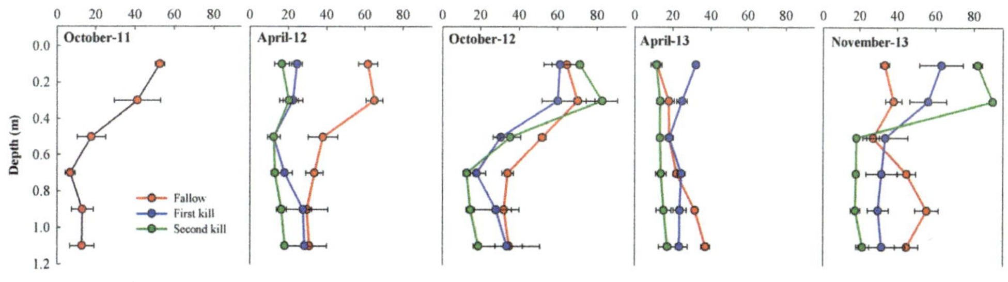

Figure 5. Soil inorganic nitrogen (kg N ha^{-1}) in the upper 1.2 m of the soil profile for the fallow (Fa), first kill (FK) and second kill (SK) treatments at different sampling times.

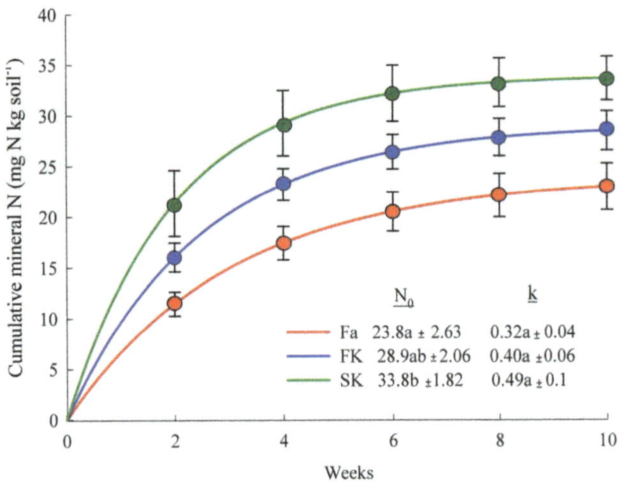

Figure 6. Cumulative N mineralization in soils from the fallow (Fa), first kill (FK) and second kill (SK) treatments during a 10-week aerobic incubation. Soil N mineralization potential (N_0) and N mineralization rate (k) were calculated by fitting a non-linear regression model ($Nt = N_0 \exp(-k\ t)$). Values followed by different letters are significantly different between treatments at ≤0.05 by Tukey's test. Bars represent the standard error.

years, maximum GC was already attained by the first kill date. The ground continued to be covered by the CC residue mulch more than six months after CC killing. Above-ground biomass increased from the FK to the SK by ~2000 kg ha^{-1}, and most of the increase was due to barley (Figure 3). Although barley biomass was greater at the SK than at the FK, no differences were found in vetch biomass between kill dates. Vetch above-ground biomass was double in 2013 than in 2012 whereas barley above-ground biomass did not vary by year. At the end of the mulch period, residues left on the soil surface were greater in the SK (2214 kg ha^{-1} in 2012, and 4014 in 2013) than in the FK treatment (966 kg ha^{-1} in 2012, 2151 in 2013. The ground was fully covered by the residues in both treatments.

Nitrogen concentration in the above-ground biomass was higher in the FK than in the SK. Nitrogen concentration in the mixture decreased from 2.7 to 2.35% in 2012 and from 3.2 to 2.6% in 2013. As expected, vetch had a higher N concentration than barley whereas the N concentration of the mixture was intermediate (Figure 3). Differences in N content between the FK and SK, calculated as the product of above-ground biomass and N concentration, were not significant for the mixture or vetch (Figure 3). However, barley N content in the SK was higher than in the FK; N uptake increased from 79 to 107 kg N ha^{-1} in 2012 and from 97 to 115 kg N ha^{-1} in 2013. The vetch reached high N_2 atmospheric fixation rates during the experiment. In the first season, N_2 atmospheric fixation was >80% of N content and ~100% in the second. No differences in N_2 atmospheric fixation were found between kill dates.

Residue quality fractions varied between kill dates (Figure 4). The NDF and ADF fractions increased when the CC was killed later. An interaction effect between treatment and year occurred for lignin content as differences in the lignin fraction between kill dates were only significant in barley and in the mixture in 2012. Lignin fraction in the mixture increased from 2.2 to 3.6% and in the barley from 1.2 to 2.9% in 2012. The C/N of barley, vetch and mixture residues was always lower for the FK than for the SK.

3. Soil inorganic nitrogen content

At the beginning of the experiment, average N_{min} of the upper layers of the soil profile was 145 kg N ha^{-1} (Figure 5, Table 2). During the first CC growing period, soil N_{min} increased in the fallow treatment, particularly in the upper 0.8 m, while it decreased in the remaining depths. During the mulch period, soil N_{min} in the fallow treatment increased slightly whereas, in the CC treatments, N_{min} increased largely in the upper soil layers, but there was no difference between the treatments by October in 2012.

During the second CC growing season, soil N_{min} was depleted in all treatments. After the CC kill in April 2013, the N_{min} was 84 kg N ha^{-1} in the SK treatment and ~140 kg N ha^{-1} in the others. The FK treatment accumulated more N_{min} in the upper layers, most likely due to the early mineralization that occurred between the CC kill and soil sampling (~4 weeks). At the end of the mulch period in November 2013, a large increase in soil N_{min} occurred in all treatments. Although no difference was observed for average N_{min} (~245 kg N ha^{-1}) over the entire profile, differences between treatments were observed at some soil depths. In the SK, most of the N_{min} (~70%) was in the top layer (0–40 cm) whereas it was ~29% in the fallow treatment. Similarly, while >40% of soil N_{min} was in the bottom layer in the fallow, < 16% was in the bottom layer in the SK. The N_{min} distribution in the FK treatment was intermediate between the others.

4. Soil nitrogen mineralization potential

The one-pool exponential model fit the cumulative mineral N from the aerobic incubation, as the F-test comparing the mean squares due to pure error and lack of fit was not significant at 0.01 level. The N_0 was higher for soils from the SK treatment (34 mg N kg $soil^{-1}$) than those from the fallow (24 mg N kg $soil^{-1}$). The N_0 from the FK treatment was intermediate. No differences between treatments were observed in the N mineralization (Figure 6).

5. Soil water content

Soil water content was affected by the presence of CC and by the kill date with differences observed in the upper layers and the whole soil profile (Figure 7, Table 3). During both CC growth periods, SWC followed a similar pattern. At sowing, the three treatments started with low SWC in the entire profile (~220 mm) in both seasons, and only small differences appeared in the top layers. The precipitation during the three months following CC sowing recharged the soil profile, reaching SWC>300 mm in 2013 due to abundant rainfall (Figure 7a). No differences were observed between treatments during this period in any season. However, during the next three months, CC extracted water from the upper layers, and by the first kill date, SWC was higher in the fallow than in the CC treatments. After the first kill date, the SWC in the FK treatment varied compared to the SK depending on the annual weather conditions. During the first season, no differences were observed between the FK and SK treatments, most likely due to low precipitation, and by the second kill date, both had similar SWC and were lower than the fallow (Table 3). During the second season, the high precipitation between both kill dates recharged the soil profile, and the FK treatment SWC was similar to the fallow except in the deeper soil layer (100–120 cm, Figure 7e) where the fallow treatment remained wetter. For the second kill date of 2013, the SK treatment SWC was lower, and differences with the FK were obvious down to 80 cm and down to 120 cm with the fallow (Table 3).

Although initial differences in SWC were observed between the two years of the study during the mulch period, such differences were not evident toward the end of the season. After water

Table 3. Soil water content (mm) in each of the soil layers and the entire profile at different sampling dates for the fallow (Fa), first kill (FK) and second kill (SK) treatments.

		0–20		20–40		40–60		60–80		80–100		100–120		0–120	
		mm													
2011–2012															
07/10/2011	Fa	23.9	a	40.3	a	38.1	a	40.8	a	37.4	a	40.4	a	221	a
CC sowing	FK	18.7	a	30.8	b	39.4	a	41.7	a	44.2	a	43.4	a	218	a
	SK	18.5	a	35.1	ab	42.6	a	41.0	a	37.0	a	41.0	a	215	a
13/03/2012	Fa	25.2	a	42.7	a	40.5	a	37.5	a	34.1	a	34.9	a	215	a
First kill date	FK	12.4	b	28.7	b	33.3	a	35.5	a	35.5	a	36.5	a	182	b
	SK	11.4	b	31.7	b	35.5	a	32.2	a	33.1	a	35.5	a	179	b
09/04/2012	Fa	33.9	a	42.2	a	40.1	a	38.3	a	35.4	ab	36.3	a	226	a
Second kill date	FK	18.3	b	28.1	b	33.2	a	36.0	a	36.3	a	37.5	a	189	b
	SK	12.6	c	28.6	b	33.8	a	29.8	a	30.8	b	36.4	a	172	c
30/04/2012	Fa	9.4	c	27.6	c	32.6	a	36.2	a	34.1	ab	36.4	a	251	a
HPD	FK	10.6	b	27.4	b	36.9	a	42.7	a	41.5	a	44.0	a	266	a
	SK	12.8	a	32.0	a	38.7	a	37.3	a	37.4	b	43.6	a	269	a
12/09/2012	Fa	9.4	a	27.6	a	32.6	a	36.2	a	34.1	b	36.4	a	176	a
Mulch period end	FK	10.6	a	27.4	a	36.9	a	42.7	a	41.5	a	44.0	a	203	a
	SK	12.8	a	32.0	a	38.7	a	37.3	a	37.4	ab	43.6	a	202	a
2012–2013															
08/10/2012	Fa	43.9	b	55.3	a	31.3	a	33.6	a	32.0	b	34.4	a	231	a
CC sowing	FK	48.5	ab	55.7	a	36.3	a	39.7	a	38.5	a	40.5	a	259	a
	SK	52.7	a	55.2	a	39.0	a	34.7	a	35.3	ab	41.0	a	258	a
15/03/2013	Fa	50.3	a	56.0	a	58.8	a	56.4	a	44.7	a	39.1	a	305	a
First kill date	FK	43.4	b	47.3	b	41.1	b	43.8	b	40.8	a	38.3	a	255	b
	SK	41.2	b	42.2	c	39.3	b	39.3	b	37.4	a	39.2	a	239	b
11/04/2013	Fa	49.4	b	56.1	a	57.7	a	60.0	a	55.9	a	47.7	a	327	a
Second kill date	FK	53.9	a	59.1	a	61.2	a	58.2	a	43.8	b	39.8	b	316	a
	SK	39.1	c	50.4	b	41.0	b	40.1	b	37.8	b	39.3	b	248	b
30/04/2013	Fa	44.2	b	52.6	a	54.6	a	57.6	a	55.6	a	51.8	a	311	a
HPD	FK	52.7	a	55.4	a	57.6	a	59.5	a	50.0	a	42.4	b	307	a
	SK	40.3	c	41.7	b	42.2	b	40.8	b	38.8	b	40.5	b	244	b
01/09/2013	Fa	7.8	a	37.7	a	47.0	a	56.2	a	56.5	a	54.9	a	315	a
Mulch period end	FK	8.5	a	32.5	ab	45.3	a	50.9	a	50.1	ab	49.2	a	318	a
	SK	8.8	a	27.5	b	41.4	a	44.0	a	43.3	b	47.2	a	244	b

Within a date and depth, means with the same letter are not significantly different at $P<0.05$.

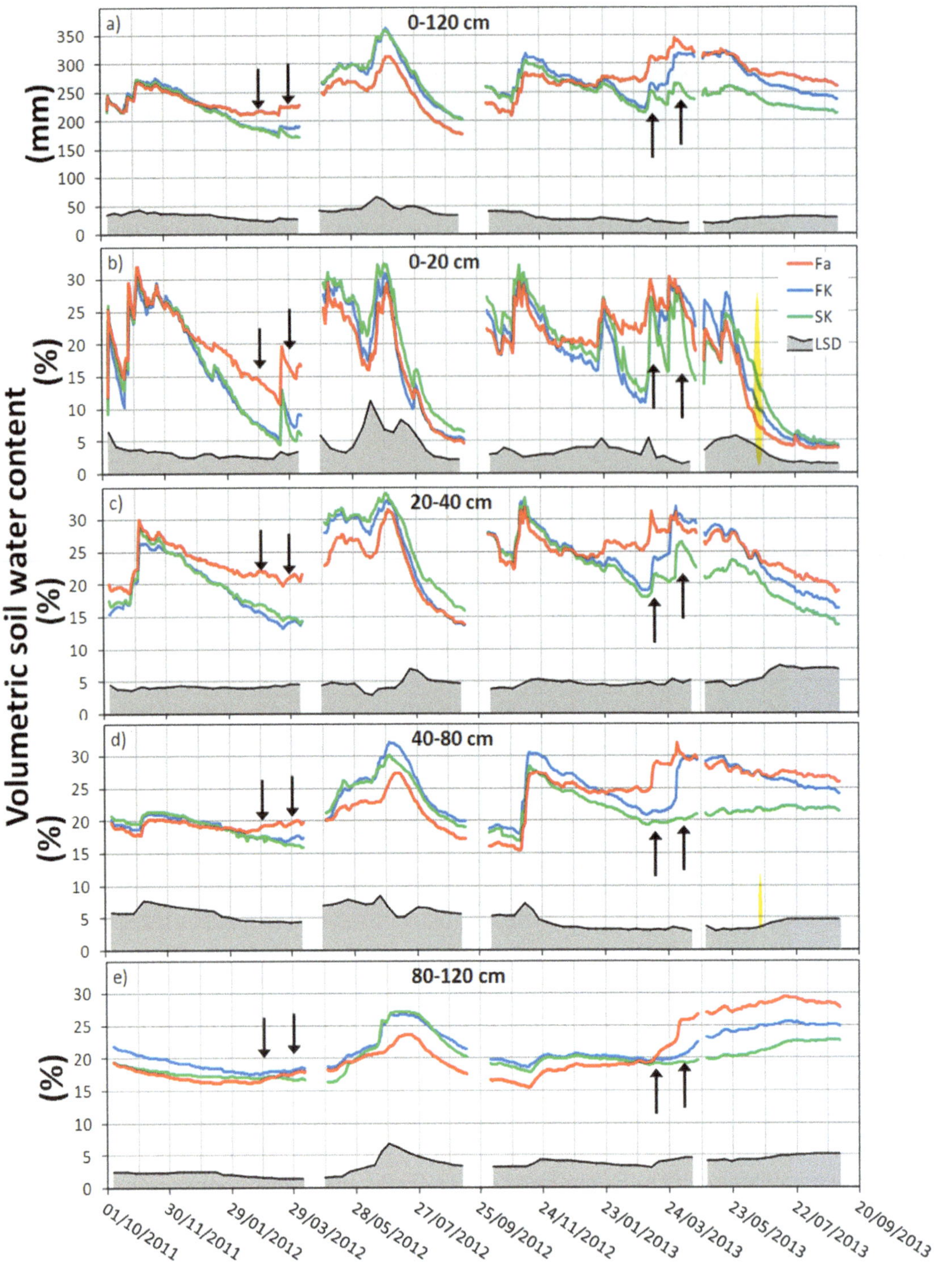

Figure 7. Soil water content monitored with the capacitance sensors at different depths for the fallow (Fa), first kill (FK) and second kill (SK) treatments during the experimental period. Arrows represent the first and second kill dates. The shadowed area in the bottom represents the LSD at 0.05 probability level.

application in April 2012, the SWC in the entire profile at the HPD was similar in all treatments (~262 mm; Table 3). During the wet/dry cycles that occurred in spring and summer, the CC treatments always retained more moisture than the fallow which showed the ability of the residue mulch to reduce soil water evaporation losses (Figure 7). After mid-June, when the soil was left to dry under the summer heat, differences in SWC between treatments disappeared, first in the upper layers and later in the entire soil profile. Even though slower drying was observed in the CC treatments, no differences were observed between treatments by the end of September (Table 3). In 2013, no water was applied, and the SWC in the entire soil profile was lower for the SK than for the FK and fallow treatments at the HPD (Table 3). At this time, no differences in SWC were observed between the FK and fallow treatments in the entire soil profile. However, the FK upper layer was wetter than the fallow because of the reduction in soil

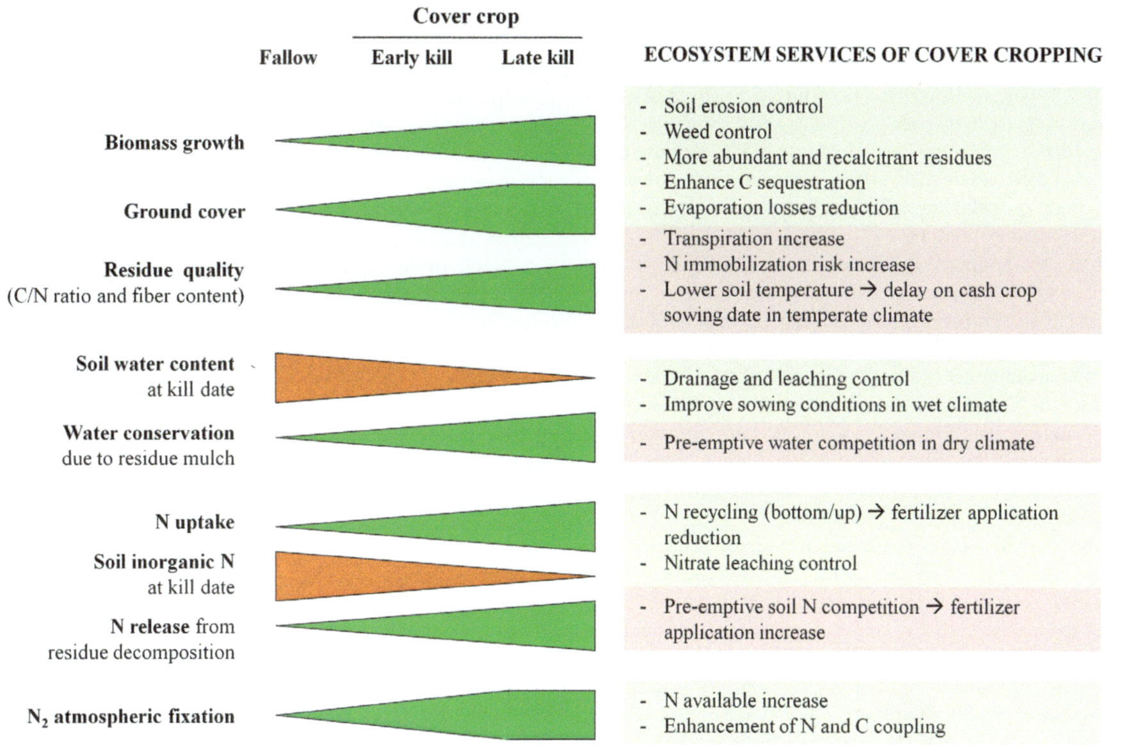

Figure 8. Schematic representation of the effect of kill date on environmental variables and ecosystem services with fallow as a reference condition.

water evaporation caused by the residue mulch (Table 3, Figure 7b). During summer, water losses in the top layer were greater in the fallow than the CC treatments, showing the effect of the mulch. By the end of the experiment, no differences in the SWC of the entire profile were observed between treatments, but some SK subsoil layers remained drier as the result of water depletion caused by the living CC.

Discussion

This study confirmed that CC kill date is an important management strategy as it affected key variables of the soil-plant system, influencing the success of the subsequent cash crop. Previous studies have focused mainly on the effect of N on the succeeding crop [6] and on the CC residue quality [14], but there is a lack of information on the combined effects of CC growth, N dynamics, and water dynamics. This information is necessary to optimize CC benefits and for development of more accurate simulation models.

Delaying kill date did not increase GC whereas the year had an effect on the maximum cover attained. We could not find data to compare the effect of kill date on GC. Nevertheless, our results agreed with those of previous researchers that showed the capacity of the asymptotic Gompertz function to capture the characteristics of the GC evolution of several CC [22,33]. Ground cover is an important variable used in studies related to soil erosion [32], evaporation [34], weed control [35,36] or radiation interception [22]. In addition, the evolution of digital technologies allows for non-destructive monitoring using reliable and efficient techniques [20]. For all of these reasons, we propose that GC should become a common variable for the characterization of CC that facilitates comparisons between species, varieties and management strategies in different regions. In the present study, GC differed with seasonal weather conditions; the dry 2011–2012 season was less favorable for CC growth than the wet and mild 2012–2013 season. In both seasons, GC>80% was already attained by the first kill date, so no additional benefits in terms of soil erosion or weed control were expected by the second kill date [32,35].

As expected, delaying kill date increased the above-ground biomass of the barley/vetch mixture. This increase was mainly due to barley as there was no difference in vetch biomass between kill dates. The barley/vetch ratio, expressed as seed weight at sowing, was 3:7, but barley was a stronger competitor due to faster establishment and growth rate [20]. The observed increase in biomass is common in many other studies [8,13], but the relationship between the legume and the non-legume differed. In Central Italy, a barley/vetch mixture with a 1:1 sowing rate yielded between 48 and 66% vetch biomass 160 d after planting which increased vetch in the sowing mixture up to 75% and produced between 53 and 76% vetch biomass at harvest [37]. A possible explanation for the large barley dominance in our study is that soil available N was not a limiting factor. Other authors reported similar superior dry matter values for the non-legume compared with the legume in mixtures, and this dominance became stronger with higher levels of N [38]. This hypothesis is corroborated by the plant and soil available N results in our study. Most of the N content in the vetch/barley mixture accumulated before the FK date. Nitrogen content of barley was greater at SK than FK (Figure 3), and in the SK treatment, the inorganic N in the soil was depleted in April, particularly from the deeper layers (Table 2, Figure 5). The high competition with barley for soil N forced the vetch to rely mainly on N_2 fixation for its N requirement, which explains the large degree of N_2 fixation observed in the legume. We were expecting the vetch to fix a significant amount of N_2 between both kill days, as observed in other experiments [10,13], but no increase in vetch N content was observed as in the study of Benincasa et al. [8]. The strong competition with barley for light explains the low vetch growth between kill dates and reinforces the need for sowing mixtures with a large proportion of vetch seeds (>50%) to favor legume activity. Most authors have found that the N content in CC and mixtures increased as kill date was delayed [10,13], but our results show that the main effect of delaying kill date increased above-ground biomass rather than N content.

The effect of postponing kill date on chemical residue composition was in agreement with previous findings in the literature [8,12,39]. Nitrogen concentration decreased from the first to the second kill date, being greater for vetch than for barley (Figure 3). These differences were reinforced in the C/N ratio as a slight increase in C content was observed in the SK samples compared with the FK (Figure 4). Vetch N concentration was similar to that found by other authors, but barley values were greater than those reported in other CC experiments [10,40,41]. In the FK samples, barley N concentration was ~2.9% and ~2.2% in the SK which agrees with data from barley cultivated as a cash crop using synthetic N fertilizer [42]. These data confirm that barley did not suffer from lack of N. The increase in fiber fractions, which is related to a decrease in the labile fraction, from delaying kill date was consistent with other studies and was expected to lead to slower C and N mineralization [43]. Increases in lignin fraction were only significant in barley in the first season, but this is not surprising as there were only three weeks between kill dates, and lignin content is less sensitive to change [39]. In the drier season (2011–2012), barley residue and the mixture had significantly lower fiber and lignin content than in the wetter season for both kill dates. A reduction in NDF, ADF and lignin cereal crop straw in dry years was previously noted in a study conducted over several years and locations in Washington State [44]. Our results confirmed that delaying kill date increased the amount of CC residue and the residues were more recalcitrant to decomposition. This is particularly interesting when the goal is preserving soil moisture [11], controlling weeds [35], or increasing the slow release of the N pool [45]. In the present experiment, a detailed study of CC residue decomposition was not conducted primarily because the lowest part of the residue layer stuck to the soil when sampling, and we decided that it was reliable for determining the amount of the remaining residue but not its quality.

The topsoil N_0 was also affected by the kill date two years after starting the experiment. Soil N mineralization potential assess the capacity of a soil to mineralize N and is a good indicator of land use as it is very sensitive to small changes in labile C and N fractions [46]. The increase of N_0 in the CC treatments proves that CC contribute to the enhancement of labile soil organic pools. The low contribution of CC to total organic C is a known fact as residues contain low quantities of recalcitrant organic matter [2]. However, CC add substrate for soil microorganisms during their growth period (root exudation, root turnover), and residue decomposition increases the organic matter labile fractions [6]. Many studies of N mineralization from CC residues have been published, but there is a lack of studies that focus on the effects on the soil in which they were cropped. In our study, no differences were observed in total soil C, but delaying kill date increased the C supplied by CC and, as a consequence, topsoil N_0. Nitrogen mineralization rate, a parameter that mainly depends on environmental conditions, was not different between treatments and was within the range found in the literature [47].

Consistent with other studies, CC absorbed most of the available soil inorganic N during their growth period, modifying the degree of pre-emptive competition and the risk of N loss with respect to the fallow (Table 2, Figure 5). This study also showed that kill date affected these parameters and could be a means to adapt soil N availability to the requirements of a particular agroecosystem. Delaying kill date increased the risk of pre-emptive competition, a negative effect that was reflected not only in the deeper layers as is common [4,48], but in the upper layers during the second season as well. The early kill date allowed more time for N release from CC residues, and by the HPD, more N was available in the topsoil. This positive effect of an early kill date on reducing pre-emptive competition has not been shown before and was evident when comparing FK with the SK and the fallow in the wet year, in which N_{min} was washed out of the soil during the fall and winter. The well-known CC effect of recycling N in the soil-plant system by uptaking inorganic N from deeper layers and subsequently releasing it to the top layers was clear in this study [6]. In the first season, the soil N_{min} depleted in April by the CC uptake was replenished by October by the N released from the residue mineralization. In the second season, CC prevented most of the available N from being lost from the soil, as happened in the fallow, and released the N by November. The difference in November was not in the amount of N_{min} but its distribution in the soil profile. In the fallow, N_{min} was mostly in the deeper layers and had greater potential for leaching during fall and winter. However, in the SK treatment, N_{min} was in the top layers making it readily available for uptake by the subsequent crop. The N_{min} profile of the FK soil was intermediate. Organic N remained in the residue layer at the end of the experiment, but as mentioned before, we could not obtain a reliable measure of N concentration.

Even if N uptake by CC depleted the soil by April in the first season and seemed to increase the risk of pre-emptive competition with respect to the fallow, we propose that this may be a positive effect in many agricultural systems as it may help control N losses. In a tomato drip irrigation study, it was observed that high soil N_{min} levels at spring planting were linked to large leaching losses as crops are often overwatered to ensure establishment [49]. Excessive irrigation during the crop establishment period may represent up to 80% of total nitrate leaching losses [34,50]. Therefore, keeping soil N_{min} at low levels and correcting fertilizer needs based on crop nutritional status during the cropping season holds promise for controlling N losses and increasing the efficiency of N use [51].

The larger above-ground biomass observed in the SK affected SWC. The CC extracted soil water by transpiration, increasing evapotranspiration losses when compared with the fallow [34]. Spring growth of the vetch/barley mixture was vigorous and soil water depletion was enhanced between the first and the second kill date. The differences in the whole profile SWC at the first kill date between CC treatments and the fallow were >35 mm in the first season and >55 in the second. At the second kill date, the SWC in the FK treatment was similar to the fallow. By the HPD and during the rest of the mulch period, the treatment effect on water availability was variable. While in the first season there was no difference, most likely because the soil was replenished by irrigation, the SWC was 60 mm lower in the SK than in the other treatments during the second season. Similar results were found in California [52,53], where reductions in spring SWC were up to 80 mm due to CC extraction. In Minnesota, the difference in SWC in the upper 0.6 m caused by delaying the rye kill date by three weeks (from April to May) was 27 mm [12]. Our results confirm that late killing of CC increases pre-emptive water competition compared with early killing, a risk that might be mitigated in rainy years or by irrigation if water is available.

A relevant aspect of the increased water uptake by CC is the effect on the upper layers' SWC. Sufficient top layer moisture at planting is crucial to ensure crop establishment and plant survival in many crops [49]. In the present study, the difference in SWC between the fallow and the CC treatments at the first kill date due to water depletion ranged between 13–19 mm in the upper 0.2 m of the soil and was enhanced between the fallow and the SK by the second kill date. However, the effect of CC residue mulching led to a higher SWC in the upper layers of the FK treatment than in the fallow at the HPD. When comparing CC treatments, the results are less clear; more water in the upper layers was retained in the SK than in the FK in the first season, but the SK was drier in the second. The FK treatment always increased water retention in the upper layers with respect to the fallow, but the result varied by year in the SK treatment. The top layer SWC was influenced by CC water uptake during winter, the residue mulch effect, and the amount of rain between the CC kill date and cash crop planting [9]. To clarify the discrepancy between the soil water depletion by CC uptake and conservation by the mulched CC residue, an experiment comparing sheltered and un-sheltered plots in Maryland showed that delaying kill date of a vetch CC produced more biomass and resulted in better soil moisture conservation in the upper 0.3 m [11]. However, recent experiments in Minnesota reported that delaying the rye kill date from the tillering to the booting stage came at the cost of SWC depletion in the top 0.3 m even if the residues continued to cover the soil after termination [12]. In Indiana, early kill of the CC resulted in better water conservation in the topsoil (0.1 m) than late kill. In drought years, early kill date produced an increase in maize yield but had a negative impact in normal precipitation years due to a lowering of soil temperature and workability at the maize planting date [7]. Our results indicate that early kill of the CC enhanced spring topsoil water moisture with respect to late kill, particularly by preserving soil water replenished by precipitation between kill dates.

A limitation found when comparing our study with others that focus on CC killing was that kill dates were always expressed as days after sowing. Given the variety of climatic regions in which cover cropping is practiced, the number of days after sowing is not a valid variable. We propose using degree-days accumulated and growth stage as more suitable variables for comparison between different regions. Soil type also has a major influence on the water balance and N cycling. Simulation models that can take multiple environmental factors into account might allow to overcome these limitations and increase the applicability of this and others studies to different locations and years.

One criticism of this study might be that a crop was not planted after the CC to obtain real data on crop water uptake. We are aware of this situation, but we think that this type of dataset is necessary to advance our understanding of the complex effect of CC kill date on crop resources. Separating the mulching and the effect of transpiration might allow for the calibration of key parameters, proper simulation of both processes, and finally, of the whole system. Replacing fallows by CC is known to provide many benefits, but the agronomic application needs to account for the requirements of specific agroecosystems. Kill date is a management tool to regulate the effects of CC on many environmental variables as presented in Figure 8. In dry environments, residue moisture conservation is the main goal of CC [9] whereas the risk of delaying sowing because of soils that are too wet is a major concern in cooler climates [7]. Depleting soil N enhances pre-emptive competition with the subsequent crop [6] but also controls N losses and creates opportunities for rational fertilizer application [34]. Late termination of CC maximizes their biomass production and the formation of a thick mulch, enhancing C sequestration, water conservation and weed control [36]. Our work does not close the debate on kill date but contributes to more rational decision-making for the successful use of CC.

Conclusions

Kill date is a major management variable with CC that may lead to increasing water and N use efficiency and controlling N losses and pre-emptive competition with the subsequent crop. The ground cover of CC was >80% by the first kill date so no more benefits in terms of soil erosion or weed control derived from the living CC were expected by the second kill date. However, delaying kill date increased above-ground biomass, providing more CC residues with a higher fiber content and C/N ratio that were more recalcitrant to decomposition and, therefore, more suitable to protect the soil and enhance the slow release of the N pool.

The early kill date decreased the risk of pre-emptive competition by diminishing the N uptake of CC, and allowing more time for N release from CC residues. Cover crops generally uptake inorganic N from deeper layers to be released on the topsoil through residue mineralization. Delaying kill date enhanced this recycling effect that prevented losses of available N by keeping soil inorganic N at low levels. Kill date was a means of controlling this process and showed promise in mitigating N losses and increasing N use efficiency. The N mineralization potential in topsoil was enhanced by the presence of CC and by delaying kill date, proving that CC contribute to the enhancement of labile organic pools.

Delaying kill date increased pre-emptive competition of water, as the water extracted by CC from the first to the second kill date was greater than the water conservation due to the extra residue generated by the SK treatment. The soil water content in the upper layers at the time of planting the subsequent crop increased by the first kill date with respect to the fallow by preserving rain harvested between kill dates, but the soil water content in the late kill could be even lower than in the fallow depending on the climatic conditions of the year.

Supporting Information

Table S1 Biomass, N content and residue left. Above-ground biomass and N content of cover crops at the first kill (FK) and second kill (SK) dates and residue covering the soil at the end of the mulch period (~7 months after cover crop killing). Means with standard error in parentheses. Within a row, means with the same letter are not significantly different between kill dates at P<0.05.

Table S2 Chemical composition of cover crops. Cover crop residue neutral detergent fiber (NDF), acid detergent fiber (ADF) and lignin (L) fractions and C/N ratio for the first kill (FK) and second kill (SK) dates. Means with standard error in parentheses. Within a row, means with the same letter are not significantly different between kill dates at P<0.05.

Image S1 Aerial view of the experimental site. Plots are marked as fallow (FA), first kill (FK) nd second kill (SK) plus the replication number. Image taken between kill dates (March 29, 2012).

Image S2 Ground level view of the experimental site. Image taken between kill dates (April 4, 2012).

Acknowledgments

We would like to thank the staff of La Chimenea field station (IMIDRA) for their helpful assistance.

Author Contributions

Conceived and designed the experiments: MQ. Performed the experiments: MA JLG MQ. Analyzed the data: MA JLG MQ. Contributed to the writing of the manuscript: MA JLG MQ.

References

1. Tonitto C, David MB, Drinkwater LE (2006) Replacing bare fallows with cover crops in fertilizer-intensive cropping systems: A meta-analysis of crop yield and N dynamics. Agriculture, Ecosystems & Environment 112: 58–72.
2. Kuo S, Sainju UM, Jellum EJ (1997) Winter cover crop effects on soil organic carbon and carbohydrate in soil. Soil Science Society of America Journal 61: 145–152.
3. Quemada M, Cabrera ML (2002) Characteristic moisture curves and maximum water content of two crop residues. Plant and Soil 238: 295–299.
4. Gabriel JL, Quemada M (2011) Replacing bare fallow with cover crops in a maize cropping system: Yield, N uptake and fertiliser fate. European Journal of Agronomy 34: 133–143.
5. Abawi GS, Widmer TL (2000) Impact of soil health management practices on soilborne pathogens, nematodes and root diseases of vegetable crops. Applied Soil Ecology 15: 37–47.
6. Thorup-Kristensen K, Magid J, Jensen LS (2003) Catch crops and green manures as biological tools in nitrogen management in temperate zones. Advances in Agronomy 79: 227–302.
7. Stipesevic B, Kladivko EJ (2005) Effects of winter wheat cover crop desiccation times on soil moisture, temperature and early maize growth. Plant Soil and Environment 51: 255–261.
8. Benincasa P, Tosti G, Tei F, Guiducci M (2010) Actual N Availability from Winter Catch Crops Used for Green Manuring in Maize Cultivation. Journal of Sustainable Agriculture 34: 705–723.
9. Clark AJ, Meisinger JJ, Decker AM, Mulford FR (2007) Effects of a grass-selective herbicide in a vetch-rye cover crop system on corn grain yield and soil moisture. Agronomy Journal 99: 43–48.
10. Tosti G, Benincasa P, Farneselli M, Pace R, Tei F, et al. (2012) Green manuring effect of pure and mixed barley-hairy vetch winter cover crops on maize and processing tomato N nutrition. European Journal of Agronomy 43: 136–146.
11. Clark AJ, Decker AM, Meisinger JJ, McIntosh MS (1997) Kill date of vetch, rye, and a vetch-rye mixture: 2. Soil moisture and corn yield. Agronomy Journal 89: 434–441.
12. Krueger ES, Ochsner TE, Porter PM, Baker JM (2011) Winter Rye Cover Crop Management Influences on Soil Water, Soil Nitrate, and Corn Development. Agronomy Journal 103: 316–323.
13. Clark AJ, Decker AM, Meisinger JJ, McIntosh MS (1997) Kill date of vetch, rye, and a vetch-rye mixture: 1. Cover crop and corn nitrogen. Agronomy Journal 89: 427–434.
14. Wagger MG, Cabrera ML, Ranells NN (1998) Nitrogen and carbon cycling in relation to cover crop residue quality. Journal of Soil and Water Conservation 53: 214–218.
15. Tosti G, Benincasa P, Farneselli M, Tei F, Guiducci M (2014) Barley-hairy vetch mixture as cover crop for green manuring and the mitigation of N leaching risk. European Journal of Agronomy 54: 34–39.
16. Soil Survey Staff (2014) Keys to Soil Taxonomy. Washington, DC, USA: USDA, Natural Resources Conservation Service.
17. Nelson DW, Sommers LE (1996) Total carbon, organic carbon and organic matter. In: Sparks D, editor. Methods of soil analysis, part 3: chemical methods. Madison, WI, USA: ASSA and SSSA. 961–1010.
18. Loeppert RD, Suarez DL (1996) Carbonate and Gypsum. In: Sparks D, editor. Methods of soil analysis, part 3: chemical methods. Madison, WI, USA: ASA and SSSA. 437–474.
19. Papadakis J (1966) Climates of the world and their agricultural potentialities. Rome, Italy: DAPCO.
20. Ramirez-Garcia J, Almendros P, Quemada M (2012) Ground cover and leaf area index relationship in a grass, legume and crucifer crop. Plant Soil and Environment 58: 385–390.
21. Richardson MD, Karcher DE, Purcell LC (2001) Quantifying turfgrass cover using digital image analysis. Crop Science 41: 1884–1888.
22. Bodner G, Himmelbauer M, Loiskandl W, Kaul HP (2010) Improved evaluation of cover crop species by growth and root factors. Agronomy for Sustainable Development 30: 455–464.
23. Unkovich M, Herridge D, Peoples M, Cadisch G, Boddey B, et al. (2008) Measuring plant-associated nitrogen fixation in agricultural systems. Camberra, Australia: Australian Centre for International Agricultural Research. 258 p.
24. Goering HK, Van Soest PJ (1970) Forage fiber analysis: apparatus, reagents, procedures, and some applications. Washington, DC: Agricultural Research Service. Handbook no. 379. US Government Printing Office.
25. Keeney DR, Nelson DW (1982) Nitrogen–inorganic forms. In: Page AL, editor. Methods of soil analysis Part 2: Chemical and microbiological properties. Madison, WI, USA: ASA and SSSA. 643–698.
26. Crooke WM, Simpson WE (1971) Determination of ammonium in Kjeldahl digests of crops by an automated procedure. Journal of the Science of Food and Agriculture 22: 9–10.
27. Stanford G, Smith SJ (1972) Nitrogen mineralization potentials of soils. Soil Science Society of America Proceedings 36: 465–472.
28. Paltineanu IC, Starr JL (1997) Real-time soil water dynamics using multisensor capacitance probes: Laboratory calibration. Soil Science Society of America Journal 61: 1576–1585.
29. Gabriel JL, Lizaso JI, Quemada M (2010) Laboratory versus field calibration of capacitance probes. Soil Science Society of America Journal 74: 593–601.
30. Quemada M, Cabrera ML (1995) CERES-N model predictions of nitrogen mineralized from cover crop residues. Soil Science Society of American Journal 59: 1059–1065.
31. Whitmore AP (1991) Method for assessing the goodness of computer simulation of soil processes. Journal of Soil Science 42: 289–299.
32. Quinton JN, Edwards GM, Morgan RPC (1997) The influence of vegetation species and plant properties on runoff and soil erosion: results from a rainfall simulation study in south east Spain. Soil Use and Management 13: 143–148.
33. Ramirez-Garcia J, Gabriel JL, Alonso-Ayuso M, Quemada M (2014) Quantitative characterization of five cover crop species. Journal of Agricultural Science. doi:10.1017/S0021859614000811.
34. Gabriel JL, Muñoz-Carpena R, Quemada M (2012) The role of cover crops in irrigated systems: Water balance, nitrate leaching and soil mineral nitrogen accumulation. Agriculture Ecosystems & Environment 155: 50–61.
35. Teasdale JR, Mohler CL (1993) Light transmittance, soil-temperature, and soil-moisture under residue of hairy vetch and rye. Agronomy Journal 85: 673–680.
36. Saini M, Price AJ, Van Santen E, Arriaga FJ, Balkcom KS, Raper RL (2008) Planting and termination dates affect winter cover crop biomass in a conservation-tillage corn-cotton rotation: Implications for weed control and yield. In: Endale DM editor. Tifton, GA, USA. Southern Conservation Agricultural Systems. 137–141.
37. Tosti G, Benincasa P, Guiducci M (2010) Competition and Facilitation in Hairy Vetch-Barley Intercrops. Italian Journal of Agronomy 5: 239–248.
38. Ofori F, Stern WR (1987) The combined effects of nitrogen-fertilizer and density of the legume component on production efficiency in a maize cowpea intercrop system. Field Crops Research 16: 43–52.
39. Wagger MG (1989) Time of desiccation effects on plant composition and subsequent Nitrogen release from several winter annual cover crops. Agronomy Journal 81: 236–241.
40. Clark AJ, Decker AM, Meisinger JJ (1994) Seeding rate and kill date effects on hairy vetch cereal rye cover crop mixtures for corn production. Agronomy Journal 86: 1065–1070.
41. Benincasa P, Guiducci M, Tei F (2011) The nitrogen use efficiency: Meaning and sources of variation-case studies on three vegetable crops in Central Italy. HortTechnology 21: 266–273.
42. Arregui LM, Quemada M (2008) Strategies to improve nitrogen use efficiency in winter cereal crops under rainfed conditions. Agronomy Journal 100: 277–284.
43. Quemada M, Cabrera ML (1995) Carbon and nitrogen mineralized from leaves and stems of four cover crops. Soil Science Society of America Journal 59: 471–477.
44. Stubbs TL, Kennedy AC, Reisenauer PE, Burns JW (2009) Chemical composition of residue from cereal crops and cultivars in dryland ecosystems. Agronomy Journal 101: 538–545.
45. Quemada M, Cabrera ML, McCracken DV (1997) Nitrogen release from surface-applied cover crop residues: Evaluating the CERES-N submodel. Agronomy Journal 89: 723–729.
46. Cookson WR, Abaye DA, Marschner P, Murphy DV, Stockdale EA, et al. (2005) The contribution of soil organic matter fractions to carbon and nitrogen mineralization and microbial community size and structure. Soil Biology and Biochemistry 37: 1726–1737.
47. Quemada M, Diez JA (2007) Available nitrogen for corn and winter cereal in Spanish soils measured by electro-ultrafiltration, calcium chloride, and incubation methods. Communications in Soil Science and Plant Analysis 38: 2061–2075.
48. Thorup-Kristensen K (1994) The effect of nitrogen catch crop species on the nitrogen nutrition of succeeding crops. Fertilizer Research 37: 227–234.
49. Vázquez N, Pardo A, Suso ML, Quemada M (2005) A methodology for measuring drainage and nitrate leaching in unevenly irrigated vegetable crops. Plant and Soil 269: 297–308.
50. Salmerón M, Isla R, Cavero J (2011) Effect of winter cover crop species and planting methods on maize yield and N availability under irrigated Mediterranean conditions. Field Crops Research 123: 89–99.
51. Vázquez N, Pardo A, Suso ML, Quemada M (2006) Drainage and nitrate leaching under processing tomato growth with drip irrigation and plastic mulching. Agriculture Ecosystems & Environment 112: 313–323.
52. McGuire AM, Bryant DC, Denison RF (1998) Wheat yields, nitrogen uptake, and soil moisture following winter legume cover crop vs. fallow. Agronomy Journal 90: 404–410.
53. Mitchell JP, Peters DW, Shennan C (1999) Changes in soil water storage in winter fallowed and cover cropped soils. Journal of Sustainable Agriculture 15: 19–31.

24

The Value of Countryside Elements in the Conservation of a Threatened Arboreal Marsupial *Petaurus norfolcensis* in Agricultural Landscapes of South-Eastern Australia—The Disproportional Value of Scattered Trees

Mason J. Crane*, David B. Lindenmayer, Ross B. Cunningham

Fenner School of Environment and Society, The Australian National University, Canberra, ACT, Australia

Abstract

Human activities, particularly agriculture, have transformed much of the world's terrestrial environment. Within these anthropogenic landscapes, a variety of relictual and semi-natural habitats exist, which we term countryside elements. The habitat value of countryside elements (hereafter termed 'elements') is increasingly recognised. We quantify the relative value of four kinds of such 'elements' (*linear roadside remnants, native vegetation patches, scattered trees* and *tree plantings*) used by a threatened Australian arboreal marsupial, the squirrel glider (*Petaurus norfolcensis*). We examined relationships between home range size and the availability of each 'element' and whether the usage was relative to predicted levels of use. The use of 'elements' by gliders was largely explained by their availability, but there was a preference for *native vegetation patches* and *scattered trees*. We found home range size was significantly smaller with increasing area of *scattered trees* and a contrasting effect with increasing area of *linear roadside remnants* or *native vegetation patches*. Our work showed that each 'element' was used and as such had a role in the conservation of the squirrel glider, but their relative value varied. We illustrate the need to assess the conservation value of countryside elements so they can be incorporated into the holistic management of agricultural landscapes. This work demonstrates the disproportional value of *scattered trees*, underscoring the need to specifically incorporate and/or enhance the protection and recruitment of *scattered trees* in biodiversity conservation policy and management.

Editor: Paul Adam, University of New South Wales, Australia

Funding: The authors have no support or funding to report.

Competing Interests: The authors have declared that no competing interests exist.

* Email: mason.crane@anu.edu.au

Introduction

Humans have transformed most of the earth's terrestrial biosphere into highly modified biomes [1], resulting in the loss and decline of many species [2]. To counter this, much focus has been on the establishment of formal conservation reserves. While these are a critical component of biodiversity conservation worldwide, reserve networks rarely provide comprehensive, adequate and representative coverage of ecosystems [3–6]. It is often the case that those natural ecosystems associated with agricultural land are the most poorly represented in conservation reserve networks. Australian box gum woodlands are a classic example of this [7]. Despite comparatively low levels of reservation in agricultural landscapes, many species associated with such ecosystems still persist there. Such species have been shown to use the relictual and semi-natural habitats of these landscapes (e.g. [8,9–15]). These habitats or countryside elements ('elements') [16] often have very different and distinct characteristics, each offering a different range of resources for wildlife.

The role that countryside elements and matrix habitats play in biodiversity conservation is increasingly appreciated. This has been demonstrated by work conducted in Australia [16], Central and South America [10,17,18] and West Africa [19]. However, to date there have been few works comparing different kinds of 'elements' and quantifying their relative usage for a single species. In this study, we examined the relative value of four kinds of wooded countryside elements in the conservation of a threatened arboreal marsupial, the squirrel glider *Petaurus norfolcensis*, in agricultural landscapes of south-eastern Australia. These 'elements' were: *scattered trees*, *tree plantings*, *linear roadside remnants* and small *native vegetation patches*. These four broad kinds of 'elements' can be found in many agricultural landscapes around the world (e.g. [20,21–24]). While their use as wildlife habitat has been reported (e.g. [12,20,21,25–28]), there are few unequivocal examples of their relative value in the conservation of different species or taxa [16,29]. To achieve this we address the following questions:

1. Is the use of an 'element' for denning and feeding proportional to their availability? If not, is there evidence of preferential usage of some kinds of 'elements'?

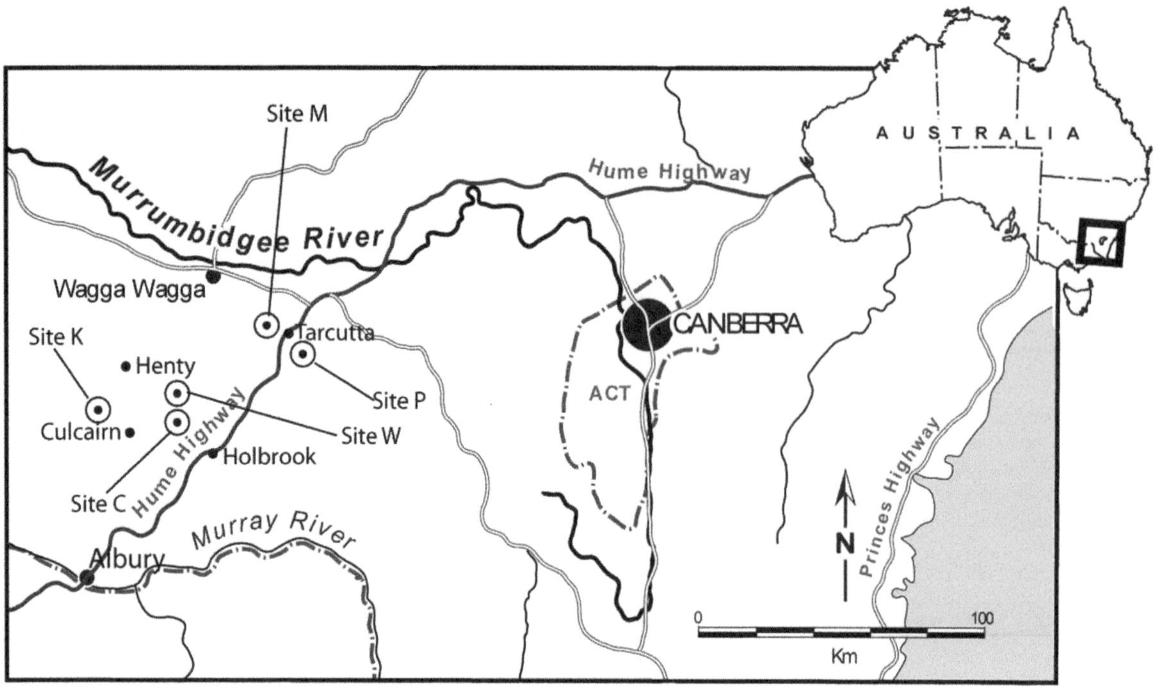

Figure 1. Location of the five study areas (C, K, M, P and W) used in radio-tracking the Squirrel Glider.

2. Does home range size depend on the proportional availability of an 'element'? Differences in home range size within species have often been attributed to habitat quality [30–32], with smaller home range indicating better quality habitat. If an 'element' offers a significantly higher or lower quality habitat relative to others, it would be expected that their availability would have a significant influence on home range size.

Our work is the first to attempt to quantify the relative contribution of different countryside elements for the squirrel glider and is one of the few that attempt to compare the ecological value of countryside elements, using empirical data. An understanding of the value of such 'elements' for biodiversity is essential for better integration of biodiversity conservation and broader management of agricultural landscapes. Without such data, managers are in danger of undervaluing certain 'elements'. This may, in turn, lead to the loss of critical habitats and further threaten associated species in these landscapes.

Methods

The squirrel glider

The squirrel glider is a nocturnal, arboreal, gliding marsupial in the Family Petauridae. It is a medium-sized possum weighing between 190–300 g and which feeds on invertebrates, insect exudates, sap of trees and shrubs, and pollen and nectar [33–35]. The species is listed as threatened in three of the four Australian states in which it is found (Victoria – *Flora and Fauna Guarantee Act 1988*; South Australia – *National Parks and Wildlife Act 1972*; New South Wales – *Threatened Species Conservation Act 1995*; and Queensland – no formal listing).

Study area

Our investigation encompassed five study areas within the south-west slopes of New South Wales, Australia [36] (Figure 1). The region is the most extensively and intensively disturbed of the 13 botanical regions of NSW, with an estimated 85% of the original cover of native vegetation removed in the past 200 years [37]. The five study areas were located in heavily modified agricultural landscapes, used predominantly for livestock grazing and dryland cropping. Study areas were approximately 3 km×3 km. Woody vegetation occurred primarily as relictual scattered paddock trees, native vegetation plantings and remnant temperate *Eucalyptus* woodlands on private lands, road reserves and travelling stock reserves.

Ethics statement

We conducted trapping and radio tracking under The Australian National University Animal Ethics Committee protocol number C.RE.39.05. Squirrel Gliders are a native species and therefore protected. Relevant permits to handle the animals were obtained from New South Wales Government agencies. Animals were captured in wire mesh cage traps covered with non-transparent, heavy-duty plastic sleeves to minimise stress on animals and protect from cold and wet weather. Qualified wildlife veterinarians anaesthetised captured animals using isofluorane gas delivered via a portable gas anaesthesia machine. Isofluorane anaesthesia ensures recovery of animals within minutes, which is considerably faster than injectable agents, thus minimising holding time and stress. Anaesthetising animals ensured that accurate measurements of body size and reproductive status could be made without undue stress to the animals. It also enabled the veterinarians to properly fit radio-collars and microchip each animal. Gliders not fitted with collars were ear tagged (hamster ear tags, Sieper & Co., Sydney, Australia) in each ear. When animals had recovered from the anaesthetic, they were released at the exact point of capture.

Land accessed was a mix of privately owned farmland, local government managed road reserves, and travelling stock reserves managed by the Livestock Health and Pest Authority and relevant access permissions were obtained from the respective land managers.

Radio-tracking

We captured gliders using drop-door, wire mesh cage traps (170 mm×200 mm×500 mm) over a three night period at each site in March 2005 [38]. We fitted 32 gliders with a single stage brass loop radio transmitter, weighing 4.5 grams (Sirtrack, New Zealand). When selecting which gliders were to be collared, we preferred adult gliders and attempted to achieve an equal sex ratio and an equal spatial coverage of animals within and between sites. We tracked gliders to their diurnal denning site at least twice a week and to a nocturnal location at least 1–3 times every 14 days, over a 4–5 month period [38,39]. For each fix, we recorded the countryside element in which the glider was located.

Home range

We derived home range estimates by using both diurnal and nocturnal fixes. The woody vegetation was scattered or represented in irregularly shaped patches, surrounded by cleared agricultural matrix. Common parametric approaches such as the minimum convex polygon method were therefore unsuitable. This is because they would have (inappropriately) included large areas of unused cleared agricultural land – as has been found by Martin et al. [40] and van der Ree and Bennett [15]. We estimated home range by using the non-parametric, grid cell method [41]. The size of the grid cell is arbitrary and can have a major influence on either underestimating or over-estimating the den range size [41]. Three sized cells were tested 40×40 m, 50×50 m and 60×60 m and compared to the minimum convex polygon (MCP) method (where MCP could be appropriately used, i.e. for some animals in areas dominated by remnant vegetation patches or large clusters of scattered trees). We calculated the minimum convex polygon using Home Range extension within Arc View GIS (ESRI, California, USA). We selected a 50×50 m grid cell as it aligned more closely with the commonly used MCP method. We connected disjointed cells by including cells that were crossed by the most direct line joining to consecutive locations, taking into consideration gap-crossing ability ([i.e. gaps in canopy <70 m; see [42]). We estimated the home range size of each squirrel glider on 95% of fixes. This was done to give an objective, repeatable method of comparison of normal home range [41]. We deleted 5% of fixes at the extremities of each animal's range to reduce the influence of exploratory movements or outlying fixes outside the 'normal' home range [15].

Countryside elements

Within our study area, we recognised four categories of countryside elements that contained woody vegetation.

(1) Linear roadside remnants. These were linear strips of remnant vegetation along roads. Remnant roadside vegetation is a major feature across agricultural landscapes providing a network of remnant vegetation corridors across what are otherwise generally heavily cleared landscapes. The width of the roadside reserves in this study ranged from 40–60 m. Vegetation along road reserves is subject to high levels of disturbance such as road construction and maintenance. However, grazing pressure by domestic livestock is often low and irregular. These areas regularly contain regeneration of overstorey species. Native understorey species are generally present but their dominance and diversity varies.

(2) Native vegetation patches. These were patches of remnant vegetation where the understorey was dominated by a diverse array of native plants. In this study, these areas were mostly on travelling stock reserves, and some patches of remnant native vegetation on freehold land. The travelling stock route network was established more than 150 years ago to facilitate the movement of domestic livestock between properties and to markets [43]. The network is made up of travelling stock routes (which today are often incorporated into the road reserves) and holding paddocks which are generally referred to as travelling stock reserves (TSRs). Because of their reservation for these purposes, these areas generally escaped clearing and continuous high-intensity livestock grazing. The *native vegetation patches* varied in size with the largest patches occurring on TSRs (≈100 ha) and the smallest patches on freehold land (>5 ha).

(3) Scattered trees. These were scattered, (mostly large, old) relictual trees remaining on land used for grazing or cropping. They include dead and living trees, are often widely spaced (more widely spaced than expected in their natural state) and contain a simple understorey, generally dominated by introduced grasses and forbs, with a low diversity of native plant species.

(4) Tree plantings. These were Australian native vegetation plantings, generally containing dense stands of trees (predominately *Eucalyptus*) and shrubs (such as *Acacia* and *Melaleuca*). The species composition was predominantly locally indigenous species, but often included species naturally found outside the region. *Tree plantings* vary in their shape and size as they were planted for various purposes such as shelterbelts or to reduce rising water tables [11]. The level of grazing by livestock within them also varied.

The vegetation was assigned to the countryside element of best fit, using the above descriptions as a guide. Where elements were adjacent to each other, boundaries were defined by differences in vegetation structure and composition, and/or management practices. We calculated the area of each countryside element available to an individual squirrel glider by measuring the total area of woody vegetation attributed to that 'element', within a 1000 m radius of the centre point of all fixes for each individual glider. We used a 1000 m radius, as 2000 m is approximately the maximum home range length that has been reported for our study species [15]. We measured the area of woody vegetation using geographical information systems software (ArcGIS 9.2-esri) to draw polygons over the canopy of woody vegetation interpreted from satellite imagery (spot 5-Astrium). We deemed that woody vegetation isolated by a gap distance of greater than 70 metres was unavailable to gliders [42].

Table 1. Number of squirrel gliders captured/collared at each site.

	Site C	Site K	Site M	Site P	Site W	Total
No. captures	5	8	25	3	11	52
No. collars fitted	5	4	12	3	8	32

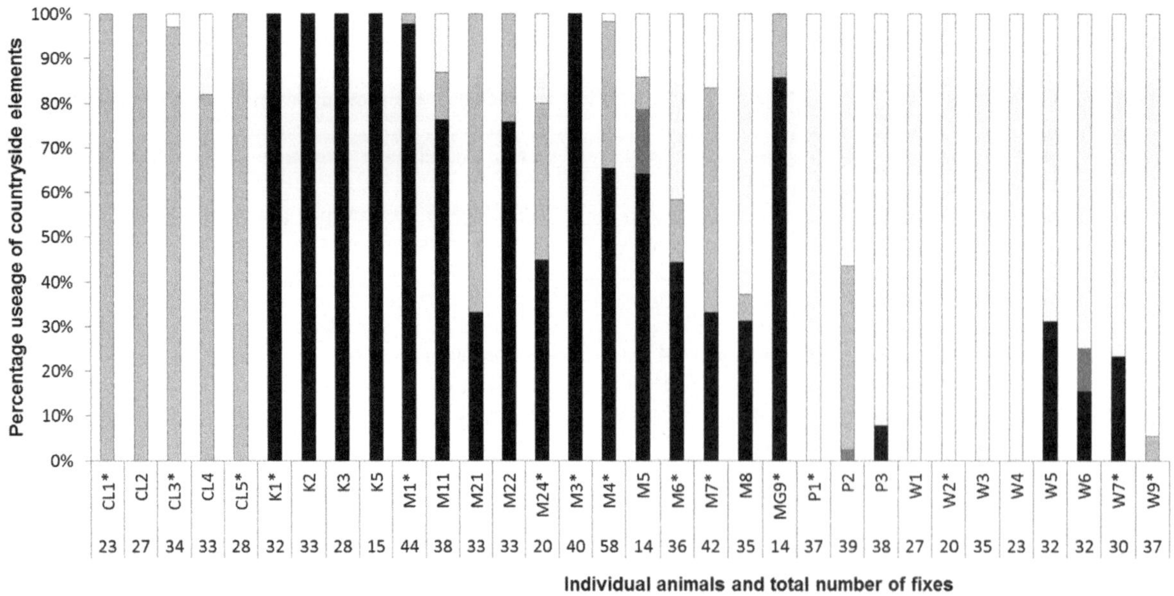

Figure 2. The percentage usage of countryside elements by individual squirrel gliders (* = female). *Remnant vegetation patches* (black shading), *tree planting* (dark grey shading), *linear roadside remnant* (intermediate grey shading) and *scattered trees* (light grey shading).

Data analyses

For each squirrel glider, our data consisted of a count of the number of fixes in each 'element', classified by circadian time - day or night. Associated data were the total area of woody vegetation in each 'element', available within a 1000 m radius of the centre point of all fixes for each individual glider. From these data, we can obtain the expected number of fixes by apportioning total fixes according to the relative area of each 'element'. Thus, for day and night data separately, we have a 2-way contingency table cross-classified by animal ID and 'element' type, where the cells are the observed count of the number of fixes and a concomitant variable is the expected number of fixes based on the relative availability of each of the 'elements'.

The Model

Considered as a two-way contingency table, our data can be modelled by

$$E[\log(y_{ij})] = u + b\log(f_{ij}) + animal_i + CSE_j$$

where y_{ij} are the expected frequencies, u is the grand mean, f_{ij} are the 'expected' frequencies derived from the percentage are occupied by *'element'*$_j$ $animal_i$ and *'element'*$_j$ are constants to account for the marginal distribution of the two-way animal by 'element' table.

The above model is a particular case of a class of models commonly known as log-linear models, and which are often used in the analyses of multiway contingency tables. The log-linear model belongs to the class of generalised linear models [45]. These models can be fitted by the use of maximum likelihood methods available in many statistical software packages, such as GENSTAT Version 15. The goodness-of-fit of the model and the significance of individual terms can be assessed by examining an analysis of deviance.

We completed further linear regression analysis to quantify relationships between home range size and total woody vegetation and each of the countryside elements. We avoided statistical issues associated with intrinsic collinearity of countryside elements by considering each of the 'elements' separately.

Results

Radio-tracking

We captured 52 individual gliders and fitted 32 with radio-transmitting collars (see Table 1). The numbers of fixes for individuals varied as signals were lost for some animals throughout

Table 2. Accumulated analysis of deviance Table after fitting a sequence of log-linear Poisson models (see section 'The Model').

Term	d.f.	Night use	Day use
+ Animal	31	27	76
+ CSE	3	186	391
Residual	91	576	949
+ log(f)	1	367	675
Residual (final)	92	209	286
Total	127	789	1416

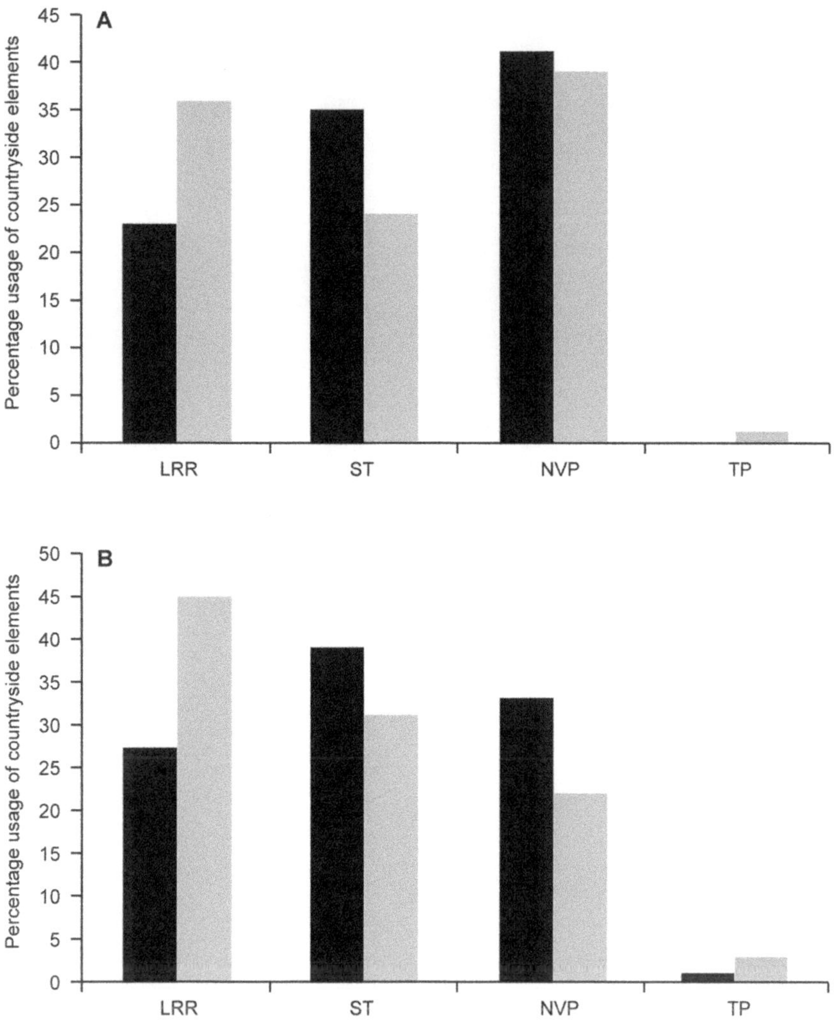

Figure 3. Actual usage verse the predicted usage of countryside elements. LRR = Linear roadside remnants, ST = Scattered trees, NVP = Native vegetation patches, TP = Tree plantings. Actual percentage usage is shown as black shading and the predicted percentage usage is shown as grey shading. Panel (a) is based on diurnal fixes, and Panel (b) is based on nocturnal fixes.

the study. Over a five month period, we tracked gliders to 1027 independent locations (655 diurnal and 372 nocturnal locations). We tracked individual gliders to an average of 21±1.16 (mean ± s.e.) diurnal locations and to 12±0.6 (mean ± s.e.) nocturnal locations.

Usage of countryside elements

We found that gliders used all four wooded countryside elements nocturnally and all but one element (*tree plantings*) for diurnal denning. Some individual gliders exclusively used one category of 'element', *scattered trees*, *linear roadside remnants* or *native vegetation* (Figure 2).

The fit of the model as defined above is summarised by the analysis of deviance (Table 2). As can be seen from Table 2, 64% (367/576) and 71% of the residual deviance was accounted for by the 'expected' count based on the percentage of available 'elements' for each animal, for night and day use respectively. The final residual deviance, based on 92 degrees of freedom after fitting the terms for animal, 'element' and expected frequency was 209 and 286, for night and day, respectively. This suggested that there remains non-random unexplained variation.

A further breakdown of the components of this residual variation revealed evidence of preferential selection of *native vegetation patches* and *scattered trees* nocturnally, and a strong preference to *scattered trees* for diurnal use, as is shown in Figure 3.

Relationship between home range size and availability of CSEs

Home ranges varied from 2.5 to 12 ha and averaged 4.9±0.45 ha (mean ±1 s.e.). Home range size increased with addition fixes, but plateaued towards the end of our study (Figure 4).

We found no significant relationship between total woody vegetation cover and home range size. However, there were significant relationships between home range size and area of woody vegetation within three 'elements'. We found a significant ($p<0.001$) negative relationship between home range size and the available area of *scattered trees* and significant ($p=0.009$ and $p=0.022$) positive relationships with the available area of *native vegetation patches* (Figure 5) and *linear roadside remnants*.

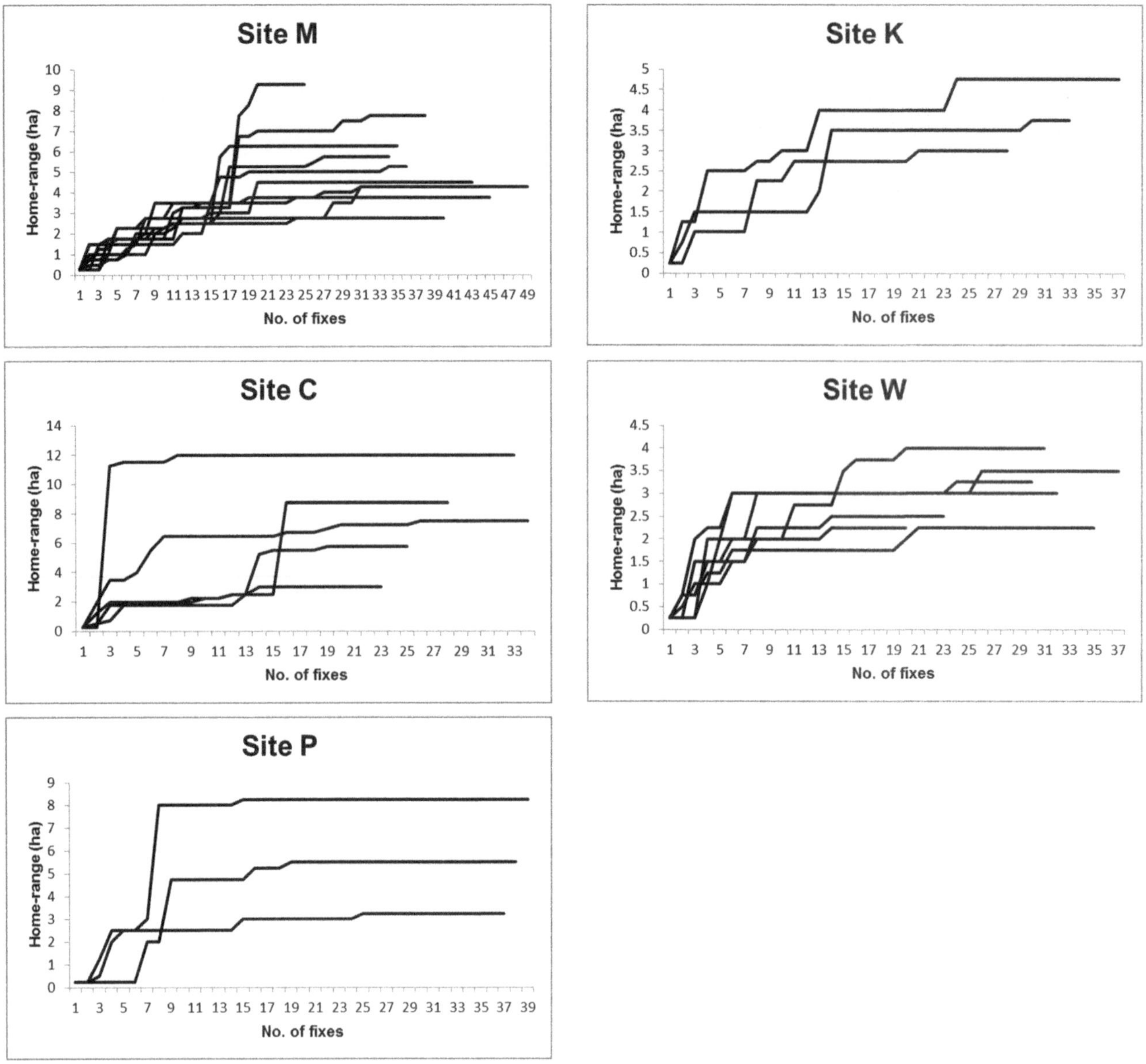

Figure 4. Home range estimates plotted against the number of fixes in sequential tracking order. The fixes are for all animals (with >20 fixes) at each of the five study sites.

Discussion

The contribution that countryside elements make towards biodiversity conservation is increasingly recognised in many ecosystems around the world [16–19]. The challenge for conservationists is to recognise different 'elements' and to understand the role each play in the conservation of different species. Our work is one of the few studies that quantify the relative value of different countryside elements for a single species. We provide empirical evidence of the disproportionate value of *scattered trees* in agricultural landscapes. Our work also highlights the need to examine the roles and values of countryside elements in biodiversity conservation, particular for species such as the squirrel glider which has a distribution largely confined to highly modified agricultural landscapes.

Across our five study sites, gliders relied entirely on countryside elements (*scattered trees*, *linear remnant vegetation*, *native vegetation patches* and *tree plantings*) located within road reserves, traveling stock reserves and freehold land. Our study demonstrates that the squirrel glider will use all of these four wooded 'elements', with availability being a key factor in determining usage.

In agricultural landscapes, roadside vegetation provides important habitat for many species of wildlife, particularly in facilitating migration and dispersal [20,46–48]. We found that the squirrel glider would commonly use *linear roadside remnants*, with some individuals relying entirely on this 'element'. The ability of *linear roadside remnants* to support squirrel glider populations has been previously reported [28]. While it is clear this kind of countryside element is of conservation importance to this species, we found evidence that suggests *linear roadside remnants* may provide inferior quality habitat compared *to native vegetation patches* and *scattered trees*. Larger home range sizes associated with increased area *of linear roadside remnants* and the under-utilisation of this 'element' is evidence of this. This may be explained by structural

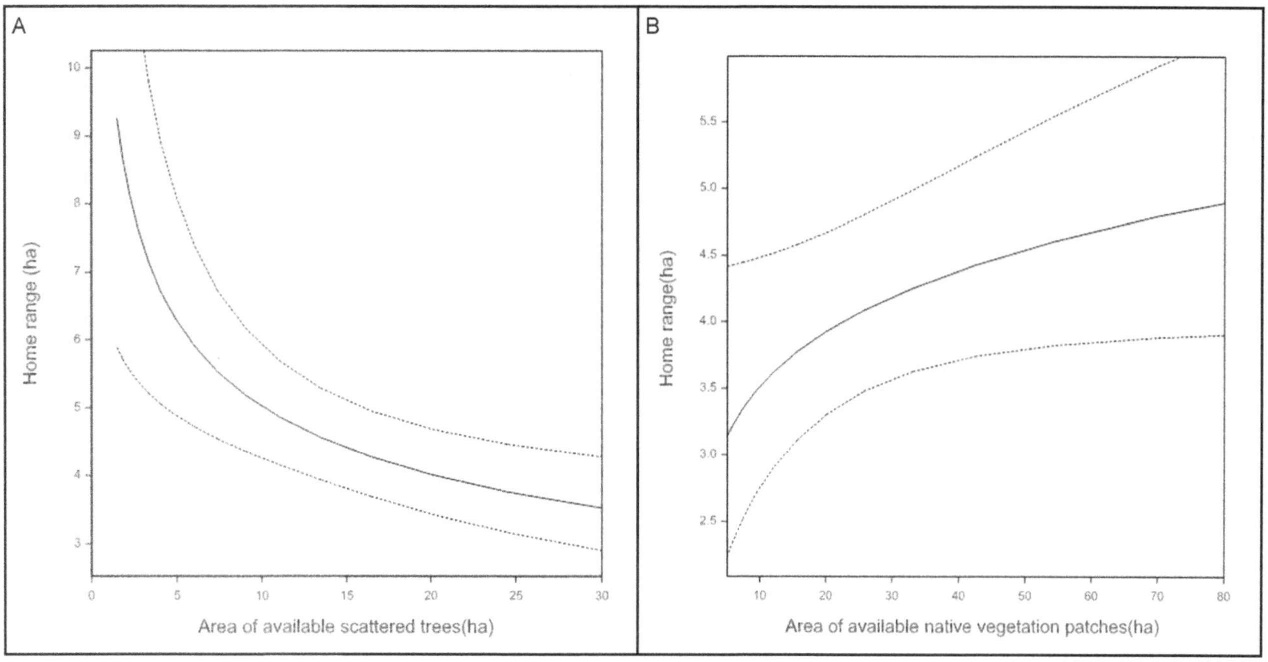

Figure 5. Influence of available area of three countryside elements on squirrel glider home range size. Panel (a) scattered trees, Panel (b) native vegetation patches. Based on 95% of fixes and associated 95% confidence intervals.

differences in the vegetation between the 'elements', such as a *linear roadside remnants* containing a higher proportion of small regrowth trees (which have been shown to offer poorer quality habitat than large trees [34,38,39]). Issues associated with the geometry of linear habitats also may explain our findings. Lindenmayer et al. [49] and Recher et al. [50] highlight the problems for species inhabiting linear habitats, such as disrupted social behaviour and additional expenditure of energy in obtaining food.

While *native vegetation patches* in agricultural landscapes are often small and highly fragmented, they have been shown to be important habitat for many species [24,51,52]. In our study, the majority (over 90%) of *native vegetation patches* occurred on traveling stock reserves. These reserves themselves have been shown to have high conservation value (higher than remnant vegetation on private land, particularly for arboreal marsupials) [53]. We found that *native vegetation patches* were heavily used by gliders, with a number of individuals using this 'element' exclusively. The use of *native vegetation patches* was higher than predicted for both diurnal and nocturnal activity. While *native vegetation patches* were preferentially used, gliders in areas dominated by *native vegetation patches* had larger home ranges. This result may suggest that while quality habitat exists within *native vegetation patches*, it is often dispersed among poorer habitats, resulting in gliders having to cover larger areas to gather resources.

Tree plantings have been shown to play an important role in the conservation of some species in agricultural landscapes, particularly birds [9,27]. For arboreal marsupial conservation, *tree plantings* may have a more limited role, at least in the short to medium term [54]. Our data indicated that *tree plantings* were used less than predicted based on availability. When compared to other treed 'elements', plantings would seem to be of minor importance to the species. However, *tree plantings* were nevertheless still used and should not be discounted as their value may increase over time as trees mature, given squirrel gliders' preference for large trees [34,38,39]. *Tree plantings* are the only countryside element examined here that can be readily introduced into these highly modified landscape. Our data suggests that for *tree plantings* to be used, they must be located in association with other countryside elements containing remnant trees.

Scattered trees are widely recognised as a critically important countryside element in many agricultural landscapes globally and are considered a keystone structure in such landscapes [23]. Numerous studies have highlighted the ecological value of *scattered trees* for various taxa of wildlife [12,16,25,26,52]. Fischer et al. [29] have shown that *scattered trees* have a disproportionate value for birds and bats. Our work provides further evidence of this. The home ranges of gliders were significantly smaller with increasing coverage of *scattered trees*, an opposite pattern to all other 'elements'. We also found that *scattered trees* were used at higher rates than predicted, particularly as diurnal den sites. Both results indicate that scattered trees have a comparatively higher habitat value for the squirrel glider than the other three kinds of countryside elements we examined.

Scattered trees are often the oldest living structures in disturbed landscapes [23]. This was the case in our study, with *scattered trees* represented by predominately large mature relictual *Eucalyptus* trees. These trees generally had well developed hollows and a large canopy, both kinds of resources have been shown to be preferentially sought by the squirrel glider [34,38,39]. Despite this, *scattered trees* are arguably under the greatest threat of all countryside elements and populations of such trees are rapidly diminishing resource in many agricultural landscapes [55,56]. It is estimated that tens of millions of *scattered trees* will be lost in grazed landscapes in Australia over the next 50 years, due to factors such as natural attrition, clearing, and a lack of regeneration [55] – a problem common to many agricultural landscapes globally [56]. Current conservation strategies offer little protection for *scattered trees* [55], and they are cleared for cropping and irrigation [57]. Without a concerted shift in conservation policy and awareness in general of the value of *scattered trees*, there is a real risk of losing this critical countryside element.

Conclusions

Agricultural landscapes are dynamic. Changing technologies, land uses and attitudes continually transform these systems [58]. There are significant threats to the integrity and extent of various kinds of countryside elements within agricultural landscapes. Therefore, understanding the contribution they make to biodiversity conservation is essential. In the case of the squirrel glider, we found that four kinds of countryside elements are important, but their relative values vary. We suggest that these countryside elements need to be better integrated into landscape management.

Acknowledgments

We thank the Hume Livestock Health and Pest Authority (formally Hume and Wagga Wagga Rural Lands Protection Boards), Wagga Wagga City and Greater Hume Shire (formerly Culcairn Shire Council) Councils, and private landholders for allowing us access to their properties, Cody Keys for assistance with trapping and radio tracking, and Damian Michael, Chris MacGregor, Ben MacDonald, Nicki Munro, and Jake Gillen for assistance with trapping. We thank our two wildlife veterinarians Karen Viggers and Arianne Lowe. We also thank Paul Adams (Academic Editor) and the two anonymous reviewers for their thoughtful contributions.

Author Contributions

Conceived and designed the experiments: MJC DBL RBC. Performed the experiments: MJC. Analyzed the data: MJC DBL RBC. Contributed to the writing of the manuscript: MJC DBL RBC.

References

1. Ellis CE (2011) Anthropogenic transformation of the terrestrial biosphere. Philos T R Soc B 369: 1010–1035.
2. Chapin FS III, Zavaleta ES, Eviner VT, Naylor RL, Vitousek PM, et al. (2000) Consequences of changing biodiversity. Nature 405: 234–242.
3. Aung UM (2007) Policy and practice in Myanmar's protected area system. J Environ Manage 84: 188–203.
4. Ioja CI, Patroescu M, Rozylowicz L, Popescu VD, Verghelet M, et al. (2010) The efficacy of Romania's protected areas network in conserving biodiversity. Biol Conserv 143: 2468–2476.
5. Pressey RL, Tully SL (1994) The cost of adhoc reservation: A case study in western New South Wales. Aust J Ecol 19: 375–384.
6. Rouget M, Richardson DM, Cowling RM (2003) The current configuration of protected areas in the Cape Floristic Region, South Africa - reservation bias and representation of biodiversity patterns and processes. Biol Conserv 112: 129–145.
7. Prober SM (1996) Conservation of the grassy white box woodlands: Rangewide floristic variation and implications for reserve design. Aust J Bot 44: 57–77.
8. Burel F (1996) Hedgerows and their role in agricultural landscapes. Critical Rev Plant Sci 15: 169–190.
9. Cunningham RB, Lindenmayer DB, Crane M, Michael DR, MacGregor C, et al. (2008) The combined effects of remnant vegetation and tree planting on farmland birds. Conserv Biol 22: 742–752.
10. Daily GC, Ceballos G, Pacheco J, Suzan G, Sanchez-Azofeifa A (2003) Countryside biogeography of neotropical mammals: Conservation opportunities in agricultural landscapes of Costa Rica. Conserv Biol 17: 1814–1826.
11. Hazell D, Cunnningham R, Lindenmayer D, Mackey B, Osborne W (2001) Use of farm dams as frog habitat in an Australian agricultural landscape: factors affecting species richness and distribution. Biol Conserv 102: 155–169.
12. Lumsden LF, Bennett AF (2005) Scattered trees in rural landscapes: foraging habitat for insectivorous bats in south-eastern Australia. Biol Conserv 122: 205–222.
13. Michael DR, Cunningham RB, Lindenmayer DB (2008) A forgotten habitat? Granite inselbergs conserve reptile diversity in fragmented agricultural landscapes. J Appl Ecol 45: 1742–1752.
14. Öckinger E, Smith HG (2007) Semi-natural grasslands as population sources for pollinating insects in agricultural landscapes. J Appl Ecol 44: 50–59.
15. van der Ree R, Bennett AF (2003) Home range of the squirrel glider (*Petaurus norfolcensis*) in a network of remnant linear habitats. J Zool 259: 327–336.
16. Haslem A, Bennett AF (2008) Countryside elements and the conservation of birds in agricultural environments. Agr Ecosyst Environ 125: 191–203.
17. Daily GC, Ehrlich PR, Sanchez-Azofeifa GA (2001) Countryside biogeography: use of human dominated habitats by the avifauna of southern Costa Rica. Ecol Appl 11: 1–13.
18. Renjifo LM (2001) Effect of natural and anthropogenic landscape matrices on the abundance of sub-Andean bird species. Ecol Appl 11: 14–31.
19. Söderström B, Kiema S, Reid RS (2003) Intensified agricultural landuse and bird conservation in Burkina Faso. Agr Ecosyst Environ 99: 113–124.
20. Bennett AF (1999) Linkages in the landscape: the role of corridors and connectivity in wildlife conservation. Gland, Switzerland: IUCN.
21. Fischer J, Lindenmayer DB (2002) Small patches can be valuable for biodiversity conservation: two case studies on birds in southeastern Australia. Biol Conserv 106: 129–136.
22. Forman RTT (1995) Land Mosaics: The Ecology of Landscapes and Regions. New York: Cambridge University Press.
23. Manning AD, Fischer J, Lindenmayer DB (2006) Scattered trees are keystone structures - implications for conservation. Biol Conserv 132: 311–321.
24. Pacheco R, Vasconcelos HL, Groc S, Camacho GP, Frizzo TLM (2013) Importance of remnants of natural vegetation for maintain ant diversity in Brazilian agricultural landscapes. Biodivers Conserv 22: 983–997.
25. Fischer J, Lindenmayer DB (2002) The conservation value of paddock trees for birds in a variegated landscape in southern New South Wales. 1. Species composition and site occupancy patterns. Biodivers Conserv 11: 807–832.
26. Manning AD, Lindenmayer DB, Barry SC (2004) The conservation implications of bird reproduction in the agricultural "matrix": a case study of the vulnerable superb parrot of south-eastern Australia. Biol Conserv 120: 363–374.
27. Munro N, Fischer J, Barrett G, Wood J, Leavesley A, et al. (2011) Bird's response to revegetation of different structure and floristics – are "restoration plantings" restoring bird communities? Restor Ecol 19: 223–235.
28. van der Ree R (2002) The population ecology of the squirrel glider (Petaurus norfolcensis) within a network of remnant linear habitats. Wildlife Res 29: 329–340.
29. Fischer J, Stott J, Law BS (2010) The disproportionate value of scattered trees. Biol Conserv 143: 1564–1567.
30. McLoughlin PD, Cluff HD, Gau RJ, Mulders R, Case RL, et al. (2003) Effect of special differences in habitat on home ranges of grizzly bears. Ecoscience 10: 11–16.
31. Mitchell MS, Powell RA (2004) A mechanistic home range model for optimal use of spatially distributed resources. Ecol Model 15: 209–232.
32. van Beest FM, Rivrud IM, Loe LE, Milner JM, Mysterud A (2011) What determines variation in home range size across spatiotemporal scales in a large browsing herbivore. J Anim Ecol 80: 771–785.
33. Ball T, Adams A, Goldingay RL (2009) Diet of the squirrel glider in a fragmented landscape near Mackay, central Queensland. Aust J Zool 57: 295–304.
34. Holland GJ, Bennett AF, van der Ree R (2007) Time-budget and feeding behaviour of the squirrel glider (*Petaurus norfolcensis*) in remnant linear habitat. Wildlife Res 34: 288–295.
35. Sharpe DJ, Goldingay RL (1998) Feeding behaviour of the squirrel glider at Bungawalbin Nature Reserve north-eastern New South Wales. Wildlife Res 25: 243–254.
36. Anderson RH (1961) Introduction, flora series Nos. 1–8: 1–6. Contributions from the New South Wales National Herbarium. Sydney: Royal Botanic Gardens.
37. Benson JS (2008) New South Wales vegetation classification and assessment: Part 2 Plant communities of the NSW South-western Slopes bioregion and update of the NSW Western Plains plant communities, version 2 of the NSWVCA database. Cunninghamia 10: 599–673.
38. Crane M, Montague-Drake RM, Cunningham RB, Lindenmayer DB (2008) The characteristics of den trees used by the Squirrel Glider (*Petaurus norfolcensis*) in temperate Australian woodlands. Wildlife Res 35: 663–675.
39. Crane MJ, Lindenmayer DB, Cunningham RB (2012) Use and characteristics of nocturnal habitats of the squirrel glider (*Petaurus norfocensis*) in Australian temperate woodlands. Aust J Zool 60: 320–329.
40. Martin JK, Handasyde KA, Taylor AC (2007) Linear roadside remnants: Their influence on the den-use, home range and mating system in bobucks (*Trichosurus cunninghami*). Austral Ecol 32: 686–696.
41. White GC, Garrott RA (1990) Analysis of Wildlife Radio-tracking Data. San Diego, USA: Academic Press.
42. van der Ree R, Bennett AF, Gilmore DC (2004) Gap-crossing by gliding marsupials: thresholds for use of isolated woodland patches in an agricultural landscape. Biol Conserv 115: 241–249.
43. Spooner PG (2005) On squatters, settlers and early surveyors: Historical development of country road reserves in southern New South Wales. Australian Geographer 36: 55–73.
44. Stirzaker R, Vertessey R, Sarre A, editors (2002) Trees, Water and Salt. An Australian Guide to Using Trees for Healthy Catchments and Productive Farms. Canberra: Joint Venture Agroforestry Program.
45. McCullagh P, Nelder JA (1989) Generalised Linear Models. New York: Chapman and Hall.

46. Arnaud J (2003) Metapopulation genetic structure and migration pathways in the land snail *Helix aspersa*: influence of landscape heterogeneity. Landscape Ecol 18: 333–346.
47. Cummings JR, Vessey SH (1994) Agricultural influences on the movement patterns of white-footed mice (*Peromyscus leucopus*). Am Midl Nat 132: 209–218.
48. Getz LL, Cole FR, Gates DL (1978) Interstate roadside as dispersal routes for *Microtus pennnsylvanicus*. J Mammal 59: 208–212.
49. Lindenmayer DB, Cunningham RB, Donnelly CF (1993) The conservation of arboreal marsupials in the montane ash forests of the Central Highlands of Victoria, south-east Australia: IV. The presence and abundance of arboreal marsupials in retained linear habitats (wildlife corridors) within logged forest. Biol Conserv 66: 207–221.
50. Recher HF, Shields J, Kavanagh RP, Webb G (1987) Retaining remnant mature forest for nature conservation at Eden, New South Wales: A review of theory and practice. In: Saunders DA, Arnold GW, Burbidge AA, Hopkins AJM, editors. Nature Conservation: the Role of Remnants of Vegetation. Chipping Norton, N.S.W.: Surrey Beatty and Sons. pp.177–194.
51. Collard S, Le Broque A, Zammit C (2009) Bird assemblages in fragmented agricultural landscapes: the role of a small brigalow remnants and adjoining land uses. Biodivers Conserv 18: 1649–1670.
52. Fischer J, Lindenmayer DB (2002) The conservation value of paddock trees for birds in a variegated landscape in southern New South Wales. 2. Paddock trees as stepping stones. Biodivers Conserv 11: 833–849.
53. Lindenmayer DB, Cunningham RB, Crane M, Montague-Drake R, Michael D (2010) The importance of temperate woodland in travelling stock reserves for vertebrate biodiversity conservation. Ecol Manage Restor 11: 27–30.
54. Cunningham RB, Lindenmayer DB, Crane M, Michael D, MacGregor C (2007) Reptile and arboreal marsupial response to replanted vegetation in agricultural landscapes. Ecol Appl 17: 609–619.
55. Gibbons P, Lindenmayer DB, Fischer J, Manning AD, Weinberg A, et al. (2008) The future of scattered trees in agricultural landscapes. Conserv Biol 22: 1309–1319.
56. Lindenmayer DB, Laurance W, Franklin WF, Likens GE, Banks SC, et al. (2014) New policies for old trees: averting a global crisis in a keystone ecological structure. Conserv Lett 7: 61–69.
57. Maron M, Fitzsimons JA (2007) Agricultural intensification and loss of matrix habitat over 23 years in the West Wimmera, south-eastern Australia. Biol Conserv 135: 587–593.
58. Lindenmayer DB, Cunningham SA, Young A, editors (2012) Land Use Intensification: Effects on Agrioculture, Biodiversity and Ecological Processes. Melbourne: CSIRO Publishing.

Alfalfa (*Medicago sativa* L.)/Maize (*Zea mays* L.) Intercropping Provides a Feasible Way to Improve Yield and Economic Incomes in Farming and Pastoral Areas of Northeast China

Baoru Sun, Yi Peng, Hongyu Yang, Zhijian Li*, Yingzhi Gao*, Chao Wang, Yuli Yan, Yanmei Liu

Key Laboratory of Vegetation Ecology, Northeast Normal University, Changchun, China

Abstract

Given the growing challenges to food and eco-environmental security as well as sustainable development of animal husbandry in the farming and pastoral areas of northeast China, it is crucial to identify advantageous intercropping modes and some constraints limiting its popularization. In order to assess the performance of various intercropping modes of maize and alfalfa, a field experiment was conducted in a completely randomized block design with five treatments: maize monoculture in even rows, maize monoculture in alternating wide and narrow rows, alfalfa monoculture, maize intercropped with one row of alfalfa in wide rows and maize intercropped with two rows of alfalfa in wide rows. Results demonstrate that maize monoculture in alternating wide and narrow rows performed best for light transmission, grain yield and output value, compared to in even rows. When intercropped, maize intercropped with one row of alfalfa in wide rows was identified as the optimal strategy and the largely complementary ecological niches of alfalfa and maize were shown to account for the intercropping advantages, optimizing resource utilization and improving yield and economic incomes. These findings suggest that alfalfa/maize intercropping has obvious advantages over monoculture and is applicable to the farming and pastoral areas of northeast China.

Editor: Wen-Xiong Lin, Agroecological Institute, China

Funding: This work was supported by the National Natural Science Foundation of China (NSFC, 31072080 and 31270444), the program for New Century Excellent Talents in University (NCET-13-0717), Key Science and Technology Program of Jilin Province (20100212, 2012ZDGG008), National Key Technology R&D Program (2011BAD17B04-3-2), and the National Program on Key Basic Research Project (2012CB722202). The funders had no role in study design, data collection and analysis, decision to publish, or preparation of the manuscript.

Competing Interests: The authors have declared that no competing interests exist.

* Email: lizj004@nenu.edu.cn (ZL); gaoyz108@nenu.edu.cn (YG)

Introduction

The farming and pastoral area (FPA) of northeast China (NEC) is an agriculture-based ecozone combining forestry and animal husbandry, and it is an important grain commodity and animal husbandry base. However, it is also a vulnerable eco-environmental zone owing to low vegetation cover, fine sandy soil and strong wind [1–2]. Wind erosion, water erosion and unsustainable production activities (e.g., single cropping and multiple-year continuous cropping) have made the land subject to dust storms in winter and spring. Moreover, cropland soil is being increasingly eroded, causing low soil fertility and reduced crop productivity and quality [3–5]. Together with the growing use of agricultural chemicals, such as fertilizers and herbicides, sustainable agricultural development is now facing many serious challenges [6].

In addition to the cropland, grassland has been degraded due to overgrazing and excessive agricultural reclamation, and grassland degradation is reflected by the reduced grassland production and forb quality and low carrying capacity [7]. "Grain for green" initiatives including reseeding of forage grasses have long been considered as effective ways to restore grassland vegetation and help to balance the ecological system [8]. Moreover, with increasing demand for meat products and high quality forage grass, China continues to give substantial support and increased financial investment to the development of the forage industry. Additionally, the Chinese government has strongly endorsed a proposal for boosting the development of the alfalfa industry so as to ensure the production, processing and sustainable supply of high quality forage grass [9]. Therefore, in the context of food and eco-environmental security and animal husbandry sustainable development, traditional farming patterns in the northeast FPA are being altered; with a tendency to adjust agricultural structure, introduce forage grass into the main crop farming system and establish an intercropping pattern between crops and forage grass, resulting in efficient resource utilization, a friendly ecological environment and good economic benefits [10–11].

Intercropping, the practice of growing two or more crops in proximity, is advantageous due to the differences in ecological characteristics and growth of the intercropped varieties. This can establish a composite population, producing complementary effects and increasing yield and economic incomes per unit area [12]. Furthermore, intercropping can improve soil fertility, alleviate disease and insect harm, and inhibit the growth of weeds [13–16]. Maize, as a principal crop of the northeast FPA, is an important food and forage crop. Its grain is an important fodder with high energy, known as "queen feed" [17]. Alfalfa is a

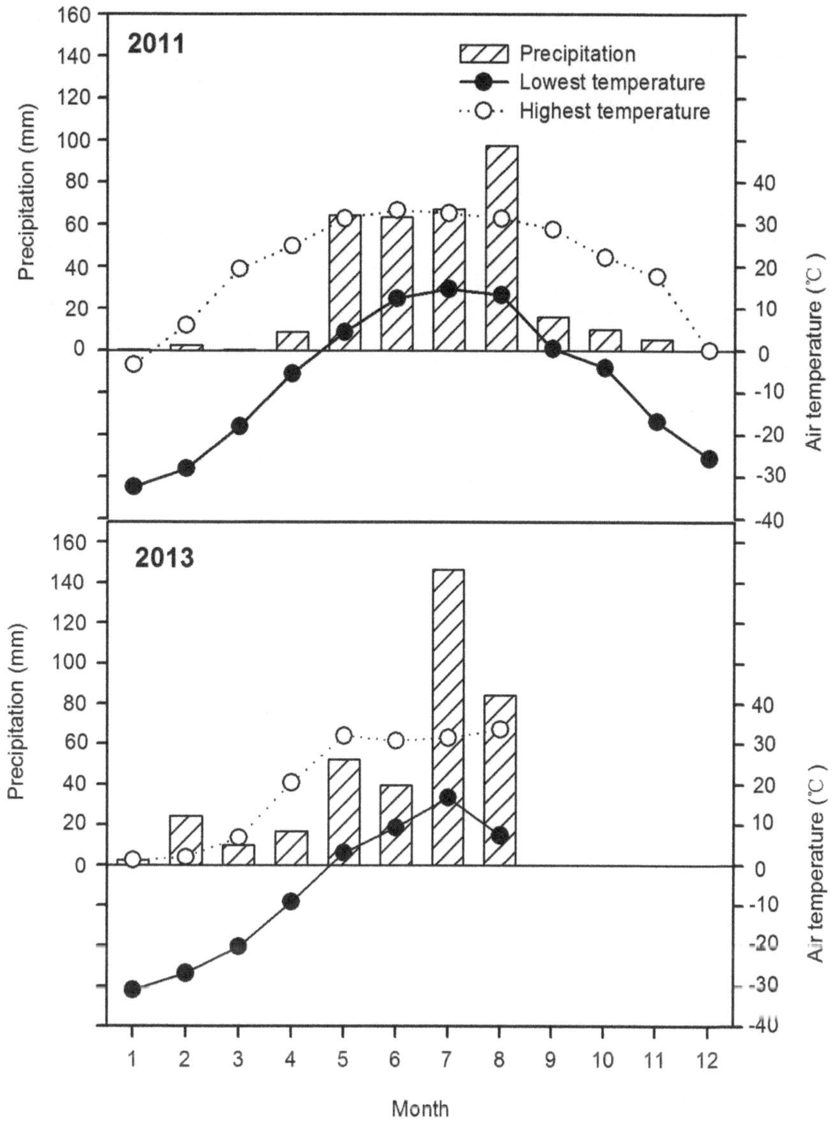

Figure 1. Monthly precipitation (bar) and air temperature (curve) of the experimental site in 2011 and 2013.

leguminous forb that has been prioritized in the development of the PFA of NEC; suitable because of its high yield, rich content and high quality of protein, abundant vitamins and minerals, good palatability and high digestibility [18]. Moreover, as a perennial, alfalfa supplies soil cover throughout the year; providing wind resistance, fixing soil and improving the environment of the planting area. Its large root system can significantly improve soil fertility and physico-chemical properties, leading to the win-win relationship between utilization and conservation [19]. Therefore, it is predictable that intercropping alfalfa with maize can not only guarantee regional food security and meet the nutritional requirements of forage industry, but also provide eco-environmental protections, and it is a promising cropping pattern in the future development of this region.

In China, maize has been traditionally cultivated in even rows. More recently, in attempts to improve maize yield agricultural scientists have experimented with alternative cultivation strategies. Indeed, a study has shown that in the main agricultural region of Jilin Province, planting maize in alternating wide and narrow rows can achieve up to 10% improved yield over the conventional pattern [20]. Previous intercropping studies focused mainly on food crop combinations [21–22], but few studies have investigated the possibility of intercropping alfalfa with maize, a strategy combining annual food crop with perennial forage crop together. Those studies that have addressed alfalfa/maize intercropping focused on improving group yield, forage quality, soil fertility and the environment [23–24]. To our knowledge, no study has investigated whether intercropping alfalfa with maize in alternating wide and narrow rows could provide sustained high yield and economic incomes while taking investment and environmental factors into account, and whether there are some constraints limiting its popularization such as crop management and the acceptability of local farmers. Hence, a field experiment was conducted to explore the following questions: (1) Compared to an even row approach, does an alternating wide and narrow row planting approach in maize improve yield in the FPA of NEC? (2) Can intercropping alfalfa with maize provide sustained high yield and economic incomes? Which intercropping mode is best, considering farmer incomes and land management?

Materials and Methods

Experimental site

The study was conducted between 2007 and 2013 at the Grassland Ecosystem Field Station of the Northeast Normal University at Songnen Grassland (123° 44′ E and 44° 40′ N, 137.8–144.8 masl), a typical FPA of NEC. This area is characterized by a semi-arid and temperate continental monsoon climate with a mean annual temperature of 4.6–6.4°C, an annual accumulated temperature (≥10°C) of 2546–3375°C, mean annual precipitation of 300–400 mm (86% of precipitation occurring from May to September) and mean annual evaporation of 1500–2000 mm. The frost-free period lasts approximately 140 days, from the end of April to early October. The two experimental years contrasted each other in terms of precipitation. In 2011, the annual total precipitation was 335 mm and mostly occurred in the growing season (308 mm), whereas the year 2013 was a year with a higher precipitation amount (376 mm from January to August), better seasonal distribution and pronounced peak in July. Air temperature in 2011 and 2013 showed a similar dynamic with the maximum and minimum air temperatures of 33°C and −32°C, respectively (Figure 1). The soil type at the site is light chernozem with deep soil layers. The plough layer consists of organic C (17.24±1.76 g kg^{-1}), total N (0.98±0.15 g kg^{-1}), rapidly available P (5.88±0.65 mg kg^{-1}) and rapidly available K (140.70±11.75 mg kg^{-1}), with an initial soil pH of 7.46±0.04.

Experiment materials and design

Alfalfa variety *Medicago sativa* L. cv. Dongmu No. 1 was used throughout this study. This variety was bred by Northeast Normal University to adapt readily to drought and cold and is now the principally cultivated variety in the study area. Alfalfa generally turns green in mid-April, continues to grow until end of October and can be harvested three times per year. The hay yield can reach up to 7500–10,000 kg ha^{-1} in the second year under the rainfed condition. *Zea mays* L. cv. Zhengdan 958 was chosen as the maize test variety. This variety matures at about 128 days and is widely cultivated by local farmers due to the high and stable yield and a great resistance to lodging and disease. The differences of alfalfa and maize in growth dynamics (Figure S4) make them easy to form temporal and spatial complementarity and promote the efficient utilization of light, water and nutrients.

At the beginning of the experiment, the whole field was fully ploughed to ensure uniform soil conditions. The experiment was conducted in a completely randomized block design with four blocks each containing five cropping patterns: maize monoculture in even (65 cm) rows (MME); maize monoculture in alternating wide (90 cm) and narrow rows (40 cm) (MMW); alfalfa monoculture in even (30 cm) rows (MA); MMW intercropped with alfalfa, with one row of alfalfa in the wide rows (23.1% alfalfa in intercropping area) (IMA1); MMW intercropped with alfalfa, with two rows of alfalfa in the wide rows (46.2% alfalfa in intercropping area) (IMA2). In the maize planting patterns, every three rows of maize were defined as one belt (1.3 m wide) with three belts in each treatment. Each plot had an area of 46.8 m^2 (3.9 m × 12 m), with 50 cm spacing between each plot and 1 m separating each block.

To establish the intercropping system, alfalfa was sown in early July 2007 at a seeding rate of 15 kg ha^{-1} and had been allowed to grow for 4 years before data collection in 2011. Every year maize was sown in early May with 26 cm separating each plant, and was irrigated with 75 mm before its sowing to ensure good germination. In all planting patterns, 135 kg P ha^{-1} and 90 kg K ha^{-1} were applied for alfalfa, and 225 kg N ha^{-1}, 120 kg P ha^{-1} and 60 kg K ha^{-1} were applied for maize. The commercial fertilizer used were: nitrogen, urea (46% nitrogen content); phosphate, diammonium phosphate (46% phosphorus content, 18% nitrogen content); potash, potassium chloride (60% potassium content). All fertilizers required by alfalfa were spread in the soil at the time of sowing. For maize, all of the phosphate and potash, and half of the nitrogen fertilizers were spread in the soil at the time of sowing and the remaining nitrogen fertilizers applied during the big flare opening period. Weeds were regularly controlled using a hand hoe, and pest and disease of alfalfa or maize were separately controlled timely with the idea of minimizing the pesticide application effects on the non-target crop.

Data collection

The data was collected from 2011. Unfortunately, there was considerably small snowfall and the air temperature was relatively high in the winter of 2011 (Figure 1), which weakened alfalfa resistance to cold and freezing [25]. In the early spring of 2012, the sprout of alfalfa was promoted due to the continuously high air temperature from 21th to 30th in March with the maximum 16.8°C, whereas the air temperature dramatically decreased from 31th March to 6th in April with the minimum −9.0°C (Figure S1). This unexpected cold snap made a serious freezing injury to the sprouting alfalfa due to its lowest cold resistence at that time [26]. Consequently, alfalfa achieved a low turning green rate. In order to ensure the sustainable production of alfalfa, we stopped data collection in 2012 and restarted sampling in 2013 when alfalfa turned a good recovery in its growth and development.

Light intensity was determined using a ST-80C illuminometer (Photoelectric Instrument Factory of Beijing Normal University, China) in 2011. Four layers for maize or alfalfa were selected in each treatment: (1) the reference layer, above the canopy; (2) the bottom layer, close to the soil surface; (3) the intermediate layer, at the point of 1/2 plant height and (4) the top layer, 10 cm below the top of the plant. Based on the light intensity, light transmission was calculated for each of the layers. Maize and alfalfa leaf area index (LAI) was measured using a LAI-2000 plant canopy analyzer (LI-COR, Inc., USA) in 2011 and 2013. From the first flowering stage of alfalfa, both light intensity and LAI were tested from 10:00–11:00 on a sunny day. The measurement was repeated every 15 days at three different positions within each testing belt.

Using time domain reflectometry (TDR 100, Campbell Scientific Inc., Logan, Utah, USA), soil water content (SWC) at depth of 0–20 cm was measured four times for all the treatments in each year (2011: 10th June, 24th July, 10th September and 4th October; 2013: 6th June, 12th July, 26th August and 30th September). The measuring time was corresponding to different developing stages of crop, that was the first, second and third flowering stage of alfalfa and the maturity stage of maize, respectively. Specifically, SWC was tested at ten different representative positions between the alfalfa or maize rows in the monoculture treatments for each plot. As to intercropping treatments, SWC was measured between the alfalfa rows (for IMA2), between the alfalfa and maize rows and between the maize rows with ten different representative positions in each belt of plots and then all measured data were averaged as the SWC condition of the intercropping system.

The final harvest of maize was taken in early October (2011: 5th October; 2013: 1th October). In each planting pattern, the second belt was selected for grain yield determination. Fresh weight was recorded before maize grains were oven-dried at 65°C to a constant weight and dry weight recorded. Water content and grain yield were calculated based on dry weight. Alfalfa was cut three times (2011: 11th June, 25th July and 11th September; 2013: 7th

June, 13th July and 27th August). For each alfalfa cut, fresh weight and dry weight were determined, and water content and hay yield calculated.

Data calculations

Light transmission. Light transmission (LT) was calculated using equation (1);

$$LT_m = \frac{LI_m}{LI_a} \tag{1}$$

where m is the canopy layer (including the top, intermediate and bottom layers) of either alfalfa or maize, LT_m is the light transmission at layer m, LI_m is the light intensity at layer m, and LI_a is the light intensity above the canopy [27].

Output value per unit area. The output value per unit area (OVPUA) was calculated according to equation (2);

$$OVPUA = P_m \times Y_m + P_a \times Y_a \tag{2}$$

where for each planting pattern, P_m and P_a denote the price of maize grain and alfalfa hay respectively. The price expressed in USD is based on the exchange rate of 630 ¥ 100 USD^{-1} in 2011 and 613 ¥ 100 USD^{-1} in 2013, thus P_m and P_a are respectively 361.90 USD t^{-1} and 380.95 USD t^{-1} in 2011 and 342.58 USD t^{-1} and 358.89 USD t^{-1} in 2013 based on local market values; Y_m and Y_a denote the yield of maize grain and alfalfa hay respectively [28].

Land equivalent ratio. Land equivalent ratio (LER) is an index that adopts yield as a comparison parameter to evaluate land use efficiency of different cultivated patterns relative to monoculture. However, the value of LER is not necessarily related to yield. The equation is defined as follows;

$$LER = (Y_{ia}/Y_{ma}) + (Y_{im}/Y_{mm}) \tag{3}$$

where Y_{ia} and Y_{im} are the respective yields of alfalfa and maize in the total intercropped area, and Y_{ma} and Y_{mm} are the yields of monocultured alfalfa and maize. An LER greater than 1.0 reveals an intercropping advantage and the favors of intercropping on crops growth and yield, while an LER less than 1.0 indicates an intercropping disadvantage and the negative affections of intercropping on crops growth and yield [29–31].

Aggressivity. Aggressivity (A_{ac}) measures the relative resource competitiveness of two intercropped species;

$$A_{ac} = Y_{ia}/(Y_{ma} \times P_a) - Y_{im}/(Y_{mm} \times P_c) \tag{4}$$

where A_{ac} is the aggressivity of alfalfa relative to maize in the intercropping system, P_a and P_c are the intercropping area proportions occupied by alfalfa and maize respectively, while the meanings of other symbols are the same as those used for the LER equation. If A_{ac} is greater than 0, the competitive ability of alfalfa exceeds that of maize in intercropping; otherwise, maize has greater competitiveness [12,23].

Statistical analysis

Normal distribution and homogeneous variances were tested for all the data with Shapiro-Wilk test [32] using SPSS 17.0 software (SPSS Inc., Chicago, IL, USA), and light transmission was biquadrate after reciprocal transformed to achieve normal distribution. One-way ANOVA was performed to examine the effects of cropping patterns on light transmission and soil water content (SWC). Repeated measures ANOVA in a general linear model (GLM) were conducted to assess the effects of planting modes on LAI, yield and output value per unit area (OVPUA), with year and alfalfa flowering stage as the repeated measures. The results were reported using the Greenhouse-Geisser correction when Mauchly's test of sphericity was violated. If the interaction between factors was significant, one-way ANOVA was conducted to evaluate the effects of cropping pattern or alfalfa flowering stage and significant differences of means were compared with Duncan's multiple-compare range test; while the effects of year were tested by independent-samples t test. Significant level was set at $P < 0.05$.

Results

Maize light transmission

With the exception of the first flowering stage of alfalfa (early June), no significant difference was found in light transmission at the top of maize among treatments (Figure 2A). At the stage of the first flowering of alfalfa, light transmission of monoculture maize (MME and MMW) was significantly higher than that of intercropped maize (IMA1 and IMA2) ($P < 0.0001$), whereas there was no significant difference between monoculture modes or between intercropping modes. This can be accounted for by the fact that during flowering, alfalfa grew taller than maize, which had an overshadowing effect. On the whole, however, no significant difference was observed for average light transmission at the top of maize, regardless of cropping pattern (Figure 2A inset).

During continuous growth of maize, there was a tendency for a reduction in both intermediate and bottom light transmission for all treatments. This was particularly evident before maize entered the big flare opening stage (24th July), with little change in light transmission occurring after this point (Figure 2B and C). In the first flowering stage of alfalfa, both intermediate and bottom light transmission of monoculture maize (MME and MMW) were higher than for intercropped maize (IMA1 and IMA2); however, the opposite pattern was observed in the resting period, when light transmission of IMA1 and IMA2 maize was higher than monoculture maize (Figure 2B and C). In addition, there were significant differences for intermediate and bottom light transmission between intercropping and monoculture maize at the second alfalfa flowering stage (24th July) (intermediate: $P < 0.0001$; bottom: $P = 0.003$) and after the second cutting of alfalfa (8th August) (intermediate: $P = 0.002$; bottom: $P = 0.005$). Significant differences were also found for intermediate and bottom light transmission between intercropped maize (IMA1 and IMA2) at the second flowering stage of alfalfa, but differences between monocultures of maize were never significant. However, for maize intermediate or bottom canopy layers, no significant difference in average light transmission was observed, regardless of planting strategy (Figure 2A inset).

Alfalfa light transmission

Throughout the alfalfa growing season, there was no evident variance in top light transmission in any treatment, whereas the intermediate and bottom light transmission dynamics of all treatments showed a bimodal curve (Figure 2D, E and F), which can be attributed to alfalfa being cut twice (11th June and 25th July). Before cutting, alfalfa was flowering and therefore the canopy had high closure and lower light transmission. Upon

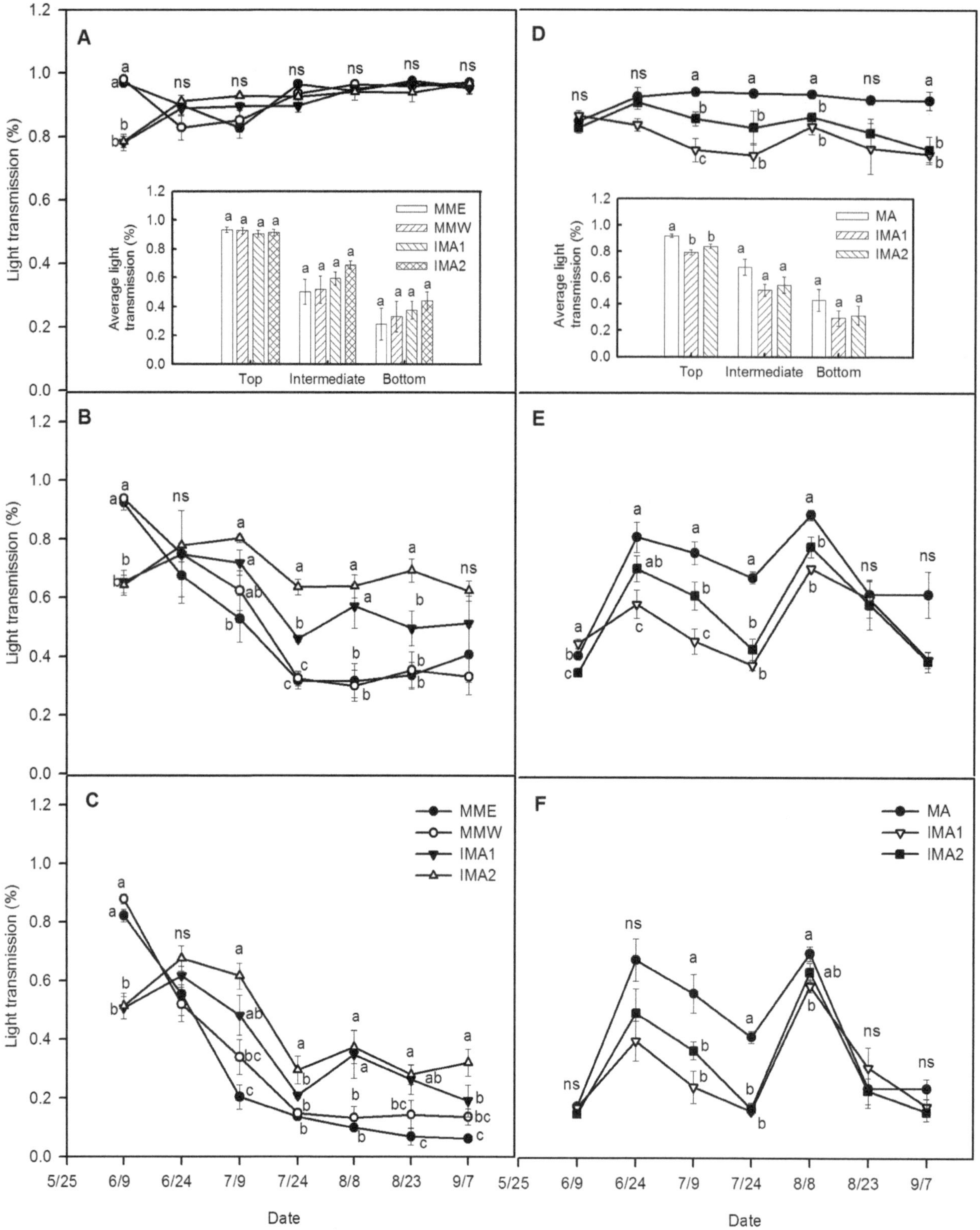

Figure 2. Light transmission dynamics of maize and alfalfa at different layers under monoculture and intercropping. The inset figures show the average light transmission of maize and alfalfa in different planting patterns during the vegetation period. A = top light transmission of maize, B = intermediate light transmission of maize, C = bottom light transmission of maize, D = top light transmission of alfalfa, E = intermediate light transmission of alfalfa, F = bottom light transmission of alfalfa. MME = monoculture maize in even rows, MMW = monoculture maize in alternating wide and narrow rows, MA = alfalfa monoculture, IMA1 = maize intercropped with one row of alfalfa in the wide rows, IMA2 = maize intercropped with two rows of alfalfa in the wide rows. Different letters for the same date indicate significant difference at $P < 0.05$ probability level, and ns represents no difference between treatments. Values = means ± SE.

cutting, light transmission increased. Subsequent alfalfa regrowth produced new canopy closure and decreased light transmission.

With the exception of the first, second and sixth measuring times, the top layer light transmission of monoculture alfalfa (MA) was significantly higher than that of intercropped alfalfa at all recorded time points (third: $P = 0.003$; forth: $P < 0.0001$; fifth: $P = 0.004$; seventh: $P = 0.010$). The season average light transmission of MA was significantly higher than that of intercropped alfalfa. There was no significant difference between the two intercropping modes (Figure 2D and inset). In the intermediate and bottom layers, light transmission showed complex patterns with time. In the intermediate layer, light transmission of MA was significantly higher than in the two intercropping models at the third, fourth and fifth measuring times (third: $P = 0.003$; forth: $P < 0.0001$; fifth: $P = 0.003$). With the exception for the first and third measuring times, there was no significant difference between the light transmissions of the two intercropping modes (Figure 2E). In the bottom layer, for the third and fourth measuring times, light transmission of MA was significantly higher than the two intercropping modes (third: $P = 0.006$; forth: $P < 0.0001$) and light transmission of the two intercropping modes showed no significant difference (Figure 2F). For intermediate and bottom layers, there were no significant differences in the season average light transmission among treatments (Figure 2D inset).

Soil water content (SWC)

SWC of all cropping patterns displayed a strong seasonal dynamic, with a peak in July and August (Figure 3). Irrespective of the growing stage, the difference of SWC between MME and MMW was not significant in both 2011 and 2013. Compared to monoculture, intercropping significantly reduced the SWC, and it was more evident in 2013 than 2011 (Figure 3). In 2011, with the exception of 24th July and 4th October, the SWC of IMA1 and IMA2 was significantly decreased compared to MMW, but with no significant difference compared with MA (both: $P < 0.0001$) (Figure 3A); while in 2013, the SWC of IMA1 and IMA2 was significantly lower than that of MMW as well as MA except for 12th July (all: $P < 0.0001$) (Figure 3B). The differences between treatments were also reflected by seasonal average SWC: the values of IMA1 and IMA2 were significantly lower than that of MMW and MME in both 2011 and 2013, while there was no significant difference between IMA1 and IMA2 in both years. For the MA treatment, the seasonal average SWC was significantly lower than that of MMW and MME in 2011, while there was no significant difference among treatments MMW, MME and MA in 2013 (Figure 3 3A inset and 3B inset).

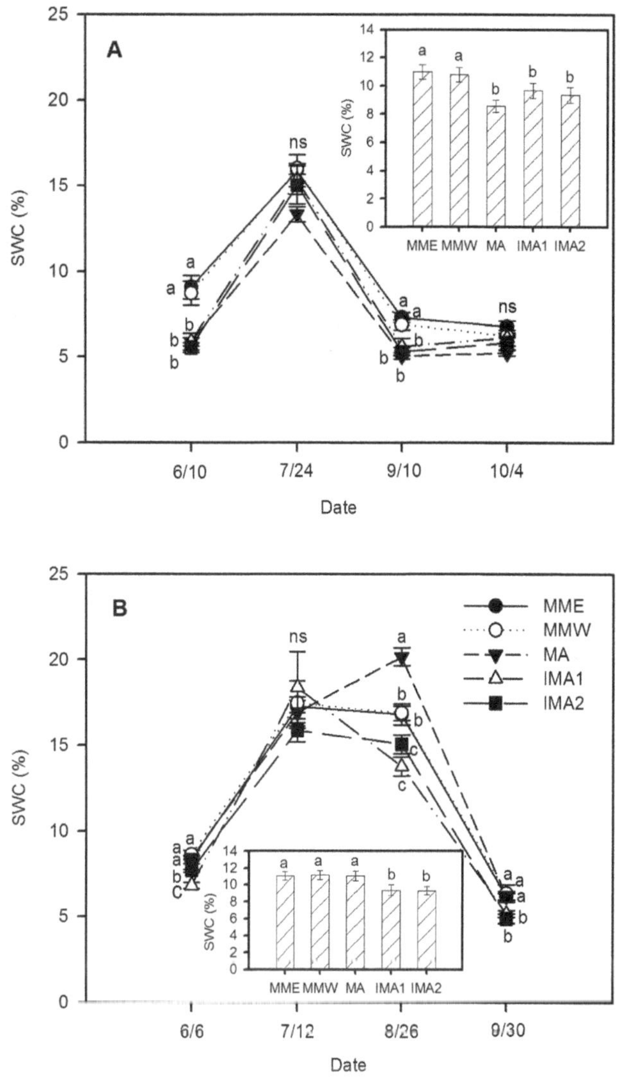

Figure 3. Soil water content comparisons of different cropping patterns in 2011 (A) and 2013 (B). The inset figures show the average soil water content in different planting patterns during the vegetation period. SWC = soil water content. The other symbols are the same as for Figure 2.

Leaf area index (LAI)

There was no significant difference in maize LAI between MME and MMW (Figure 4). Compared to MMW, the LAI of intercropped maize (IMA1 and IMA2) was significantly reduced ($P < 0.0001$), and no significant difference was observed between the two intercropping patterns (Table 1; Figure 4). Regarding alfalfa LAI, the values of IMA1 and IMA2 were significantly higher than that of MA ($P = 0.009$), but with no significant difference between IMA1 and IMA2 (Table 1; Figure 5A). Meanwhile, alfalfa LAI was significantly affected by its flowering stage ($F = 32.648$, $P < 0.0001$), LAI in the first flowering stage was significantly higher than that in the second and third flowering stages ($P < 0.0001$) (Figure 5B).

Yield and output value per unit area (OVPUA)

Although both grain yield and OVPUA of MMW were higher (6.8% in 2011 and 6.5% in 2013) than the corresponding values of MME, the difference between the two cultivated patterns was not statistically significant (Table 2). Compared to monoculture, alfalfa hay yield in the intercropping treatments increased significantly, while maize grain yield in the same treatments was reduced dramatically in both years (all: $P < 0.0001$). The corresponding parameters in the IMA1 cropping pattern were altered in a greater extent than that in IMA2. Additionally, maize grain yield of IMA1 in 2013 was significantly higher than that in 2011 ($P = 0.028$) (Table 1 and 2).

The comprehensive benefits for total yield and OVPUA were significantly affected by the interaction between cropping pattern and year (Table 1). In 2011, both total yield and OVPUA of IMA1 and IMA2 were significantly enhanced compared to MA (both: $P < 0.0001$), while no significant increase was found compared to MMW. In 2013, both total yield and OVPUA of the two intercropping patterns were significantly higher than that of MA as well as MMW (both: $P < 0.0001$). In terms of year effects, both total yield and OVPUA of MA in 2013 was significantly

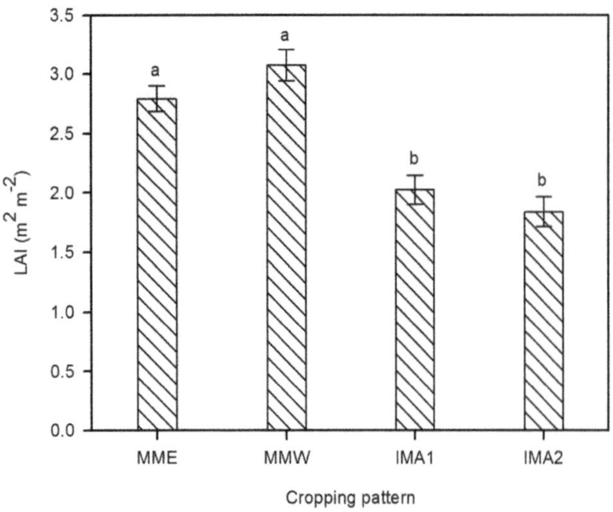

Figure 4. Leaf area index comparisons of maize at the harvest stage under monoculture and intercropping. Significant differences between different cropping patterns are indicated by lower case letters (P <0.05). The other symbols are the same as for Figure 2.

higher than that in 2011 (total yield: $P = 0.038$, OVPUA: $P <$ 0.0001), and there was also a significant increase in the total yield of IMA1 in 2013, as compared to 2011 (P <0.0001) (Table 2).

Furthermore, land equivalent ratios (LERs) of IMA1 and IMA2 were 1.27 and 1.23 in 2011 and 1.12 and 1.08 in 2013, respectively, demonstrating that both intercropping strategies were advantageous, and that the IMA1 pattern was superior to IMA2. The calculated aggressivity (A_{ac}) values for IMA1 and IMA2 were 2.71 and 1.40 in 2011 and 1.31 and 0.65 in 2013, respectively, demonstrating that the resource competitiveness of alfalfa was greater than that of maize in the two intercropping systems (Table 2).

Discussion

Maize monoculture in wide and narrow rows vs. in even rows

Liu et al. [20] reported that in the main agricultural region of Jilin Province, compared to even rows, maize planted in alternating wide and narrow rows increased group light transmission, photosynthetic potential, leaf area and grain yield by more than 10%. The results presented here are in agreement with these findings; both grain yield and output value of MMW were enhanced by 6.8% in 2011 and 6.5% in 2013 relative to MME (Table 2). This was mainly attributable to the improved spatial structure of the group, which increased light transmission (Figure 2), improved maize growth conditions and promoted the formation of edge effect [33]. At harvest time, maize achieved a greater LAI and dry matter accumulation (Figure 4 and S3). Therefore, in the PFA of NEC, maize cropped in alternating wide and narrow rows also had a superior economic benefit compared to maize planted in even rows. We recommend that this approach should be popularized and put into widespread use.

The advantages of intercropping alfalfa with maize

Intercropping plays an important part in traditionally intensive agriculture and has captured attention for its efficient utilization of limited resources [21,34]. Among numerous agricultural intercropping modes, legume/cereal intercropping has been most

Table 1. Results of repeated measures ANOVA on maize leaf area index (LAI) and yield, alfalfa yield and comprehensive benefit analysis of total yield and output value per unit area (OVPUA), with cropping pattern (CP) as the independent variable and year (Y) as the repeated measure.

Factors	Maize			Alfalfa		Comprehensive benefit analysis		
	Df	LAI	Yield	Df	Yield	Df	Total yield	OVPUA
CP	3	26.691*	49.157**	2	430.879**	4	16.343**	15.063**
Y	1	0.349 ns	21.174**	1	0.248 ns	1	32.515**	8.052*
CP × Y	3	0.846 ns	9.024**	2	3.985 ns	4	10.850**	10.699**

Df = degrees of freedom, ns = no significant difference, * p <0.05, ** p <0.01

successful, with a long history and several apparent advantages [12,35–36]. Common patterns include intercropping peanuts [37], soybeans [38] or cowpeas [39] with maize. However, few studies have investigated the potential advantages of intercropping alfalfa with maize [24,40–41], and there has been no systematic study attempting to identify whether it can provide sustained high yield and economic incomes while taking investments and environmental factors into consideration or analyze the constraints limiting its popularization. Furthermore, previous intercropping studies have planted their maize in even rows [23,28]. To our knowledge, no field data are available for intercropping alfalfa with maize in alternating wide and narrow rows.

Owing to differences in the traits, growth and development characteristics of alfalfa and maize (Figure S4), the alfalfa/maize intercropping system resulted in temporal and spatial complementarity, which optimized resource utilization and promoted intercropping advantages [27]. The details are presented as follows:

First, as a perennial forb, alfalfa turns green in early spring, grows fast and mainly covers the soil by late April, whereas maize is sown in early or mid-May. We found that when alfalfa was entering the first flowering stage, maize was still a seedling with canopy lower than that of alfalfa. Thus alfalfa and maize occupy complementary spatial and temporal niches, resulting in complementarity in light interception (Figure 2) and facilitating the circulation and diffusion of air (especially CO_2) in the composite population. This result is consistent with many other studies [42–43]. The increase of light transmission of intercropped alfalfa produces an edge effect and enhances the LAI of alfalfa (Figure 5), thereby significantly improving the hay yield of alfalfa in the first cut, especially when using the IMA1 strategy (Figure S2). Furthermore, the hay yield of alfalfa in the first cut accounted for more than 50% of the total hay yield [19]; therefore, the increase in hay yield in the first cut made a great contribution to total hay yield (Figure S2).

Second, between alfalfa first cut and third flowering stage, maize achieves a period of peak growth, became taller than alfalfa; occupying a more advantageous position in the intercropping system so as to make full use of light, heat and other resources. This effect is particularly evident immediately after the second cutting of alfalfa, when intermediate and bottom light transmission of intercropped maize is dramatically enhanced (Figure 2B and C); producing better growing conditions for intercropped maize and accelerating dry matter accumulation (Figure S3). As a result, intercropped maize has an opportunity for recovery [44], partly compensating for the reduction in the maximum dry matter and grain yield caused by competition with alfalfa. In addition, many studies have demonstrated that there is nitrogen transfer from legumes to cereals in intercropping systems [45–46]. Especially, alfalfa, a perennial leguminous forb, has a strong ability to fix nitrogen, and its fixed and transferred nitrogen contributed as much as 30% to the total N accumulated in the associated grass [47–48]. Therefore, we speculate that there would be nitrogen transfer from alfalfa to maize in the alfalfa/maize intercropping system, which can enhance soil nitrogen availability, improve soil physical and chemical properties and be responsible for the improved growth and development of intercropped maize. It should be noted that the relatively favorable precipitation and allocation in the growing season in 2011 and 2013 was also beneficial in alleviating the growth inhibition of intercropped maize caused by water deficit (Figure 1). Consequently, the combined effects of these factors narrowed the grain yield gap between intercropped maize and monoculture maize and promoted the formation of intercropping advantages.

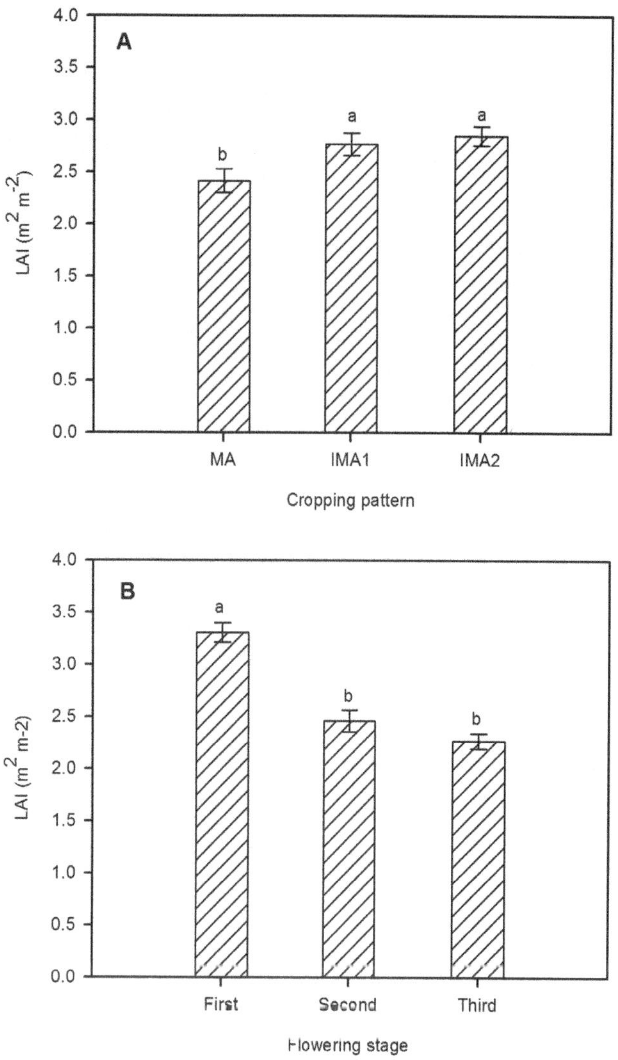

Figure 5. Leaf area index comparisons of alfalfa under monoculture and intercropping (A) and at different flowering stages (B). Significant differences between different alfalfa flowering stages are indicated by lower case letters ($P < 0.05$). The other symbols are the same as for Figure 2 and 4.

Third, alfalfa can continue to grow for around one month after harvesting of maize (Figure S4), providing enough time for alfalfa roots to store adequate carbohydrates to overwinter [23]. Thus, the temporal and spatial differentiation in alfalfa/maize intercropping system avoided wasting light, heat, water, air and other natural resources, prolonged photosynthetic effective time, increased photosynthetic effective area, and ultimately enhanced group yield (Table 2).

Although intercropping alfalfa with maize occupies largely complementary aboveground ecological niches, there is belowground competition. Alfalfa has deep strong roots, that can penetrate>10 m into the soil but mainly proliferate at a soil depth of 0–60 cm. The roots of maize are shallow and mostly distributed at a soil depth of 0–40 cm [49]. Despite some differentiation in root distribution, alfalfa and maize will compete in the shallow soil layers where most water and nutrients are distributed. Moreover, alfalfa has a higher evaporation coefficient and requires more water than maize [18,50]. Therefore, when water is limited, intercropped alfalfa competes with maize for water in the shallow

Table 2. Comparisons of total yield and OVPUA of different planting patterns.

Treatment	Yield (t ha^{-1})			OVPUA (USD ha^{-1})			LER	A_{ac}
	Maize	Alfalfa	Total	Maize	Alfalfa	Total		
2011								
MME	11.00±0.33a	–	11.00±0.33a	3982.31±120.85	–	3982.31±120.85a	–	–
MMW	11.75±0.33a	–	11.75±0.33a	4253.31±120.38	–	4253.31±120.38a	–	–
MA	–	7.85±0.31a	7.85±0.31b	–	2991.92±118.56	2991.92±118.56b	–	–
IMA1	5.87±0.34 (7.63±0.44b)	6.08±0.53 (26.35±2.29b)	11.95 ± 0.30a	2124.17±122.92	2316.91±201.23	4441.08±116.95a	1.27	2.70
IMA2	3.71±0.64 (6.89 ± 1.19b)	7.21±0.34 (15.63±0.74c)	10.92±0.67a	1342.17±231.66	2747.93±129.75	4090.10±246.77a	1.23	1.40
2013								
MME	10.68±0.29a	–	10.68±0.29a	3656.77±100.28	–	3656.77±100.28a	–	–
MMW	11.37±0.23a	–	11.37±0.23ab	3894.69±79.08	–	3894.69±79.08a	–	–
MA	–	11.76±0.05a	11.76 ± 0.05b*	–	4221.72±18.90	4221.72±18.90b*	–	–
IMA1	7.18±0.18 (9.34±0.23b*)	5.79±0.27 (25.06±1.17b)	12.97±0.19c*	2460.38 ± 61.59	2075.89±97.10	4536.26±67.64c	1.12	1.31
IMA2	4.77±0.15 (8.87 ± 0.27b)	7.79±0.11 (16.88±0.23c)	12.56 ± 0.05c	1633.92±50.89	2795.73±38.22	4429.65±14.12c	1.08	0.65

MME = monoculture maize in even rows, MMW = monoculture maize in alternating wide and narrow rows, MA = alfalfa monoculture, IMA1 = maize intercropped with one row of alfalfa in the wide rows, IMA2 = maize intercropped with two rows of alfalfa in the wide rows. Values in the parentheses are yields based on the whole of the intercropping area, including the areas occupied by both maize and alfalfa, and are equal to the yields of maize or alfalfa divided by their respective area proportion. The intercropping area proportions of maize and alfalfa were respectively 76.9% and 23.1% in the IMA1 treatment, while the intercropping area ratios occupied by alfalfa and maize were 53.8% and 46.2% in the IMA2 treatment. Different letters in the same column following the values indicate significant difference between different cropping patterns, and * denotes significant difference between years (P <0.05). Value = mean ± S.

layers to meet its growth and development besides utilizing deep groundwater [28], which contributed to the promotion of LAI and hay yield of intercropped alfalfa at the flowering stage (Figure 5 and S1; Table 2).

However, the strong competition of alfalfa with maize significantly reduced SWC of the intercropping system, especially at the first flowering stage of alfalfa and between its second and third flowering stage when maize was in the period of seedling and peak growth with high water demand (Figure 3). The water stress restrained the growth and development of intercropped maize, particularly in the early growth stage (first 80 days post-emergence) (Figure S3), and significantly declined the LAI and grain yield at harvest time (Figure 4; Table 2).

The competitiveness for resources of intercropped species differs due to competition and complementary in the intercropping system. Generally, cereals are considered to have a competitive advantage over legumes and have a decisive influence on the total yield in annual legume/cereal intercropping systems [31,51–52]. However, Zhang et al. [23] studied an alfalfa/maize intercropping system and found that alfalfa had much greater competitiveness than maize, and that alfalfa yield dramatically influenced the total yield of the intercropping system. Our results are consistent with this study. In our two intercropping modes (IMA1 and IMA2), the competitiveness for resources of alfalfa was much stronger than that of maize, and alfalfa hay yield was significantly improved whereas maize grain yield was significantly reduced. In addition, any reduction in maize yield was more than offset by increased alfalfa yield (Table 2). Thus both intercropping systems have consistently accomplished a successful tradeoff between complementarity and competition, enhanced land use capability and achieved a significant yield and output value advantage over monoculture, except that in 2011 the two intercropping patterns achieved a similar yield and OVPUA compared to MMW (Table 2).

When investment is considered, both IMA1 and IMA2 can improve economic benefits in a great extent relative to MMW in both years. This is because alfalfa is a perennial legume that does not require ploughing, sowing and fertilization after the establishment. Compared to maize, planting alfalfa can reduce costs by at least 584 USD ha^{-1} (based on the fertilizer level and seeding rate in our study and in which fertilizer, seed and farming labor savings were 376, 113 and 95 USD ha^{-1}). In this way, relative to MMW, the IMA1 mode not only increased output value (+187.77 USD ha^{-1} in 2011 and +641.57 USD ha^{-1} in 2013), but also reduced investment (−135 USD ha^{-1}), thus enhanced economic benefits by 322.77 USD ha^{-1} and 776.57 USD ha^{-1} in 2011 and 2013, respectively. Similarly, the final economic benefit of IMA2 was also enhanced by 106.79 USD ha^{-1} in 2011 and 804.96 USD ha^{-1} in 2013, as compared to MMW.

The higher total yield and output value of alfalfa/maize intercropping in 2013 than 2011 (Table 2) could be attributed to the following two aspects. First, the favorable higher amount and better distribution of precipitation in 2013, especially in the growing season, improved the SWC (Figure 3). Thus water competition between alfalfa and maize was minimized, nutrient availability to crops was enhanced [53] and crops obtained a superior condition for growth and development. Meanwhile, the improved growth conditions made alfalfa reach the flowering stages earlier, especially for the second and third times. Thus the co-growth time of alfalfa and maize with strong competition was shortened, and maize obtained a more unconstrained condition to utilize resource and grow. Second, with the intercropping year's increase, alfalfa roots descend into deeper soil layers [49] and could extract water at greater depths when the upper soil horizons get drier, which would improve the growth conditions of intercropped maize; at the same time, it has been proved that alfalfa can fix and transfer more nitrogen to the associated grass as time goes on [54], thus, it is likely that alfalfa could transfer more nitrogen to intercropped maize and improve its growth and development.

Therefore, the comprehensive benefit of alfalfa intercropped with maize was not completely stable and had a certain variation with the changes of rainfall and planting years. It should be noted that the economic benefit of intercropping is also strongly correlated with local market prices of alfalfa and maize [28]. In addition, environmental stress (e.g. pest, disease and the freezing injury to alfalfa in early spring) should also be taken into account to evaluate the valorization of the comprehensive benefits of the intercropping system, and it could severely reduce crops productivity and the economic returns. However, compared to monoculture, intercropping enhanced crops resistance to stress and reduced the management costs and economic losses per unit area [14–16]. Overall, both of the two intercropping strategies have potential to improve economic incomes and are superior to monoculture. This result is in accordance with many other studies and manifests the advantages of intercropping [16,28,55]. Moreover, the intercropping mode of IMA1 was superior to that of IMA2, regardless of land use efficiency, total yield or output value.

Nevertheless, the maize/alfalfa intercropping system still remains relatively unpopular for a number of reasons. First, because of the distinct cultural requirements of each crop, it is difficult to manage the two crops together with existing farm machinery. In this study, all crop management was performed using manual labor. Sowing, fertilizer application, weed control and harvest were conducted separately due to growth and management differences between the two crops. It was also required to manage pest and disease control independently for each crop - when either crop suffered from pests or disease, appropriate pesticides and safeguards were selected to minimize influence on the other crop. It is clear that managing the intercropping system is more complicated and inconvenient than management of a monoculture system. In order to simplify management of this intercropping system, it will be necessary to integrate a multidisciplinary body of knowledge to develop efficient machinery and cautiously advocate the popularization of this intercropping system on a moderate scale to realize yield and economic advantages [56]. Secondly, although the 383,000 km^2 FPA and NEC areas [1] are well suited for the popularization of this intercropping system, local farmers have traditionally utilized a single cropping system and are likely unaware of intercropping systems, especially an intercropping system mixing a food crop and forage legume. Under current conditions, these farmers are more likely to select traditional planting strategies with simple and convenient management schedules and are little concerned about sustainable food production, animal husbandry or eco-environmental security (local investigation and personal communications). Therefore, the dissemination of efficient intercropping technologies and expert technical guidance, as well as financial support of government will necessarily play an important role in putting alfalfa/maize intercropping into practice in the FPA of NEC [56]. Finally, further research is required to assess the long-term benefits of the composite crop population and its responses to rainfall, planting years and environmental stresses (pest, disease and freezing injury), in order to avoid agronomic risks and economic loss.

Conclusions

Based on the above analyses, we conclude that alfalfa/maize intercropping has obvious advantages in grain yield and economic incomes; it guarantees regional food security and provides superior forage. Therefore, this intercropping strategy can help maximize use of limited land resources and promote sustainable development of agriculture and animal husbandry. Moreover, nitrogen fixation and transfer from alfalfa to maize can improve soil fertility and reduce fertilizer investment [19]. Furthermore, alfalfa hay yield increases continuously throughout the first five years [18], providing sustainable economic benefits.

With rising demand for meat, egg, milk and nutrient balance, China is giving increasing importance to the development of animal husbandry and investing the forage industry [9]. Therefore, alfalfa market prices are likely to increase. In this way, planting alfalfa will not only meet the demands of the animal husbandry industry, but also promote the rapid development of local economies. In addition, the multiple-year coverage of alfalfa on the soil can alleviate wind erosion and water erosion, improving the environment of planting areas [19]. In conclusion, there are clear and significant economic, social and ecological benefits in alfalfa/maize intercropping, and maize intercropped with one row of alfalfa was identified as the optimal strategy.

Supporting Information

Figure S1 Daily variation dynamics of air temperature from February to April in 2012. T-highest = highest temperature in a day, T-lowest = lowest temperature in a day.

Figure S2 Alfalfa hay yield at three different flowering stages under monoculture and intercropping. MA = alfalfa monoculture, IMA1 = maize intercropped with one row of alfalfa in the wide rows, IMA2 = maize intercropped with two rows of alfalfa in the wide rows. Different lower case letters for the same flowering stage in one year indicate significant difference, and significant differences of alfalfa total hay yield for one year between different cropping patterns are indicated by different capital letters ($P < 0.05$). Values = means ± SE.

Figure S3 Accumulation dynamics of maize above-ground dry matter under monoculture and intercropping. MME = monoculture maize in even rows, MMW = monoculture maize in alternating wide and narrow rows. The other symbols are the same as for Figure S2.

Figure S4 Growth dynamics of alfalfa and maize in the intercropping system.

Acknowledgments

Special thanks to Baotian Zhang for field help and Scott Diloreto for English check.

Author Contributions

Conceived and designed the experiments: ZL YG. Performed the experiments: BS CW YY YL. Analyzed the data: BS YP HY YG. Contributed reagents/materials/analysis tools: ZL. Wrote the paper: BS YG.

References

1. Zhang HX, Shao MA, Zhang XC (2004) The resuming of weak ecological environment and sustainable development in farming-pasture zone of north-eastern China. J Arid Land Resour Env 18: 129–134.
2. Luo CP, Xue JY (1995) Ecologically vulnerable characteristics of the farming-pastoral zigzag zone in northern China. J Arid Land Resour Env 9: 1–7.
3. Zhou DW, Lu WX, Xia LH, Wu FZ, Li JD, et al. (1999) Grassland degradation and soil erosion in the eastern ecotone between agriculture and animal husbandry in northern China. Resour Sci 21: 57–61.
4. Zhao LP, Wang HB, Liu HQ, Wang YL, Liu SX, et al. (2006) Mechanism of fertility degradation of black soil in corn belt of Song Liao Plain. Acta Pedol Sinica 43: 79–84.
5. He LN, Liang YL, Gao J, Xiong YM, Zhou MJ, et al. (2008) The effect of continuous cropping on yield, quality of cucumber and soil enzymes activities in solar green house. J Northwest A F Univ (Nat. Sci. Ed.) 36: 155–159.
6. Zhen Z, Bo WJ, Wu GL, Luo XC, Zheng YH (2012) Important effect of the organic fertilizer on soil fertility and yield of crop: a case study in Zhende organic farm, Henan, China. J Eng Stud 4: 19–25.
7. Lin JX, Wang JF, Li XY, Zhang YT, Xu JT, et al. (2011) Effects of saline and alkaline stresses in varying temperature regimes on seed germination of *Leymus chinensis* from the Songnen Grassland of China. Grass Forage Sci 66: 578–584.
8. Tian X, Yang YF (2009) Current situation of grassland degradation and its management options in farming-pasturing ecotone in western Jilin Province and eastern Inner Mongolia. Chinese J Ecol 28: 152–157.
9. Lu XS (2012) The opportunity, connotation and future of Chinese grassland agriculture. The Second China Grassland Agriculture Conference. pp. 7–10.
10. Ren JZ (2002) Establishment of on agro-grassland systems for grain storage—A thought on restructure of agricultural framework in Western China. Acta Prataculturae Sinica 11: 1–3.
11. Zhu TC, Li ZJ, Zhang WZ, Liang CZ, Yang HJ, et al. (2002) A preliminary report on the cereal-forage rotation system in the plain of Northeast China. Acta Prataculturae Sinica 12: 34–43.
12. Li L, Sun JH, Zhang FS, Li XL, Yang SC, et al. (2001) Wheat/maize or wheat/soybean strip intercropping I. Yield advantage and interspecific interactions on nutrients. Field Crop Res 71: 123–137.
13. Hauggaard-Nielsen H, Ambus P, Jensen ES (2001) Interspecific competition, N use and interference with weeds in pea-barley intercropping. Field Crop Res 70: 101–109.
14. Miriti JM, Kironchi G, Esilaba AO, Heng LK, Gachene CKK, et al. (2012) Yield and water use efficiencies of maize and cowpea as affected by tillage and cropping systems in semi-arid Eastern Kenya. Agr Water Manage 115: 148–155.
15. Trenbath BR (1993) Intercropping for the management of pests and diseases. Field Crops Res 34: 381–405.
16. Rusinamhodzi L, Corbeels M, Nyamangara J, Giller KE (2012) Maize-grain legume intercropping is an attractive option for ecological intensification that reduces climatic risk for smallholder farmers in central Mozambique. Field Crop Res 136: 12–22.
17. Chen YX, Zhang X, Chen J, Zhou DW (2009) The maize proper harvesting methods in ecotone between agriculture and animal husbandry in Northeast China. J Agr Mech Res 31: 113–117.
18. Wang X, Ma YX, Li J (2003) The nutrient content and main biological characteristics of alfalfa. Pratacultural Sci 10: 39–41.
19. Li ZJ, Guo JX, Zhang YS, Wu ZY (2003) The role and status of alfalfa industry in the structure adjustment of planting industry in Jilin province. J Jilin Agr Sci 28: 40–46.
20. Liu WR, Feng YC, Zheng JY, Liu FC, Zhu XL, et al. (2003) Yield and benefit analysis of maize planted in wide and narrow rows. J Maize Sci 11: 63–65.
21. Lesoing GW, Francis CA (1999) Strip intercropping effects on yield and yield components of corn, grain sorghum, and soybean. Agron J 91: 807–813.
22. Gilbert RA, Heilman JL, Juo ASR (2003) Diurnal and seasonal light transmission to cowpea in sorghum-cowpea intercrops in Mali. Agron Crop Sci 189: 21–29.
23. Zhang GG, Yang ZB, Dong ST (2011) Interspecific competitiveness affects the total biomass yield in an alfalfa and corn intercropping system. Field Crop Res 124: 66–73.
24. Wang T, Zhu B, Xia LZ (2012) Effects of contour hedgerow intercropping on nutrient losses from the sloping farmland in the Three Gorges Area, China. J Mt Sci 9: 105–114.
25. Yin XJ, Cui GW (2006) Causes and preventing techniques of alfalfa freezing injury in northern cold regions. Feed Review 4:31–33.
26. Sun QZ, Wang YQ, Hou XY (2004) Alfalfa winter survival research summary. Pratacultural Sci 21:21–25.
27. Bedoussac L, Justes E (2010) Dynamic analysis of competition and complementarity for light and N use to understand the yield and the protein content of a durum wheat-winter pea intercrop. Plant Soil 330: 37–54.
28. Smith MA, Carter PR (1989) Strip intercropping corn and alfalfa. J Prod Agric 11: 345–353.
29. Anil L, Park J, Phipps RH, Miller FA (1998) Temperate intercropping of cereals for forage: a review of the potential for growth and utilization with particular reference to the UK. Grass Forage Sci 53: 301–317.

30. Chu GX, Shen QR, Gao JL (2004) Nitrogen fixation and N transfer from peanut to rice cultivated in aerobic soil in an intercropping system and its effect on soil N fertility. Plant Soil 263: 17–27.
31. Lithourgidis AS, Viachostergios DN, Dordas CA, Damalas CA (2011) Dry matter yield, nitrogen content, and competition in pea-cereal intercropping systems. Eur J Agron 34: 287–294.
32. Shapiro SS, Wilk MB (1965) An analysis of variance test for normality (complete samples). Biometrika 52:591–611.
33. Liu AN, Liu ZG, Zhou XG, Meng ZJ, Chen JP (2005) Study on edge effect and ecological effect in system of winter wheat intercropping with cotton. J Mt Agr Biol 24: 471–476.
34. Trenbath BR (1976) Plant interactions in mixed crop communities. In: Papendick RI, Sanchez PA, Triplett GB, editors. Multiple cropping. ASSA, CSSA, and SSSA. Madison Wis. pp. 129–170.
35. Willey RW (1990) Resource use in intercropping systems. Agr Water Manage 17: 215–231.
36. Mandal BK, Das D, Saha A, Mohasin M (1996) Yield advantage of wheat (*Triticum aestivum*) and chickpea (*Cicer arietinum*) under different spatial arrangements in intercropping. Ind J Agron 41: 17–21.
37. Inal A, Gunes A, Zhang F, Cakmak I (2007) Peanut/maize intercropping induced changes in rhizosphere and nutrient concentrations in shoots. Plant Physiol Biochem 45: 350–356.
38. Prasad RB, Brook RM (2005) Effect of varying maize densities on intercropped maize and soybean in Nepal. Exp Agr 41: 365–382.
39. Ghanbari A, Dahmardeh M, Siahsar BA, Ramroudi M (2010) Effect of maize (*Zea mays* L.)-cowpea (*Vigna unguiculata* L.) intercropping on light distribution, soil temperature and soil moisture in arid environment. J Food Agr Environ 8: 102–108.
40. Liebman M, Graef RL, Nettleton D, Cambardella CA (2011) Use of legume green manures as nitrogen sources for corn production. Renew Agr Food Syst 27: 180–191.
41. Guldan SJ, May T, Martin CA, Steiner RL (1998) Yield and forage quality of interseeded legumes in a high-desert environment. J Sustain Agr 12: 85–97.
42. Tsubo M, Walker S (2002) A model of radiation interception and use by a maize-bean intercrop canopy. Agr Forest Meteorol 110: 203–215.
43. Awal MA, Koshi H, Ikeda T (2006) Radiation interception and use by maize/peanut intercrop canopy. Agr For Meteorol 139: 74–83.
44. Li L, Sun JH, Zhang FS, Li XL, Rengel Z, et al. (2001) Wheat/maize or wheat/soybean strip intercropping II. Recovery or compensation of maize and soybean after wheat harvesting. Field Crop Res 71: 173–181.
45. Shen QR, Chu GX (2004) Bi-directional nitrogen transfer in an intercropping system of peanut with rice cultivated in aerobic soil. Biol Fertil Soils 40: 81–87.
46. Li YF, Ran W, Zhang RP, Sun SB, Xu GH (2009) Facilitated legume nodulation, phosphate uptake and nitrogen transfer by arbuscular inoculation in an upland rice and mung bean intercropping system. Plant Soil 315: 285–296.
47. Ta TC, Faris MA (1987) Species variation in the fixation and transfer of nitrogen from legumes to associated grasses. Plant Soil 98: 265–274.
48. Yang SX, Yang ZZ (1992) A study on superiorities in mixed cropping of alfalfa and Siberian wildrye. Sci Agr Sin 25: 63–68.
49. Zhang GG, Zhang CY, Yang ZB, Dong ST (2013) Root distribution and N acquisition in an alfalfa and corn intercropping system. J Agr Sci 5: 128–142.
50. Sun HR (2003) Alfalfa's pre-blossoming transpiration coefficients and comparison of the water consumption coefficients between alfalfa and maize in terms of economic yield. Acta Agrestia Sinica 4: 346–349
51. Misra AK, Acharya CL, Rao AS (2006) Interspecific interaction and nutrient use in soybean/sorghum intercropping system. Agron J 98: 1097–1108.
52. Wahla IH, Ahmad R, Ehsanullah AA, Jabbar A (2009) Competitive functions of components crops in some barley based intercropping systems. Intl J Agr Biol (Pakistan) 11: 69–71.
53. Jensen JR, Bernhard RH, Hansen S, McDonagh J, MØberg JP, et al. (2003) Productivity in maize based cropping systems under various soil-water-nutrient management strategies in semi-arid Alfisol environment in East Africa. Agr Water Manage 59: 217–237.
54. Goodman PJ (1988) Nitrogen fixation transfer and turnover in upland and lowland grass-clover awards, using ^{15}N isotope dislution. Plant Soil 112: 247–254.
55. Mucheru-Muna M, Pypers P, Mugendi D, Kung'u J, Mugwe J, et al. (2009) A staggered maize-legume intercrop arrangement robustly increases crop yields and economic returns in the highlands of Central Kenya. Field Crop Res 115: 132–139.
56. Yu Y, He Y (2009) The investigation and popularization of the efficient mode of grain-economic intercropping. Shanxi J Agr Sci 5: 90–93.

26

Responses of Super Rice (*Oryza sativa* L.) to Different Planting Methods for Grain Yield and Nitrogen-Use Efficiency in the Single Cropping Season

Song Chen[1], Danying Wang[1], Chunmei Xu[1], Chenglin Ji[1], Xiaoguo Zhang[1], Xia Zhao[1], Xiufu Zhang[1]*, Bhagirath Singh Chauhan[2]*

1 China National Rice Research Institute, Chinese Academy of Agricultural Sciences, Hangzhou, Zhejiang, China, **2** Queensland Alliance for Agriculture and Food Innovation (QAAFI), University of Queensland, Toowoomba, Queensland, Australia

Abstract

To break the yield ceiling of rice production, a super rice project was developed in 1996 to breed rice varieties with super high yield. A two-year experiment was conducted to evaluate yield and nitrogen (N)-use response of super rice to different planting methods in the single cropping season. A total of 17 rice varieties, including 13 super rice and four non-super checks (CK), were grown under three N levels [0 (N0), 150 (N150), and 225 (N225) kg ha^{-1}] and two planting methods [transplanting (TP) and direct-seeding in wet conditions (WDS)]. Grain yield under WDS (7.69 t ha^{-1}) was generally lower than TP (8.58 t ha^{-1}). However, grain yield under different planting methods was affected by N rates as well as variety groups. In both years, there was no difference in grain yield between super and CK varieties at N150, irrespective of planting methods. However, grain yield difference was dramatic in japonica groups at N225, that is, there was an 11.3% and 14.1% average increase in super rice than in CK varieties in WDS and TP, respectively. This suggests that high N input contributes to narrowing the yield gap in super rice varieties, which also indicates that super rice was bred for high fertility conditions. In the japonica group, more N was accumulated in super rice than in CK at N225, but no difference was found between super and CK varieties at N0 and N150. Similar results were also found for N agronomic efficiency. The results suggest that super rice varieties have an advantage for N-use efficiency when high N is applied. The response of super rice was greater under TP than under WDS. The results suggest that the need to further improve agronomic and other management practices to achieve high yield and N-use efficiency for super rice varieties in WDS.

Editor: Guoping Zhang, Zhejiang University, China

Funding: This research was partly supported by grants from the 'Five-twelfth' National Science and Technology Support Program (2012BAD04B00), the National Natural Science Foundation of China (31301255&3117150156), and the MOA Special Fund for Agro-Scientific Research in the Public Interest of China (201203096). The funders had no role in study design, data collection and analysis, decision to publish, or preparation of the manuscript.

Competing Interests: The authors have declared that no competing interests exist.

* Email: b.chauhan@uq.edu.au (BSC); chais.zju@gmail.com (XZ)

Introduction

More than 90% of the world's rice (*Oryza sativa* L.) is grown and consumed in Asia. In China, it is the staple food for about 60% of the population [1]. To meet the growing demand for food that will result from population growth and economic development in the next decade [2], great efforts should be made to breed new rice varieties with higher yield potential in order to enhance average farm yield [3,4]. In the past decade, programs for super or ideotype rice breeding, aiming at dramatically increasing yield potential, have greatly progressed at the International Rice Research Institute (IRRI) and some countries, including Japan, Korea, and China [5–7]. In 1996, China established a nationwide mega-project on the development of super rice based on the ideotype concept [7]. A rice variety could be recognized as a super rice if it meets the yield target at two pilot sites in two successive years, or if it meets the goal of yield advantages over the control variety in regional yield trials. The criteria of super rice cultivars vary with both production area and subspecies type [7]. Up to 2012, 96 commercially released super rice varieties were grown on over a total of 80 million ha in China [8]. A number of super-high yield records of over 12 t ha^{-1} have been reported [4,9,10]. In addition, it was also reported that super hybrid rice varieties, such as Liangyoupeijiu and Xieyou 9308, produced 8–20% higher grain yield than check varieties, such as Shanyou 63 and Xieyou 63, in a similar environment [11,12]. On the other hand, there was little increase in terms of large-scale rice production, indicating that the yield potential of newly released super rice varieties has not been fully realized, mainly because of unreasonable agronomic practices. Thus, it is imperative to understand yield responses of super rice to various agronomic practices, such as N rates or planting methods.

Traditionally, rice is cultivated using the transplanting method, consisting of raising nurseries, uprooting and picking seedlings, and transplanting [13], which cost labor and energy [14,15]. Paradoxically, labor availability is limited in China because an increasing number of young farmers have left for jobs in the cities [16]. Direct-seeded rice is an alternative production system that

Table 1. Details on the varieties used in the experiment.

Variety	Type	Year of release	Growth period (d)	Plant height (cm)
II-you838	[a]Hybrid Indica -CK	1995	140–145	98–102
Shanyou63	Hybrid Indica -CK	1984	138.0	105–110
II-Youhang1	Hybrid Indica	2005	135.8	127.5
II-Youhang2	Hybrid Indica	2007	134.5	129.9
Fengliangyou6	Hybrid Indica	2008	145.0	121.8
Neiliangyou6	Hybrid Indica	2007	154.4	110.4
Liangyoupei9	Hybrid Indica	2001	150.0	110–120
Peiliangyou3076	Hybrid Indica	2006	135.7	119.1
Yangliangyou6	Hybrid Indica	2005	134.1	120.6
Zhongzheyou1	Hybrid Indica	2004	138.0	125.0
Xiushui09	[b]Japonica-CK	2005	159.0	92.5
Xiushui63	Japonica-CK	1997	150–154	95–100
Huaidao11	Japonica	2008	156.0	103.9
Huaidao9	Japonica	2006	152.0	100.0
Ningjing1	Japonica	2004	145.0	93.1
Ningjing3	Japonica	2008	159.0	98.6
Yangjing4038	Japonica	2008	135.0	106.0

[a]The hybrid Indica rice were hybrid varieties developed using the three-line method.
[b]The Japonica rice were inbred varieties.
The information of varieties was obtained from the China Rice Data Center (http://www.ricedata.cn/index.htm).

can help reduce labor costs [17]. Therefore, the planting pattern in the past two decades is being gradually and partially replaced by direct seeding in many countries [18,19]. This is also the case in China; the rice area for wet direct seeding (WDS) is rapidly increasing [13]. Previous studies suggest that WDS could be adopted on a large area in the Yangtze River Basin [20]. Furthermore, WDS cultivation provides rice with a completely different growth condition at the seedling stage than cultivation by transplanting does. Studies on rice under WDS indicate favorable changes for high yield formation in comparison with transplanted rice [21–23]. These changes include earlier seedling emergence [21], stronger root activity, higher seed setting rate [22], and greater biomass production at the early stage [22,23]. On the other hand, a significant improvement in grain yield has not been achieved in WDS systems because of unstable seedling establishment, sensitivity to lodging, shorter growth period, and severe competition with weeds [17,20]. Generally, most super rice varieties currently planted in China are developed by transplanting (TP) cultivation. Therefore, further research is needed to evaluate the different genotypic responses in growth and yield of super rice to planting methods.

Nitrogen (N) is the most important yield-limiting nutrient for rice [24]. Increased rice production is largely attributed to the increased use of N fertilizer. The amount of N uptake needed to produce one ton of rough rice is 15–17 kg N for an average yield of 5–6 t ha^{-1} [24]; it is 19 kg for high-yielding rice [25]. For super rice varieties, the amount of N uptake was more than 18 kg N t^{-1} of grain yield, and, in some cases, even reached as high as 20 kg N t^{-1} of grain yield [9,12,26–28]. Some super rice varieties, such as Liangyoupeijiu, were bred for high fertility conditions with high N application, achieving the yield of 10.98 t ha^{-1} with an application of 234 kg N ha^{-1} [9]. An earlier study also indicated that japonica super rice in north China reached a very high yield (11 t ha^{-1}) with the application of more than 247.5 kg N ha^{-1} [29]. In China, 150 kg N ha^{-1} applied at key growth stages is recommended for traditional varieties in TP systems [30]. Therefore, the abundant use of N fertilizer on super rice on a large scale would generally accelerate non-point source pollution. In addition, it was also reported that TP has a greater N-use efficiency than WDS in six non-super indica varieties [31]. However, beyond the yield output, as well as for relative physiological or morphological traits, the study of the effect of planting methods on N-use efficiency in high-yielding varieties is still limited.

In this study, a two-year field research was conducted to compare the yield performance of super rice varieties with three N levels (0, 150, and 225 kg ha^{-1}) and two planting methods (TP and WDS). The objectives of this study are to (1) evaluate the grain yield performance of super rice in different planting systems, and (2) evaluate the N-use efficiency of super rice varieties, switching from TP to WDS.

Materials and Methods

Experimental design and crop management

Experiments were conducted in 2011 and 2012 at the experimental farm of the China National Rice Research Institute, Fuyang County (28′09′N, 113′37′E, 43 m a.s.l.), Zhejiang Province, China. The soil was clay with the following properties: pH 6.0, 22.3 g kg^{-1} organic matter, 197 mg kg^{-1} alkali-hydrolyzale N, 102.8 mg kg^{-1} available P, and 205 mg kg^{-1} available K. The soil test was based on samples taken from the upper 20 cm of the soil.

Treatments were arranged in a split-split-plot design with planting method in the main plots, N rate in split-plots, and variety in split-split plots. The experiment was replicated three times and the sub-sub plot size was 30 m^2.

A total of 17 and 15 varieties in the single rice cropping system were used in 2011 and 2012, respectively. Four groups of varieties

Table 2. Analysis of variance for nitrogen (N), planting method, and variety group on grain yield, total N accumulation, NAE, NPE, and NRE in 2011 and 2012.

Factor	Grain yield (t ha^{-1})		Total N accumulation (kg ha^{-1})		NAE (kg grain kg^{-1} N)		NPE (kg grain kg^{-1} N)		NRE (%)	
	2011	2012	2011	2012	2011	2012	2011	2012	2011	2012
Nitrogen (N)	**	**	**	**	**	**	**	**	**	**
Planting method (P)	**	**	**	**	ns	ns	*	*	ns	ns
Variety group (V)	**	**	*	ns	**	**	**	**	*	*
P*N	ns	*	ns	ns	*	*	*	*	ns	ns
V*N	ns	ns	**	*	*	*	**	**	*	*
V*P	**	*	*	*	*	*	ns	ns	ns	ns
V*P*N	**	*	*	*	*	*	*	*	**	**

Abbreviations: *means the p value at <0.01; **means the p value at <0.001; ns, nonsignificant; NAE, N agronomic efficiency; NPE, N physiological efficiency; NRE, N recovery efficiency.

were chosen based on the subspecies and super yield: high-yielding indica hybrid varieties, high-yielding japonica inbred varieties, and indica and japonica check varieties. These varieties have been widely grown by farmers in the Yangtze River Basin in China. Details on the varieties are given in Table 1.

We confirm that our study did not involve endangered or protected species. No specific permissions were required for using the locations and varieties. The seeds of all varieties used in the study are commercially available.

Two planting methods were included: manual wet direct seeding (WDS) and transplanting (TP). For TP, pre-germinated seeds were sown on a seedbed. Thirty- and 25-d-old seedlings were transplanted on 20 June 2011 and 25 June 2012, respectively. Two seedlings were transplanted at a spacing of 30 cm×16.7 cm. In WDS, pre-germinated seeds (about 3–4 seeds per hill) were placed manually on the surface of puddled soil, with a spacing of 30×16.7 cm. To maintain similar plant density, excess seedlings were removed after emergence (about 10–15 days after seeding).

Three N treatments were included: 0 (N0), 150 (N150), and 225 (N225) kg N ha^{-1}. For N150, 75, 45, and 30 kg N ha^{-1} were applied at basal, mid-tillering, and panicle initiation, respectively. For N225, 113, 67, and 45 kg N ha^{-1} were applied at basal, mid-tillering, and panicle initiation, respectively. Phosphorus (60 kg P_2O_5 ha^{-1}) was applied and incorporated in all plots 1 day before TP or WDS. Potassium (150 kg K_2O ha^{-1}) was applied in two equal splits at basal and panicle initiation in both establishment methods.

Crop management followed the standard cultural practices. Weeds in the field were well controlled using intensive hand weedings. The experimental field was kept flooded from the day of transplanting until 10 days before maturity in both planting methods. Insects were intensively controlled by chemicals to avoid biomass and yield loss.

Sampling and measurements

At maturity, 12 hills were sampled diagonally from a 5-m^2 harvest area and were separated into panicle and straw to determine the dry weight. Dry weight was determined after oven-drying the samples at 70°C to the constant weight. Grain yield was determined from a 5-m^2 area in each plot and adjusted to the standard moisture content of 14%. The N concentrations of straw and panicles were determined by micro-Kjeldahl digestion, distillation, and titration [32]. N accumulation (panicle or straw) resulted from dry weight and N concentration (panicle or straw). Total N accumulation was calculated from the panicles and straw.

Nitrogen agronomic efficiency (NAE), nitrogen physiological efficiency (NPE), and nitrogen-recovery efficiency (NRE) were calculated as:

$$NAE = (GYNi - GYN0)/FNi$$

$$NPE = (GYNi - GYN0)/(TNi - TN0)$$

$$NRE = 100.(TNi - TN0)/(FNi)$$

Table 3. Effect of nitrogen (N) rates [N0, N150, and N225 are 0, 150, and 225 kg N ha^{-1}, respectively] and planting methods (transplanting, TP and wet direct seeding, WDS) on grain yield in 2011.

Variety	N0		N150		N225	
	WDS	TP	WDS	TP	WDS	TP
CK (Indica)						
II-you838	7.0±0.2	7.2±0.4	8.1±0.8	9.0±0.5	7.8±0.6	7.7±0.2
Shanyou63	7.5±0.6	7.9±0.5	8.5±0.6	8.5±0.4	8.1±0.3	8.5±0.6
Average	7.3b	7.6b	8.3a	8.7a	8.0ab	8.1ab
High yielding varieties (Indica hybrid)						
II-Youhang1	5.4±0.2	7.2±0.3	8.0±0.4	9.9±0.9	7.4±0.7	9.1±0.2
II-Youhang2	5.5±0.4	6.2±0.3	8.1±0.4	8.7±0.8	8.0±0.8	10.5±0.3
Fengliangyou6	5.9±0.6	6.4±0.5	7.9±0.4	8.5±0.6	7.9±0.5	8.1±0.2
Neiliangyou6	6.3±0.3	8.2±0.5	8.0±0.5	10.1±0.4	8.3±0.7	10.3±0.6
Liangyoupei9	5.7±0.2	7.3±0.5	8.3±0.4	8.4±0.6	6.5±0.5	10.2±0.3
Peiliangyou3076	5.6±0.3	6.5±0.4	8.4±0.8	8.4±0.4	6.9±0.6	7.3±0.7
Yangliangyou6	5.9±0.3	7.1±0.2	7.5±0.3	8.3±0.4	6.6±0.3	7.6±0.3
Zhongzheyou 1	6.5±0.3	7.0±0.4	8.6±0.3	9.3±0.9	8.1±0.3	9.1±0.5
Average	5.9d	7.0c	8.1b	9.0a	7.5bc	9.0a
CK (Japonica)						
Xiushui09	7.0±0.7	7.9±0.7	9.3±0.4	9.9±0.7	8.3±0.6	9.8±0.9
Xiushui63	6.1±0.2	6.8±0.5	8.2±0.7	8.2±0.6	7.6±0.2	8.0±0.3
Average	6.6c	7.4b	8.8a	9.1a	8.0b	8.9a
High yielding varieties (Japonica)						
Huaidao11	6.1±0.4	8.5±0.4	7.9±0.7	9.7±0.5	8.6±0.4	10.4±0.8
Huaidao9	6.6±0.4	7.8±0.6	9.3±0.5	10.3±0.7	9.5±0.6	10.0±0.4
Ningjing1	6.3±0.5	7.9±0.2	8.2±0.7	9.4±0.9	9.4±0.4	10.2±0.4
Ningjing3	6.6±0.5	7.6±0.6	8.5±0.8	9.9±0.4	8.6±0.7	10.3±0.3
Yangjing4038	7.2±0.2	7.6±0.5	8.2±0.6	8.6±0.7	8.7±0.7	10.9±1
Average	6.6d	7.9c	8.4bc	9.6ab	9.0b	10.4a

Yields of rice varieties were expressed as Mean ± SE (with 3 replications). Average means in rows followed by different letters were significantly different at $p<0.05$ (Tukey's HSD).

In the equations, *GYNi* is the grain yield from the plots that received N fertilizer; *GYN0* is the grain yield in the zero-N (control) plots; *FNi* is the amount of N fertilizer applied; *TNi* is the total above plant N accumulation in the plots that received N fertilizer; and *TN0* is the total above plant N accumulation in zero-N (control) plots.

Statistical analyses

Data generated from the experiments were subjected to statistical software SAS 8.0 for Windows separately for 2011 and 2012. A three-way analysis of variance (ANOVA) was conducted for all of the abovementioned parameters from three replicates at harvest, with the following effects: variety group, nitrogen rates, planting methods, variety group×N rates, variety group×planting methods, planting methods×N rates, and variety group×N rates×planting methods. Yield comparisons were made among the combination of N rates and planting methods for each variety group using Tukey's HSD; a *p* value <0.05 was considered significantly different.

Results

ANOVA results for grain yield, N accumulation, NAE, NRE, and NPE are shown in Table 2. The year effect was significant; therefore, the results are presented and discussed separately for years.

Grain yield

In 2011, grain yield at N150 ranged from 7.5 to 9.3 t ha^{-1} in WDS and 8.2 to 10.3 t ha^{-1} in TP, whereas, the corresponding values at N225 were 6.5–9.5 t ha^{-1} and 7.3–10.9 t ha^{-1}, respectively (Table 3). In 2012, grain yield at N150 ranged from 7.5 to 9.9 t ha^{-1} in WDS and 7.9 to 10.3 t ha^{-1} in TP, whereas, the corresponding values at N225 were 6.1–9.1 t ha^{-1} and 7.4–10.9 t ha^{-1}, respectively (Table 4). The interaction of N rates, planting methods, and variety groups was significant for grain yield (Table 2). In general, grain yield under WDS was lower than TP–9.1% and 10.6% lower in 2011 and 2012, respectively. However, the differences varied with N rates and planting methods. At N225, the average grain yield for super japonica, super indica, and japonica CK varieties in 2011 under TP were 9.03, 8.90, and 10.36 t ha^{-1}. These were 20.9%, 11.9%, and 15.6% higher than those under WDS, respectively. However, no

Table 4. Effect of nitrogen (N) rates [N0, N150, and N225 are 0, 150, and 225 kg N ha^{-1}, respectively] and planting methods (transplanting, TP and wet direct seeding, WDS) on grain yield in 2012.

Variety	N0		N150		N225	
	WDS	TP	WDS	TP	WDS	TP
CK (Indica)						
II-you838	7.5±0.4	7.5±0.6	8.1±0.7	9.5±0.9	8.2±0.8	8.0±0.5
Shanyou63	7.3±0.6	7.8±0.7	8.3±0.3	8.1±0.6	8.7±0.2	8.8±0.3
Average	7.4b	7.7b	8.2ab	8.8a	8.4a	8.4a
High yielding varieties (Indica hybrid)						
II-Youhang1	5.0±0.4	6.9±0.6	8.5±0.4	9.9±0.8	7.7±0.6	8.5±0.4
II-Youhang2	5.8±0.2	6.6±0.3	7.5±0.2	8.5±0.4	7.6±0.6	10.2±0.4
Neiliangyou6	6.2±0.6	7.8±0.5	7.9±0.7	9.6±0.7	8.5±0.7	9.8±0.4
Liangyoupei9	5.4±0.3	7.6±0.4	8.2±0.7	7.9±0.6	6.7±0.4	9.7±0.6
Peiliangyou3076	5.6±0.2	6.8±0.1	8.7±0.2	8.4±0.8	7.4±0.3	7.4±0.2
Zhongzheyou1	6.6±0.4	7.2±0.5	8.5±0.7	9.1±0.9	8.0±0.2	9.3±0.5
Average	5.8c	7.2b	8.2ab	8.9a	7.7b	9.1a
CK (Japonica)						
Xiushui09	7.8±0.8	9.9±0.7	10.2±0.6	8.0±0.7	9.5±0.7	7.8±0.8
Xiushui63	6.7±0.4	7.7±0.8	8.7±0.4	7.2±0.6	8.0±0.4	6.7±0.4
Average	6.5c	7.2b	8.8ab	9.5a	7.6b	8.7ab
High yielding varieties (Japonica)						
Huaidao11	8.6±0.4	7.8±0.4	9.2±0.7	9.1±0.4	10.9±0.3	8.6±0.4
Huaidao9	7.3±0.2	9.2±0.4	10.3±0.5	8.8±0.5	10.5±0.9	7.3±0.2
Ningjing1	8.2±0.5	8±0.3	9.5±0.3	9.0±0.8	10.7±0.4	8.2±0.5
Ningjing3	7.5±0.5	8.2±0.5	9.7±0.4	8.8±0.6	9.9±0.8	7.5±0.5
Yangjing4038	7.5±0.6	8.7±0.5	8.8±0.7	8.1±0.5	10.8±0.7	7.5±0.6
Average	6.6c	7.8b	8.4b	9.5ab	8.8b	10.6a

Yields of rice varieties were expressed as Mean ± SE (with 3 replications). Average means in rows followed by different letters were significantly different at $p<0.05$ (Tukey's HSD).

difference was found between TP and WDS for indica CK varieties. Similar results were also found in 2012. In 2011, at N150, yield increases of 0.85 and 1.18 t ha^{-1} were obtained under TP than under WDS for super varieties japonica and indica, respectively; there was no difference found for CK varieties. In 2012, yield increase in transplanted rice was found only for super japonica varieties, and no difference was found for the rest of the variety groups.

These results indicate that super rice has a greater yield potential at high N rates in the TP system. The optimum N rate for wet direct-seeded rice was 150 kg ha^{-1} in the current condition. In both years, grain yield at N150 and N225 was

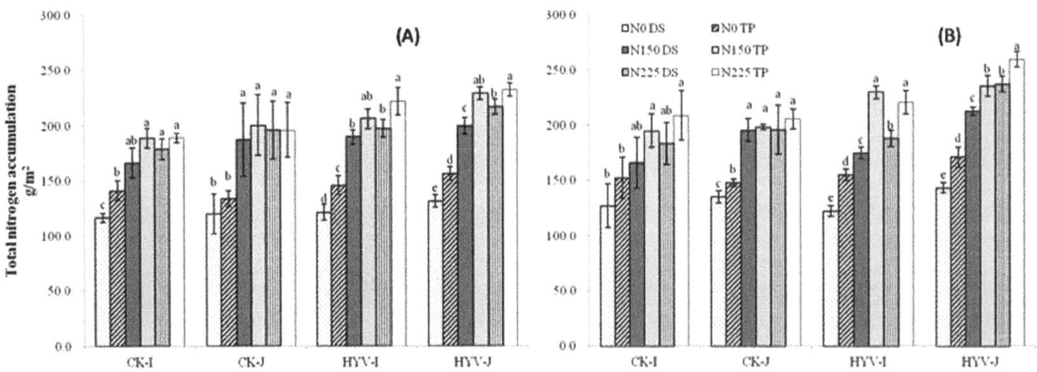

Figure 1. Effect of nitrogen (N) rates and planting methods on total nitrogen accumulation in above biomass in 2011 (A) and 2012 (B). WDS, wet direct-seeded rice; TP, transplanting rice; N0, 0 kg N ha^{-1}; N150, 150 kg N ha^{-1}; N225, 225 kg N ha^{-1}. The vertical bars stands for standard error of means. Average means followed by different letters were significantly different at $p<0.05$ (Tukey's HSD) in the same variety group.

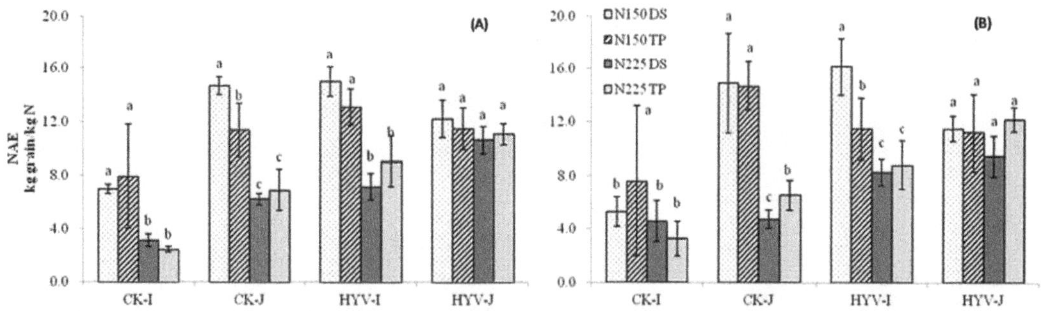

Figure 2. Effect of nitrogen (N) rates and planting methods on N agronomic efficiency (NAE) in 2011 (A) and 2012 (B). WDS, wet direct-seeded rice; TP, transplanting rice; N0, 0 kg N ha^{-1}; N150, 150 kg N ha^{-1}; N225, 225 kg N ha^{-1}. The vertical bars stands for standard error of means. Average means followed by different letters were significantly different at $p<0.05$ (Tukey's HSD) in the same variety group.

similar, except for super indica and japonica CK varieties under the WDS system, in which grain yield was 8.5% and 10.1% higher at N150 than at N225, respectively. However, yield at N150 and N225 were always greater than that at N0. At N0, the average grain yield of super varieties was relatively lower than that of CK varieties. For indica varieties under WDS, average grain yield of the checks (7.25 t ha^{-1} and 7.69 t ha^{-1}) was 23.9% and 27.6% higher than that of super varieties in 2011 and 2012, respectively. In both years, irrespective of planting method, no difference was found in grain yield between super and CK varieties at N150. At N225, however, grain yield difference was dramatic for the japonica group; there was an 11.3% and 14.1% average increase in grain yield for super varieties under WDS and TP, respectively, in comparison with the checks. In both years, for indica, the grain yield of super varieties (9.1 t ha^{-1}) was 9.2% higher than CK varieties (7.5 t ha^{-1}) under the TP system, but 8.4% lower under the WDS system. Such cases of yield decline for super indica varieties at N225 under WDS might lead to the negative effect of super varieties (e.g., Liangyoupei9 and Peiliangyou3076). The yield decline was generally due to the excessive growth of plants and/or lodging during panicle initiation and/or grain filling (visual observations). The results imply a potential risk to super varieties applied with high N doses in the WDS system.

Total nitrogen accumulation

N×planting method×variety group interaction was significant. However, N×planting method interaction was not significant (Table 2), suggesting the consistent effect of planting methods on total nitrogen accumulation (TNA) at different N rates (Fig. 1). In 2011, average TNA values under WDS across N rates were 153.9, 169.4, and 182.6 g m^{-2} for indica CK, super indica, and super japonica varieties, respectively, which were 14.6%, 12.0%, and 11.8% lower than those under TP, respectively. However, no difference was found for japonica CK. Similar results were found in 2012, except for japonica CK at N0, in which TNA values increased under WDS than under TP.

Differences among variety groups varied with N rates and planting methods. In 2011, no difference was found between super and CK varieties for the indica group, except for plants grown at N150 in the TP system, in which TNA was 17.9% higher in super rice than in CK varieties. In japonica, TNA in the TP system was 15.9% and 18.6% higher in super rice than in CK varieties at N0 and N150, respectively. However, no difference was found in the WDS system. A significant increase in TNA between super and CK varieties was found at N225 in both planting methods–21.1% and 26.3% greater in super rice than in CK varieties. In 2012, similar results were found for both japonica and indica groups, except at N0 under TP, in which no difference was detected between super and CK varieties.

Nitrogen agronomic efficiency, N-recovery efficiency, and N physiological efficiency

In both years, irrespective of N rate and variety group, no difference was found in NAE between the TP and WDS systems, except for the super indica varieties and japonica CK varieties at N150 in 2011, in which NAE was greater in WDS than in TP (Fig. 2). In the indica group in 2011, NAE was greater (66–270%)

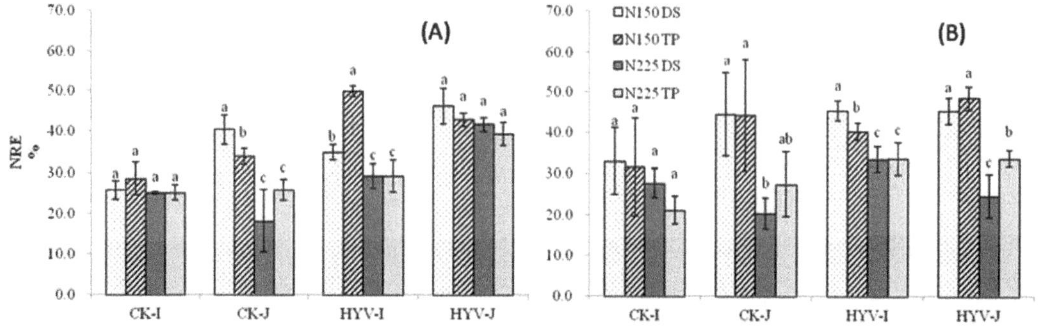

Figure 3. Effect of nitrogen (N) rates and planting methods on N recovery efficiency (NRE) in 2011 (A) and 2012 (B). WDS, wet direct-seeded rice; TP, transplanting rice; N0, 0 kg N ha^{-1}; N150, 150 kg N ha^{-1}; N225, 225 kg N ha^{-1}. The vertical bars stands for standard error of means. Average means followed by different letters were significantly different at $p<0.05$ (Tukey's HSD) in the same variety group.

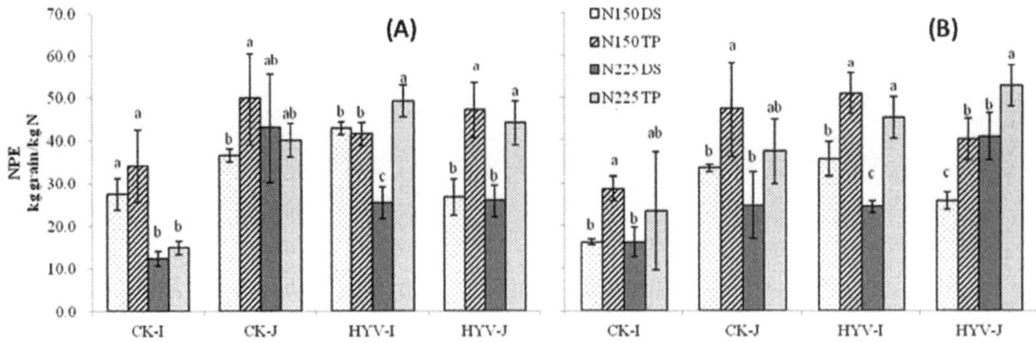

Figure 4. Effect of nitrogen (N) rates and planting methods on N physiological efficiency (NPE) in 2011 (A) and 2012 (B). WDS, wet direct-seeded rice; TP, transplanting rice; N0, 0 kg N ha^{-1}; N150, 150 kg N ha^{-1}; N225, 225 kg N ha^{-1}. The vertical bars stands for standard error of means. Average means followed by different letters were significantly different at $p<0.05$ (Tukey's HSD) in the same variety group.

in super rice (11.1 kg grain kg^{-1} N on average) than in CK varieties (5.1 kg grain kg^{-1} N on average). However, the difference in japonica varieties varied with N rate. No difference was found for NAE at N150, whereas super rice varieties (10.9 kg grain kg^{-1} N on average) had greater NAE than CK varieties (6.5 kg grain kg^{-1} N on average) at N225–71% and 61% greater under WDS and TP, respectively. In 2012, irrespective of planting method and variety group, the difference between super and CK varieties was not significant at N150, except for indica varieties under WDS, in which the average NAE in super varieties was 206% greater than that in CK varieties. Regardless of planting method and variety group, super varieties had greater NAE values than CK varieties at N225. A higher NAE in super varieties at N150 and N225 for indica and only at N225 for japonica suggests that indica super varieties might be more sensitive to N than japonica super varieties.

The difference in NRE was not significant between TP and WDS, except for the japonica CK variety at N225, in which the NRE was 42% and 36% greater in TP than in WDS (Fig. 3). In both years, irrespective of N rate and planting method, no difference was found for NRE between super and CK varieties, except for japonica at N225 in 2011, in which the NRE was 130% and 53% greater in the super rice varieties than in the CK varieties under WDS and TP, respectively (Fig. 3).

In both years, irrespective of N rate and variety group, plants under TP had generally greater NPE than those under WDS, except for super indica varieties at N150 and japonica varieties at N225 in 2011, in which no difference was found between TP and WDS (Fig. 4). The average NPE in the TP system was 43 and 42 kg grain kg^{-1} N in 2011 and 2012, respectively, which were 33% and 55% higher than in the WDS system. Among varieties, NPE was generally higher in super rice than in CK varieties for the indica group, but it was similar for the japonica group. In both years, the average NPE of super indica varieties was 39 kg grain kg^{-1} N, which was 82% higher than that of CK varieties.

Discussion

Grain yield response to planting method

In current rice production systems with high economic outputs, seedling establishment has become more important than ever. Direct seeding methods have been suggested to replace the traditional transplanting method because of their advantage in saving labor without yield loss for high-yielding rice varieties [20]. In general, rice yield under direct seeding in farmers' fields is lower than under the transplanting method [19]. However, there is a genotypic and environmental interaction for the yield of direct-seeded rice. No distinct difference in grain yield was found between direct seeding and transplanting under flooded conditions [20,33]. A previous study reported that direct-seeded rice had superior grain yield than transplanted rice when short-duration varieties were used, but had equal or lower grain yield when medium- and long-duration varieties were used [34].

In our study, grain yield under WDS (7.69 t ha^{-1}) was generally lower (9.9%) than under TP (8.58 t ha^{-1}). However, the effect of planting methods on grain yield varied with N rate and variety group. No difference was found between TP and WDS at N150 in all the variety groups, except for the super japonica varieties. These results indicate that the optimum N rate for rice under WDS is 150 kg ha^{-1} in current conditions. In addition, the reduced growth period, ranged from 6–12 days across nitrogen levels and years, was also observed in the WDS system (data not shown), which might have reduced the photosynthetic duration during grain filling, reduced the biomass accumulation, and finally resulted in the yield decline [33]. Interestingly, Liangyoupei9 and Peiliangyou2046 had greater grain yield in WDS than in TP, indicating the possibility of breeding varieties suitable for direct seeding at optimum N rates.

Grain yield response to super rice varieties

A total of 13 and 11 super high-yielding rice varieties (hybrid indica and inbred japonica) were used in 2011 and 2012, respectively. These varieties have been approved as super rice and widely spread in farmers' fields in China (www.moa.gov.cn). Many field experiments have shown that super rice can achieve a grain yield of around 12 t ha^{-1} [7,9,10]; however, the yield output of farmers' fields is still arguable. One of the important results of our study is that 10.9 t ha^{-1} was the highest yield observed under the environment at Fuyang, China. The high-yielding varieties did not meet the grain yield criteria for super rice, which was a minimum of 11.7 t ha^{-1} for a single season of indica/japonica rice in the Yangtze River Basin. This could partly be the effect of the environment on yield behavior and indicates a need for further work on improving yield performance of super rice varieties through agronomic and other management practices. On the other hand, relative grain yield was still superior for super rice varieties than for CK varieties. Current results show that the effect of super rice varieties on grain yield varied with the variety group. In both years, no difference was found for grain yield between super rice and CK varieties at N150 regardless of planting method. However, grain yield difference was dramatic in the japonica group at N225. Averaged over years, super rice varieties had 11.3% and 14.1% higher yield than CK varieties in the WDS and TP systems, respectively. The results suggest that high N input could contribute to narrowing the yield gap in super rice varieties and also provide evidence that super rice was bred for high fertility conditions.

Nitrogen-use efficiency traits of super rice

Nitrogen is one of the most active elements for rice growth and yield performance. To achieve improved yield over traditional varieties, super rice varieties are generally bred in fertile conditions through transplanting [9,26,27]. During the last decade, the effect of N application on the NUE of super rice has been well-studied in the TP system; however, contradictory results still exist. In a recent study, the NUE of super japonica rice is greater at 246 kg N ha^{-1} than at 214 kg N ha^{-1} [35]. Another study compared the NPE and NAE of super rice Shennong265 and CK Liaojing294 at two N levels (150 and 250 kg ha^{-1}) and reported that the NPE and NAE values of the super rice variety were greater than that of the CK variety at higher N application and lesser at lower N application [36]. Moreover, the physiological parameters affecting NUE also changes when the planting method shifts from TP to WDS. In general, panicle density of direct-seeded rice was greater than that of transplanted rice, but spikelet number per panicle was reduced because of the internal compensation [37]. It was also reported that the lower tissue N concentration during spikelet differentiation in direct-seeded rice may limit spikelet numbers compared with transplanted rice [21,37]. In our study, more N was accumulated in super rice than in the check at N225 in japonica, but no difference was found between super and CK varieties at N0 and N150, indicating that the yield increase of japonica super rice resulted from high N inputs. Similar results were also found for NAE. These results suggest that super rice varieties have an advantage at NUE when high N is applied.

Acknowledgments

The authors would like to thank Ms. Priscilla Grace Cañas for providing comments on the manuscript.

Author Contributions

Conceived and designed the experiments: SC Xiufu Zhang. Performed the experiments: DYW CMX CLJ Xiaoguo Zhang. Analyzed the data: SC DYW. Contributed reagents/materials/analysis tools: X. Zhao. Contributed to the writing of the manuscript: SC Xiufu Zhang BSC.

References

1. Zhu DF (2000) Bridging the rice yield gap in China, In Bridging the Rice Yield Gap in the Asia, MK Papademetriou, FJ Dent, EM Herath, Editors. Pacific Region, Food and Agriculture Organization of the United Nations Regional Office for Asia and the Pacific: Bangkok, Thailand. p. 69–83.
2. Normile D (2008) Reinventing rice to feed the word. Sci 321: 330–333.
3. Peng SB, Khush GS, Virk P, Tang QY, Zou YB (2008) Progress in ideotype breeding to increase rice yield potential. Field Crops Res 108: 32–38.
4. Yuan L (1996) Prospects for yield potential in rice through plant breeding. Hybrid Rice 11: 1–3.
5. Amano T, Shi C, Qin D, Tsuda MYM (1996) High yielding performance of paddy rice achieved in Yunnan Province, China: I. High yielding ability of japonica F1 hybrid rice, Yu-Za 29. Jpn J Crop Sci 65: 16–21.
6. Ying JF, Peng S, He Q, Yang H, Yang CD, et al. (1998) Comparison of high-yield rice in tropical and subtropical environments: I. Determinants of grain and dry matter yields. Field Crops Res 57: 71–84.
7. Cheng SH, Zhuang JY, Fan Y, Du JH, Cao LY (2007) Progress in research and development on hybrid rice: a super-domesticate in China. Ann Bot 100: 959–966.
8. Cheng SH, Fang FP, Zhu ZW, Zhu D, Xiang J (2013) Reports on China's rice production, China National Rice Research Institution: Beijing, China. p. 33.
9. Zou YB, Zhou SY, Tang QY (2003) Status and prospect of high yielding cultivation researches on China super hybrid rice. J Hunan Agric Univ (Nat Sci) 29: 78–84.
10. Peng SB, Laza RC, Visperas RM, Khush GS, Virk P, et al. (2004) Rice: progress in breaking the yield ceiling. In Proceedings of the 4th International Crop Science Congress.
11. Zong S, Lu C, Zhao L, Wang C, Dai Q, et al. (2000) Physiological basis of high yield of an intersubspecific hybrid rice, Liangyoupeijiu. J. Nanjing Agric. Tech. College, 16 p. 8–12 (in Chinese, with English abstract).
12. Wang X, Tao L, Yu M, Huang X (2002) Physiological characteristics of super hybrid variety Xieyou 9308. Chin J Rice Sci 16: 38–44 (in Chinese with English abstract).
13. Chen S, Cai SG, Chen X, Zhang GP (2009) Genotypic differences in growth and physiological responses to transplanting and direct seeding cultivation in rice. Rice Sci 16: 143–150.
14. Bhushan L, Ladha JK, Gupta RK, Singh S, Tirol-Padre A, et al. (2007) Saving of water and labor in a rice–wheat system with no-tillage and direct seeding technologies. Agron J 99: 1288–1296.
15. Chen S, Xia GM, Zhao WM, Wu FB, Zhang GP (2007) Characterization of leaf photosynthetic properties for no-tillage rice. Rice Sci 14: 283–288.
16. Derpsch R, Friedrich T (2009) Global overview of conservation agriculture no-till adoption. *In* 4th World Congress on Conservation Agriculture New Delhi, India.
17. Farooq M, Siddique KHM, Rehman H, Aziz T, Lee DJ, et al. (2011) Rice direct seeding: experiences, challenges and opportunities. Soil Till Res 111: 87–98.
18. Dawe D (2005) Increasing water productivity in rice-based systems in Asia-past trends, current problems, and future prospects. Plant Prod Sci 8: 221–230.
19. Pandey S, Velasco LE, Suphanchaimat N (2002) Economics of direct seeding in Northeast Thailand, in Direct Seeding: Research Strategies and Opportunities, Pandey S, et al., Editors. IRRI: Manila, Philippines. p. 139–160.
20. Huang M, Zou Y, Feng Y, Cheng Z, Mo Y, et al. (2011) No-tillage and direct seeding for super hybrid rice production in rice–oilseed rape cropping system. Eur J Agron 34: 278–286.
21. Dingkuhn M, Schnier H, De Datta S, Dörffling K, Javellana C, et al. (1990) Nitrogen fertilization of direct-seeded flooded vs. transplanted rice: II. Interactions among canopy properties. Crop Sci 30: 1284–1292.
22. Naklang K, Shu F, Nathabut K (1996) Growth of rice cultivars by direct seeding and transplanting under upland and lowland conditions. Field Crops Res 48: 115–123.
23. Chen S, Zhang X, Zhang G, Wang D, Xu C (2012) Grain yield and dry matter accumulation response to enhanced panicle nitrogen application under different planting methods (*Oryza sativa* L.). Aust J Crop Sci 6: 1630–1636.
24. Mae T (1997) Physiological nitrogen efficiency in rice: Nitrogen utilization, photosynthesis, and yield potential, In Plant nutrition for sustainable food production and environment, Springer. p. 51–60.
25. Wada G, Shoji S, Mae T (1986) Relationship between nitrogen absorption and growth and yield of rice plants. Jpn Agric Res Q 20: 135–145.
26. Lin X, Zhu D, Zhang Y (2002) Achieving high yielding plant type in a "super" rice variety by optimising crop management. China Rice 2: 10–12 (in Chinese with English abstract).
27. Cheng SH, Cao LY, Zhuang JY, Chen SG, Zhan XD, et al. (2007) Super hybrid rice breeding in China: achievements and prospects. J Integrative Plant Biol 49: 805–810.
28. Katsura K, Maeda S, Horie T, Shiraiwa T (2007) Analysis of yield attributes and crop physiological traits of Liangyoupeijiu, a hybrid rice recently bred in China. Field Crops Res 103: 170–177.
29. Shi HR, Zhang WZ, Xie WX, Yang Q, Zhang ZY, et al. (2008) Analysis of matter production characteristics under different nitrogen application patterns of japonica super rice in north China. Acta Agronomica Sinica 34: 1985–1993 (in Chinese with English abstract).
30. Peng SB, Huang JL, Zhong XH, Yang JC, Wang GH, et al. (2002) Research strategy in improving fertilizer nitrogen use efficiency of irrigated rice in China. Scientia Agricultura Sinica 35: 1095–1103 (in Chinese with English abstract).
31. Chen S, Zhou WH, Zeng FR, Zhang GP (2011) Effect of planting method on grain quality and nutrient utilization for no-tillage rice. Comm Soil Sci Plant Analysis 42: 1324–1335.
32. Bremner JM, Mulvaney CS (1982) Nitrogen total, In Agronomy: A Series of Monographs, Methods of Soil Analysis, Part 2. Chemical and Microbiological Properties (second ed.), AL Page, Editor. p. 595–624.
33. Arromratana U, Na Nagara T, Kaewkhongkha T, Uraipong B, Morita HA, et al. (2002) Development of no-tillage direct seeding cultivation of rice. JIRCAS Working Rep 30: 129–138.
34. Peng S, Cassman KG, Virmani SS, Sheehy J, Khush GS (1999) Yield potential trends of tropical rice since the release of IR8 and the challenge of increasing rice yield potential. Crop Sci 39: 1552–1559.
35. Shi LH, Ji XH, Zhu XQ, Li HS, Peng H, et al. (2010) A preliminary study on optimizing nitrogen fertilization amount at different phases to enhance the storage capacity of super hybrid rice. Scientia Agricultura Sinica 43: 1274–1281 (in Chinese with English abstract).

36. Song GY, Xu ZJ, He M, Zhang XJ, Chen WF, et al. (2007) Influences of nitrogen fertilizer on the assimilation and use efficiency of nitrogen in rice. Soil Fert Sci China 4: 44–48.

37. Schnier H, Dingkuhn M, De Datta S, Mengel K, Faronilo J (1990) Nitrogen fertilization of direct-seeded flooded vs. transplanted rice: I. Nitrogen uptake, photosynthesis, growth, and yield. Crop Sci 30: 1276–1284.

27

Designing Mixed Species Tree Plantations for the Tropics: Balancing Ecological Attributes of Species with Landholder Preferences in the Philippines

Huong Nguyen[1,2]*, David Lamb[1,3], John Herbohn[1,2], Jennifer Firn[4]

1 School of Agriculture and Food Sciences, The University of Queensland, St Lucia, Australia, **2** Forest Industries Research Centre, The University of the Sunshine Coast, Sippy Downs, Australia, **3** Centre for Mined Land Research, The University of Queensland, St Lucia, Australia, **4** Faculty of Science and Technology, School of Earth, Environmental and Biological Sciences, Queensland University of Technology, Brisbane, Australia

Abstract

A mixed species reforestation program known as the Rainforestation Farming system was undertaken in the Philippines to develop forms of farm forestry more suitable for smallholders than the simple monocultural plantations commonly used then. In this study, we describe the subsequent changes in stand structure and floristic composition of these plantations in order to learn from the experience and develop improved prescriptions for reforestation systems likely to be attractive to smallholders. We investigated stands aged from 6 to 11 years old on three successive occasions over a 6 year period. We found the number of species originally present in the plots as trees >5 cm dbh decreased from an initial total of 76 species to 65 species at the end of study period. But, at the same time, some new species reached the size class threshold and were recruited into the canopy layer. There was a substantial decline in tree density from an estimated stocking of about 5000 trees per ha at the time of planting to 1380 trees per ha at the time of the first measurement; the density declined by a further 4.9% per year. Changes in composition and stand structure were indicated by a marked shift in the Importance Value Index of species. Over six years, shade-intolerant species became less important and the native shade-tolerant species (often Dipterocarps) increased in importance. Based on how the Rainforestation Farming plantations developed in these early years, we suggest that mixed-species plantations elsewhere in the humid tropics should be around 1000 trees per ha or less, that the proportion of fast growing (and hence early maturing) trees should be about 30–40% of this initial density and that any fruit tree component should only be planted on the plantation margin where more light and space are available for crowns to develop.

Editor: Han Y.H. Chen, Lakehead University, Canada

Funding: The study was funded from ACIAR Smallholer Forest projects ASEM/2003/052 and ASEM/2006/091. The funders had no role in study design, data collection and analysis, decision to publish, or preparation of the manuscript.

Competing Interests: The authors have declared that no competing interests exist.

* E-mail: h.nguyen22@uq.edu.au

Introduction

Over the past two decades, there has been rising interest in planting mixtures of tree species to establish plantations that provide multiple services including production and improved nutrient cycling and also to provide more biodiversity at the landscape level [1]. Mixed-species plantations have the potential to generate a variety of forest products, as well as a range of ecosystem services [2–4]. Mixed species plantations are often established using just two or three species but sometimes far more diverse mixtures have been used which include representatives of a variety of successional stages [5].

A major issue facing those wishing to establish mixed species plantations is that knowledge of the silvicultural attributes of most species is usually so limited that it is difficult to predict how these will grow when planted into novel combinations [6]. The growth strategies of species growing in natural forests can provide an indication of the role they might play in a plantation including whether they are shade-intolerant, a canopy dominant species or whether they can grow in sub-canopy strata. The interactions among species will strongly influence the productivity of mixtures [7,8]. Species attributes that could be used as possible indicators of the performance of species grown in a mixed-species stand include shade tolerance/intolerance, height growth rate, crown structure, foliar phenology, and root depth and phenology [9,10]. Combinations of species with complementary traits can reduce competition and allow for the most efficient use of limiting resources like water, nutrients and light in plant communities [6].

Identifying complementary species is a difficult task, particularly in the tropics when native tree species are preferred, as there is a limited knowledge of growth strategies of most native species. A number of challenges are likely to take place in newly established mixed-species stands because of competition and differences in the species growth rates. Trees grown in mixed species stands can sometimes suffer both intra-specific competition and inter-specific competition. Both of these are influenced by tree density [11–13]. High initial planting density may facilitate early site capture, reduce weed control, improve form and lower stem taper [14]. On the other hand, higher densities mean costs are greater because more seedlings are needed and stand thinning is necessary at an earlier stage of stand development. This problem may not be as great as it seems if the thinned trees have some economic value

Table 1. Site characteristics and planting history of 18 of the mixed-species sites in Leyte Province, the Philippines*.

Site	Site location	Year planted	Area (ha)	Soil type	Topography	No. of plots
02	Marcos, Baybay	1995	0.61	Clay loam	Slightly to moderately rolling	2
03	Catmon, Ormoc	1998	1.4	Clay loam	Flat	12
04	Patag, Baybay	1998	1.0	Clay loam	Slightly rolling	5
05	Cienda, Baybay	1996	0.9707	Clay to clay loam	Flat	9
06	Pomponan, Baybay	1997	0.38	Clay loam	Slightly rolling	2
07	Punta, Baybay	1996	5.442	Limestone	Moderately rolling	9
08	Maitum, Baybay	1996	0.478	Clay loam	Slightly to moderately rolling	2
09	Mailhi, Baybay	1996	3.22	Clay to clay loam	Slightly to moderately rolling	10
10	Vila Solidasidad, Baybay	1995	0.4377	Clay	Flat	4
11	Maitum, Baybay	1996	0.4686	Limestone	Moderately rolling	2
12	Maitum, Baybay	1996	0.9862	Limestone	Slightly rolling	2
13	Maitum, Baybay	1996	0.2518	Clay loam	Moderately rolling	2
14	Pomponan, Baybay	1996	0.9518	Clay loam	Slightly to moderately rolling	2
15	Pomponan, Baybay	1997	0.438	Clay	Moderately rolling	2
16	Pomponan, Baybay	1997	0.41	Clay loam	Moderately rolling	2
17	Pomponan, Baybay	1999	0.8475	Sandy loam	Flat	2
19	Licoma, Ormoc	2000	0.25	Clay loam	Moderately rolling	2
22	Milagro, Ormoc	1996	1.5	Clay loam	Flat	7

*Milan et al. 2004.

and there are usually some fast growing species that can be harvested at an early age.

The Rainforestation Farming plantations in the Philippines present an ideal opportunity to explore basic questions concerning forest dynamics, optimal planting densities and human-use patterns of tropical polycultures [15–19]. We hypothesise that these mixed species plantations could balance the ecological attributes of species with landholder preferences in the Philippines. Here, over three time periods, we measure changes in the structure and species composition of these mixtures. We consider the following questions:

(i) How has the composition and structure of these mixed-species plantings changed over time?

(ii) How have patterns of species loss (due to mortality and harvesting) and recruitment been affected by the design of these species mixtures?

(iii) What are the implications for the future design and management of tropical polycultures used by smallholders?

Materials and Methods

Ethics Statement

The study was permitted by the owners of the lands (for more details see Table 1) to be conducted in 18 sites. It was not possible to sample from all 28 sites that were established under the Rainforestation Farming system because several plantations had been detrimentally affected by fire, harvesting, clearing for other agricultural activities; because access was not granted by the land owners; or did not meet minimum requirements for measurements (e.g. trees greater than 5 cm diameter).

Study Area

The study was conducted in Leyte province, which is one of two provinces located in Leyte Island (Figure S1 in File S1). This is the eighth largest island in the middle of the Philippines archipelago and covers about 800,000 ha [20,21]. Leyte Province has a humid monsoon climate and the average rainfall in the study area for the years 1980–2000 was 2,686 mm with an annual variation of between 1,775 mm in 1987 and 3,697 mm in 1999 [22]. Although there is no pronounced dry season, the region experiences its lowest rainfall of less than 100 mm between March and May [23]. Dry periods of several months duration with rainfall of less than 100 mm can sometimes occur as was the case during the 'El Nino' year of 1993, the year in which the project commenced. The average annual temperature is 27.5°C and ranges from 26.3 to 28.7°C. The relative humidity is always high and the average monthly level for the years 1980–2000 ranged from 75.1% in March to 80.1% in October [22]. The soils are derived from volcanic parent material and were strong acidic with a pH 4.1–4.9 [24]. The natural vegetation in the region is a species-rich, lowland dipterocarp forest but natural forest now only remains on the less accessible slopes of the Leyte cordillera [25].

A unique polyculture reforestation program was started in the Philippines in 1992 called the Rainforestation Farming system [26–29]. It involved 28 small-scale mixed-species plantations on private farms on Leyte Island. These plantings were established to provide smallholders with a form of reforestation that generated income from an early age while also providing the benefits of increased biodiversity in the highly cleared and mainly agricultural landscape. The intent was to create resilient and sustainable forms of reforestation that would also be financially attractive to local farmers [27,28]. The program focussed on native species and used a large number of species to establish mixtures to resemble a few natural forests in the region. In response to demands by landholders some exotic species were also used in the mixtures.

Table 2. Equations to calculate Important value index (IVI) of species.

Index	Equation
Importance value index (IVI)	Relativedensity + Relativefrequency + Relativedominance
Relative density	$\frac{\text{Density of a species}}{\text{Total density of all species}} \times 100$
Relative frequency	$\frac{\text{Frequency of a species}}{\text{Total frequency of all species}} \times 100$
Relative dominance	$\frac{\text{Dominance of a species}}{\text{Total dominance of all species}} \times 100$
Density	$\frac{\text{Number of a species}}{\text{Total area sampled}}$
Frequency	$\frac{\text{Area of plots in which a species occurs}}{\text{Total area sampled}}$
Dominance	$\frac{\text{Total basal area of a species}}{\text{Total area sampled}}$

The system used approximately 100 endemic pioneer and shade-intolerant tree species, longer-lived species of Dipterocarp, fruit trees and a limited number of exotic timber species. These were planted to create a series of small scale plantations in the average area of about 1 ha [29].

At the beginning of the Rainforestation Farming project, species assessed as shade-intolerant pioneers were planted in a spacing of 2 m×2 m to provide an environment for supposedly shade-tolerant timber and fruit tree species to be established in the following year. These were inter-planted at a general spacing of 2 m×1 m [28]. The estimated density at planting time was about 5000 trees/ha at sites. From 7 to 40 species were planted at each farm. The aim was to create a three storied structure with pioneer and shade-intolerant trees in the upper canopy layer, shade-tolerant trees and fruit trees in the second storey and shade-tolerant crops in the lowest layer. Supposedly there would be roughly equal numbers of shade-tolerant and shade-intolerant species. In fact the numbers and identity of species planted at each farm were not consistent but depended upon seedling availability at the time, and the preferences of the local landholder. As a consequence detailed records of plantings at each farm were not kept. Sites were developed over a period of several years from 1995 to 2000. Further details of the Rainforestation Farming methodology are given by Kolb [22].

Data Collection and Analysis

Data were collected from 80 plots distributed across 18 of the mixed-species plantations (Table 1). These plots have also been used in a recent biodiversity-productivity study [18]. The first measurement of species composition and stand structural characteristics was undertaken from February to March 2006 when the trees were aged between six and 11 years. Measurements of trees and site properties were collected from randomly located circular plots with a radius of 5 m (78 m^2 area) within the plantations. The number of plots sampled at each farm ranged from 1 to 12 plots on the size of the farm's plantings, with the number being determined by the size of site so that the sampling area occupied at least 5% of plantation [30,31]. All plots and trees within them were permanently marked in the field. In each plot, all trees were counted and identified to the species level and the height (H) and diameter at breast height (dbh) were measured. Each plot contained at least seven trees greater than 5 cm in diameter. All the plots were re-measured during 2008 and again in 2012. On each occasion new recruits were identified and measured and tree deaths were recorded. A distinction was made between deaths and by harvesting (i.e. evidence in the form of stumps) and natural tree deaths.

The species were grouped into two categories based on provenance (i.e. native species and exotic species) and three categories based on ecological types. These included shade-intolerant species (pioneer as well as longer lived secondary forest species), shade-tolerant species (species from the Dipterocarpaceae plus other shade tolerant species) and fruit tree species. All of the shade-tolerant species but especially the Dipterocarps are regarded locally as having highly valued timbers.

Annual rates of mortality and recruitment were calculated as the mean annual proportion of lost trees or new recruits using the total number of stems at the time of the first measurement as the base line.

Table 3. Changes of species composition of canopy trees in the Rainforetation plantations over time.

Category	Remaining number of species		
	2006	2008	2012
Shade-intolerant species	43	38	35
Shade-tolerant species	19	18	19
Fruit-tree species	14	12	11
Native species	57	50	49
Exotic species	19	18	16
Total	76	68	65*

*includes a new species recruited from outside plot.

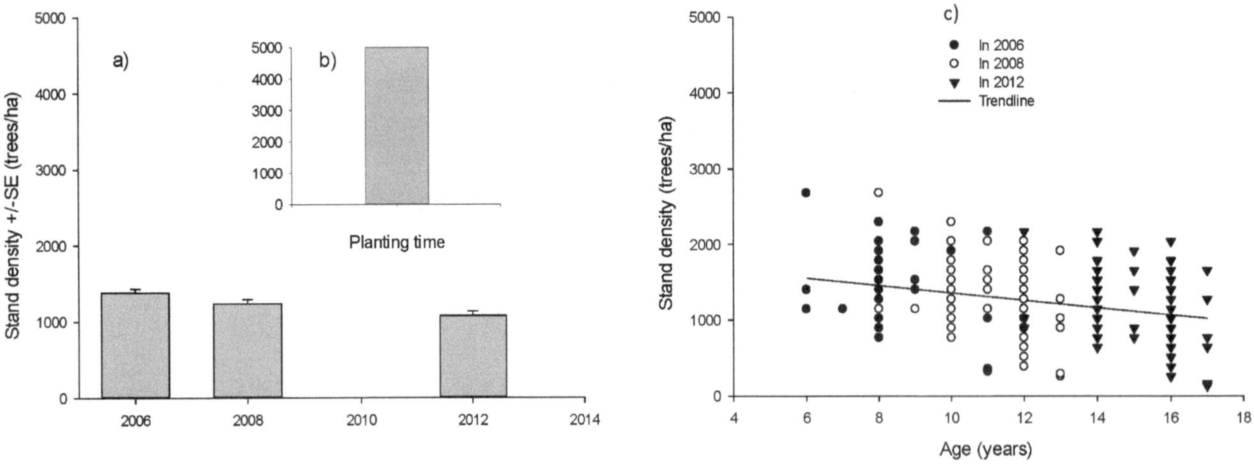

Figure 1. Changes in stand density of Rainforestation plantations over time. Mean stand density at measurements (a), estimated stand density at the planting time (b), and stand density of plots at different ages (c).

The estimated times needed for species growing in Rainforestation stands to reach threshhold sizes of 10 cm and 30 cm were calculated based on relative growth rates of the 32 common species across sites.

The Importance Value Index (IVI; see Table 2) was used to assess the importance of different species in each plantation [32–36].

One-way ANOVA test was used to compare stand densities and the Importance Value Indices between the three measurements. The differences in tree loss and recruitment between species groups were also analysed using a one-way ANOVA combined with LSD post-hoc tests for all pairwise comparisions between group means.

Statistical analysis was undertaken using SPSS 21 and Excel 2010 and data was visually presented using SigmaPlot 12.3.

Results

Species Compositional Changes

A total of 76 canopy species of trees were recorded at 80 plots distributed across the 18 sites at the time of the first measurement in 2006 (Table 3). These represented 33 families and 58 genera. They included 43 shade-intolerant species, 8 shade-tolerant non-Dipterocarp species, 11 Dipterocarp species (all shade-tolerant) and 14 fruit tree species. Of these, 57 are native species and 19 are exotic species.

Each of the different species types were present at most sites, but the proportion of trees represented by the different types of species varied: exotic species represented 36% of all trees, while shade intolerant species represented 78% of all trees. Fruit trees were present in only 21 of the 80 plots and represented 8% of all trees.

Overtime, we found the number of species present decreased from 76 in 2006 to 65 in 2012 (Table 3). The species lost included eight that were shade-intolerant, three fruit tree species and one species of Dipterocarpaceae. On the other hand an additional species (*Strombosia philippinensis*) was recruited from trees growing outside the plots in the period between 2008 and 2012 and grew up to exceed the 5 cm dbh size threshold above which trees were assessed.

Changes in Stand Density Caused by Mortality and Harvesting

The average stand density across all sites decreased from an estimated planting density of 5000 trees ha^{-1} to 1383±50 trees ha^{-1} at the time of first measurement in 2006 when the sites were aged between six and 11 years. From 2006 the average stand density then decreased further to 1145±56 trees ha^{-1} in 2012 (Figure 1). These changes were not uniform and density was constant in 19 plots (24% total plots) during the whole period over which measurements were made. The average site stand density differed significantly over time ($F_{18} = 4.720$, $p = 0.013$).

Part of the decline was due to between-tree competition but some was also caused by tree harvesting by farmers. Loss due to competition (i.e. mortality) over the period differed between the various categories of species. Overall, shade-intolerant species and fruit trees lost around 5% of trees each year but only 0.7% of the shade-tolerant trees were lost (Table 4). Around 5.4% and 8.6% of individuals of native and exotic species respectively were lost each year due to competition and harvesting (Table 4, Figure 2). There was no significant difference in the loss of stems between native and exotic species ($F_{18} = 0.450$, $p = 0.507$) but a difference between functional groups of shade-intolerant, shade-tolerant and fruit trees ($F_{18} = 3.111$, $p = 0.054$) across sites. We found that the shade-tolerant group had lost significantly fewer stems than the other groups ($p = 0.030$ and 0.043 between shade-tolerant and shade-intolerant, and between shade-tolerant and fruit trees, respectively). We found no significant difference in terms of losing individuals between shade-intolerant species and fruit trees ($p = 0.881$). Most of the deaths occurred in trees less than 10 cm dbh although a few larger trees had also died by the time of the final measurement (Figure 3). Most of the losses occurred in smaller size classes and were more common in the 2006–2008 period than in the 2008–2012 period (Figure 3).

Evidence of harvesting and the identity of these trees could be seen in the form of old stumps. Harvesting commenced when the trees were in the 5–10 cm dbh class but harvesting was common in trees up to around 25 cm dbh (Figure 3). During the six years, 20.4% of all trees >5 cm dbh recorded in 2006 were harvested. The smaller sized trees harvested belonged to 38 species of all shade-intolerant, shade-intolerant species and fruit trees while the larger trees belonged to only 11 shade-intolerant species (e.g. *Melia dubia, Gmelina arborea, Leucaena leucocephala, Terminala macrocarpa* and

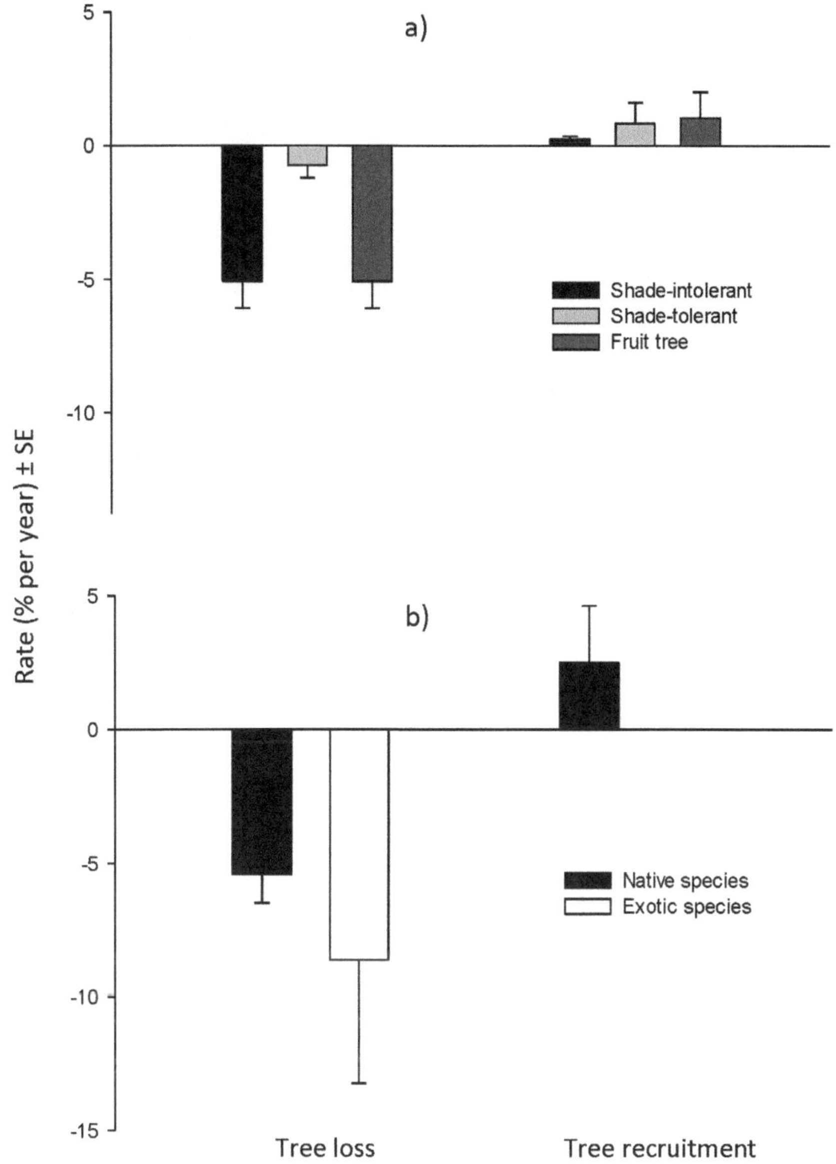

Figure 2. Annual rates of mortality and recruitment at sites during the period of 2006–2012. In categories of species ecology (a) and in categories of species provenance (b).

Swietenia macrophylla) and only one individual of fruit tree (i.e. *Sandroricum koetjape*). Some trees of these species were also harvested at dbh < = 25 cm. The other common species that were harvested at dbh <25 cm included *Vitex parviflora*, *Gymnostoma rumphianum*, *Artocarpus heterophyllus*, *Samanea saman*, *Pterocymbium tinctorium* and *Tectona grandis*.

Most of trees which were died due to competition or disease had dbh smaller than 25 cm, including shade-intolerant, shade-tolerant and fruit trees; only one death of decayed tree of *Melia dubia* had dbh around 30 cm.

In contrast to these losses, a number of trees were also added to the stands as trees grew larger and exceeded the 5 cm dbh size class. These were found in 23 of the 80 plots. At the time of the first measurement in 2008 trees belonging to 9 species had been added to the stands and by 2012 trees from 20 native species were being added. They included Dipterocarp species (e.g. *Parashorea plicata*, *Shorea contorta*, *Hopea malibato*, *Hopea plagata*) and some individuals of other shade-tolerant, shade-intolerant and fruit trees. There were no exotic species found in the recruitment (Table 4, Figure 2).

Growth Rates

By 2006 about 2% of all trees exceeded 30 cm dbh but this has increased to 6.7% by 2012 (see Figure S2 in File S1). The results showed that three shade-intolerant species (i.e. *Leucaena leucocephala*, *Melia dubia* and *Gmelina arborea*) could reach the dbh threshold of 30 cm by age 10 years. Most these species could achieve 30 cm dbh before 20 years old of planting. The estimated time it takes for particular species to reach a harvestable age is shown in Table 5.

Importance Value Index

Although Importance Value Index of species was not significantly different between measurements across sites for all species ($F_{77} = -1.1E\text{-}13$, p>1), mortality and early harvesting by land-

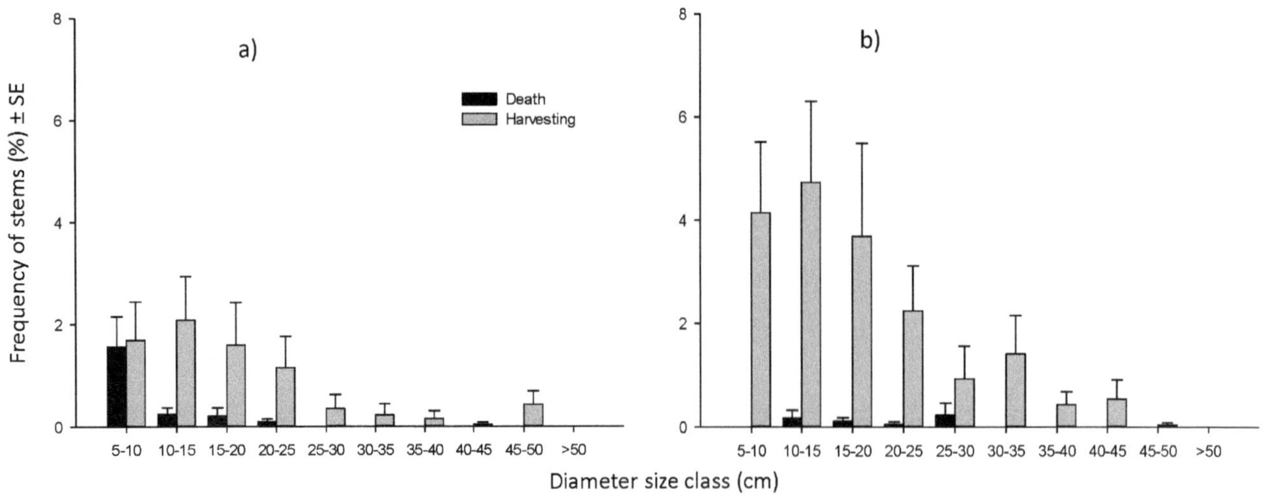

Figure 3. Size class distributions of lost trees (death and harvesting) at the Rainforestation sites. In the periods of 2006–2008 (a) and 2008–2012 (b).

holders caused a change in the Importance Value Index of some species in specific plantations (Figure 4). At the time of the first measurement shade-intolerant species had the highest Importance Value Indices due to their density, relative common and uniform distribution and their faster-growth (e.g. *Swietenia macrophylla*, *Gmelina arborea*, *Terminalia macrocarpa*, *Vitex parviflora* and *Gymnostoma rumphianum*). However, their Importance Values declined because of harvesting and the shade-tolerant species gradually acquired greater Importance Values (e.g. *Gmelina arborea*, *Gymnostoma rumphianum*, *Melia dubia* and *Leucaena leucocephala*) (Figure 4 and Figure S3 in File S1).

Discussion

Our results indicate that the Rainforestation Farming plantations are a highly dynamic system. Changes in stand structure and species composition have been brought about by a combination of anthropogenic factors and natural mortality. It appears that the harvesting of trees by farmers was carried out in an entirely opportunistic way according to their particular circumstances; it was not based on silvicultural prescriptions that specified the timing or intensity of tree removals. This means it is unlikely that production was optimised in the way that managers of industrial plantations usually seek to achieve. On the other hand, this was exactly the intention of the designers of the Rainforestation Farming system; optimal productivity has been traded-off for the sake of flexibility. Mortality has been hastened by the very high initial planting densities. Not surprisingly this has mostly occurred amongst the shade-intolerant species. The net effect of these two processes has meant that over the six years of the study, slower growing native species have become more dominant while faster growing (mainly shade-intolerant) species have become less dominant. At this point it is not clear how these stands should be now managed because the timing of any future harvesting (i.e. financial benefit to landholders) is difficult to specify. It is likely that farmers will continue to remove trees once they reach some threshold size chosen by the landholder to suit their purpose irrespective of the value of the timbers or market for the logs.

Fruit trees were added to the species mixtures in the expectation that once mature, they would yield a large annual crop of fruit. Experience has shown however that this has not occurred even though some fruit trees are now more than 15 years old. Typically fruit trees require wide spacing and full sun to bear productively [15,35–40]. But in the Rainforestation Farming design they were incorporated into a closely spaced system and were quickly shaded by faster growing pioneer species. As a consequence there appears

Table 4. Mortality and recruitment of species groups in Rainforestation plantations over time during the period of 2006–2012.

Category	Proportion of trees	
	Mortality (% per year)± SE	**Recruitment (% per year)± SE**
Shade-intolerant species	5.08±1.01	0.24±0.11
Shade-tolerant species	0.73±0.49	0.83±0.78
Fruit-tree species	4.79±2.11	1.03±0.98
Native species	5.43±1.08	2.51±2.13
Exotic species	8.62±4.63	0
New species in plots (but planted species in farms)		0.004±0.004
Total	4.89±1.01	0.49±0.32

Table 5. The estimated time needed for species growing in Rainforestation Farming stands to reach threshold sizes for firewood (10 cm dbh) and lumber (30 cm dbh).

Species	Provenance	Estimated time (years) to reach	
		10 cm dbh (Mean ± SD)	30 cm dbh (Mean ± SD)
Shade-intolerant species:			
Ipil-Ipil (*Leucaena laucocephala*)	Exotic	3.3±1.0	9.8±3.1
Bagalunga (*Melia dubia*)	Native	5.1±2.4	15.2±7.3
Gmelina (*Gmelina arborea*)	Exotic	5.9±3.5	17.6±10.5
Taluto (*Pterocymbium tinctorium*)	Native	7.0±2.6	20.9±7.7
Teak (*Tectona grandis*)	Exotic	7.5±3.6	22.6±10.8
Santol (*Sandoricum koetjape*)	Native	7.7±3.0	23.2±8.9
Mt Agoho (*Gymnostoma rumphianum*)	Native	7.8±2.8	23.5±8.3
Kalumpit (*Terminalia macrocarpa*)	Native	7.9±3.9	23.7±11.6
Raintree (*Samanea saman*)	Exotic	7.9±5.0	23.7±15.1
Dao (*Dracontamelon dao*)	Exotic	8.6±3.7	25.7±11.0
Nangka (*Artocarpus heterophyllus*)	Native	8.9±2.7	26.7±8.2
Thailand acacia (*Senna siamea*)	Exotic	9.3±2.5	28.0±7.4
Mahogany (*Swietenia macrophylla*)	Exotic	9.3±4.2	27.9±12.6
Molave (*Vitex parviflora*)	Native	9.4±3.1	25.2±9.3
Antipolo (*Artocarpus blancoi*)	Native	9.9±4.4	29.8±13.3
Bitanghol sibat (*Calophyllum lancifolium*)	Native	10.4±3.4	31.2±10.3
Narra (*Pterocarpus indicus*)	Native	10.6±3.7	31.7±11.2
Hindang laparan (*Myrica javanica*)	Native	10.9±4.7	32.6±14.2
Lanipga (*Toona ciliate*)	Exotic	10.3±2.8	30.9±8.5
Malakawayan (*Podocarpus rumphii*)	Native	18.2±4.2	54.7±12.5
Shade-tolerant species:			
Mayapis (*Shorea palosapis*)	Native	5.3±1.6	16.0±4.9
Tangeli (*Shorea polysperma*)	Native	6.5±2.7	19.4±8.1
Apitong hagakhak (*Dipterocarpus kunstleri*)	Native	8.0±3.6	24.1±10.8
Marang banguhan (*Artocarpus odoratissimus*)	Native	8.3±3.0	24.8±9.0
Bagtikan (*Parashorea plicata*)	Native	8.7±3.8	26.2±11.4
Rambutan (*Nephelium lappaceum*)	Native	8.8±2.4	26.3±7.2
Cacao (*Theobroma cacao*)	Exotic	8.9±2.6	26.7±7.9
White lauan (*Shorea contorta*)	Native	8.9±3.7	26.8±11.1
Almaciga (*Agathis philippinensis*)	Native	9.5±2.9	28.6±9.5
Durian (*Durio zibethinus*)	Exotic	9.8±3.7	29.5±11.2
Yakal saplungan (*Hopea plagata*)	Native	10.4±3.2	31.2±9.6
Yakal kaliot (*Hopea malibato*)	Native	12.0±2.7	35.9±8.2

to be high mortality of fruit trees. There are reports of some fruit harvesting at some sites by Milan et al. [26] but overall productivity has been very low. A better outcome might have been achieved by plantings fruit trees along plantation edges where they would receive more light and be easier to harvest.

Based on this experience we suggest a modified set of silvicultural prescriptions for smallholder and community tree plantations in this part of the Philippines, along with elsewhere in SE Asia. The suggestions are outlined in Table 6 and show recommended planting densities and constituent species. The planting density is set at around 1100 trees per ha rather than the 5000 trees per ha used in the Rainforestation Farming plantations. This reduces the cost of buying and planting seedlings and, while there may be slight increase in the time needed for weed control until the seedlings are established, this cost should be modest. The numbers of species used is reduced from that originally established in Rainforestation Farming plantings (at least 76 species assessed in the plots in this study) to between 11–23 species. This is because the type of species (i.e. fast growing and/or exotic versus slow growing/native) drives productivity in these plantings rather than increased biodiversity [18].

The identity of these species has been based on their performance in the field (See Herbohn et al. [15] for the initial assessment) and the prospective markets likely to be available to farmers in this region. They include fast-growing species currently preferred by most farmers as well as slow-growing species likely to

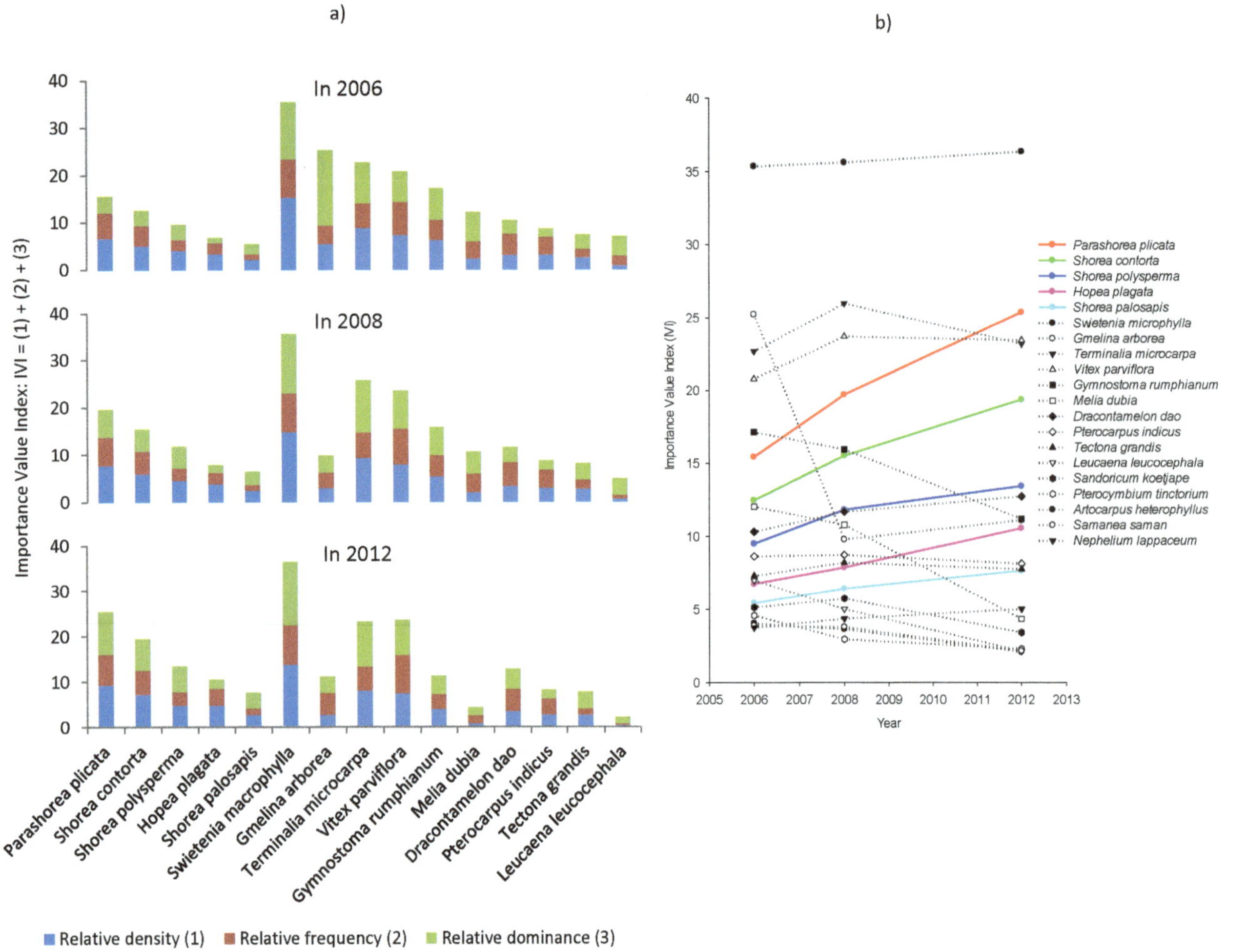

Figure 4. Importance Value Indices (IVIs) of the 15 common species with the highest IVI in rainforestation plantations in period of 2006–2012. IVI at measurements (a) and changing trend of IVI between 2006 and 2012 (b).

generate higher value timbers in the long term [26]. As was the case in the original Rainforestation Farming design, the shade intolerant species will be planted one year before the other species to facilitate their establishment. The intent of these prescriptions is to provide a steady flow of trees of sufficient size and with properties making them suitable for various markets. A significant proportion of the species planted are destined quickly capture the site and then be removed at an early age (6–10 years) in order to generate an income. This harvest will also act as a thinning to prevent the stands stagnating and to hasten the development of larger and more valuable residual trees. Most of these will be faster-growing species such as *Gmelina arborea, Casuarina equestifolia* or *Acacia* sp but some slower-growing species having trees with poor form might also be felled at this time as well. *Casuarina equisetifolia* was not found in our plots but observed in surrounding areas, we, therefore, suggest this species for plantations in the future based on our observation and the local farmers and experts during our research periods. At this age these are likely to be mostly used for fuelwood or pulp [41]. A second harvesting period could occur when the stands are 8–12 years old and a number of species are capable of being large enough for a pole market. Based on species growth rates a third harvesting period would develop around 14–18 years leaving a final harvest of slower growing but higher value trees after 20 years. The selective harvesting could be applied for different species at different times and for different products. Trees of some fast-growing species such as *Leucaena leucocephala, Melia dubia* and *Gmelina arborea* could be early harvested for household consumption or firewood while most of Dipterocarp species could be late harvested for high valued products (e.g. lumber). By this time a number of additional seedlings may have regenerated and begun to growing using light from canopy gaps. These seedlings might even be supplemented by enrichment plantings using species favoured by the market. The plantations could then gradually move to become a selectively managed forest or it could be clear-felled and replanted. Throughout this sequence the timbers produced by the plantations are becoming progressively more valuable. Similarly, the market for these timbers is increasing from a largely local market or on-farm use for fuelwood to a regional or national market for higher value sawlogs.

Finally, these mixtures could also achieve the ecological goal of the Rainforestation Farming project i.e. to achieve a planted forest with a physical structure and species composition and succession similar to the original local rainforest ecosystem after around 14 – 16 years. A three-storey structural complexity is now beginning to develop these plantings resembling natural rainforests inthearea.Thechangesindensityandspeciescompositions inthese

Table 6. Recommended design for smallholder tree plantations in Leyte, Philippines based on performance of Rainforestation Farming plantations.

Product	Time of thinning (yrs)	Number of species	Density (trees/ha)	Typical species
Firewood	6–10	3–5	450	Gmelina (*Gmelina arborea*)
				Bagalunga (*Melia dubia*)
				Ipil-Ipil (*Leucaena laucocephala*)
				Raintree (*Samanea saman*)
				Mt agoho (*Gymnostoma rumphianum*)
				Agoho (*Casuarina equisetifolia*)
				Thailand acacia (*Senna siamea*)
Pole	8–12	2–3	200	Gmelina (*Gmelina arborea*)
				Mahogany (*Swietenia macrophylla*)
				Kalumpit (*Terminalia macrocarpa*)
				Dao (*Dracontamelon dao*)
				Narra (*Pterocarpusindicus*)
				Mt agoho (*Gymnostoma rumphianum*)
Fast-growing timber	14–18	3–5	250	Gmelina (*Gmelina arborea*)
				Mahogany (*Swietenia macrophylla*)
				Teak (*Tectona grandis*)
				Mayapis (*Shorea palosapis*)
				White lauan (*Shorea contorta*)
				Almaciga (*Agathis philippinensis*)
				Tangeli (*Shorea polysperma*)
				Bagtikan (*Parashorea plicata*)
Slower-growing timber	>20	3–10	200	Yakal saplungan (*Hopea plagata*)
				Yakal kaliot (*Hopea mallbato*)
				Malakawayan (*Podocarpus rumphii*)
				Molave (*Vitex parviflora*)
				Narra (*Pterocarpus indicus*)
				Malapanau (*Dipterocarpus kerrii*)
				Tangeli (*Shorea polysperma*)
				Apitong (*Dipterocarp grandiflorus*)
				Apitong hagakhak (*Dipterocarpus kunstleri*)
				Dalingdingan (*Hopea dalingdingan*)
				Red lauan (*Shorea negrosensis*)
				Bitanghol (*Calophyllum blancoi*)
				Bitanghol sibat (*Calophyllum lancifolium*)
Total		11–23	1100	Fruit trees: Durian (*Durio zibethinus*), Mango (*Mangifera indica*), Rambutan (*Nephelium lappaceum*), Nangka (*Artocarpus heterophyllus*)

plantings over time indicated a characteristic often found in uneven-aged forests, old-growth forest or natural forest that is a continuous recruitment and mortality in the forest succession [42]. The ratio of Dipterocarp trees to total trees was approximately 1:4 across sites, which is similar to the species proportion in the three canopy strata at Mt. Pangasugan in Leyte, the Philippines [29].

Conclusions

It is clear from our results that the composition and structure of the Rainforestation plantations have changed over time, with the plantations being a highly dynamic system. There has been a decrease in the relative importance of shade-intolerant species, especially exotic species and a corresponding increase in the relative importance of shade-tolerant native species. The design of the species mixtures affected the patterns of species loss in that the very high initial stocking rates resulted in high rates of mortality, mostly among the shade-intolerant species and fruit trees. Other changes in species loss were largely due to ad hoc harvesting decisions by the land owners. We draw on our current results to recommend a modified set of prescriptions for smallholder and community tree plantations. These recommendations include a lower initial stocking rate of 1100 trees per hectare; the proportion of fast growing species should be around 30 to 40 per cent of this initial density; and any fruit trees should only be planted on the plantation margin where more space and light are available for crowns to develop.

Supporting Information

File S1 Figure S1. Map of the Rainforestation sites in Leyte province, the Philippines. The numbers in the map refer to the names of sites. **Figure S2. Size class distributions of species groups in the Rainforestation sites.** Depending on species provenance (i.e. exotic and native) or ecological characteristics of species (i.e. Shade-intolerant, Shade-tolerant and Fruit tree). 1a, 2a & 3a: all size classes of provenance groups; 1b, 2b & 3b: three largest size classes of provenance groups; 4a, 5a & 6a: all size classes of ecological groups; and 4b, 5b & 6b: three largest size classes of ecological groups. **Figure S3. The survival and mortality of the most common species at 80 plots of 18 rainforestation sites in period of 2006–2012.**

Acknowledgments

Thank you to the staff of the Visayas State University, in particular the College of Forestry and Natural Resources and Institute of Tropical Ecology, for supporting the work and also to the people in Leyte province, for allowing us to conduct our study on their land and for their great assistance during the field work. The authors would like to acknowledge to Grant Wardell-Johnson for contributing to the design of the sampling procedure. We thank Jack Baynes for helping with the map of study sites.

Author Contributions

Conceived and designed the experiments: HN JH. Performed the experiments: HN. Analyzed the data: HN DL. Wrote the paper: HN DL JH JF.

References

1. Lamb D, Erskine PD, Parrotta JA (2005) Restoration of degraded tropical forest landscapes. Science 310: 1628.
2. Debell DS, Cole TG, Whitesell CD (1997) Growth, Development, and Yield in Pure and Mixed Stands of Eucalyptus and Albizia. Forest Science 43: 286–298.
3. Montagnini F, Gonzalez E, Porras C, Rheingans R (1995) Mixed and pure forest plantations in the humid neotropics: a comparison of early growth, pest management and establishment costs. Commonwealth Forestry Review 74: 306–314.
4. Piotto D, Víquez E, Montagnini F, Kanninen M (2004) Pure and mixed forest plantations with native species of the dry tropics of Costa Rica: a comparison of growth and productivity. Forest Ecology and Management 190: 359–372.
5. Kelty MJ (2006) The role of species mixtures in plantation forestry. Forest Ecology and Management 233: 195–204.
6. Lamb D (2011) Regreening the bare hills: Tropical forest restoration in the Asia-Pacific region; Palo M, editor. Dordrecht: Springer.
7. Hooper DU, Chapin FS III, Ewel JJ, Hector A, Inchausti P, et al. (2005) Effects of biodiversity on ecosystem functioning: A consensus of current knowledge. Ecological Monographs 75: 3–35.
8. Wormald TJ (1992) Mixed and pure forest plantations in the tropics and subtropics. Rome: FAO. 152 p.
9. Kelty MJ (1992) Comparative productivity of monocultures and mixedspecies stands. In: Kelty MJ, Larson BC, Oliver CD, editors. The Ecology and Silviculture of Mixed-Species Forests. Dordrecht, Boston: Kluwer Academic Publishers. 125–141.
10. Haggar JP, Ewel JJ (1997) Primary Productivity and Resource Partitioning in Model Tropical Ecosystems. Ecology 78: 1211–1221.
11. Bullock BP, Burkhart HE (2005) An evaluation of spatial dependency in juvenile loblolly pine stands using stem diameter. Forest Science 51: 102–108.
12. Grant JC, Nichols JD, Pelletier M-C, Glencross K, Bell R (2006) Five year results from a mixed-species spacing trial with six subtropical rainforest tree species. Forest Ecology and Management 233: 309–314.
13. Kooyman RM (1996) Growing rainforest: rainforest restoration and regeneration - recommendations for the humid sub-tropical region of northern New South Wales and south east Queensland. Brisbane: Greening Australia - Queensland. 79 p.
14. Lamprecht H (1989) Silviculture in the tropics. Tropical forest ecosystems and their tree species - Possibilities and methods for their long-term utilization. Eschborn, Federal Republic of Germany: Technical Co-operation.
15. Herbohn JL, Vanclay J, Nguyen H, Le HD, Baynes J, et al. (2014) Inventory Procedures for Smallholder and Community Woodlots in the Philippines: Methods, Initial Findings and Insights. Small-Scale Forestry 13: 79–100.
16. Le HD, Smith C, Herbohn J (2014) What drives the success of reforestation projects in tropical developing countries? The case of the Philippines. Global Environmental Change, 24:334–348.
17. Le HD, Smith C, Herbohn J, Harrison S (2012) More than just trees: Assessing reforestation success in tropical developing countries. Journal of Rural Studies 28: 5–19.
18. Nguyen H, Herbohn J, Firn J, Lamb D (2012) Biodiversity–productivity relationships in small-scale mixed-species plantations using native species in Leyte province, Philippines. Forest Ecology and Management 274: 81–90.
19. Nguyen HTT (2011) Performance of mixed-species plantaitons using the Rainforestation Farming system in Leyte province, the Philippines [PhD thesis]. The University of Queensland.
20. Milan PP (1997) Appraisal mission of rainforestation farming in Leyte. Baybay, Leyte: Visayas State College of Agriculture (ViSCA). 16 p.
21. Milan PP (1997) Strategy for community involvement in Rainforestation farming. In: Margraf J, Goltenboth F, Milan PP, editors; 1997 3–6 March; Leyte. Visayas College of Agriculture (ViSCA) - GTZ. 336.
22. Kolb M (2003) Silvicultural analysis of "Rainforestation Farming" areas on Leyte island, Philippines [PhD thesis]. Deutschland: Universitat Gottingen. 116 p.
23. Jahn R, Asio VB (2001) Climate, geology, geomorphology and soils of the tropics with special reference to Leyte islands (Philippines). In: Goltenboth F, Asio VB, editors; 2001 9–20 Apr; Baybay, Leyte. Visaya State College of Agroculture. 25–43.
24. Marohn C (2007) Rainforestation farming on Leyte island, Philippines - aspects of soil fertility and carbon sequestration potential [PhD thesis]. Stuttgart: University of Hohenheim.
25. Hussain I (2008) Plant Physiology. 1 ed. New Delhi: Oxford Book Co.
26. Milan PP, Ceniza MJC, Asio VB, Bulayog SB, Napiza MD (2004) Evaluation of silvicultural management, ecological change and market study of products of existing rainforestation demonstration and cooperators' farms. Baybay: Institute of Tropical Ecology.
27. Schulte A (1998) Community-based rainforestation options for the Visayas, Philippines: a review. Hoxter, Germany: GTZ.
28. Schulte A (2002) Rainforestation Farming: Option for rural development and biodiversity conservation in the humid tropics of Southeast Asia. A review of major issues on community-based rehabilitation silviculture and guide to recommended native tree species for the Visayas/Philippines; Goltenboth F, Milan P, editors. Aachen: Shaker. 312 p.
29. Margraf J, Milan P (1996) Ecology of Dipterocarp forests and its relevance for island rehabilitation in Leyte, Philippines. In: Schulte A, Schone D, editors. Dipterocarp forest ecosystems: Towards sustainable management. Berlin: World Scientific. 124–154.
30. Barbour MG (1999) Terrestrial plant ecology. Menlo Park, Calif.: Benjamin/Cummings. 1 v. (various pagings).
31. Barbour MG, Burk JH, Pitts WD (1980) Terrestrial plant ecology. Menlo Park, Calif.: Benjamin/Cummings Pub. Co. xi, 604 p.
32. Arroyo-Rodríguez V, Mandujano S (2006) The Importance of Tropical Rain Forest Fragments to the Conservation of Plant Species Diversity in Los Tuxtlas, Mexico. Biodiversity & Conservation 15: 4159–4179.
33. Dangol DR, Shivakoti GP (2001) Species composition and dominance of plant communities in Westren Chitwan, Nepal. Napal Journal of Science and Technology 3: 69–78.
34. Shrestha R, Karmacharya SB, JHA PK (2000) Vegetation analysis of natural and degraded forests in Chitrepani in Siwalik region of Central Nepal. Tropical Ecology 41: 111–114.
35. Guariguata M, Chazdon R, Denslow J, Dupuy J, Anderson L (1997) Structure and floristics of secondary and old-growth forest stands in lowland Costa Rica. Plant Ecology 132: 107–120.
36. Nebel G, Kvist LP, Vanclay JK, Christensen H, Freitas L, et al. (2001) Structure and floristic composition of flood plain forests in the Peruvian Amazon: I. Overstorey. Forest Ecology and Management 150: 27–57.
37. Crane JH (2004) Selected cultural techniques to improve production of some subtropical and tropical fruit crops. In: Albrigo LG, Sauco VG, editors. Acta Horticulturae; 2004 29 Feb 2004; Toronto. 179–187.
38. Ferguson L, Krueger WH, Reyes H, Metheney P (2002) Effect of mechanical pruning on California black ripe (Olea europea L.) cv. 'Manzanillo' table olive yield. In: Vitagliano C, Martelli GP, editors. Proceedings of the Fourth International Symposium on Olive Growing. Leuven. 281–284.
39. Medina-Urrutia VM, Nunez-Elisea R (1997) Mechanical pruning to control tree size, flowering, and yield of mature "Tommy Atkins" mango trees. In: Lavi U, Degani C, Gazit S, Lahav E, Pesis E et al., editors. 5th International Mango Symposium. Israel. 305–314.

40. Schupp JR, Baugher TA, Miller SS, Harsh RM, Lesser KM (2008) Mechanical thinning of peach and apple trees reduces labor input and increases fruit size. Horttechnology 18: 660–670.
41. Tolentino EL (2008) Restoration of Philippine Native Forest by Smallholder Tree Farmers. In: Snelder DJ, Lasco RD, editors. Smallholder Tree Growing for Rural Development and Environmental Services. Dordrecht; London: Springer. 319–346.
42. Oliver CD, Larson BC (1996) Forest stand dynamics. Updated edition. Wiley, New York.

PERMISSIONS

All chapters in this book were first published in PLOS ONE, by The Public Library of Science; hereby published with permission under the Creative Commons Attribution License or equivalent. Every chapter published in this book has been scrutinized by our experts. Their significance has been extensively debated. The topics covered herein carry significant findings which will fuel the growth of the discipline. They may even be implemented as practical applications or may be referred to as a beginning point for another development.

The contributors of this book come from diverse backgrounds, making this book a truly international effort. This book will bring forth new frontiers with its revolutionizing research information and detailed analysis of the nascent developments around the world.

We would like to thank all the contributing authors for lending their expertise to make the book truly unique. They have played a crucial role in the development of this book. Without their invaluable contributions this book wouldn't have been possible. They have made vital efforts to compile up to date information on the varied aspects of this subject to make this book a valuable addition to the collection of many professionals and students.

This book was conceptualized with the vision of imparting up-to-date information and advanced data in this field. To ensure the same, a matchless editorial board was set up. Every individual on the board went through rigorous rounds of assessment to prove their worth. After which they invested a large part of their time researching and compiling the most relevant data for our readers.

The editorial board has been involved in producing this book since its inception. They have spent rigorous hours researching and exploring the diverse topics which have resulted in the successful publishing of this book. They have passed on their knowledge of decades through this book. To expedite this challenging task, the publisher supported the team at every step. A small team of assistant editors was also appointed to further simplify the editing procedure and attain best results for the readers.

Apart from the editorial board, the designing team has also invested a significant amount of their time in understanding the subject and creating the most relevant covers. They scrutinized every image to scout for the most suitable representation of the subject and create an appropriate cover for the book.

The publishing team has been an ardent support to the editorial, designing and production team. Their endless efforts to recruit the best for this project, has resulted in the accomplishment of this book. They are a veteran in the field of academics and their pool of knowledge is as vast as their experience in printing. Their expertise and guidance has proved useful at every step. Their uncompromising quality standards have made this book an exceptional effort. Their encouragement from time to time has been an inspiration for everyone.

The publisher and the editorial board hope that this book will prove to be a valuable piece of knowledge for researchers, students, practitioners and scholars across the globe.

LIST OF CONTRIBUTORS

Aaron J. Gassmann, Jennifer L. Petzold-Maxwell, Ryan S. Keweshan and Mike W. Dunbar
Department of Entomology, Iowa State University, Ames, Iowa, United States of America

Sikander Khan Tanveer, Xiaoxia Wen, Xing Li Lu, Junli Zhang and Yuncheng Liao
College of Agronomy, Northwest A&F University Yangling, Shaanxi, P.R. China

Xiao Yang and Bao-Rong Lu
Ministry of Education Key Laboratory for Biodiversity Science and Ecological Engineering, Department of Ecology and Evolutionary Biology, Fudan University, Shanghai, China

Feng Wang and Jun Su
Fujian Province Key Laboratory of Genetic Engineering for Agriculture, Fujian Academy of Agricultural Sciences, Fuzhou, China

Joost van Heerwaarden
Biometris, Plant Sciences Group, Wageningen University and Research Center, Wageningen, The Netherlands
Instituto de Ecología, Universidad Nacional Auto´noma de México, México D.F., México,

Diego Ortega Del Vecchyo
Interdepartmental Ph.D. Program in Bioinformatics, University of California Los Angeles, Los Angeles, California, United States of America

Elena R. Alvarez-Buylla
Instituto de Ecología, Universidad Nacional Autónoma de México, México D.F., México
Centro de Ciencias de la Complejidad, Universidad Nacional Autónoma de México, México D.F., México

Mauricio R. Bellon
Bioversity International, Rome, Italy

Hugh Roderick and Howard J. Atkinson
Centre for Plant Sciences, University of Leeds, Leeds, United Kingdom

Elvis Mbiru, Danny Coyne and Leena Tripathi
International Institute of Tropical Agriculture, Kampala, Uganda

Christian H. Krupke, Greg J. Hunt, Gladys Andino and Krispn Given
Department of Entomology, Purdue University, West Lafayette, Indiana, United States of America

Brian D. Eitzer
Department of Analytical Chemistry, The Connecticut Agricultural Experiment Station, New Haven, Connecticut, United States of America

Amita Mohan, William F. Schillinger and Kulvinder S. Gill
Department of Crop and Soil Sciences, Washington State University, Pullman, Washington, United States of America

Shujuan Chang
Institute of Grassland Science, China Agricultural University, Beijing, People's Republic of China
Inner Mongolia Rangeland Survey & Design Institute, Hohhot, Inner Mongolia, People's Republic of China

Nan Liu, Xiaoya Wang, Yingjun Zhang and Yue Xie
Institute of Grassland Science, China Agricultural University, Beijing, People's Republic of China

Matthew H. Meisner
Department of Evolution and Ecology, University of California Davis, Davis, California, United States of America
Department of Statistics, University of California Davis, Davis, California, United States of America

Jay A. Rosenheim
Department of Entomology and Nematology, University of California Davis, Davis, California, United States of America

Ling Liu, Qian Zhang, Liangliang Hu, Jianjun Tang, Ligen Xu and Xin Chen
College of Life Science, Zhejiang University, Hangzhou, China

Xiantian Yang
Taizhou Academy of Agriculture Sciences, Taizhou, China

Jean W. H. Yong
Singapore University of Technology and Design, Singapore, Singapore

Yuejian Mao
Energy Biosciences Institute, University of Illinois, Urbana, Illinois, United States of America
Institute for Genomic Biology, University of Illinois, Urbana, Illinois, United States of America

Anthony C. Yannarell
Energy Biosciences Institute, University of Illinois, Urbana, Illinois, United States of America
Institute for Genomic Biology, University of Illinois, Urbana, Illinois, United States of America
Department of Natural Resources and Environmental Sciences, University of Illinois, Urbana, Illinois, United States of America

Roderick I. Mackie
Energy Biosciences Institute, University of Illinois, Urbana, Illinois, United States of America
Institute for Genomic Biology, University of Illinois, Urbana, Illinois, United States of America
Department of Animal Sciences, University of Illinois, Urbana, Illinois, United States of America

Xiu-liang Jin and Sen-nan Song
Key Laboratory of Crop Genetics and Physiology of Jiangsu Province, Yangzhou University, Yangzhou, China
Beijing Research Center for Information Technology inAgriculture, Beijing Academy of Agriculture and Forestry Sciences, Beijing, China
National Engineering Research Center for Information Technology in Agriculture, Beijing, China

Hai-kuan Feng, Xin-kai Zhu1, Zhen-hai Li, Xiao-yu Song, Guijun,Yang and Xin-gang Xu
Beijing Research Center for Information Technology in Agriculture, Beijing Academy of Agriculture and Forestry Sciences, Beijing, China
National Engineering Research Center for Information Technology in Agriculture, Beijing, China

Wen-shan Guo
Key Laboratory of Crop Genetics and Physiology of Jiangsu Province, Yangzhou University, Yangzhou, China

James Angus Fraser
Departamentos de Antropología e Historia, Universidad de los Andes, Bogotá, Cundinamarca, Colombia
Department of Anthropology, University of Sussex, East Sussex, United Kingdom

Alessandro Alves-Pereira
Post-graduate Program in Genetics and Plant Breeding, Universidade de São Paulo, Piracicaba, Estado de São Paulo, Brazil

Laboratório de Evolução Aplicada, Universidade Federal do Amazonas, Manaus, Amazonas, Brazil

André Braga Junqueira
Coordenação de Tecnologia e Inovação, Instituto Nacional de Pesquisas da Amazônia, Manaus, Amazonas, Brazil
Centre for Crop Systems Analysis/Technology and Agrarian Development Group Wageningen University, Wageningen, Gelderland, The Netherlands

Nivaldo Peroni
Departamento de Ecologia e Zoologia, Universidade Federal de Santa Catarina, Florianópolis, Santa Catarina, Brazil

Charles Roland Clement
Laboratório de Evolução Aplicada, Universidade Federal do Amazonas, Manaus, Amazonas, Brazil
Coordenação de Tecnologia e Inovação, Instituto Nacional de Pesquisas da Amazônia, Manaus, Amazonas

Hamid R. Asghari
Faculty of Agriculture, Shahrood University of Technology, Shahrood, Iran

Timothy R. Cavagnaro
School of Biological Sciences, Monash University, Clayton, Victoria, Australia
AustralianCentre for Biodiversity, Monash University, Clayton, Victoria, Australia

Jingfeng Huang and Xinxing Li
Institute of Agricultural Remote Sensing & Information Application, Zijingang Campus, Zhejiang University, Hangzhou, China
China Ministry of Education KeyLaboratory of Environmental Remediation and Ecological Health, Zhejiang University, Hangzhou, China
Key Laboratory of Agricultural Remote Sensing and Information System, Zhejiang Province, China

Xiuzhen Wang
Institute of Remote Sensing and Earth Sciences, Hangzhou Normal University, Hangzhou, China

Hanqin Tian
International Center for Climate and Global Change Research, Auburn University, Auburn, Alabama, United States of America
Ecosystem Dynamics and Global Ecology (EDGE) Laboratory, School of Forestry and Wildlife Sciences, Auburn University, Auburn, Alabama, United States of America

Zhuokun Pan
Institute of Agricultural Remote Sensing & Information Application, Zijingang Campus, Zhejiang University, Hangzhou, China
Key Laboratory of Agricultural Remote Sensing and Information System, Zhejiang Province, China

David B. Lindenmayer, Amanda R. Northrop-Mackie, Rebecca Montague-Drake, Mason Crane,
Damian Michael, Sachiko Okada and Philip Gibbons
Fenner School of Environment and Society, ARC Centre of Excellence for Environmental Decisions, and National Environment Research Program, The Australian National University, Canberra, Australia

Yulin Gao, Qingbo Wei, Wenyan Yu, Zhi Zhang and Zhongren Lei
State Key Laboratory for Biology of Plant Diseases and Insect Pests, Institute of Plant Protection, Chinese Academy of Agricultural Sciences, Beijing, P. R. China

Stuart R. Reitz
Department of Crop and Soil Sciences, Malheur County Extension, Oregon State University, Ontario, Oregon, United States of America

Danying Wang, Song Chen, Chenglin Ji, Chunmei Xu and Xiufu Zhang
State Key Laboratory of Rice Biology, China National Rice Research Institute, Hangzhou, Zhejiang, China

Bhagirath Singh Chauhan
Queensland Alliance for Agriculture and Food Innovation, University of Queensland, Queensland, Australia

Zaiman Wang
Key Laboratory of Key Technology on Agricultural Machine and Equipment, South China Agricultural University, Guangzhou, Guangdong, China

Rose Gageler, Mark Bonner, Gunnar Kirchhof, Nicole Robinson and Susanne Schmidt
School of Agriculture and Food Sciences, The University of Queensland, St Lucia, Australia

Mark Amos
Lake Baroon Catchment Care Group, Maleny, Australia

Luke P. Shoo
School of Biological Sciences, The University of Queensland, St Lucia, Australia

Ji Chen, Fengjuan Lv, Jingran Liu, Yina Ma, Youhua Wang, Binglin Chen, Yali Meng and Zhiguo Zhou
Key Laboratory of Crop Physiology & Ecology, Ministry of Agriculture, Nanjing Agricultural University, Nanjing, Jiangsu Province, PR China

Derrick M. Oosterhuis
Department of Crop, Soil, and Environmental Sciences, University of Arkansas, Fayetteville, Arkansas, United States of America

Anja Schmithals and Norbert Kühn
Department of Landscape Architecture and Environmental Planning, Chair of Vegetation Technology and Planting Design, Berlin Institute of Technology, Berlin, Germany

Thomas S. Adams and David M. Eissenstat
Department of Ecosystem Science and Management and the Ecology Graduate Program, the Pennsylvania State University, University Park, Pennsylvania, United States of America

María Alonso-Ayuso, José Luis Gabriel and Miguel Quemada
School of Agriculture Engineering, Technical University of Madrid, Madrid, Spain

Mason J. Crane, David B. Lindenmayer and Ross B. Cunningham
Fenner School of Environment and Society, The Australian National University, Canberra, ACT, Australia

Baoru Sun, Yi Peng, Hongyu Yang, Zhijian Li, Yingzhi Gao, Chao Wang, Yuli Yan and Yanmei Liu
Key Laboratory of Vegetation Ecology, Northeast Normal University, Changchun, China

Song Chen, Danying Wang, Chunmei Xu, Chenglin Ji, Xiaoguo Zhang, Xia Zhao and Xiufu Zhang
China National Rice Research Institute, Chinese Academy of Agricultural Sciences, Hangzhou, Zhejiang, China

Bhagirath Singh Chauhan
Queensland Alliance for Agriculture and Food Innovation (QAAFI), University of Queensland, Toowoomba, Queensland, Australia

Huong Nguyen and John Herbohn
School of Agriculture and Food Sciences, The University of Queensland, St Lucia, Australia
Forest Industries Research Centre, The University of the Sunshine Coast, Sippy Downs, Australia

David Lamb
School of Agriculture and Food Sciences, The University of Queensland, St Lucia, Australia
Centre for Mined Land Research, The University of Queensland, St Lucia, Australia,

Jennifer Firn
Faculty of Science and Technology, School of Earth, Environmental and Biological Sciences, Queensland University of Technology, Brisbane, Australia

Index